Crystalline Metal Oxide Catalysts

Wataru Ueda
Editor

Crystalline Metal Oxide Catalysts

Springer

Editor
Wataru Ueda
Department of Material and Life
Chemistry, Faculty of Engineering
Kanagawa University
Yokohama, Kanagawa, Japan

ISBN 978-981-19-5012-4 ISBN 978-981-19-5013-1 (eBook)
https://doi.org/10.1007/978-981-19-5013-1

This Springer imprint is published by the registered company Springer Nature Singapore Pte Ltd.
The registered company address is: 152 Beach Road, #21-01/04 Gateway East, Singapore 189721, Singapore

Preface

Invitation to Crystalline Catalysts that have Not Yet been Found

People are fascinated by glitter materials and are strongly attracted to them. A typical example is a jewel. The elements in jewels are lined up in an orderly fashion to create the sparkle, but this is not apparent to a human's naked eye. However, through the shining light of jewels, people have intuitively sensed the arrangement of the elements and felt the power generated by them. This kind of daily occurrence is also true in the world of science. Scientists are amazed by the various functions of materials and are impressed by the functions that arise from the ordered but complex arrangement of elements, thinking about the state of the elements that make up crystals. Although crystalline metal oxide catalysts, the subject of this book, do not shine like jewels, chemical reactions are triggered and promoted by the crystal surface fields over the solid crystalline material. Scientists are fascinated by this and are keenly interested in the subtleties of nature's weaving. They feel that there must be an infinite number of crystalline solid materials in nature with unprecedented catalytic capabilities. Nature is always waiting for scientists to approach and discover. Today, human society is facing various problems with energy and materials. One effective way to deal with these problems is to use catalysts, and in fact, there are high expectations for solid catalysts, especially those with catalytic functions generated by crystals. Now is the time to dive into the world of undiscovered crystalline materials and bring new catalytic functions to our science and technology world. I hope that this book will provide an opportunity to start this process.

Organizations of human societies are naturally composed as a result of the daily activity of various people, and likewise, catalytic material structures should be naturally composed of various elements arrangement. However, "naturally" is not "totally free", but through the process of exchanging information between individuals, recognizing and identifying everyone make fit in a predetermined place and the structure of individuals. In human society, various works are started under necessity, and then an organized society is naturally established in order to maximize society's effectiveness. If the overall policy is decided in society, a social structure will be created in that direction. In the case of catalysts, a series of chemical reactions must be controlled by various catalytic elements in solid catalysts, and then the constituting elements are organized in a particular structure in order to optimize the catalytic promotion and work harmonically. However, even if the overall catalytic function is decided, it will not be possible to create the catalyst structure that meets the uniquely required function.

Working on complex things can have complex consequences, but a fascinating nature behind this will motivate a lively challenge to complex things.

Yokohama, Kanagawa, Japan Wataru Ueda

Contents

Contributors

Abe Ryu Graduate School of Engineering, Kyoto University, Katsura, Kyoto, Japan

Aihara Takeshi Laboratory for Materials and Structures, Institute of Innovative Research, Tokyo Institute of Technology, Yokohama, Kanagawa, Japan

Fujima Jun Department of Chemistry, Hokkaido University, Sapporo, Hokkaido, Japan

Hamamoto Nobutsugu Faculty of Engineering, Department of Applied Chemistry, Sanyo-Onoda City University, Yamaguchi, Japan

Hensen Emiel J. M. Department of Chemical Engineering and Chemistry, Laboratory of Inorganic Materials and Catalysis, Eindhoven University of Technology, Eindhoven, The Netherlands

Hinuma Yoyo Department of Energy and Environment, National Institute of Advanced Industrial Science and Technology (AIST), Ikeda, Osaka, Japan

Hosokawa Saburo Faculty of Materials Science and Engineering, Kyoto Institute of Technology, Kyoto, Japan

Hosono Hideo Materials Research Center for Element Strategy, Tokyo Institute of Technology, Yokohama, Kanagawa, Japan

Ishikawa Satoshi Department of Material Chemistry and Life, Kanagawa University, Yokohama, Kanagawa, Japan

Kamachi Takashi Department of Life, Environment and Applied Chemistry, Fukuoka Institute of Technology, Fukuoka, Fukuoka, Japan

Kamata Keigo Laboratory for Materials and Structures, Institute of Innovative Research, Tokyo Institute of Technology, Yokohama, Kanagawa, Japan

Kitano Masaaki Materials Research Center for Element Strategy, Tokyo Institute of Technology, Yokohama, Kanagawa, Japan

Motohashi Teruki Department of Material and Life Chemistry, Kanagawa University, Yokohama, Kanagawa, Japan

Nakajima Kiyotaka Institute for Catalysis, Hokkaido University Kita, Sapporo, Hokkaido, Japan

Nishimura Shun Graduate School of Advanced Science and Technology, Japan Advanced Institute of Science and Technology (JAIST), Nomi, Japan

Ohyama Junya Faculty of Advanced Science and Technology, Kumamoto University, Chuo-ku, Kumamoto, Japan

Ooyama Haruka E. Department of Applied Chemistry, Graduate School of Advanced Science and Engineering, Hiroshima University, Higashi-Hiroshima, Hiroshima, Japan

Osuga Ryota Institute of Multidisciplinary Research for Advanced Materials, Tohoku University, Aoba-ku, Sendai, Miyagi, Japan

Padovan Daniele Institute for Catalysis, Hokkaido University Kita, Sapporo, Hokkaido, Japan

Sadakane Masahiro Department of Applied Chemistry, Graduate School of Advanced Science and Engineering, Hiroshima University, Higashi-Hiroshima, Hiroshima, Japan

Suzuki Hajime Graduate School of Engineering, Kyoto University, Katsura, Kyoto, Japan

Takahashi Keisuke Department of Chemistry, Hokkaido University, Sapporo, Hokkaido, Japan

Takahashi Lauren Department of Chemistry, Hokkaido University, Sapporo, Hokkaido, Japan

Ueda Wataru Department of Material Chemistry and Life, Kanagawa University, Yokohama, Kanagawa, Japan

Yokoi Toshiyuki Institute for Innovative Research, Tokyo Institute of Technology, Midori-ku, Yokohama, Kanagawa, Japan

Zhang Zhenxin School of Materials Science and Chemical Engineering, Ningbo University, Ningbo, Zhejiang, P. R. China

Chapter 1
Overview of Crystalline Metal Oxide Catalysts

Wataru Ueda

1.1 General Introduction

Catalysts are used in many modern chemical industrial processes. Recently, CO_2 conversion catalysts, bio-conversion catalysts, electrolysis catalysts, fuel cell catalysts, photocatalysts, etc. have largely progressed, and new use targets of catalysts have expanded along with the sophisticated development of catalyst systems. It is no exaggeration to say that catalysts are a key technology that can essentially contribute to solving the environmental, resource, and energy problems that are important issues to be solved for humankind in this century and even the coming century. This is because catalysts inherently have no load on the surrounding nature system, and further development of catalysis technologies is desired for the realization of a sustainable society.

Catalytic ammonia synthesis and catalytic polymerization are the most famous examples of how catalysts have contributed significantly to human society's improvement by supplying chemical products profoundly. Although iron catalysts for ammonia synthesis were developed through trial-and-error-type research in response to the demands, the catalyst performance has made remarkable progress with practical application from time to time. At present, the energy efficiency of ammonia synthesis from hydrogen and nitrogen performed at optimized temperature and pressure is said to be at a level comparable to that of nitrogen fixation by the biological enzyme nitrogenase. This improvement has been attained, thanks to the advanced understanding of catalytic phenomena elucidated by sophisticated analysis, surface science, and quantum-chemistry at the elementary reaction step level, and also thanks that it has become possible to effectively perform catalytic material synthesis and reaction control based on the understanding. This is exactly the result of the high

W. Ueda (✉)
Department of Material Chemistry and Life, Kanagawa University, 3-27-1, Rokkakubashi, Kanagawa-ku, Yokohama 221-8686, Japan
e-mail: uedaw@kanagawa-u.ac.jp

W. Ueda (ed.), *Crystalline Metal Oxide Catalysts*,
https://doi.org/10.1007/978-981-19-5013-1_1

degree of chemical understanding of catalytic reactions and the advancement of catalytic reaction engineering.

However, we must notice that the accumulation of scientific results by practical research took more than 100 years from the time when the ammonia synthesis catalyst was put into practical use until the current level of catalytic performance and energy efficiency was reached. Needless to say, time is necessary to steadily build basic catalytic science and chemistry, but under the current strong demand for solving environmental and energy problems, we might not be allowed to conduct such long-time research for building a new catalyst system. Undoubtedly, prompt and timely development of an advanced ultra-high performance catalyst system is necessary to meet the demands of human society. However, the ultra-high performance catalyst systems must be inevitably complex and highly sophisticated, and thus it must be very difficult to intentionally construct and rapidly synthesize catalysts in those states. We are still facing such difficulties, but must overcome the difficult task of developing catalysts by introducing a new dimension of catalytic science and technology [1].

One of the keywords in the above context is the introduction of "complexity" into catalysts. Catalysis researchers have long noticed that high-level complexation is often seen in bio-enzymatic systems and experienced that introduction of a complex situation into solid-state catalysts, for example, complexity in constituting elements, structure, and electronic state of solid-state catalysts mostly brings positive results in catalytic performance, although the researchers felt that there was a lot of uncertainty behind complexity. In other words, controlling complexity is controlling catalysis and ultimately bringing high-performance catalysts.

This book focuses on complex metal oxide catalysts in crystalline form because this type of catalyst has made great success in achieving high catalytic performance in various catalytic reactions by the introduction of "complexity".

1.1.1 Progress of Metal Oxide Catalyst Development: Toward Complexation and Crystalline State

Significant progress has been made in the use of complex metal oxides as catalysts for various chemical reactions [2]. The history of their development is summarized according to catalytic reactions in Table 1.1. First, the early development of catalysts was mostly based on single metal oxides that could be easily obtained or synthesized in those days. It is easy to imagine that the researchers were surprised when they saw the phenomenon of catalytic functionality even over such simple metal oxides. This phenomenon is seen in almost all the listed catalytic reactions. From the opposite point of view, the idea of combining elements did not fully exist among the researchers at the beginning of the catalyst development. The synthesis technologies of metal oxide catalysts were gradually progressed and at the same time, natural crystalline minerals were also utilized in catalyst development. Under the trend of

Table 1.1 Progress of metal oxide catalysts from simple oxide to crystalline complex metal oxides

Reaction type	Representative reaction	Progress of metal oxide catalysts				
		From	To			
		Simple oxide	Oxide-supported oxide	Mixed oxide (amorphous)	Mixed oxide (multi-phasic)	Single crystalline complex oxide
Oxidation	Methane coupling	MgO	Li/MgO, Mn/KWO_4/SiO_2		$LaAlO_3$	Li_2CaSiO_4
		La_2O_3				
	Methane to oxygenates	MoO_3, V_2O_5	MoO_3/SiO_2, V_2O_5/SiO_2			
		Fe_2O_3	Fe_2O_3/SiO_2			$FePO_4$, Fe-zeolite
		CuO				Cu_xO_y-zeolite
	Methanol to HCHO	MoO_3	MoO_3/TiO_2	Mo–Sn–O	Fe–Mo–O/Al_2O_3	$Fe_2(MoO_4)_3$
	Ethane to ethene (to acetic acid)	NiO			NiO–Nb_2O_5	
		V_2O_5	V_2O_5–MoO_3/SiO_2	Mo–V–Nb–O	Mo–V–Nb–Ln–O	$Mo_3VO_{11.2}$
	Propene (amm) oxidation	Cu_2O		Cu–Se–O		
		MoO_3	Mo–Bi–P–O/SiO_2	Mo–Bi–Fe–O	Mo–Bi–Fe–Co–K–O	$Bi_2(MoO_4)_2$, Bi_2MoO_6
		Sb_2O_3		Fe(Sn, U)–Sb–O	Fe–Sb–Te–Mo–O	$FeSbO_4$
	Propane (amm) oxidation	V_2O_5	V_2O_5/SiO_2, V_2O_5/Nb_2O_5			$MoV_{0.24}Te_{0.12}Nb_{0.08}O_x$, $Mg_2V_2O_7$, $VSbO_4$, $MgMoO_4$
		MoO_3				
	Acrolein to acrylic acid	MoO_3, V_2O_5		Mo–V–O	Mo–V–Cu–W–Sb–O	MMo_3VOx (M: Cu, W, Sb)
		TeO_2			TeO_2–$CuMoO_4$	$CoTeMoO_6$
	Methacrolein to acid	MoO_3				$MPMo_{11}VO_{40}$ (M: Cs, Cu, Sb)

(continued)

Table 1.1 (continued)

Reaction type	Representative reaction	Progress of metal oxide catalysts				
		From	To			
		Simple oxide	Oxide-supported oxide	Mixed oxide (amorphous)	Mixed oxide (multi-phasic)	Single crystalline complex oxide
	C4, C6 to $C_4H_2O_3$	V_2O_5	V_2O_5–MoO_3/SiO_2	V(Mo)–P–O		$(VO)_2P_2O_7$
	Epoxidation, ammoximation	TiO_2	TiO_2/SiO_2			Ti-zeolite
	HCl oxidation	RuO_2	RuO_2/TiO_2			
	Complete oxidation	Fe_2O_3, CoO			$CuZnO_2$	$LaSrFeCoO_3$,
		MnO_2, CuO			$CuMn_2O_4$,	$BaMnAl_{11}O_{19}$
		CeO_2, V_2O_5	CeO_2/ZrO_2, V_2O_5/SiO_2			
	NH_3–SCR	V_2O_5	V_2O_5/TiO_2,			
			V_2O_5–WO_3/TiO_2			$W_4V_3O_{19}$
		CuO				Cu-zeolite
Acid–base	Cracking	Al_2O_3		Activated clay	SiO_2–Al_2O_3	Zeolite
	Isomerization	ZrO_2	WO_3/ZrO_2, SO_4/ZrO_2		WO_3–ZrO_2	Layered
		B_2O_3	B_2O_3/Al_2O_3			B-zeolite
	Hydration-dehydration	Al_2O_3, WO_3		Mg–Si–O	SiO_2–TiO_2	Ti–zeolite, $PW_{12}O_{40}$
	Methanol conversion	P_2O_5	P_2O_5/SiO_2	Al–P–O		$AlPO_4$, SAPO, $PW_{12}O_{40}$
		WO_3				Metal phosphate

(continued)

Table 1.1 (continued)

Reaction type	Representative reaction	Progress of metal oxide catalysts				
		From	To			
		Simple oxide	Oxide-supported oxide	Mixed oxide (amorphous)	Mixed oxide (multi-phasic)	Single crystalline complex oxide
	Alkylation	Nb_2O_5, TiO_2		Nb–P–O, Nb–W–O	P_2O_5–TiO_2	HNb_3O_8, $TiNb_3O_8$
	Condensation (base-catalyzed)	MgO, CaO				$Ca_3(PO_4)_2$
		Na, K, SrO	M/Al_2O_3(M:Na, K)		Na–MgO	M-zeolite (M: Na, K, Cs)
		LnO_3, Y_2O_3				
Hydrogenation	CO, CO_2 hydrogenation	CuO, ZnO, Fe_2O_3, $Cr_2O_3ZrO_2$, In_2O_3	Cu–ZnO/Al_2O_3		ZnO–Cr_2O_3	$CuCr_2O_4$, $CuMn_2O_4$
Dehydrogenation			FeO_x/Al_2O_3 In_2O_3/ZrO_2		ZnO–ZrO_2	$Cu_{1-x}Zn_xFe_2O_4$, $CuAl_2O_4$
Electro-catalysis	Water gas shift reaction		Fe_2O_3–Cr_2O_3/Al_2O_3			$La_{1.85}Ca_{0.15}CuO_3$
	Dehydrogenation to olefin	Fe_2O_3, Cr_2O_3	Cr_2O_3/Al_2O_3		Fe–K–Mg(Ce, Mo)–O	$K_2Fe_2O_4$
		ZrO_2, Ga_2O_3				
	Alcohol dehydrogenation	CuO, ZnO, MgO		MgO–SiO_2	CuO–ZnO	
Photo-catalysis	Water decomposition	TiO_2, WO_3, Nb_2O_5				$SrTiO_3$, $K_4Nb_6O_{17}$, $BiVO_4$, Zn_2GeO_4
	Hydrocarbon oxidation	V_2O_5	V_2O_5/SiO_2			

this movement, researchers have soon been interested in using familiar natural materials such as clay as a catalyst. Clay has multiple constituent elements, like Si and Al. In practical application, acid and heat-treatments of clay were implemented for improving catalytic activity. As the success was achieved on clay materials, the catalytic function generated by constituting multiple elements seems to have been obvious but at the same time mysterious. Eventually combining multiple elements to construct solid-state oxide catalysts became a conscious effort. As for the solid-acid catalysts, the development of mixed oxide catalysts had directed to the artificial synthesis of a composite of Si and Al oxides. These were the advent of composite oxide catalysts. Although the composite metal oxide catalysts synthesized in this way were mostly amorphous or simply mixed oxides, the combined catalysts have long been used in industrial processes as catalysts with high performance. At the same time, much fundamental research on catalysis itself and the catalytic materials have progressed. In the further development process, researchers again paid their attention to natural minerals and found zeolite catalysts. This is a composite oxide of Si and Al but is a crystalline solid acid with a porous structure able to adsorb small organic reactant molecules. This was one of the major milestones in the development of solid metal oxide catalysts. This crystalline material was immediately applied as a catalyst to industrial processes and has been used to date. In addition, many new zeolites having different porous structures have been artificially synthesized. Nowadays, a great number of new structural zeolites have emerged, and moreover, not only Si and Al, but also zeolitic materials of other elements such as phosphorus have recently emerged. Therefore, the development of Si- and Al-based complex oxides as solid-acid catalysts is the synergistic result of the direction toward crystalline materials, the addition of porosity, and the complexation of constituting elements. Recent advancement in controlling the structural position of constituting elements in zeolite chemistry is introduced in Chap. 6. With the advent of the zeolite catalysts, the horizontal development to expand the catalytic materials category has been widely undertaken.

Exactly the same trend can be seen in the history of the development of metal oxide catalysts for oxidation [3–7]. The firstly developed catalysts were here again simple metal oxides. We can easily notice that almost all kinds of simple metal oxides were tested and used practically. Based on these findings, the characteristics of each simple metal oxide catalyst were serially organized for the understanding of catalytic functions by element. This approach strongly promoted a fundamental understanding of the catalytic chemistry of simple metal oxides. However, simple metal oxide catalysts had limited catalytic functionality and gave an unsatisfactory catalytic performance. Methods improving catalytic performance much more were strongly demanded. Under such circumstances, methods to immobilize metal oxide catalysts, which is a metal oxide supported on other metal oxides, were developed in order to improve their performance. Supported catalysts were widely used because they have the advantage of easy synthesis and are superior in terms of catalytic reaction engineering. At the same time, the methods were also possible to easily combine more than two single metal oxides to bring multiple functions. This is the starting point of elemental complexation and eventually, this approach brought about complex metal oxide catalysts that consist of more than two metal elements with completely different elemental properties and catalytic functions. One of the most prominent examples is the Fe_2O_3–MoO_3 catalyst for gas-phase oxidation of methanol to formaldehyde. Fe_2O_3 catalyst was active but non-selective and MoO_3 was selective but poorly active, and then combined oxide showed catalytically extremely active and selective. This is absolutely the combination effect. This approach achieved a remarkable improvement in catalytic performance and was a very important turning point in the history of the development of metal oxide catalysts for oxidation catalysis. This remarkable complexation effect accelerated the development of catalysts as shown in Table 1.1, and the elemental compositions have become increasingly more complex [8]. However, with such elemental diversification, catalysts have become a mixed form of multiple crystalline phases or amorphous forms, which has adversely hindered a deeper understanding of catalysts through further performance improvement. At this stage, complex metal oxides had the inevitable consequence of crystal formation, shedding new light on catalytic chemistry. Along with the development of a wide variety of crystalline complex metal oxide catalysts, the approach of extraction of structural materials truly responsible for catalytic activity has progressed, and it is now possible to achieve better catalytic performance than before with a single crystalline phase while maintaining the multi-element state. What is unique here is that catalysts mostly exhibit complex structures with an appreciation of the effects of multi-element complexation. This direction must be correct, and indeed, the importance of crystalline complex metal oxide catalysts has become apparent as they have evolved into catalysts for the selective oxidation of alkanes which has been very difficult to achieve in the past.

Metal oxide catalysts for hydrogenation [9], dehydrogenation [10], photocatalysis, and electrocatalysis [11] are not exceptional. Elemental complexation and crystallization in complex metal oxide catalysts were widely used for these reactions, but extreme complexation in the catalyst for these reactions is a very recent trend in development. The reason, for instance, is that the reducing atmosphere under catalytic hydrogenation and dehydrogenation conditions causes unclear valence state and structural arrangement of the main constituent elements, preventing the aggressive application of crystallization and element complexation. Along with gradual clarification of the detailed character of the catalysts, dramatic progress using complexation by keeping solid crystal structure can be made in photocatalysis and electrocatalysis on the basis of solid-state chemistry such as band structure, defect structure, electro-conductivity, oxygen migration, and so on. Well-crystallized state is directly related to the catalytic activity, so crystallization and element complexation have been considered from the beginning of this catalyst development.

It should be worthy to consider why many catalytic reactions needed the direction of crystallization and element complexation of metal oxide catalysts. The first reason is that although the reaction equation itself is usually simple, the actual elemental catalytic reaction step is "complex," and the catalyst must be appropriately "complex" to precisely control all catalytic elemental reaction steps. Taking the selective oxidation reaction of organic compounds with molecular oxygen as an example, oxygen molecules to react with organic substances need to be activated by receiving electrons from oxide catalysts to weaken or dissociate O–O bonds. Oxygen then assumes a variety of activated states depending on electron donation ability from oxide catalysts. This obviously leads to the formation of various products depending on the state of oxygen activation. The role of the catalyst that brings the oxygen to a specific activated state is very important for the desired reaction to occur. The activated state of the organic reactant is also important, which in turn changes the reaction between the activated organic intermediate and the oxygen species. Usually, several steps are required to reach a stable product, therefore high selectivity to the desired product cannot be achieved without eliminating the possibility of undesired reaction pathways. In other words, the spatial structure and electronic structure of the surface of the metal oxide catalyst must be controlled for the appropriate oxygen activation and the direction of the organic reaction. It is theoretically possible to control the various reaction steps by using an appropriate catalyst, but a simple catalytic function made of a single catalytic element, like as simple metal oxide catalysts, is practically unable to control many reaction steps at once perfectly for attaining high selectivity. The structural environment in which multiple catalytic functions made of multiple catalytic elements can work cooperatively is most necessary to achieve selective oxidation. It is not surprising that composite oxides, which can be composed of many metal elements, are the most suitable solid functional materials as catalysts for selective oxidation and in fact are used in many gas-phase selective oxidation reactions. This is also valid for other catalytic reaction systems such as organic syntheses using metal oxide catalysts (see Chap. 8).

1.1.2 Structural Classification of Complex Metal Oxide Catalysts

The history of the development of metal oxide catalysts, particularly of industrial catalysis, shown in Table 1.1, clearly shows a trend toward increasing complexity, with crystallization and elemental complexation as the keywords. The same keywords can also be applied to other metal–oxygen bonded materials, as long as they are composed of a clear structure of metal–oxygen bonding. Expanding the scope in this way, we can see a variety of catalysts. Examples of such catalysts are summarized in Table 1.2, and examples of metal oxide catalysts with their respective characteristics are shown in Figs. 1.1, 1.2, 1.3, 1.4, 1.5, 1.6, 1.7, 1.8, 1.9, 1.10, 1.11 and 1.12, organized in a catalyst structure diagram.

The catalyst examples in Table 1.2 are classified by the spatial dimension of the catalytically active site structure. In the first case, the structure is not limited to a 3D periodic structure like a crystalline oxide but forms a minimum structural organization necessary for completing a catalytic cycle. The requirement to form a minimal local structural active site naturally leads to a molecular and discrete structural state of the catalytic material. The most basic example is the single atomic catalyst. There are many examples of single atomic catalysts in the liquid phase,

Table 1.2 Various states of metal oxide-based complex catalysts

Classification	Catalytic material form	Principal catalyst structure unit	Elemental co-unit for creating the principal catalyst	Brief description of complexity in constituting elements and structure	Example and structural view in figure
Single atom	Single atom in metal oxide lattice matrix	Atomic	Metal oxide lattice matrix	Single atomic metal element incorporated in surface metal oxide lattice matrix	Fe–Zeolite Ag–MnO_2 (1.1)
Molecular	Multinuclear metal oxide cluster	Molecular assembly with metal–oxygen bonds	Organic moiety stabilizing metal oxide cluster, or zeolite framework	3D complex metal oxide cluster stabilized with surrounding organic moiety or with inorganic framework	$CaMn_4O_5$ cluster in bio-system (1.2) Cu_3Ox/Zeolite (1.3)
	High-dimensional polyoxometalates	Polyoxometalate anions	Ionic linker	Orderly condensed oxo anions of different metal elements in salt form or with surrounding organic ions	POM with organic complex or metal-substituted POM (1.4)
	Metal oxide clusters and polyoxometalates	Metal oxide clusters, poly-oxometalates	Organic linker	Metal oxide clusters and polyoxometalates 3-dimensionally connected with organic linker in porous framework structure	MOF-type (1.5)
Surface layer or interface	Interface materials formed in metal-on-metal oxide surface	Nano-sized metal particles	Solid-state metal oxide crystals	Interface state formed in the close contact of nano-sized metal particles with crystalline metal oxide surface	Au/TiO_2 (1.6) Pd–$LaCoO_3$ Ni/$LaCoMnO_3$

(continued)

Table 1.2 (continued)

Classification	Catalytic material form	Principal catalyst structure unit	Elemental co-unit for creating the principal catalyst	Brief description of complexity in constituting elements and structure	Example and structural view in figure
	Interface materials formed in metal oxide-on-metal surface	2D metal oxide	Metallic board Metal particles	Surface materials formed either by covering metal particles with metal oxide or by interaction of metal oxide with metallic surface	FeOx/Pt (1.7) TiO_2/Pt CrOx/Ru, NiOx/Au
	Interface materials formed in metal oxide-on-metal oxide surface	Metal oxide nano-particles	Crystalline metal oxides	1D metal oxide, like oxide clusters, 2D oxide, like layered oxide formed by epitaxy, and 3D oxide formed on basal metal oxide supports	WO_3/ZrO_2 RuO_2/TiO_2 (1.8) Re_2O_7/Al_2O_3 V_2O_5/TiO_2
3D solid (layer structure)	Layer (slab) structure of metal oxide with additional elements in interlayer spacing	2D structural metal oxide	Elements in interlayer spacing	Layered multi-element structuring controlled by 3D lattice packing with accommodating elements in interlayer space	Sheet silicate (1.9.a) Metal phosphate (1.9.b) Metal niobate (1.9.c)
(non-porous)	Mixed metal oxide	3D metal oxides	Another 3D metal oxides or	Tight or loose mixture of metal oxides, crystal phase mixture, additional impurity phase, etc.	
	Complex metal oxides in amorphous form	3D metal oxides	Occluded cations, etc.	Uniform distribution of constituting elements in amorphous state	

(continued)

Table 1.2 (continued)

Classification	Catalytic material form	Principal catalyst structure unit	Elemental co-unit for creating the principal catalyst	Brief description of complexity in constituting elements and structure	Example and structural view in figure
	Metal oxides in condensed crystal form with elemental complexity	3D metal oxides	Counter cations, etc.	Uniform distribution of multi-elements constructing infinite crystal structure	Scheelite (110.a) Perovskite (1.10.b) Spinel (1.10.c) Hexaaluminate (1.10.d)
(porous)	Metal oxide framework crystal structure with pore	Metal oxide	Counter cations, etc.	Organized porous materials, like high-dimensional porous metal oxides and/or added metal cation position-controlled zeolite	Metalosilicates (1.11.a) Metalophosphates (1.11.b) Organosilicates (1.11.c)
	High-dimensional crystalline complex metal oxide with high elemental complexity	Polyoxoanion of metals	Linker and counter cations	Framework arrangement with molecular polyoxoanions and linkers, forming high-dimensional complex metal oxides structure with pore and constituting multi-elements	$Mo_3V(M)O_{11.2}$ (1.12.a) All-inorganic porous POM (1.12.b) $W_4V_3O_{19}$ (1.12.c)

Fig. 1.1 Isolated single Ag atoms on MnO_2 with hollandite structure as an example of single atom catalyst. Reproduced with permission from Ref. [13]. Copyright 2015 John Wiley and Sons

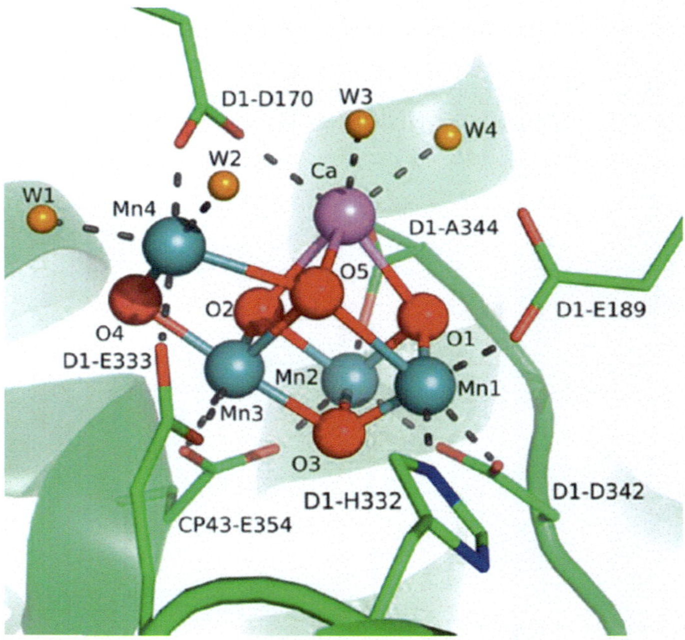

Fig. 1.2 Local structure of oxygen-evolving complex with $CaMn_3O_4$ cubane oxide cluster in biological catalyst. (Adapted with permission from Ref. [14]. Copyright 2013 American Chemical Society)

including organic metal complex catalysts, so this is not particularly new. However, in solid catalysts, it is essential to construct a peripheral structure to maintain the single atomic state, and therefore, discrete structural states are required [12]. An example is an Ag atom surrounded by a metal oxide framework shown in Fig. 1.1 [13]. This example has the periodic structure of the metal oxide to support single atoms, indicating even single atom catalyst needs local structuring based on multi-element. Slightly more complicated is the Mn oxide cluster [14, 15] which is the

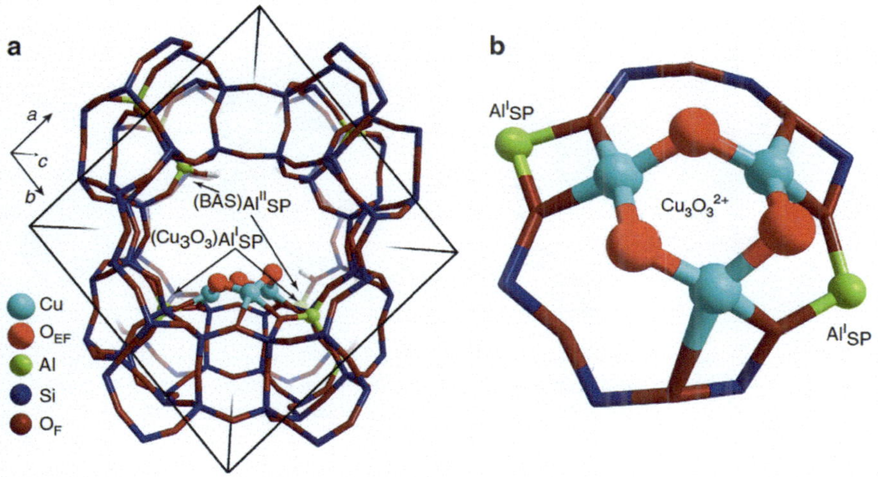

Fig. 1.3 Structure model. **a** of $[Cu_3(\mu\text{–}O)_3]^{2+}$ oxide cluster stabilized by two anionic centers **b** at the mordenite side pocket for methane oxidation to methanol from Ref. [15]

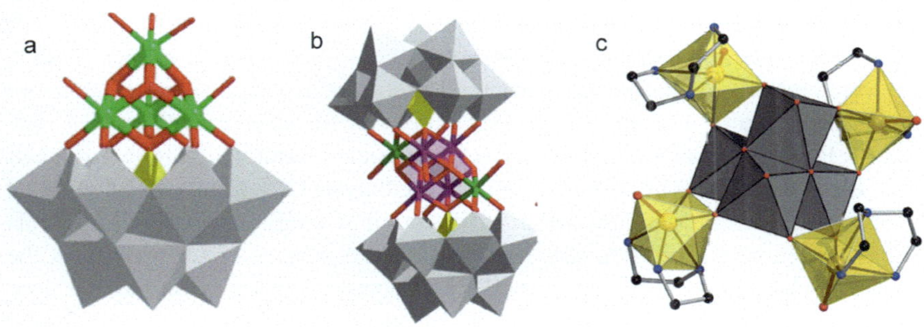

Fig. 1.4 Cubane oxide cluster in discrete structure. **a** Ni oxide cubane coordinated by a lacunary α-SiW_9O_{34}, **b** face-shared double cubane of Mn oxide sandwiched by two lacunary α-SiW_9O_{34} (Adapted with permission from Ref. [19]. Copyright 2016 Elsevier) and **c** W oxide cubane coordinated by four vanadyls with organic ligands (Adapted with permission from Ref. [20]. Copyright 2016 Royal Society of Chemistry)

oxygen-evolving site of photosynthesis in the bio-system (Fig. 1.2). The catalytic site is a discrete metal oxide cluster constructed with four Mn and one Ca and surrounded by amino acid groups. In a very similar system, instead of amino acids playing the role of structuring ligands, there is a catalyst in which Cu di (or tri)-nuclear oxide clusters are formed in the cavity of a zeolite structure as an inorganic ligand (Fig. 1.3). This is the catalyst that has attracted much attention because it can catalyze the oxidation of methane to methanol with molecular oxygen [16]. This is partially mimicking methane mono-oxidase (MMO) [17]. These examples introduced above are the catalysts with a minimal catalytic element, but if we include

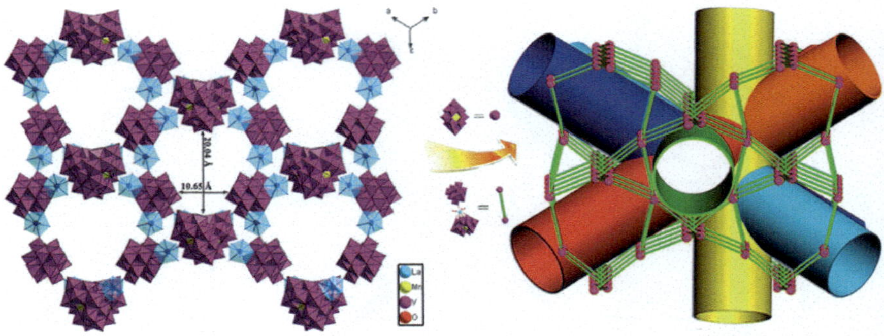

Fig. 1.5 3D framework structure with covalent connection of heteropolyvanadate and trivalent lanthanide ions and topological representation of six directional channels in the 3D framework. (Adapted with permission from Ref. [21]. Copyright 2013 Royal Society of Chemistry)

the structural environment around active elements in the catalytically active structure, it is classified as a complex catalyst in a discrete form. Perhaps further control of complexity around the catalytically active structure will be needed to tackle highly challenging methane oxidation reaction, because this direction is considered to be the right way for the evolution of new catalysts.

Polyoxometalates are the next crystalline materials with discrete structural units, i.e., oxide clusters, as the primary structural unit. The discrete structural units are arranged in a regular manner via organic cations or metal cations (Figs. 1.4, 1.5, and 1.6). Polyoxometalate is a well-known catalytic material with a long history of catalysis research and is important as industrial catalysts [18]. Catalyst design based on polyoxometalates has progressed in the direction of complexing various elements while maintaining the discrete structure [19–22]. Polyoxometalates are one of the most designable crystalline complex oxide catalysts and in fact are continuously used in catalyst development [23]. This material has also contributed significantly to the development of other solid metal oxide catalysts. An interesting example in a historical background is that the development of the composite metal oxide catalysts was initially driven by polyoxometalate materials as the basic starting materials. They were transformed into high-performance composite metal oxide catalysts through the steps of elemental complexation and thermal decomposition of the starting polyoxometalates. This is because the starting polyoxometalates can be synthesized with a wide variety of constituting elements.

Another completely different progress that has been made in the polyoxometalate system is the evolution of metal–organic framework (MOF) materials with polyoxometalate units shown in Fig. 1.5. The materials have various 3D structures with porous framework structures [21, 24] and have been tried to use as solid catalysts aiming at the utilization of the pore as a catalytic reaction field. At the same time, comparisons between metal oxide clusters in MOFs and bulk crystalline oxide will promote a better understanding of the catalytic property of metal oxides. Furthermore, there seem links between artificial metal oxide clusters supported in the organic

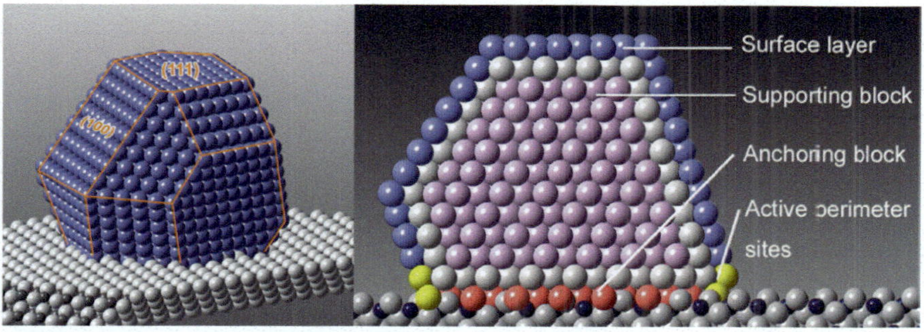

Fig. 1.6 Gold nanoparticle anchoring onto CeO_2 surface, low temperature CO oxidation catalyst, with schematic illustration of the functions of different domains. (Adapted with permission from Ref. [22]. Copyright 2012 American Chemical Society)

moiety of MOF and natural enzyme catalysts occluding molecular clusters as shown in Fig. 1.2 [25].

2D materials refer here to the group of materials formed on the surface layer [26]. Examples in this category are Au nano-particles supported on TiO_2 (Fig. 1.6), discrete metal oxide clusters held on metal surfaces (Fig. 1.7), and a monolayer of RuO_2 having rutile type structure epitaxially formed on a crystalline rutile type TiO_2 surface (Fig. 1.8) [22, 27, 28]. The support substrate only provides a spatial field over its surface, but also strongly influences the metal–oxygen bonding of catalytically active elements, discrete structure like monodisperse $(WO_3)_3$ oxide clusters on $TiO_2(110)$ [29], and electronic state of the discrete material through strong structural interactions [30] and electron transfer [31] between the support and catalytic elements. These result in the evolution of catalytic functionality. Metal oxide catalysts in this category have an old history as supported catalysts, so to speak, but as analysis progresses, they are increasingly recognized as catalysts of complex systems with a 2D local structure as depicted in Figs. 1.6 and 1.8.

As the number of dimensions further increases to three, the crystalline metal oxide catalyst cases that have been introduced in Table 1.1 come into play majorly. These are the usual crystalline complex oxides with a uniform periodic arrangement of elements. The first typical examples are complex metal oxides with layered structures as shown in Fig. 1.9, followed by 3-dimensionally condensed complex metal oxides like perovskite as representative (Fig. 1.10) [32–35]. These are very common as catalysts because of their wide constituent elemental diversity. Various constituting elements are represented as colors in these figures.

With the increase in dimensionality, zeolite-type materials with porous nature, like aluminosilicates and aluminophosphates, and functional zeolites with metal-substituted type in the structure framework [36], like titanosilicates and heteroatom-containing aluminophasphates [37], so-called single site catalysts, appear as shown in Fig. 1.11. Zeolitic material itself shows prominent solid-acid property and heteroatoms incorporated in the framework show additional catalytic functions like oxidation. In this silicate material system, MOF-type approaches using organic

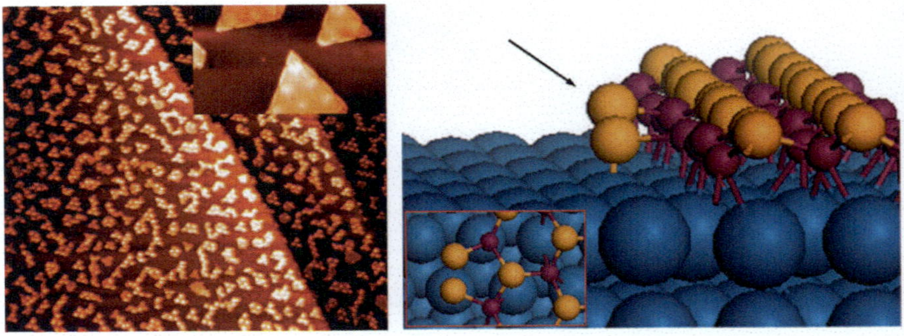

Fig. 1.7 STM images of Fe-oxide clusters on Pt surface and schematic structure of the boundary between Fe–O and Pt active for CO oxidation. From Ref. [27]. Reprinted with permission from AAAS

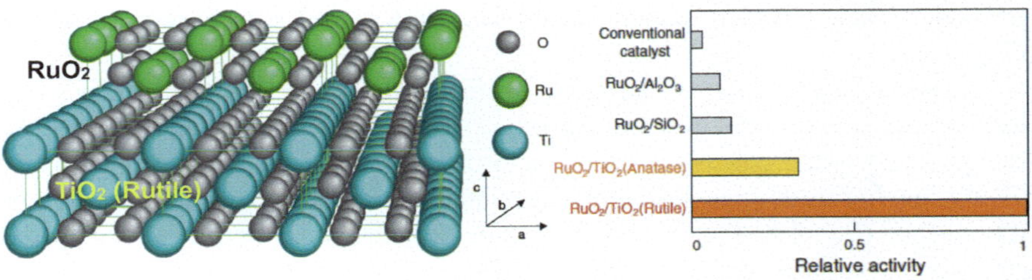

Fig. 1.8 Crystalographically well-fitted RuO_2 on TiO_2 with rutile structure and superior catalytic activity for HCl oxidation compared to other supported catalysts. (Adapted with permission from Ref. [28]. Copyright 2010 Springer)

linkers were also progressed, resulting in porous organosilicate materials [38] as one of the successful examples shown in Fig. 1.11.

The next to emerge after zeolite are catalytic materials of a crystalline complex metal oxide in which a periodic crystal structure is composed of 3D oxide discrete clusters connected 3-dimensionally via metal–oxygen polyhedral and accompanied by the formation of additional local structures (Fig. 1.12) [39–41]. The additional local structures here are pore structures and structural environment that are different from the 3D oxide discrete clusters of the basic unit. Details about this structural category of the complex oxide catalyst are described in Chaps. 3, 4, and 5. In this material group, many elements are orderly arranged in crystalline units, creating an environment in which many catalytic functions can be realized. Therefore, it is considered that a system for the future development of catalysts will be based on such crystalline complex metal oxides. The evolution of structural dimensions that can be led by increased complexity will be important for solid catalysts.

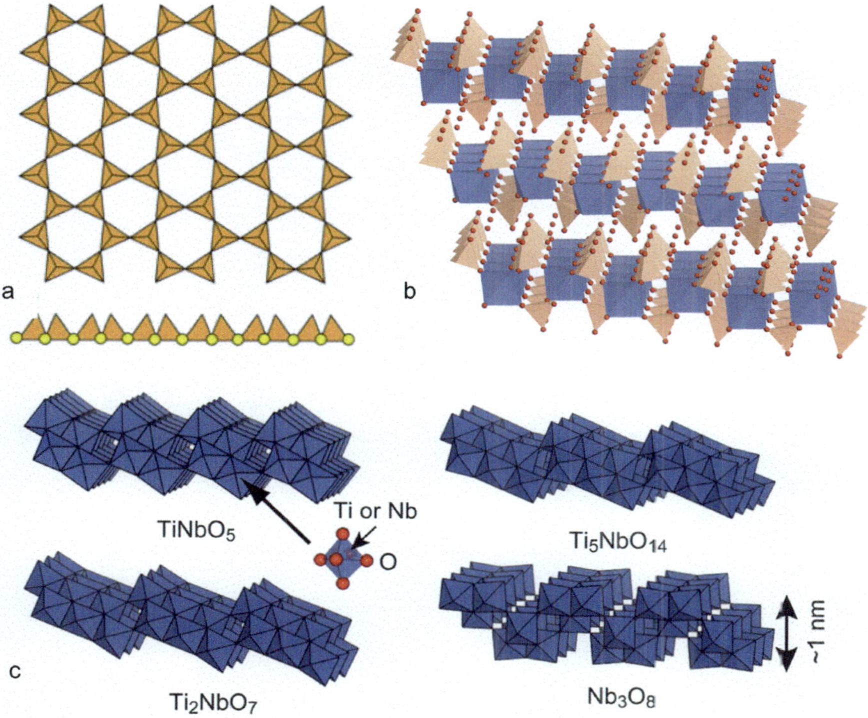

Fig. 1.9 Various layer-structured complex metal oxides. **a** sheet silicate (From Ref. [32]) Reprinted with permission from Cambridge University Press), **b** zirconium phosphate, and **c**; various niobates (Adapted with permission from Ref. [33]. Copyright 2011 Royal Society of Chemistry)

1.2 Structure Chemistry and Catalysis Chemistry of Crystalline Complex Metal Oxide Catalysts

1.2.1 Simple Metal Oxides with Catalytic Function

Oxides are compounds of oxygen with other elements and are known for most elements. Among them, oxides of typical metal elements, transition metal elements, and rare earth oxides such as lanthanides are often used as basic materials for oxide catalysts because of their thermodynamic and structural stability in the solid state. These oxides can be made by direct reaction of metals with oxygen molecules or other oxidants, but in catalyst preparation they are synthesized by more homogeneous and mass-synthesizable methods, such as dehydration-condensation in acids, dehydration of hydroxides, decomposition of salts, and reaction of compounds with oxygen. However, the structures of oxides are diverse, and the catalytic ability varies greatly accordingly, so precise catalyst preparation is important.

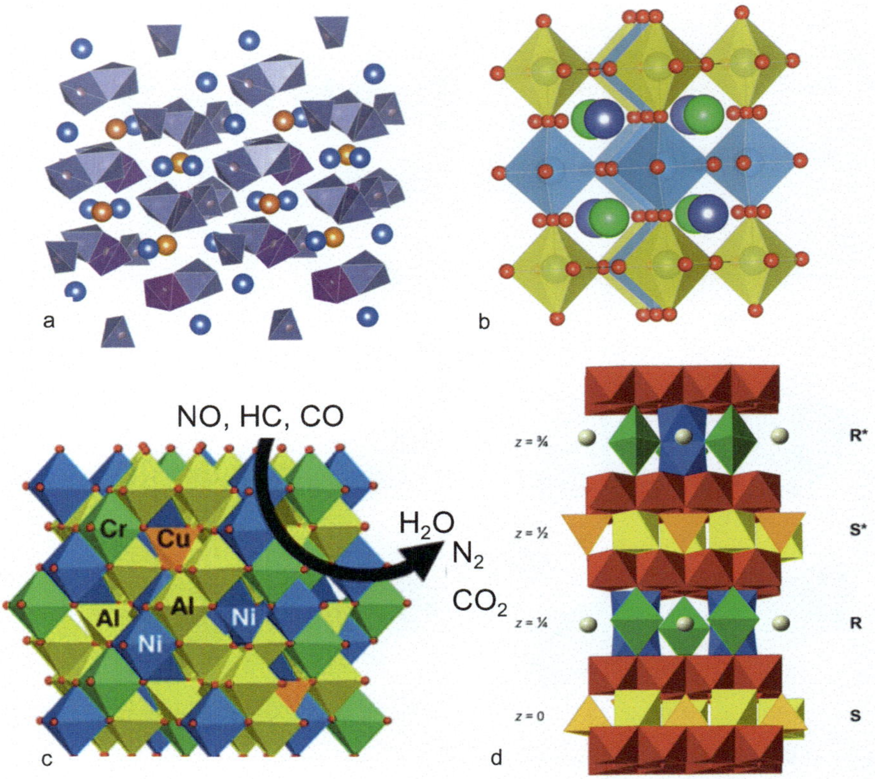

Fig. 1.10 Various condensed crystalline complex metal oxides having constituting element complexity. **a** scheelite structure ($(Bi_{2-x}Fe_x(MoO_4)_3$ or $Bi_2(Mo_{1-x}V_xO_4)_3$, **b** perovskite structure, **c** spinel structure (Adapted with permission from Ref. [10]. Copyright 2019 American Chemical Society), and **d** magnetoplumbite-type structure with various polyhedra. (From Ref. [35] Reprinted with permission from Cambridge University Press)

Table 1.3 summarizes the catalytic properties of the simple metal oxides related to the properties of each element along the periodic table. For oxides in groups 1 and 2, BeO is an amphoteric oxide with covalent bonding properties, and for elements after Li, the alkaline and alkali-earth oxides are ionically bonded and basic due to low electronegativity. Elements of group 13, B, Al, Ga, and In are strongly covalent and give acidic or amphoteric oxides. In the 14th oxides of group 4, only carbon gives molecular compounds, followed by covalent oxides. SiO_2 forms single covalent bonds due to their more extended orbitals compared to carbon, giving rise to poly-silicates with linked oxygen tetrahedra. SiO_2 is the oxide forming zeolites and is an essential material for catalysts. GeO_2 is similar in structure to SiO_2. SnO_2 is amphoteric, while the oxide of divalent Sn is basic. Divalent Pb is stable and its oxide is basic. In the 15th group, only nitrogen gives molecular oxides. P exists as an oxoacid ion and tends to condense to form covalent polymeric solids. As and

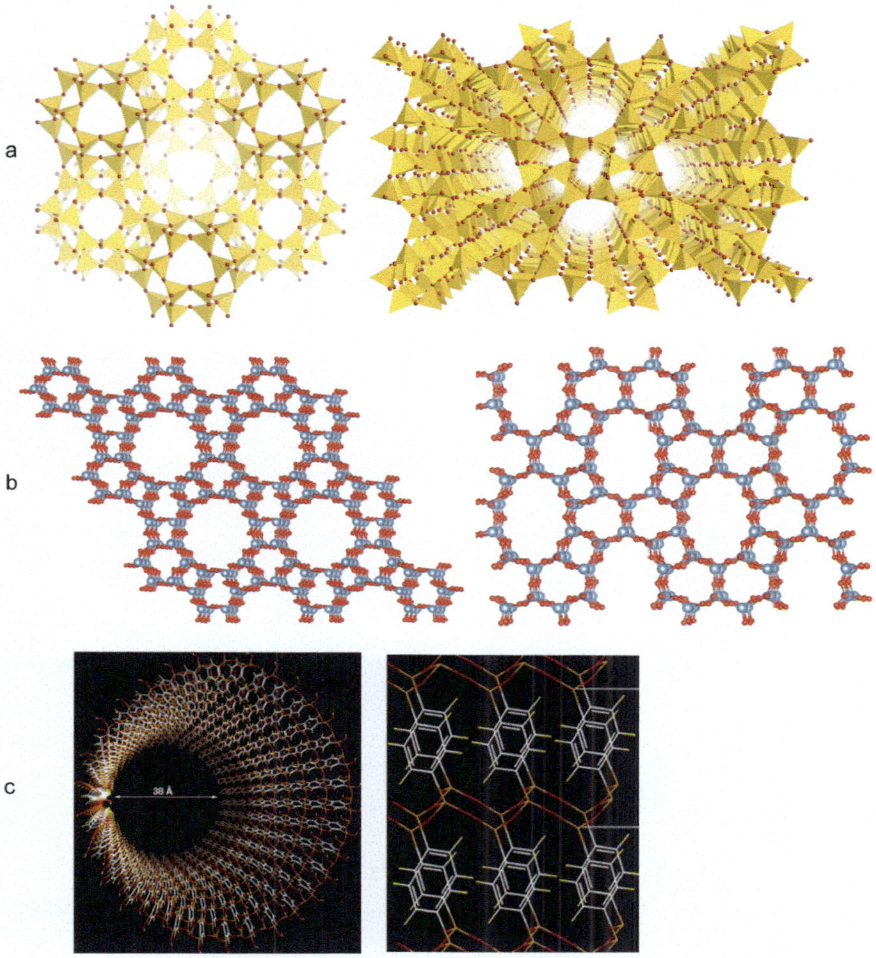

Fig. 1.11 Various crystalline complex metal oxides with porous framework structures based on tetrahedra arrangement. **a** aluminosilicates (zeolite), **b** metalophosphate (AlPO, SAPO) [37], and **c** organosilicate (From Ref. [38]. Reprinted with permission from Springer Nature)

Sb are trivalent oxides, but in an oxidizing atmosphere their valence increases, and the unstable $As_2O_5$and Sb_2O_5 oxides are formed. Since Bi has an inert electron pair effect, the trivalent oxide is extremely stable and becomes a basic oxide with strong ionic bonding. In group 16, oxygen and sulfur form molecular oxides, Se and Te take both tetravalent and hexavalent oxides, and the tetravalent oxides are relatively stable. When the oxidation number is high, it forms a molecular and acidic oligomer. In group 12 Zn, Cd, and Hg elements, *s* electrons in the two outermost shells dominate the reactivity, and in Zn and Cd, the divalent oxide has a stable ionic bond crystal structure and is basic.

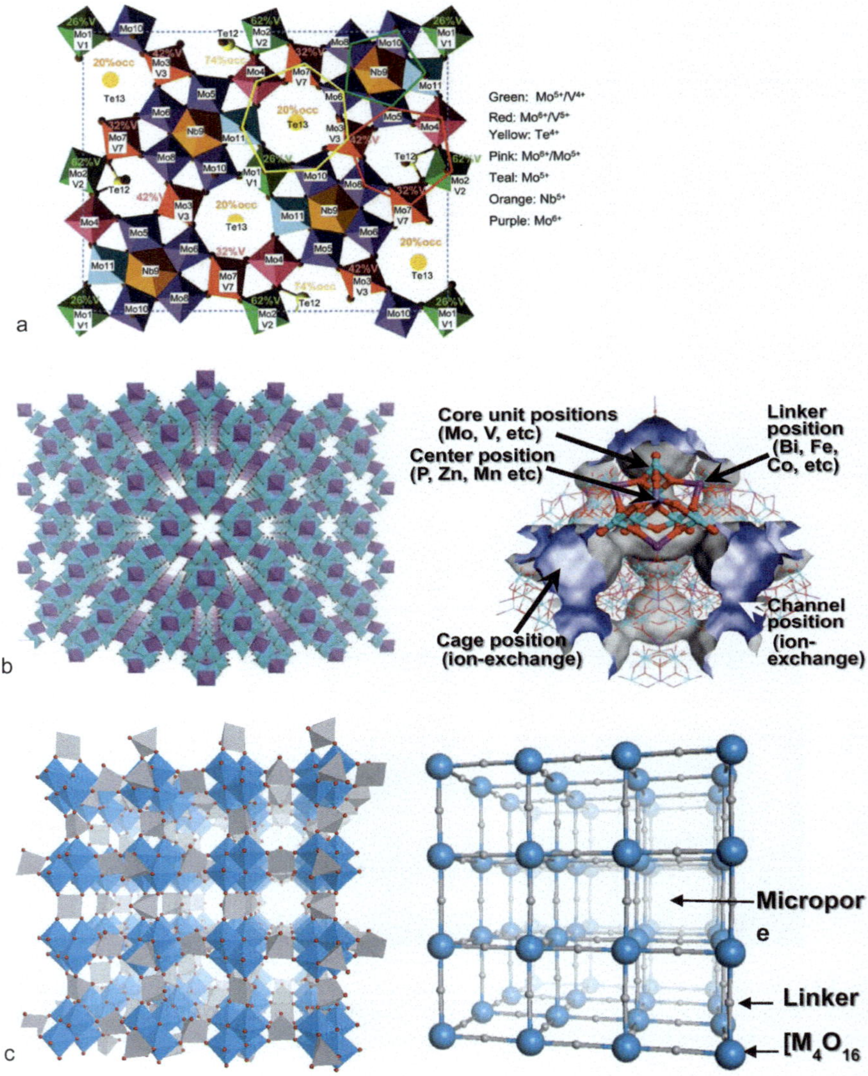

Fig. 1.12 Various crystalline complex metal oxides with porous framework structures based on octahedra arrangement and with high elemental complexity. **a** $Mo_1V_{0.23}Te_{0.09}Nb_{0.24}O_x$ (Adapted with permission from Ref. [39]. Copyright 2013 American Chemical Society), **b** ε-Keggin-based porous polyoxometalate [40], and **c** porous framework structure based on cubane-type oxide clusters [41]

Table 1.3 Classification of simple metal oxides with catalytic function

Type	O/M	Structural character	Crystal structure (coordination number)		Example	Catalytic function[a]
MO_4	4	Molecular		(4)	RuO_4, OsO_4	Oxidation
M_2O_7	3.5	Molecular		(6)	Mn_2O_7, Te_2O_7, Re_2O_7	Oxidation
MO_3	3	Chain		(4)	CrO_3, SO_3	Oxidation, Acid
		3D crystal		(6)	MoO_3	Redox, Acid
			ReO_3	(6)	WO_3, MoO_3	Redox, Acid
M_5O_{14}	2.8	3D crystal	Bronze	(6,7)	Mo_5O_{14}	Redox
M_2O_5	2.5	Chain		(6)	P_2O_5, As_2O_5, Sb_2O_5	Acid, Oxidation
		Layer		(5,6)	V_2O_5,	Redox, Acid
		3D crystal	Block	(4,6)	Nb_2O_5, Ta_2O_5	Acid, oxidation
MO_2	2		Chain	(6)	SeO_2	Redox
		3D crystal	Fluorite	(8)	ZrO_2, HfO_2,CeO_2, UO_2	Redox, Neutral
			Rutile	(6)	TiO_2, VO_2, NbO_2, TaO_2	Reduction
					CrO_2, MoO_2, WO_2, MnO_2	Acid–base
					ReO_2, RuO_2, GeO_2, SnO_2	Oxidation
					PbO_2, TeO_2, TiO_2, PtO_2	IrO_2 photo
			Anatase	(6)	TiO_2	Photo
			Brookite	(6)	TiO_2	Photo
			Hollandite	(6)	MnO_2	Acid–base
			Silica	(4)	SiO_2, GeO_2	Neutral
M_2O_3	1.5	Chain			Sb_2O_3	Redox
		Layer	Sillen	(6)	As_2O_3, Bi_2O_3	Redox, Base
		3D crystal	Corundum	(6)	Al_2O_3, Ti_2O_3, V_2O_3, Cr_2O_3	Acid–base
					Fe_2O_3, Ga_2O_3, In_2O_3	Reduction
				(>6)	Sc_2O_3, Y_2O_3, Ln_2O_3	Redox, Base
M_3O_4	1.3	Chain		(6,5)	Pb_3O_4	Base

(continued)

Table 1.3 (continued)

Type	O/M	Structural character	Crystal structure (coordination number)		Example	Catalytic function[a]
		3D crystal	Spinel	(6,4)	Fe_3O_4, Co_3O_4, Mn_3O_4	Redox, Oxidation
MO	1	Layer			PbO, SnO	
		3D crystal	NaCl	(6)	MgO, CaO, SrO, BaO, TiO	Reduction, Base
					VO, MnO, FeO, CoO, NiO	
			Wurtzite	(4)	BeO, ZnO	Reduction, Base
M_2O	0.5	3D crystal	Antifluorite	(4)	Li_2O, Na_2O, K_2O, Pb_2O	Base
			Cuprite	(2,4)	Cu_2O, Ag_2O	Redox

[a] Redox; reduction–oxidation property during catalysis and OER and ORR properties
Oxidation: property oxidizing molecules; Reduction: dehydrogenation or hydrogenation
Photo: photocatalysis
Neutral: property showing very weak acid or base

Oxides of transition elements are important as catalysts. In group 4, Ti oxides are used as catalysts in many systems such as photocatalysts and as supports. The tetravalent of Ti is the most stable and forms TiO_2, and there are three types of crystal transformations composed of oxygen octahedrons called rutile, anatase, and brookite. The oxides are acidic and form titanates with basic cations. Since Zr and Hf of the same family have larger ionic radii, the oxide is neutral or basic. The thermal stability is high, and the crystal phase of ZrO_2 includes tetragonal and cubic crystals. In group 5 V is an indispensable element as an oxidation catalyst. This is because a variety of oxidation states can be taken, so that various oxides with different oxidation state and different structures are formed, and non-stoichiometric oxides are also formed. Nb_2O_5 and Ta_2O_5 are chemically inactive against oxidation and reduction, but show characteristic acid and base properties, for example, water-tolerance solid-acid (see more examples in Chap. 11). Since they form niobate and tantalate, it is classified as weakly acidic oxide. Up to Cr and Mn, the maximum oxidation number equal to the group number can be taken, and the oxide becomes a powerful oxidizing agent. Mo in the next row has small ionization energy, and forms oxide with a high oxidation number in various crystal structures. The oxidizing power of Mo is lower than Cr, but acid properties increase. W does not show oxidizing power, but its acidity is strong. Like V oxides, many types of Mo and W oxides are known, including non-stoichiometric oxides. The characteristic of these oxides is that the oxygen octahedron shares corners, edges, and face, and condenses to form oxides in various structures. One of the characteristic structures is a shear structure which can flexibly change with the composition of an indefinite ratio. Mo and W are also the main elements of polyoxometalate.

There is a big difference between Mn and Re in Group 7. Mn has the most stable divalent oxide and Mn in a higher oxidation state forms a variety of oxide structures based on edge and face sharing of octahedral (Chap. 8). On the other hand, Re has mostly 4+ and 6+ valence and forms ReO_3 structure with corner sharing of octahedral like as perovskite structure. The catalytic function of Re oxide is not well-clarified but Re oxide shows redox performance and acid properties like Mo. In group 8, Fe has the maximum oxidation number of 6+. Co and Ni have mostly a low valence like 2+ in the stable simple oxides and 4+ is the maximum oxidation number for these elements. Fe does not form FeO_4, but as the period decreases, the ionization energy decreases, and oxides of RuO_4 and OsO_4 can be formed and show strong oxidizing ability. Rh and Ir differ from Ru and Os and they do not form oxoacid or high oxidation number oxides.

Cu possesses mostly a low valence, 1+ and 2+, in the stable simple oxides, and the monovalent oxide of Cu, Cu_2O easily changes to CuO. The stability of oxide decreases to Ag, Pd, and Pt, and becomes minimal at Au. Group 3 and lanthanoid element oxides are characterized by taking the M_2O_3 type. For this reason, the valence hardly changes, but some elements become divalent or tetravalent oxides or non-stoichiometric oxides. The oxides are thermally stable and mostly exhibit basic properties.

Most of the single metal oxides are ionic crystals and some are those with covalent metal–oxygen bond character. Even single metal oxides have a wide variety of crystal structures depending on the O/M ratio as shown in Table 1.3. In stably coordinated structures, metal ions are surrounded by oxygen octahedra and/or tetrahedra, and the crystal structure depends on how the metal ions are arranged in the tetrahedral and/or octahedral gaps that are formed in the closely packed structures (hexagonal and cubic) created by the arrangement of oxygen ions. The metal–oxygen bond distances and bond angles, or polyhedral distortions also vary with the splitting of the *d* orbitals in the crystal field, leading to changes in the overall structure. In the first transition element series, the bonding to the coordinating oxygen is weak, leading to a variety of structures, but from the second transition element series onward, the bonding is strong, leading to a polyhedron with high symmetry. This is one reason for the difference in structural properties between molybdenum oxide and tungsten oxide. Molybdenum oxide produces distorted oxygen octahedra, resulting in metal–oxygen bonds with different bonding strengths in a single octahedral, which has a significant effect on its properties as an oxidation catalyst. On the other hand, tungsten oxide has a highly symmetric and structurally stable oxygen octahedron but can allow a coordinatively unsaturated state on its surface and provide a catalytic reaction field. Thus, the coordination structure and local structure of oxides are strongly related to their catalytic properties.

Another important factor affecting catalysis is the electronic state of the metal elements and oxide ions that make up the oxide. Figure 1.13 shows the electronegativity of the lattice oxygen of the metal oxide, expressed as Pauling's scale based on Sanderson's rule. The metal oxides are arranged according to the calculated values [42]. The lower the value, the closer the lattice oxygen charge is to the formal charge of −2, i.e., nucleophilic, and the higher the value, the closer it is to the neutral oxygen atom, i.e., electrophilic. This trend is also observed for metal elements that have a high valence and a large oxygen-to-metal ratio (O/M), while those with a low valence and a small O/M ratio are categorized at the bottom. Comparing the catalytic functions of these oxides, the oxide catalysts belonging to the upper group promote oxygen addition reactions, while those belonging to the lower group give priority to dehydrogenation or C–H bond cleavage. Similarly, the higher the O/M ratio the oxide has, the more acidic the catalyst shows. Clearly, element-dependent catalytic functions can be seen in a systematic manner. This ordering is based on the average metal–oxygen bonds, whereas in actual metal oxides, the bonding mode is more complex depending on changes in the coordination structure. In particular, the upper oxides can take a variety of bonding configurations and thus can perform a variety of catalytic functions, e.g., both oxygen addition and dehydrogenation depending on these configurations. In fact, the oxides of the upper elements exhibit a variety of catalytic activities. The lower oxides tend to be more catalytically oriented toward dehydrogenation.

Therefore, even single metal oxides exhibit certain catalytic functions, and in fact, in the early stages of catalytic evolution, single metal oxides were used (Table 1.1). Moreover, controlling the bulk crystal structure and exposed facet makes it possible to possess multiple catalytic functions even for a single metal oxide, and many studies have been conducted. However, since catalytic reactions of ordinary organic reactants always consist of multiple elementary reaction steps, it is impossible to satisfy all necessary catalytic functions with a single oxide in a well-balanced manner, as already discussed and a composite of metal oxides with different properties will work to advantage. This is exactly what is happening, and further improvement can be achieved with crystallization, which creates an inevitability toward crystalline complex oxide catalysts.

1.2.2 Crystalline Complex Metal Oxides with Catalytic Function

If we look at crystalline complex oxides from the viewpoint of solid-state chemistry, we notice that they form a truly large group of materials, and the diversity of their crystal structures is immeasurable. New crystals are continuously synthesized, and new physical properties are created. From the viewpoint of solid catalysts, on the other hand, it is hard to say that this wide library of complex metal oxide crystals has been fully utilized for creating solid catalysts. Rather the utilization seems to be limited to several crystal structures. Although many crystalline materials must have been tested as catalysts in the history of catalyst development, the successful examples of crystalline complex metal oxides as catalysts are limited to common solid-state structural materials, as shown in Table 1.4. Also, the number of newly developed crystalline metal oxide catalysts is not increasing steadily even at present. The major reason for this is that, as mentioned in the history of catalyst development, there was

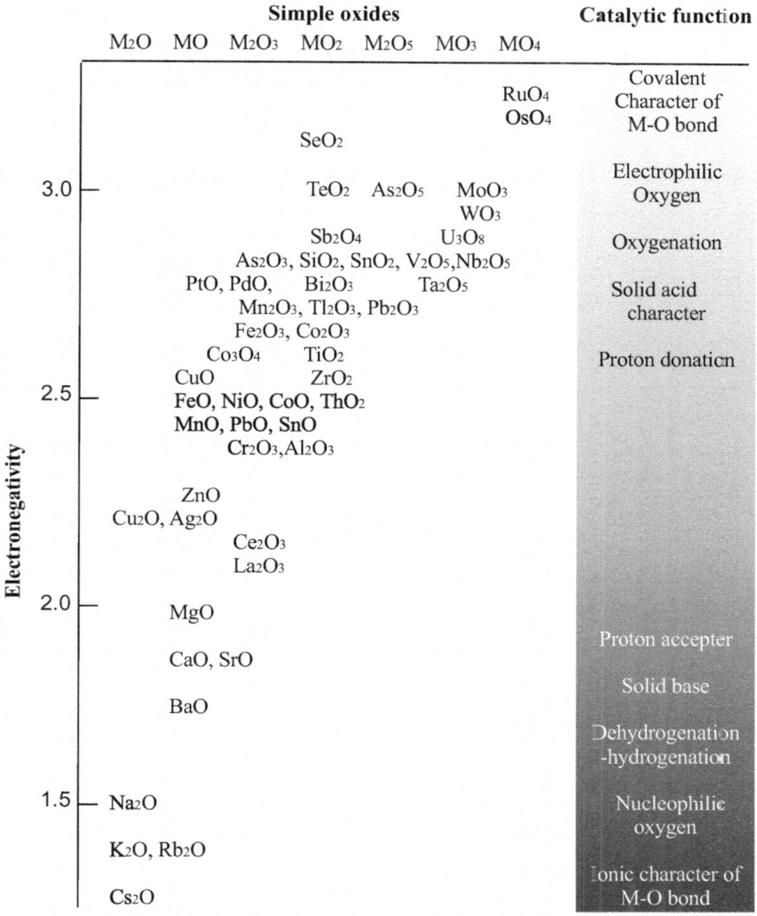

Fig. 1.13 Electronegativity of lattice oxygen and catalytic function of simple metal oxides

a technological background in the transition from simple composite metal oxides to a crystalline form, namely satisfactory achievement of catalytic performance was a priority rather than the state of catalytic materials. For example, crystallization by high-temperature calcination, which is a normal solid synthesis process, usually forms materials with extremely low surface area. This is obviously disadvantageous from the viewpoint of catalytic function generation and therefore was avoided to use, except in special cases such as combustion reactions where high reaction temperature is needed.

Meantime, even though catalysts do not have the state of crystal form in whole materials prepared by the normal methods, the resulting materials have been considered to possess catalytically active local structures on the surface as far as the resulting materials show appreciable catalytic properties. This situation in turn

Table 1.4 Type of crystalline complex metal oxides (A–M–O) and catalyst examples with their function

Type	O/(A+M) ratio	Structural character	Crystal Coordination structure number (A, M)[a]		Example	Catalytic function	Structure in figures
$[AM_xO_y]$	3.43 3.14 3.0	Discrete Chain Sheet	Polyoxometalate Polyoxometalate Linked polyhedral	(4, 6) (6, 6) (6, 6)	$[PMo_{12}O_{40}]$, $[PW_{12}O_{40}]$, $[PMo_{11}VO_{40}]$, $[PW_{11}VO_{40}]$, $[TeMo_6O_{24}]$, $[P_2W_{18}O_{62}]$ $[SeW_6O_{21}]$, $[TeW_6O_{21}]$ $[TaO_3]$	Oxidation, Acid Acid Oxidation, Acid	(1.14) (1.15) (1.16)
	2.86 2.71	3D porous framework	Linked polyoxometalate Linked polyhedral	(6, 6) (6, 5)	$[Bi_2Mo_{12}O_{40}]$ $[W_4V_3O_{19}]$	Oxidation NH_3–SCR, Acid	(1.12.b) (1.12.c)
$[A_xBO_3]$	2.8	3D porous framework	Linked polyoxometalate	(6,5)	$[Mo_3VO_{11.2}]$, $[Mo_3VO_x]$	Oxidation	(1.17)
A_xBO_3	2–2.7	3D non-porous framework	ReO_3, bronze-type	(6, 5)	$K_{0.5}WO_3$, $Sb_2Mo_6O_{20}$, $Mo_{5-x}V_xO_{14}$ $Cs_{0.5}Nb_{2.5}W_{2.5}O_{14}$, $Nb_8W_9O_{47}$, $Mo_{7.8}V_{1.2}Te_{0.94}NbO_{28.9}$	Oxidation	(1.19) (1.12.a)
$[AB_2O_8]$	2.66	Open layer	Phosphate, molybdate	(6, 4)	$[ZrP_2O_8]$, $[Zr_{1+x}Mo_{2-x}O_8]$	Acid	(1.19.b)
$[AMO_5]$	2.5	Open layer	Niobate	(6, 6)	$[TiNbO_5]$	Acid	(1.19.c)
AM_3O_{10}	2.5	3D condensed	Antimonate, uranate	(8, 6)	USb_3O_{10}, UTa_3O_{10}	Oxidation	
$A_2M_2O_9$	2.25	Layer condensed	Pyrophosphate	(5, 4)	$(VO)_2P_2O_7$	Oxidation, Acid	(2.4)
$[AMO_4]$	2.0	3D porous framework	Metalosilicate Metalophosphate	(4, 4) (4, 4)	Zeolite, Titanosilicate AlPO-11, SAPO-11	Acid, Oxidation	(1.11.a) (1.11.b)
A_2MO_6	2.0	Layer condensed	Aurivillious Fluorite	(6,6)	Bi_2MoO_6 Sb_2MoO_6	Oxidation Oxidation	(1.20.c)

(continued)

Table 1.4 (continued)

Type	O/(A+M) ratio	Structural character	Crystal Coordination structure number (A, M)[a]		Example	Catalytic function	Structure in figures
AMO_4	2.0	3D condensed	Scheelite	(4, 8)	$CaWO_4$, $PbMoO_4$, $BiVO_4$ $Bi_2(MoO_4)_3$, $Fe_2(MoO_4)_3$	Oxidation, Acid, Photo	(1.20.a)
			α-$CoMoO_4$	(6, 6)	$NiMoO_4$	Oxidation, Acid	(1.21.a)
			α-$MnMoO_4$	(4, 6)	$CoMoO_4$, $MnMoO_4$	Oxidation, Acid	(1.21.b)
			Phosphate	(8, 4)	$BiPO_4$, $CePO_4$, $FePO_4$, BPO_4, $LaPO_4$	Oxidation, Acid	(1.22)
			Rutile	(6, 6)	$FeSbO_4$, $AlSbO_4$, $Ru_xTi_{1-x}O_2$	Oxidation, Acid	(1.23.a)
			Fluorite	(8, 8)	$CeZrO_4$	Oxygen carrier	(1.23.b)
$A_2M_2O_7$	1.75	3D condensed	Pyrochlore	(4, 6	$La_2Zr_2O_7$, $Pb_2Sb_2O_7$	Reforming, Redox	(1.23.c)
			Dion-Jacobson	(6, 6)	$BaSrNb_2O_7$	Photo	(1.24)
			Pyrophosphate	(6, 6)	$Cu_2P_2O_7$, $Ni_2P_2O_7$, $Mn_2P_2O_7$	Oxidation	
			Pyrovanadate	(6, 4)	$Mg_2V_2O_7$	Oxidation	(1.22.d)
$A_4M_3O_{12}$	1.71	Layer condensed	Aurivillious	(6, 6)	$Bi_4Ti_3O_{12}$	Photo	(1.24)
$A_4M_6O_{17}$	1.7	3D condensed	Niobate	(6, 6)	$K_4Nb_6O_{17}$	Photo	(1.22.e)
$A_3M_2O_8$	1.6	3D condensed	Phosphate	(6, 4)	$Ca_3(PO_4)_2$	Acid–base	(1.22.c)
AMO_3	1.5	Layer condensed	Double perovskite	(12, 6)	$LaCuMnO_6$, $LaNiTiO_6$ $La_{0.5}Sr_{0.5}FeNiO_6$,	Oxidation, OER, Photo	(1.24)
		3D condensed	Perovskite	(12, 6)	$LaCoO_3$, $BaTiO_3$, $SrTiO_3$ $LaSrCoO_3$	Redox, Electrolyte, Photo	(1.24)
			Corundum	(6, 6)	$FeVO_3$, $MnFeO_3$	Reduction, Acid–base	(1.24)

(continued)

Table 1.4 (continued)

Type	O/(A+M) ratio	Structural character	Crystal Coordination structure number (A, M)[a]		Example	Catalytic function	Structure in figures
			Ilmenite	(6, 6)	$MgSiO_3$, $FeTiO_3$	Base	(1.24)
			Defect pyrochlore	(6, 6)	$BiYO_3$	Base	
			Garnet	(6, 4)	$Fe_3Al_2(SiO_4)_3$	Redox	
			Hexaalluminate		$BaMAl_{11}O_{19}$(M = Cr, Mn, Fe, Co, Ni)	Redox	
$AM_{11}O_{18}$	1.5	Layer condensed	β-alumina		$LaFe_{11}O_{18}$	Redox	
$AM_{12}O_{19}$	1.46	Layer condensed	Magnetoplumbite		$Sr_{1-x}La_xFe_{12}O_{19}$	Redox	(1.10.d)
$A_4M_3O_{10}$	1.43	3D condensed	Ruddlesden-Popper	(12, 6)	$Sr_4Ti_3O_{10}$	Redox	(1.24)
$A_3M_2O_7$	1.4	3D condensed	Ruddlesden-Popper	(12, 6)	$Sr_3Fe_2O_7$, $La_xCa_{3-x}Mn_2O_7$	NO oxidation, SOFC	(1.24)
A_2MO_4	1.33	3D condensed	Ruddlesden-Popper	(12, 6)	$Pr_{0.5}Ca_{1.5}MnO_4$, $(La_{0.6}Sr_{0.4})_2(Co,Fe)\ O_4$	ORR, OER, SOEC CO_2 reduction	(1.24)
A_2MO_4	1.33	3D condensed	K_2NiF_4		La_2NiO_4	Redox	
AM_2O_4	1.33	3D condensed	Spinel	(4, 6)	$ZnCr_2O_4$, $CuCr_2O_4$, $NiCr_2O_4$ $MnCo_2O_4$, $CuCr_{2-x}Fe_xO_4$, $CuAl_2O_4$	Redox, Base, Photo	(1.10.c)
			Inverse spinel	(6, 4)	$CoFe_2O_4$, $NiCr_{2-x}Fe_xO_4$, $Mg_{1-x}Cu_xAl_2O_4$	Redox	
$A_2M_2O_5$	1.25	3D condensed	Brown-Millerite		$Ca_2Fe_2O_5$, $Sr_2Fe_2O_5$	Redox, Base	
AMO_2	1.0	3D condensed	NaCl	(6, 6)	$LiNiO_2$, $LiAlO_2$	Oxidation, Base	
			Phosphate	(6, 4)	Li_3PO_4	Oxidation, Base	

(continued)

Table 1.4 (continued)

Type	O/(A+M) ratio	Structural character	Crystal Coordination structure	number (A, M)[a]	Example	Catalytic function	Structure in figures
			Silicate, germanate	(6, 4)	Li_2CaSiO_4, Li_2SiO_3, Li_2GeO_3	Oxidation, Base	
			α-$NaFeO_2$	(6, 6)	$LiCoO_2$	Electrocatalysis	
			Olibin	(6, 4)	$LiFePO_4$	Electrocatalysis	
A_4MO_4	0.8	3D condensed	Silicate, germanate		Li_4SiO_3, Li_4GeO_3	Oxidation, Base	

[ ] is for samples in negatively charged materials, so that counter cations in the examples of the discrete structure and 3D open framework structure with negative charge are not taken into consideration for O/(A+M) ratio and are not shown in chemical formulae

[a] Coordination numbers are the main catalyst components of typical examples

promoted various trials of transforming ill-crystallized materials into crystalline complex oxides by developing various catalyst synthesis methods. This trial ultimately resulted in high-performance catalysts. In addition, advances in synthetic technology helped the evolution of new crystalline complex metal oxide catalysts because there is a great variety of crystalline complex oxide crystals in the materials library. Eventually, the prediction of which crystal structure is suitable for certain catalysis will be practical.

The crystalline complex metal oxide catalysts created through the above historical steps are classified on the basis of the crystal structure (Figs. 1.14, 1.15, 1.16, 1.17, 1.18, 1.19, 1.20, 1.21, 1.22, 1.23 and 1.24) and the tendency of the catalytic function in Table 1.4. If the basic body of the structure has a negative charge and other metal cations are incorporated separately into the structure to compensate the negative charge, this cationic element is not included in the calculation of the composition ratio, O/[A+M]. Instead, the basic body is only used. These cases are distinguished by using parenthesis [] from others. In addition, the structural properties are also summarized in Table 1.5. The general theory of the effects of crystalline structuring and element complexation will be discussed later, so here the catalytic characteristics of each structure are listed separately.

The classification of catalytic reactions is very similar to that of single oxides (O/M) when listed by the O/(A+M) ratio of elemental compositions. If M, the main constituent, has a high oxidation number, i.e., a high O/(A+M) ratio, the catalytic properties are dominated by M, which catalyzes oxidation and acid reactions. If the element A that can replace M is one with properties similar to M, i.e., the elements at the top region in Fig. 1.13, and if the charge of A is different from that of M, the M–O–A bond character and the acidic quality also change. The complex metal oxides in this O/(A+M) region are also characterized by the formation of multiple condensed oxygen octahedra in a cluster morphology. A typical example is a heteropoly compound like polyoxometalate shown in Fig. 1.14. Structures of the clusters are completely discrete, while some others are interconnected and develop into 3D crystals while maintaining the cluster motif. For example, discrete clusters condense each other to form an infinite chain, which further connects each other with the assistance of metal cations to form 3D solids. The examples are shown in Fig. 1.15 [43, 44]. If discrete clusters condense each other to form sheets, which further connect each other with the assistance of metal cations in interlayer space, 3D solids form like the example shown in Fig. 1.16 [45–48]. If discrete clusters interact with each other more densely, 3D solids accommodating pore structure can be formed (Figs. 1.17 and 1.18) [49–52]. The formed pore structure is usually occupied by other metal ions (Fig. 1.18) [53], resulting in non-porous properties that cannot absorb any small molecules, although some of these materials are called octahedral molecular sieves (OMS) [54] or SOMS [52] However, if materials with fully open pore can be synthesized like as $Mo_3VO_{11.2}$ in Fig. 1.17 [49, 50], the resulting material should be regarded as truly porous crystalline complex metal oxide, which can provide unique catalytic functions.

A high degree of condensation of discrete clusters by further sharing of lattice oxygen ions results in the disappearance of the pore structure (Fig. 1.19) but again in

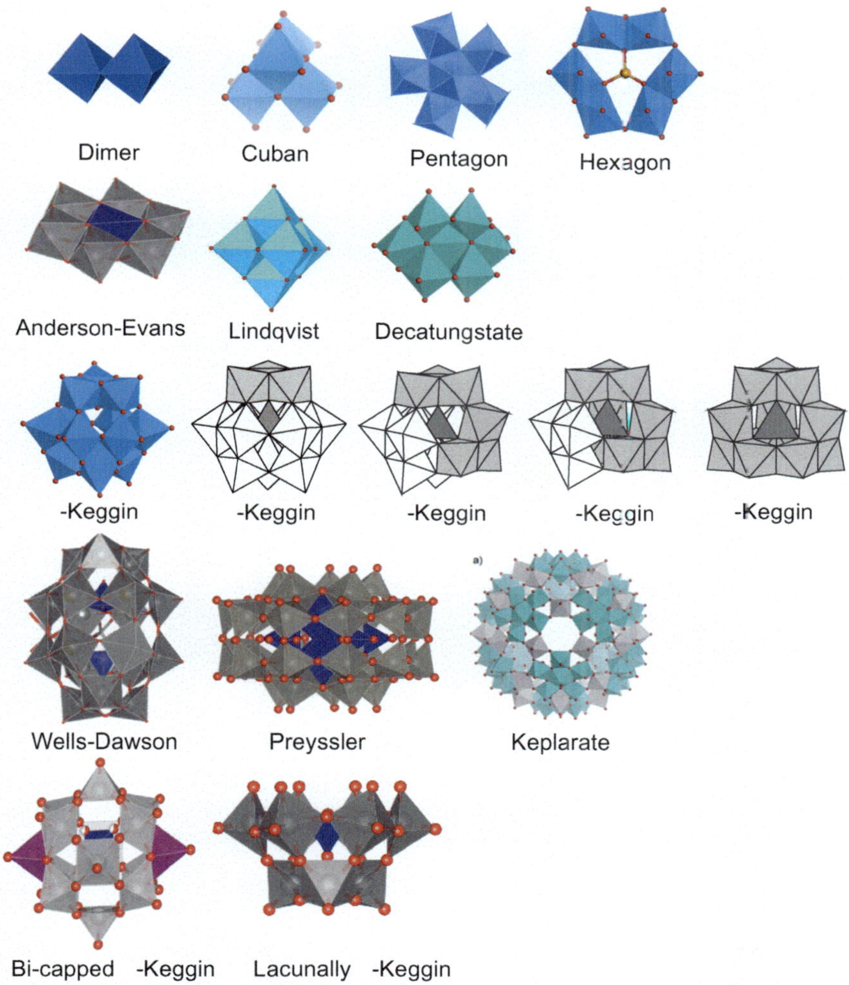

Fig. 1.14 Discrete clusters of metal oxygen octahedral assembly (see more examples in Chap. 3)

the transformation into a densely layered compound without space or voids between the layers (Figs. 1.19 and 1.20a). This type of layered compound is thermodynamically more stable than that of porous structures and is relatively easy to form. In fact, layered compounds are often obtained in the process of catalyst development and are highly practical because the protons and metal ions that exist between the layers are also functional as catalysts. As condensation proceeds further, the layered structure disappears and discrete structures consisting of multiple polyhedral begin to form. $(VO)_2P_2O_7$ is the example, in which an octahedral forms a dimer surrounded by tetrahedral as can be seen in Fig. 2.4.

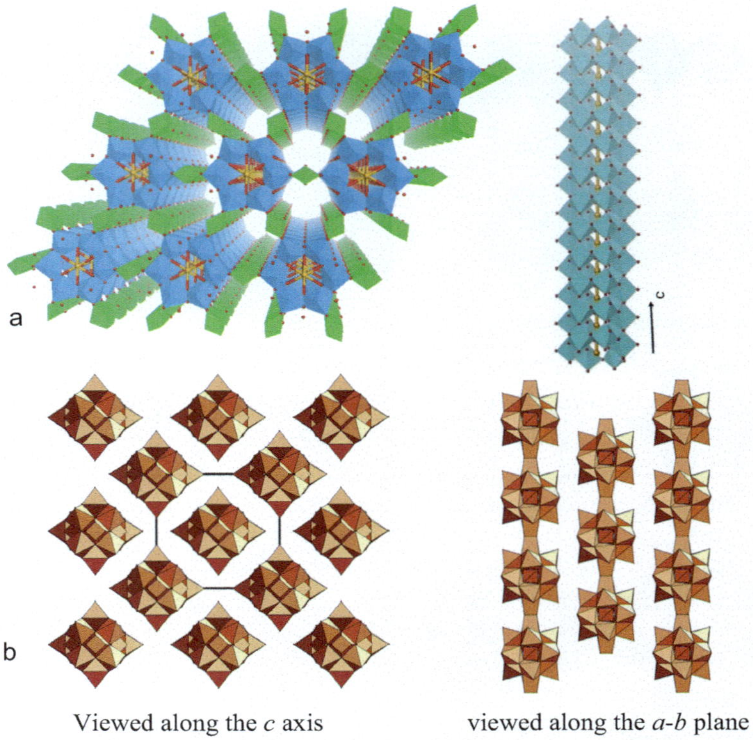

Fig. 1.15 Discrete oxide cluster assembling into chains and then linking each other into 3D structure with metal polyhedra linker. **a** $Co[TeW_6O_{20}]$ [43], **b** Na_{12} $[Ti_2O_2]$ $[SiNb_{12}O_{40}]4H_2O$ (Adapted with permission from Ref. [44]. Copyright 2005 American Chemical Society)

Fig. 1.16 Structure of $RbTaO_3$ as an example of discrete oxide cluster, tetramer of Ta–O, assembling into slab and then linking each other into 3D layer structure with accommodating metal cations (Rb) in interlayer. Adapted with permission from Ref. [48]. Copyright 2021 Royal Society of Chemistry

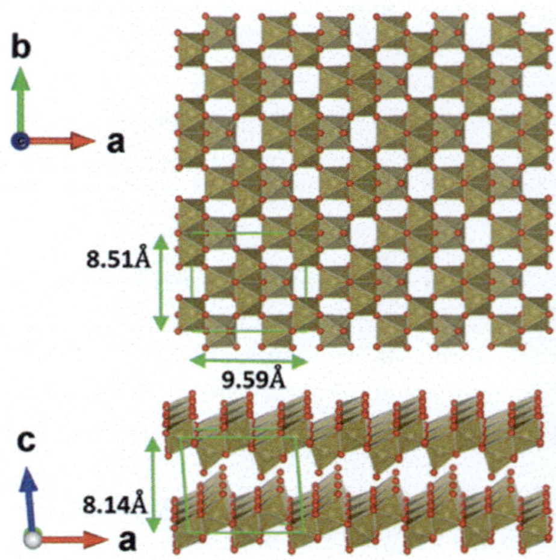

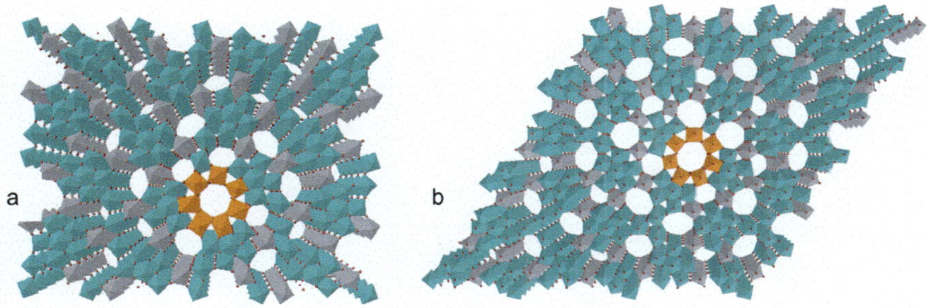

Fig. 1.17 Discrete oxide cluster assembling with metal polyhedra linker into open porous 3D framework structure. **a**: orthorhombic $Mo_3VO_{11.2}$ and **b**; trigonal $Mo_3VO_{11.2}$ [49, 40]

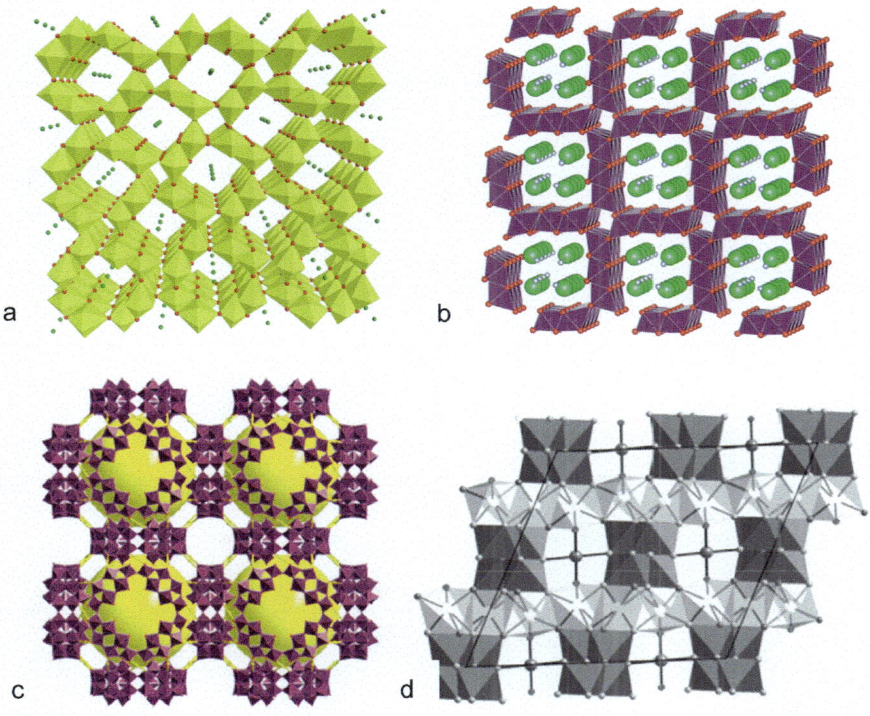

Fig. 1.18 Discrete oxide cluster assembling with metal polyhedra linker into porous 3D framework structure. The pore is occupied with additional cations and water. **a**:hollandite $Ba[MnO_2]$, **b** Todorokite, $Sr_x[Mn_6O_{12}]$, **c** $K_{18}Li_6[Mn_8(H_2O)_{48}P_8W_{48}O_{184}]$ (Adapted with permission from Ref. [51]. Copyright 2011 American Chemical Society) **d** $Na_2[Nb_{1.6}Ti_{0.4}O_{5.6}(OH)_{0.4}]$ (Adapted with permission from Ref. [52]. Copyright 2001 American Chemical Society)

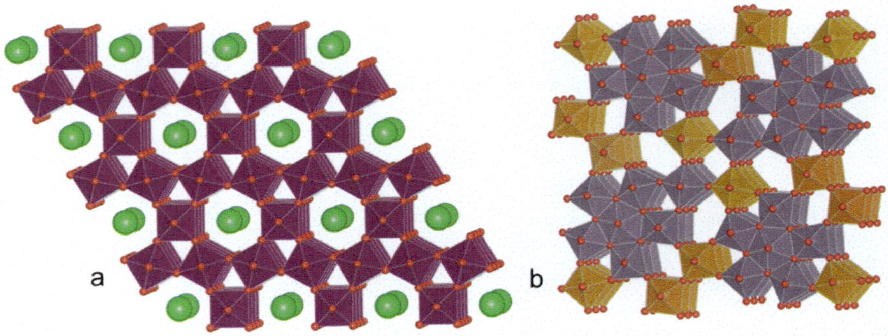

Fig. 1.19 Discrete oxide cluster (**a** trimer, **b** pentagonal) assembling with metal cations (green in the left illustration) and with metal polyhedra linker (brawn in the right illustration) into non-porous 3D framework structure. **a** $K_{0.5}WO_3$ Bronze, **b** $Mo(V)_5O_{14}$

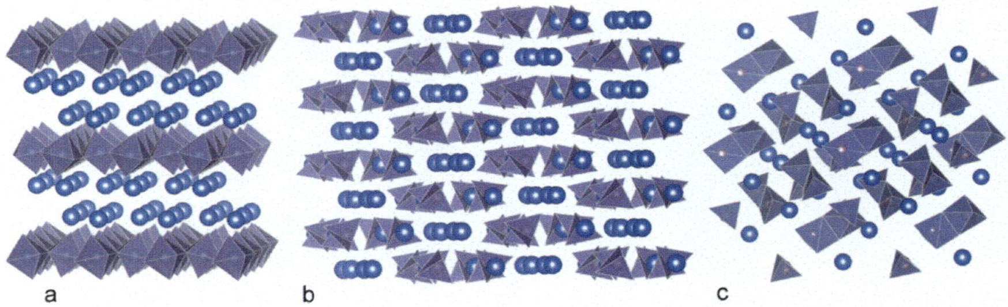

Fig. 1.20 3-Dimensionally infinite assembly of metal oxygen polyhedra into condensed solid. Bismuth molybdates from layered octahedra to isolated tetrahedral or square pyramidal, **a** γ-Bi_2MoO_6 (Aurivillious), **b** β-$Bi_2Mo_2O_9$, **c** α-$Bi_2(MoO_4)_3$ (Scheelite)

Fig. 1.21 3-Dimensionally infinite assembly of metal oxygen polyhedra into condensed solid. Cobalt molybdates from Mo octahedra to Mo tetrahedra, **a** α-$CoMoO_4$, **b** β-$CoMoO_4$. Adapted with permission from Ref. [55]. Copyright 2011 American Chemical Society

When the O/(A+M) ratio enters the region of 2, structures based on oxygen octahedrons and oxygen tetrahedrons, which are connected 3-dimensionally, become more common (Fig. 1.21) [55]. These are zeolites with a network structure based on SiO_2 in the oxygen tetrahedral configuration. Metal cations can be incorporated into the framework. This is a typical example of a solid-acid catalyst and oxidation catalyst. Similarly, aluminum phosphate and metal-substitute aluminum phosphates have a network-type structure with pores. This pore is empty and large enough to adsorb various organic molecules. In this region of 2, there are also many crystalline complex metal oxides with densely connected polyhedral structures, such as scheelite (Fig. 1.10a), metal phosphate (Fig. 1.22), and rutile and fluorite structures (Fig. 1.23), most of which form solid catalysts for various selective oxidation reactions.

The crystalline complex metal oxides in O/(A+M) ratios above 2 are extremely diverse in terms of structural features and from a catalytic point of view. As shown in Table 1.4, the catalyst examples dominate much of the listed examples. This fact indicates that complexity is necessary for crystalline complex metal oxides for solid catalysts to be viable. There is still much potential and importance in constructing new catalysts in this area of structural variety. In other words, this seems to be an area where further catalyst development should be pursued in the future.

For O/(A+M) ratios below 2, there is a great structural variety in complex metal oxides which can be used for solid catalysis, as shown in Table 1.4. Complex metal oxides belonging to this category are well known in the ceramics field of solid-state physics, and various applications and developments of these materials have recently

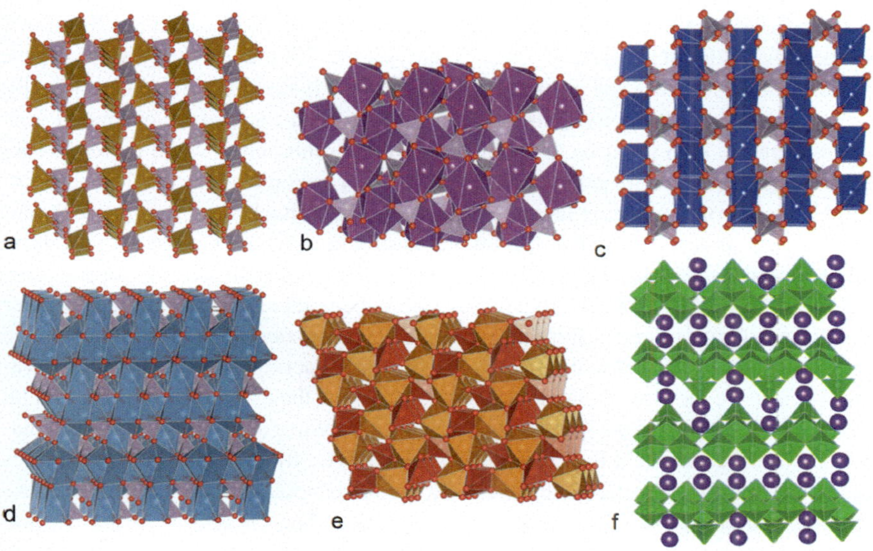

Fig. 1.22 3-Dimensionally infinite assembly of metal oxygen polyhedra into condensed solid. Metal phosphate (**a** $FePO_4$, **b** $BiPO_4$, **c** $Ca_2P_2O_7$, **d** $Ca_3(PO_4)_2$), vanadate (**e** $Mg_2V_2O_7$), and niobate (**f** $K_4Nb_6O_{17}$)

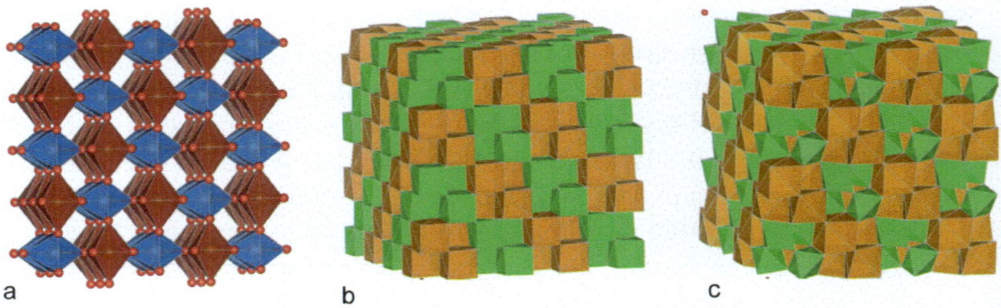

Fig. 1.23 3-Dimensionally infinite assembly of metal oxygen polyhedra into condensed solid. **a** $FeSbO_4$ (rutile structure), **b** $CeZrO_4$ (fluorite structure), **c** $La_2Zr_2O_7$ (pyrochlore structure)

been made in the fields of ionics, electronics, and energy. The crystal structure is determined by the packing of lattice oxygen ions and the arrangement of metal elements in the gaps between them, which is determined by the ionic radius ratio of lattice oxygen to metal, such as the Goldschmidt tolerance factor (t) [56, 57]. Therefore, there is a certain degree of freedom in the arrangement of the elements, and the synthesis conditions are influenced by heating, pressure, and other complex factors, which makes it valuable as a catalyst (Fig. 1.24). Among the materials in this area, perovskite structures have been the most widely applied to solid catalysts. This is because, as described later, elemental diversity, structural diversity, defect structures, and heteroatomic valence can be controlled by solid-state chemistry, and catalytic activity can be completely controlled in conjunction with these elements. Mixed oxides in this region exhibit functions such as complete oxidation reactions, photocatalysis, de-NOx catalysts, electrocatalysts, and ORR and OER [58, 59]. There is a demand to increase the surface area of solids to make them more catalytically active, and the regioselective introduction of defects of the constituent metal ions is considered to be a possible methodology. On the other hand, when the O/(A+M) ratio is slightly lower than 2, the Ruddlesden-Popper structure (Fig. 1.24) can appear, which consists of alternating layers of perovskite units stacked on top of a rock-salt type structure as layer units. This has favorable structural features in terms of catalytic functions and also electrocatalysis [60]. A similar structure is the Dion-Jacobson type.

As the O/(A+M) ratio further decreases, complex metal oxide crystals with simpler structures but with more complex constituent elements become possible, such as the spinel structure [61, 62]. Materials in this region are generally stable at high temperatures because of the strong metal–oxygen bond. Thus, the crystalline complex metal oxides in this group are used as catalysts involving hydrogen, such as dehydrogenation and hydrogenation at high temperatures. Catalytic functionality in this group is also expressed through the generation of oxygen vacancies on the surface because the structure can possess stable oxygen vacancies. The crystalline materials are, therefore, used for catalytic electrochemical reactions such as HER, OER, ORR, SOFC, and SOEC [63, 64]. In the case of HER, oxygen vacancies can facilitate water dissociation and weaken hydrogen adsorption under electrocatalytic H_2 evolution from water. In the same sense as a strong metal–oxygen bond, the complex

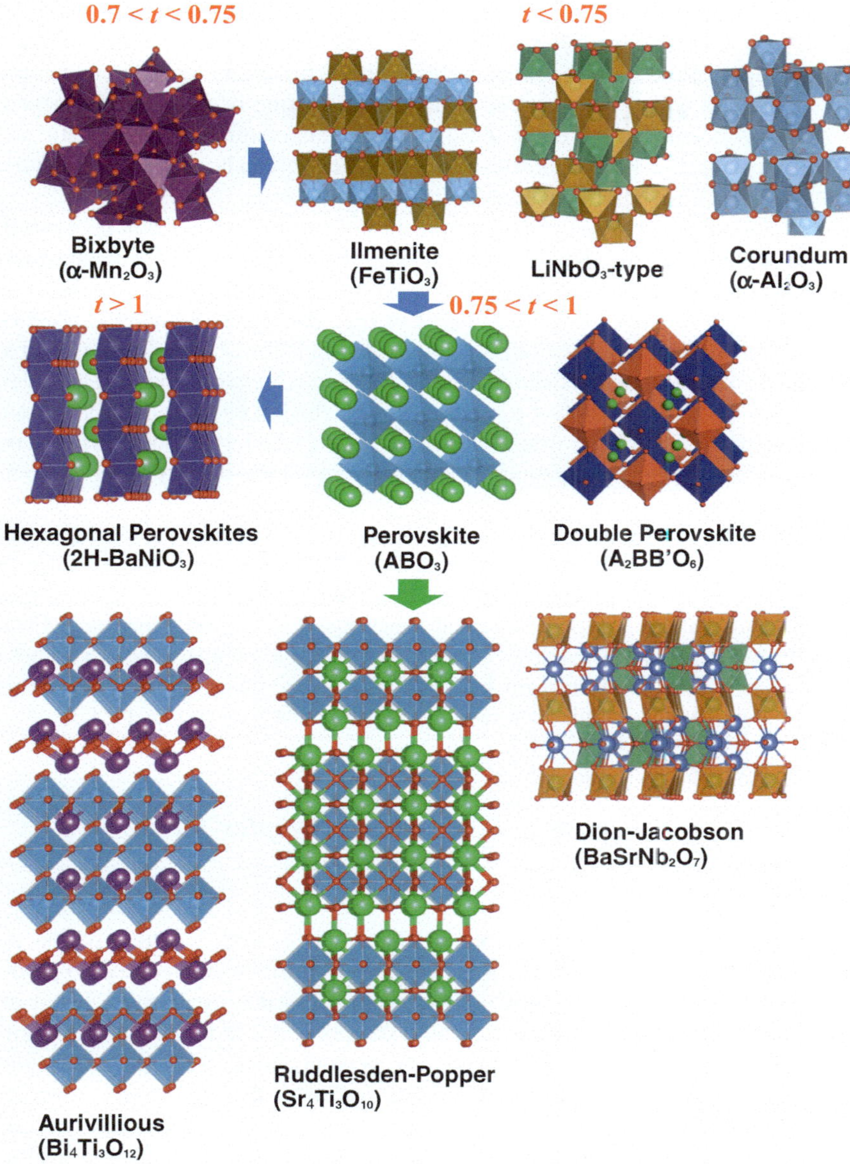

Fig. 1.24 3-Dimensionally infinite assembly of metal oxygen polyhedra into condensed solid, Perovskite family [56, 57]

oxides show a strong solid base property. Elemental complexation is also possible in this region, and it is possible to control the solid base nature by controlling the level of complexation. In many cases in this group, particularly hexaaluminate and magnetoplumbite which have layered structures consisting of alternatively tacked spinel blocks of close-packed oxide ions [65], high temperature is a prerequisite for material synthesis, but in the opposite sense, the materials can be applied to catalytic reactions that require high temperatures, such as combustion.

The additional features that can be seen in the crystal structures of complex metal oxides throughout the list is the oxygen packing. As already described, crystal structure is basically determined by the fashion of oxygen packing. The listed structures in Table 1.4 tend to have a lower density of oxygen packing in a higher O/(A+M) ratio. In other words, the degree of polyhedral sharing is lower in the crystal with a higher O/(A+M) ratio and this situation may result in a higher probability of local structure deformation with defect formation and polyhedral changes. This situation is apparently profitable for creating catalytic functions, which is a major reason why there are many examples of crystalline metal oxide catalysts in the region of higher O/(A+M) ratio.

The above low dense state of oxygen packing can be realized by forming metal element defects in crystalline oxide with a lower O/(A+M) ratio. If metal ions at specific positions can be selectively deleted, the resulting structure with a less dense state can bring a more complex state and functions such as pore structure which is suitable for an advanced catalytic system. An example of this case is a spinel-type structure which can transform into a Pharmacosiderite structure when some constituting element is selectively removed [66] as can be seen in Fig. 1.25.

1.2.3 Synthetic Chemistry of Crystalline Complex Metal Oxide Catalysts

Crystalline complex metal oxides are extremely diverse both structurally and in terms of their constituent elements. Catalyst synthesis must be tailored to each property [67]. The synthesis methods widely used for crystal formation are listed in Table 1.6 according to the catalyst structural category. While individual synthesis methods are not described in detail here, the general theory from the viewpoint of constructing catalytic reaction sites and reaction fields through crystal structure formation is outlined.

The construction of solid catalytic materials is first the establishment of the catalytic function that is essential to activate reactant molecules by the right selection of necessary elements and introducing a catalytically active structure. The second is the creation of a suitable catalyst particle structure in relation to molecular reaction. Catalytically active material must be homogeneous from the reacting molecule's point of view and all reacting molecules must be able to identically access to the catalytic active area and are equally activated. In addition, molecules produced must

Table 1.5 Classification of crystalline complex metal oxide catalysts on the basis of structuring principle and resulting crystal structures

Classification	Material form	Principal structure unit	Co-unit for constructing whole structure	Brief structure description	Catalyst example
Metal oxide cluster-based structure	Discrete	Finite cluster of metal–oxygen octahedra assembly	(poly) Oxy cations (linker or cap), counter cations, and proton	Discrete cluster units assembling with co-units into 3D structure	Metal-substituted POM Metal lacunary POM Caped POM
	Chain			Discrete cluster units assembling into chain form and then into 3D structure with co-units	SeW_6O_{24} Linear Caped POM
	Layer			Discrete cluster units assembling into slab form with co-units and then into 3D structure with accommodating co-units in interlayer	
	3D-framework (Porous)			Discrete cluster units assembling with co-units into 3D porous framework structure	$Mo_3VO_{11.2}$ $W_4V_3O_{19}$ All-inorganic porous POM
	(Non-porous)			Discrete cluster units assembling with co-units into non-porous 3D framework structure	Bronze-type

(continued)

Table 1.5 (continued)

Classification	Material form	Principal structure unit	Co-unit for constructing whole structure	Brief structure description	Catalyst example
Infinite metal oxide-based structure	Open layer	2D slab of metal–oxygen polyhedra assembly	Counter cations, metal hydroxy cations, and oxy cations in interlayer space	2D slab assembling with interlayer cations into 3D layered metal oxide	Sheet silicate Pillared clay Metal phosphate (Zr–P–O)
	Condensed layer	2D slab of metal–oxygen polyhedra assembly	Another slab of metal oxide	Alternative assembling of slabs of metal oxides into 3D layered metal oxide	Metal phosphate (Nb–P–O) $(VO)_2P_2O_7$ Hexaaluminate (BaMnAlO) Bi_2MoO_6
	3D porous framework	3D infinite assembly of metal–oxygen polyhedral	Counter cations, proton	Open porous structure with tetrahedral metal oxide network and additional metal counter cations inside the pore structure	Metalosilicates (MOR) Metalosilicates (MFI) AlPO, SAPO
	3D condensed	3D infinite assembly of metal–oxygen polyhedral	none	Condensed multi-element structuring controlled by 3D lattice oxygen packing without ordered porous structure	Perovskite Spinel ($FeSbO_4$) Scheelite ($Bi_2(MoO_4)_3$

be quickly expelled from the catalytic space. The former is material synthesis chemistry that achieves functions arising from electronic states of constituting elements under the control of the spatial mutual position among the elements by the introduction of the crystalline state in catalysts. In the case of complex metal oxide catalysts, structure formation reflects the nature of the metal elements and their bonding states with oxygen atoms. The latter also determines the structural form of the reaction field in a way that optimizes the variables of mass transfer, such as the uptake of reactant molecules, and energy transfer associated with the reaction. Although they share the same meaning of dealing with a 3D spatial factor as a catalytic reaction field, the spatial domain of the catalytic structure field with size changes ranging from the atomic level of the catalytically active site to a larger size, from several nano- to several micrometers, and at the same time the spatial domain of the catalytic

Fig. 1.25 Selective removal of constituting metal cations in octahedral from spinel structure (left) to form pharmacosiderite structure (right) with pore. Adapted with permission from Ref. [66]. Copyright 2012 Elsevier

structure field with a temporal size changes ranging from pico- or femto-seconds such as elementary reactions to several microseconds such as mass transfer between elements. The time domain of molecular motion to a catalyst structure field such as diffusion has a time axis with a width of milliseconds. In order to cope with such a space–time structure, catalyst materials must be in a state of highly ordered structural materials in which the arrangement of elements at the atomic level as well as the arrangement between catalyst particles is controlled.

However, uncertainty arises here. It is primarily necessary to create a catalytic reaction field with structure at the atomic and molecular levels to generate catalytic function. Even though catalytic materials are uniformly created to a size of several micrometers, the mutual position of primary crystal particles, even if prepared homogeneously, is not uniform but rather random. Therefore, this situation does not fully correspond to controlling the time region of molecular movement. On the other hand, in order to construct a catalytic active site field that can correspond to a wide time domain of molecular movement, it is required to prepare catalytic materials spatially homogeneous at a nano-sized level.

Our current synthetic methodologies are still uncertain and offer limited control over the above requisite. The catalyst material thus produced, therefore, will inevitably have complexity and inhomogeneity, resulting in complexity in the entire catalyst system, and making the actual performance of the system unclear and difficult to control. In addition, the space–time structure strongly affects the state of the reaction molecule and the formed molecule, changing the reaction factor under catalytic action. As a result, the space–time structure controls the effective performance of the catalyst and determines the success or failure of the catalyst. It can be said that overcoming this uncertainty is being done in every catalyst development and synthesis.

Table 1.6 Typical synthetic method depending on the type of crystalline complex metal oxides

Classification	Material form	O/(A+M)	Synthesis type	Brief description of synthetic procedure	Pretreatment conditions before catalysis
Metal oxide cluster unit-based structure	Discrete	3.4	Crystallization or precipitation from solution	Octahedral monomeric metal oxo anions in solution condense each other into discrete polyoxometalate units with other heteroatoms under carefully controlled pH condition depending on the kind of metal oxo anions, followed by crystallization or precipitation by adding counter cations with desired amounts	Usually no heat-treatment necessary but to remove ammonium cation to form proton-type catalysts, careful heat-treatment is necessary because structural phase change occurs
	Chain	3.1	Crystallization from solution Refluxing Hydrothermal	Octahedral monomeric metal oxo anions in solution condense each other into discrete polyoxometalate units with other heteroatoms under carefully controlled pH condition depending on the kind of metal oxo anions, followed by chain structure formation under refluxing or hydrothermal conditions and by crystallization or precipitation with counter cations	Usually no heat-treatment necessary but to remove water or ammonium cation to form proton-type catalysts, careful heat-treatment is necessary because structural phase change occurs
	Layer	3	Refluxing Hydrothermal	Octahedral monomeric metal oxo anions in solution condense each other into discrete polyoxometalate units with other heteroatoms under hydrothermal conditions to form 2D layer structure with accommodating heteroatoms in the layer slab as well as counter cations in interlayer spacing	Usually no heat-treatment necessary but to remove water or ammonium cation to form proton-type catalysts

(continued)

Table 1.6 (continued)

Classification	Material form	O/(A+M)	Synthesis type	Brief description of synthetic procedure	Pretreatment conditions before catalysis
	3D framework (Porous) (Non-porous	2.8	Hydrothermal	Octahedral monomeric metal oxo anions in solution condense each other into discrete polyoxometalate units with other heteroatoms under carefully controlled pH condition depending on the kind of metal oxo anions, followed by 3D porous framework formation with polyoxometalate units, the heteroatom linker, and counter cations in pore under hydrothermal conditions Post-treatment with acid to remove element from constructure structure to introduce vacancy and coordinatively unsaturated state or to replace constituting element with others	Heat-treatment is necessary to remove water or ammonium cation in the pore to make the materials catalytically active. Materials are mostly stable under heat-treatment but less stable compared to 3D condensed solids. Atmosphere during heat-treatment affects the oxidation state of constituting elements and thus catalytic properties
Infinite metal oxide-based structure	Layer (open)	2.5–2.7	Hydrothermal Sol–gel Microemulsion	Tetrahedral silicate (phosphate, germanate) anions condense each other infinitely under hydrothermal conditions to form 2D layer structure (e.g., sheet silicate) with accommodating heteroatoms in the layer slab as well as counter cations in interlayer spacing	Heat-treatment is necessary to remove interlayer water

(continued)

Table 1.6 (continued)

Classification	Material form	O/(A+M)	Synthesis type	Brief description of synthetic procedure	Pretreatment conditions before catalysis
	(condensed)	1.5–2.3	Hydrothermal Solvothermal Sol–gel Precipitation Solid-state reaction Flux synthesis Microwave assisted	Most frequently used synthetic method is precipitation in which metal elements in desired valences and compound states are mixed in aqueous media under desired pH condition and temperature, forming precipitates eventually. In this synthesis, metal(hydroxy) oxides and hydrated oxides often form and thus are subject to heat-treatment to form complex metal oxides in layer structure. Versatile method is a solid-state reaction of simple mixture of desired metal compounds at high temperature	Layered metal hydroxy oxide transforms to condensed layer structure under heat-treatment Reversely, condensed layered materials are acid treated and exfoliated to catalytically active nano-sheet
	3D (porous)	2.0	Hydrothermal Sol–gel	Tetrahedral silicate (phosphate) anions condense each other infinitely under hydrothermal conditions often in the presence of organic structure directing agents to form 3D open framework structure with accommodating heteroatoms in the framework as well as counter cations in pores. Post-treatment with acid to remove element from constructure structure to introduce vacancy and coordinatively unsaturated state	Layered hydroxy aluminosilicate transforms to 3D open structure under heat-treatment Various pretreatments, for example, acid, high temperature, and steam, are applied in order to tune catalytic properties

(continued)

Table 1.6 (continued)

Classification	Material form	O/(A+M)	Synthesis type	Brief description of synthetic procedure	Pretreatment conditions before catalysis
	3D (condensed)	1.4–2.7	Solvothermal Sol–gel Microemulsion Precipitation Mixed salt decomposition Solid-state reaction Flux synthesis Pechini	Most versatile method is solid-state reaction of simple mixture of desired metal compounds at high temperature and solvothermal method are in some case used because condensed complex metal oxide can be synthesized directly. Key points are uniformity of precursor mixture of elements. For this purpose, spray drying, reverse micelle, complexing agent method, polymerized complex method, etc. are utilized	Synthesized materials are usually heat-treated to modifying surface state but the process is not indispensable in terms of catalytic bulk structure formation Sometimes, heat-treated samples are treated with acid or base to remove surface impurity phases or to create defect structure

The above measures are inevitable particularly when introducing high-performance catalysts. To overcome or avoid these uncertainties, the reaction field region should be taken into consideration. One of the most desirable directions is that multiple catalytic elements are uniformly arranged in a higher order local structure in a nano-sized space. Apparently, new catalyst material construction engineering is needed for this aim. Table 1.6 summarizes the recent academic and technological technologies for the synthesis of crystalline complex metal oxides that have attracted attention. As described in Table 1.6, there is something in common that results from higher order structurization of different material states rather than simple elemental or molecular organization on a narrow atomic scale. Recently, hydrothermal synthesis received a lot of attention [68, 69] because the method can allow forming nanostructured catalysts like $Mo_3VO_{11.2}$ (Fig. 1.17) under a controlled structure formation process [70] as illustrated in Fig. 1.26a. Because the hydrothermal condition is usually mild and soft, this method controls progressive structuring of complex metal oxides through the formation of discrete polyoxometalates depending on pH conditions. This formation process similarly takes place even in facile but carefully controlled solution-based synthesis to form dense complex metal oxides like spinel with simultaneous tuned crystal phase and elemental compositions (Fig. 1.26c) [71, 72].

It should be, however, referred to that a bottom-up method from the atomic/molecular level still has difficulty in creating complex interactions as intended. Many other atomic-nanoscale synthetic chemistry faces these problems and still requires a lot of trials and time to create new catalytic materials. It should be added herewith, as a countermeasure, that there are automated test methods such as combinatorial methods and high-throughput synthesis. To make them meaningful, it is necessary to combine theories and cooperate with advanced analysis, and in the end, advanced judgment by humans is required. This has resulted in further temporal divergence.

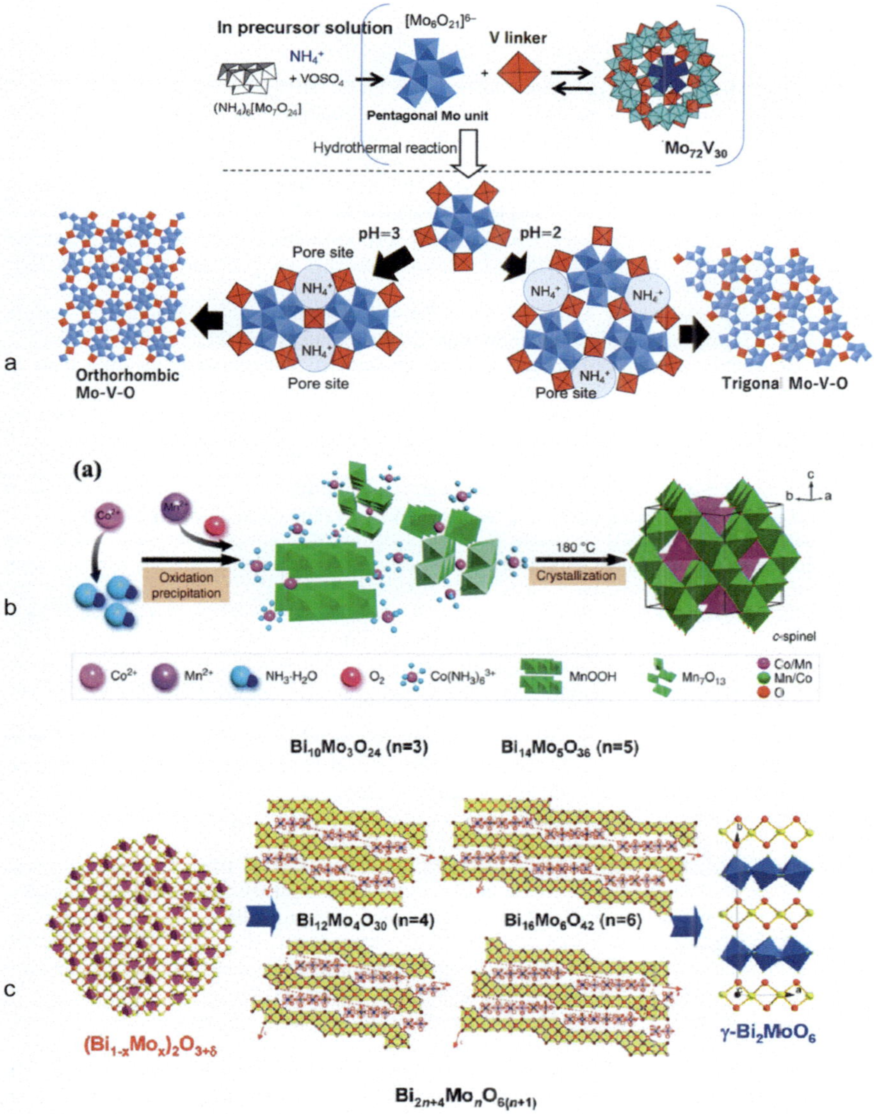

Fig. 1.26 Schematics of the crystal formation through discrete unit formation, successive condensation and transformation under mild heating condition like as hydrothermal condition. **a** the pentagonal unit supplied from precursor condenses each other with linkers [70], **b** discretely condensed octahedra reacts each other with metal cations for crystallization (From Ref. [71] (Reprinted with permission from Springer Nature), and **c** randomly distributed molybdenum polyhedra self-organize under hydrothermal condition (Adapted with permission from Ref. [72]. Copyright 2012 American Chemical Society)

References

1. Schlögl R. Heterogeneous catalysis. Angew Chem Int Ed. 2015;54:3465–520.
2. Védrine JC. Heterogeneous catalysis on metal oxides. Catalysts. 2017;7:341.
3. Getsoian AB, Zhai Z, Bell AT. Band-gap energy as a descriptor of catalytic activity for propene oxidation over mixed metal oxide catalysts. J Am Chem Soc. 2014;136:13684–97.
4. Chieregato A, Nieto JML, Cavani F. Mixed-oxide catalysts with vanadium as the key element for gas-phase reactions. Coord Chem Rev. 2015;301–302:3–23.
5. Najari S, Saeidi S, Concepcion P, Dionysiou DD, Bhargava SK, Lee AF, Wilson K. Oxidative dehydrogenation of ethane: catalytic and mechanistic aspects and future trends. Chem Soc Rev. 2021;50:4564–605.
6. Malik MI, Abatzoglou N, Achouri IE. Methanol to formaldehyde: an overview of surface studies and performance of an iron molybdate catalyst. Catalysts. 2021;11:893.
7. Lin R, Amrute AP, Pérez-Ramírez J. Halogen-mediated conversion of hydrocarbons to commodities. Chem Rev. 2017;117:4182–4247.
8. Sprenger P, Kleist W, Grunwaldt J. Recent advances in selective propylene oxidation over bismuth molybdate based catalysts: synthetic, spectroscopic, and theoretical approaches. ACS Catal. 2017;7:5628–42.
9. Zhang S, Wu Z, Liu X, Hua K, Shao Z, Wei B, Huang C, Wang H, Sun Y. A short review of recent advances in direct CO_2 hydrogenation to alcohols. Top Catal. 2021;64:371–94.
10. Hirakawa T, Shimokawa Y, Tokuzumi W, Sato T, Tsushida M, Yoshida H, Hinokuma S, Ohyama J, Machida M. Multicomponent spinel oxide solid solutions: a possible alternative to platinum group metal three-way catalysts. ACS Catal. 2019;9(12):11763–73.
11. Kim JS, Kim B, Kim H, Kang K. recent progress on multimetal oxide catalysts for the oxygen evolution reaction. Adv Energy Mater. 2018;8:1702774.
12. Lang R, Du X, Huang Y, Jiang X, Zhang Q, Guo Y, Liu K, Qiao B, Wang A, Zhang T. Single-atom catalysts based on the metal-oxide interaction. Chem Rev. 2020;120:11986–2043.
13. Chen Y, Kasama T, Huang Z, Hu P, Chen J, Liu X, Tang X. Highly dense isolated metal atom catalytic sites: dynamic formation and in situ observations. Chem Eur J. 2015;21:17397–402.
14. Tsui EY, Kanady JS, Agapie TA. Synthetic cluster models of biological and heterogeneous manganese catalysts for O_2 evolution. Inorg Chem. 2013;52(24):13833–48.
15. Umena Y, Kawakami K, Shen J, Kamiya N. Crystal structure of oxygen-evolving photosystem II at a resolution of 1.9 Å. Nature. 2011;473:55–60.
16. Grundner S, Markovits MAC, Li G, Tromp M, Pidko EA, Hensen EJM, Jentys A, Sanchez-Sanchez M, Lercher JA. Single-site trinuclear copper oxygen clusters in mordenite for selective conversion of methane to methanol. Nature Commun. 2015;(6):7546.
17. Chan SI, Lu Y, Nagababu P, Maji S, Hung M, Lee MM, Hsu I, Minh PD, Lai JC-H, Ng KY, Ramalingam S, Yu SS-F, Chan MK. Efficient Oxidation of methane to methanol by dioxygen mediated by tricopper clusters. Angew Chem Int Ed. 2013;52:3731–5.
18. Okuhara T, Mizuno N, Misono M. Catalytic chemistry of heteropoly compounds. Adv Catal. 1996;41:113–252.
19. Rasmussen M, Näther C, Leusen JV, Warzok U, Schalley CA, Kögerler P, Bensch W. Small, beautiful and magnetically exotic: $\{V_4\ W_2\}$- and $\{V_4\ W_4\}$-type polyoxometalates. Dalton trans. 2016;45:10519–22.
20. Liu Z, Wang X, Qin C, Zhang Z, Li Y, Chen W, Wang E. Polyoxometalate-assisted synthesis of transition-metal cubane clusters as artificial mimics of the oxygen-evolving center of photosystem II. Coord Chem Rev. 2016;313(15):94–110.
21. Liu D, Lu Y, Tan H, Chen W, Zhang Z, Li Y, Wang E. Polyoxometalate-based purely inorganic porous frameworks with selective adsorption and oxidative catalysis functionalities. Chem Commun. 2013;49:3673.
22. Ta N, Liu J, Chenna S, Crozier PA, Li Y, Chen A, Shen W. Stabilized gold nanoparticles on ceria nanorods by strong interfacial anchoring. J Am Chem Soc. 2012;134(51):20585–8.
23. Wang S, Yang G. Recent advances in polyoxometalate-catalyzed reactions. Chem Rev. 2015;115:4893–962.

24. Rodriguez-Albelo LM, Ruiz-Salvador AR, Sampieri A, Lewis DW, Gómez A, Nohra B, Mialane P, Marrot J, Sécheresse F, Mellot-Draznieks C, Biboum RN, Keita B, Nadjo L, Dolbecq A. Zeolitic polyoxometalate-based metal-organic frameworks (Z-POMOFs): computational evaluation of hypothetical polymorphs and the successful targeted synthesis of the redox-active Z-POMOF1. J Am Chem Soc. 2009;131:16078–16087.
25. Yang D, Babucci M, Casey WH, Gates BC. The surface chemistry of metal oxide clusters: from metal-organic frameworks to minerals. ACS Cent Sci. 2020;6:1523–33.
26. Stacchiola DJ, Senanayake SD, Liu P, Rodriguez JA. Fundamental studies of well-defined surfaces of mixed-metal oxides: special properties of $MO_x/TiO_2(110)$ {M = V, Ru, Ce, or W}. Chem Rev. 2013;113:4373–90.
27. Fu Q, Li W, Yao Y, Liu H, Su H, Ma D, Gu X, Chen L, Wang Z, Zhang H, Wang B, Bao X. Interface-confined ferrous centers for catalytic oxidation. Science. 2010;328(5982):1141–4.
28. Seki K. Development of RuO_2/Rutile-TiO_2 catalyst for industrial HCl oxidation process. Catal Surv Asia. 2010;14:168–75.
29. Bondarchuk O, Huang X, Kim J, Kay BD, Wang L, White JM, Dohnálek Z. Formation of monodisperse $(WO_3)_3$ clusters on $TiO_2(110)$. Angew Chem Int Ed. 2006;45:4786–9.
30. Wang J, Tan H, Yu S, Zhou K. Morphological effects of gold clusters on the reactivity of ceria surface oxygen. ACS Catal. 2015;5:2873–81.
31. Mohrhusen L, Grebien M, Al-Shamery K. Electron transfer in oxide-oxide cocatalysts: interaction of tungsten oxide clusters with Ti^{3+} states in rutile TiO_2. J Phys Chem C. 2020;124:23661–23673.
32. Hawthorne FC, Uvarova YA, Sokolova E. A structure hierarchy for silicate minerals: sheet silicates. Mineral Mag. 2019;83(1):3–55.
33. Shibata T, Takanashi G, Nakamura T, Fukuda K, Ebinad Y, Sasaki T. Titanoniobate and niobate nanosheet photocatalysts: superior photoinduced hydrophilicity and enhanced thermal stability of unilamellar Nb_3O_8 nanosheet. Energy Environ. 2011;4:535–42.
34. Clearfield A, Smith GD. The crystallography and structure of α-zirconium bis (Monohydrogen Orthophosphate) monohydrate. Inorg Chem. 1969;8:431–6.
35. Holtstam D, Hålenius U. Nomenclature of the magnetoplumbite group. Mineral Mag. 2020;84:376–80.
36. Thomas JM, Raja R, Lewis DW. Single-site heterogeneous catalysts. Angew Chem Int Ed. 2005;44:6456–82.
37. Li J, Yu J, Xu R. Progress in heteroatom-containing aluminophosphate molecular sieves. Proc R Soc A. 2012;468:1955–67.
38. Inagaki S, Guan S, Ohsuna T, Terasaki O. An ordered mesoporous organosilica hybrid material with a crystal-like wall structure. Nature. 2002;416:304–7.
39. Amakawa K, Kolen'ko YV, Villa A, Schuster ME, Csepei L, Weinberg G, Wrabetz S, d'Alnoncourt RN, Girgsdies F, Prati L, Schlögl R, Trunschke A. Multifunctionality of crystalline MoV(TeNb) M1 oxide catalysts in selective oxidation of propane and benzyl alcohol. ACS Catal. 2013;3:1103–1113.
40. Zhang ZX, Sadakane M, Murayama T, Izumi S, Yasuda N, Sakaguchi N, Ueda W. Tetrahedral connection of ε-keggin-type polyoxometalates to form an all-iinorganic octahedral molecular sieve with an intrinsic 3D pore system. Inorg Chem. 2014;53:903–11.
41. Zhang Z, Zhu Q, Sadakane M, Murayama T, Hiyoshi N, Yamamoto A, Hata S, Yoshida H, Ishikawa S, Hara M, Ueda W. A zeolitic vanadotungstate family with structural diversity and ultrahigh porosity for catalysis. Nature commun. 2018;9:3789.
42. Moro-oka Y. Reactivities of active oxygen species and their roles in the catalytic oxidation of inactive hydrocarbon. Catal Today. 1998;45:3–12.
43. Zhang Z, Murayama T, Sadakane M, Ariga H, Yasuda N, Sakaguchi N, Asakura K, Ueda W. ultrathin inorganic molecular nanowire based on polyoxometalates. Nature commun. 2015;6:7731.
44. Bonhomme F, Larentzos JP, Alam TM, Maginn EJ, Nyman M. Synthesis, structural characterization, and molecular modeling of dodecaniobate keggin chain materials. Inorg Chem. 2005;44(6):1774–85.

45. Elshof JET, Yuan H, Rodriguez PG. Two-dimensional metal oxide and metal hydroxide nanosheets: synthesis, controlled assembly and applications in energy conversion and storage. Adv Energy Matar. 2016;6:1600355.
46. Zhu Q, Zhang Z, Sadakane M, Matsumoto F, Hiyoshi N, Yamamoto A, Yoshida H, Yoshida A, Hara M, Ueda W. Structural characterization of 2D zirconomolybdate by atomic scale HAADF-STEM and XANES and its highly stable electrochemical properties as a li battery cathode. Inorg Chem. 2017;56:14306–14.
47. Takagaki A, Yoshida T, Lu D, Kondo JN, Hara M, Domen K, Hayashi S. Titanium niobate and titanium tantalate nanosheets as strong solid acid catalysts. J Phys Chem B. 2004;108:11549–55.
48. Timmerman MA, Xia R, Wang Y, Sotthewes K, Huijben M, Elshof JET. Long-range ordering of two-dimensional wide bandgap tantalum oxide nanosheets in printed films. J Mater Chem C. 2021;9:5699–705.
49. Sadakane M, Watanabe N, Katou T, Nodasaka Y, Ueda W. Crystalline Mo_3VO_x mixed metal oxide catalyst with trigonal symmetry. Angew Chem Int Ed. 2007;46:1493–6.
50. Sadakane M, Kodato K, Kuranishi T, Nodasaka Y, Sugawara K, Sakaguchi N, Nagai T, Matsui Y, Ueda W. Molybdenum-vanadium-based molecular sieves with microchannels of seven-membered rings of corner-sharing metal oxide octahedra. Angew Chem Int Ed. 2008;47:2493–2496.
51. Mitchell SG, Boyd T, Miras HN, Long D, Cronin L. Extended polyoxometalate framework solids: two Mn(II)-linked P8W48 network arrays. Inorg Chem. 2011;50:136–43.
52. Nyman M, Tripathi A, Parise JB, Maxwell RS, Harrison WTA, Nenoff TM. A new family of octahedral molecular sieves: sodium Ti/Zr^{IV} niobates. J Am Chem Soc. 2001;123:1529–30.
53. Mitchell SG, Boyd T, Miras HN, Long D, Cronin L. Extended polyoxometalate framework solids: two Mn(II)-linked P8W48 network arrays. Inorg Chem. 2011;50(1):136–43.
54. Shen YF, Zerger RP, Deguzman RN, Suib SL, McCurdy L, Potter DI, O'young CL. Manganese oxide octahedral molecular sieves: preparation, characterization, and applications. Science. 1993;260:511.
55. Robertson LC, Gaudon M, Jobic S, Deniard P, Demourgues A. Investigation of the first-order phase transition in the $Co_{1-x}Mg_xMoO_4$ solid solution and discussion of the associated thermochromic behavior. Inorg Chem. 2011;50:2878–84.
56. Saha-Dasgupta T. Double perovskites with 3d and 4d/5d transition metals: compounds with promises. Mater Res Express. 2020;7: 014003.
57. Dos santos-García AJ, Solana-Madruga E, Ritter C, Ávila-Brande D, Fabeloc O, Sáez-Pucheb R. Synthesis, structures and magnetic properties of the dimorphic Mn_2CrSbO_6 oxide. Dalton Trans. 2015;44:10665–10672.
58. Suntivich J, May KJ, Gasteiger HA, Goodenough JB, Shao-Horn Y. A perovskite oxide optimized for oxygen evolution catalysis from molecular orbital principles. Science. 2011;334:1383–5.
59. Cheng X, Fabbri E, Nachtegaal M, Castelli IE, Kazzi ME, Haumont R, Marzari N, Schmidt TJ. Oxygen evolution reaction on $La_{1-x}Sr_xCoO_3$ perovskites: a combined experimental and theoretical study of their structural, electronic, and electrochemical properties. Chem Mater. 2015;27:7662–72.
60. Ding P, Li W, Zhao H, Wu C, Zhao L, Dong B, Wang S. Review on ruddlesden-popper perovskites as cathode for solid oxide fuel cells. J Phys Mater. 2021;4: 022002.
61. Zhao Q, Yan Z, Chen C, Chen J. Spinels: controlled preparation, oxygen reduction/evolution reaction application, and beyond. Chem Rev. 2017;117:10121–211.
62. Ye A, Li Z, Ding J, Xiong W, Huang W. Synergistic catalysis of Al and Zn sites of spinel $ZnAl_2O_4$ catalyst for CO hydrogenation to methanol and dimethyl ether. ACS Catal. 2021;11:10014–9.
63. Zhu Y, Lin Q, Zhong Y, Tahini HA, Shao Z, Wang H. Metal oxide-based materials as an emerging family of hydrogen evolution electrocatalysts. Energy Environ Sci. 2020;13:3361–92.
64. Song F, Bai L, Moysiadou A, Lee S, Hu C, Liardet L, Hu X. Transition metal oxides as electrocatalysts for the oxygen evolution reaction in alkaline solutions: an application-inspired renaissance. J Am Chem Soc. 2018;140:7748–59.

65. Tian M, Wang XD, Zhang T. Hexaaluminates: a review of the structure, synthesis and catalytic performance. Catal Sci Technol. 2016;6:1984–2004.
66. Baur WH. Rigid frameworks of zeolite-like compounds of the pharmacosiderite structure-type. Microporous Mesoporous Mater. 2012;151:13–25.
67. Carneiro JSA, Williams J, Gryko A, Herrera LP, Nikolla E. Embracing the complexity of catalytic structures: a viewpoint on the synthesis of nonstoichiometric mixed metal oxides for catalysis. ACS Catal. 2020;10:516–27.
68. Yoshimura M, Byrappa K. Hydrothermal processing of materials: past, present and future. J Mater Sci. 2008;43:2085–103.
69. Patzke GR, Zhou Y, Kontic R, Conrad F. Oxide nanomaterials: synthetic developments, mechanistic studies, and technological innovations. Angew Chem Int Ed. 2011;50:826–59.
70. Sadakane M, Endo K, Kodato K, Ishikawa S, Murayama T, Ueda W. Assembly of a pentagonal polyoxomolybdate building block, $[Mo_6O_{21}]^{6-}$, into crystalline Mo V oxides. Eur J Inorg Chem. 2013;10–11:1731–6.
71. Li C, Han X, Cheng F, Hu Y, Chen C, Chen J. Phase and composition controllable synthesis of cobalt manganese spinel nanoparticles towards efficient oxygen electrocatalysis. Nat Commun. 2015;6:7345.
72. Kongmark C, Coulter R, Cristol S, Rubbens A, Pirovano C, Löfberg A, Sankar G, Beek WV, Bordes-Richard E, Vannier R. A comprehensive scenario of the crystal growth of γ–Bi_2MoO_6 catalyst during hydrothermal synthesis. Cryst Growth Des. 2012;12:5994–6003.

Chapter 2
Catalysis Chemistry of Crystalline Complex Metal Oxide Catalysts

Wataru Ueda

2.1 Catalysis Chemistry of Crystalline Complex Metal Oxides

2.1.1 Metastable State

A great number of complex metal oxides having crystal structures act as solid catalysts and are put to practical use in a variety of catalytic reactions. Thermal stability of solid catalysts is always a crucial issue in practical use. Even if high catalytic performance is achieved at the laboratory level, thermal distribution will inevitably occur in the catalyst layer zone of the practical large-size reactor, causing the catalytic performance degradation over a long time on stream. To prevent such problems, solid catalysts have to gain higher or improved thermal resistance. A further combination of elements is often used in many cases for this purpose. However, the combination of elements does not always improve catalytic performance, and in many cases catalytic performance is sacrificed for the purpose of improving thermal tolerance. One of the reasons is that the structural flexibility required for a stable catalytic reaction cycle at local structural sites formed on a crystalline matrix depends on the thermal stability of the whole structural of the catalyst, so that when changes are made to the whole structural matrix to increase thermal stability, the local structural sites are also changed at the same time. Another important reason is the change in the number of catalytically active local structural sites. While pretreatment at a temperature sufficiently higher than the reaction temperature naturally increases the state of crystallization and improves the stability of the catalyst material, the surface state is greatly affected by sintering, reduction in surface roughness, and changes in

W. Ueda (✉)
Department of Material Chemistry and Life, Kanagawa University, 3-27-1, Rokkakubashi, Kanagawa-ku, Yokohama 221-8686, Japan
e-mail: uedaw@kanagawa-u.ac.jp

W. Ueda (ed.), *Crystalline Metal Oxide Catalysts*,
https://doi.org/10.1007/978-981-19-5013-1_2

the state of intergranular materials, resulting in a decrease in the number of catalytically active local structural sites. Nevertheless, if the number of catalytically active local structural sites per surface area is maintained to some extent while the desired selectivity and other catalytic properties are maintained, the net effect of thermal stabilization is regarded as positive.

What emerges from the above discussion is that the surface structure of a crystalline catalyst is never as stable as the bulk state of a solid catalyst, even though the surface structure has lower surface energy and is sustained by the bulk structure. In addition, catalytic reactions can induce dynamic changes in the surface structure, which in turn induce structural changes in the bulk [1]. The surface structure of solid catalysts is thus a metastable state. The metastable state is understood in terms of solid-state chemistry as follows: If various polymorphs exist in the crystal structure of a solid material, the phase transition of one polymorph into a thermodynamically more stable one occurs much more easily at the surface than the structural transition in the bulk. This transition usually occurs at the Tamman temperature range. Even in the absence of polymorphic structures, changes in elemental composition due to migration of elements, displacement of elements, and valence change due to reduction or oxidation between elements or under the atmosphere will occur, resulting in a completely different crystalline phase in bulk, or in the generation or disappearance of defect structures at the surface. Looking at such crystallographic metastable state formation from a catalytic chemistry point of view, a metastable state is always established under various perturbations added to the surface structure, like the direct action of reactant and produced molecules on the surface during a catalytic reaction, electron transfer, oxygen atom transfer and preferential metal element migration, metal–oxygen bond breaking and regeneration, metal–oxygen bond displacement, reaction heat transfer, etc. in addition to thermal deformation during the catalytic reaction with heating [2–4], so-called irreversible changes, in situ activation of catalysts, and also surface reconstruction known as common phenomena in oxide catalysts not only for the redox-type reaction [5] but also for solid-acid type reaction [6, 7]. The stable functioning as a solid catalyst means that the catalytic material has dynamic reversibility in the local structure and electronic state while repeating the catalytic cycle under such perturbations. It is important to note, however, that such perturbations should not irreversibly cause large deformations in surface local structure. This is because if the change is irreversible, the material cannot be a catalyst. A solid catalyst is a material that can tolerate dynamic reversibility in the range of small displacements and deformations. Crystalline complex metal oxide catalysts acting at 800 °C, such as the catalysts for OCM, appear to undergo little change during the reaction, but in reality, small changes occur in the structure and electronic state at the surface through binding to the reactants during catalysis, which is why they can act as solid catalysts. In other words, it is more important for solid catalysts whether the change with a smaller energy difference is dynamically reversible rather than the change of crystallographic metastable state. This leads to the further implication, that is, reversible dynamic micro-displacements of the structure of crystalline solid catalysts do not necessarily occur over the entire structure of the crystal, but a reversible dynamic change in a narrow or limited region is enough to promote catalysis. The local structures produced or sustained over the entire crystal structure should have micro-kinetical reversibility and dynamic reversibility to show catalytic functions. Crystalline $Mo_3VO_{11.2}$ oxides are good examples. As can be seen in Fig. 1.17, the rigid ring-type framework with six pentagonal units occludes five octahedra connected to each other in the center in the case of the orthorhombic type and three octahedra in the case of the trigonal type and seems to support the dynamic change of the octahedral connection under the catalytic reaction. Crystal structure information of metal oxide catalysts always gives us better pictures of catalytically active structures. The catalytic functions of solid catalysts should be viewed in terms of such metastable states. But at the same time, it must be in mind that correlating the catalytic behaviors of complex metal oxides to their bulk structures in a simple manner without actual surface features [8] will mislead in understanding real catalysis and giving the wrong direction of catalyst design.

2.1.2 Dynamic Reversible Phase Transfer Under Catalysis (MvK Mechanism)

Dynamic reversible elemental displacement and change in electronic state in bulk structure and/or on the surface structure of crystalline complex metal oxide catalysts can be either very small at the surface, as exemplified by the complex metal oxide catalyst for OCM described above, or a major structural transition of the entire crystalline catalyst. The former is a dynamic change at a surface atomic level under a catalytic cycle of oxidizing an organic substance using the lattice oxygen in the surface region of the oxide catalyst as an oxidant, while the latter is a rather dynamic change accompanying a structural phase transition, where an oxide catalyst with a crystal structure changes to another crystal structure with releasing lattice oxygen to oxidize organic molecules and then the phase-changed catalysts return to the original structure of the oxide catalyst again by replenishment of lattice oxygen with molecular oxygen. This working mechanism is effective in making the catalytic reaction selective because catalytically active oxygen species is only lattice oxygen. This working process of the latter is known engineeringly as a chemical loop [9], which is an effective method in chemical processes. Catalytic reactions utilizing lattice oxygen of metal oxide catalysts are also carried out in a manner known as a moving-bed reaction process in which the reaction of the reactant with lattice oxygen of metal oxide catalysts and replenishment of lattice oxygen with molecular oxygen are carried out in separate reactors and the catalyst is moved around. The key to these reaction methods is to utilize lattice oxygen to the extent that the reversibility of the crystal phase of the complex metal oxide catalyst can be maintained. Maximization of the amount of available lattice oxygen and sustaining of reversibility as well as both activity and selectivity of reaction are mandatory in selecting the crystal system and adding elements in catalyst design.

The separation and cycling properties of solid catalysts as described above are very important not only from a reaction engineering point of view, but also from a catalytic chemistry point of view. This catalytic working process is well-known as the Mas van Krevelen (MvK) mechanism [10, 11]. This mechanism is applicable for a variety of catalytic reactions over solid-state oxide catalysts and the mechanism for oxidation reactions is described in Fig. 2.1 as a representative.

MvK mechanism is classified into two categories at least. First is the catalytic reaction mechanism based on a back-and-forth cycle between oxidized oxide catalyst crystals (crystal phase 1) and oxygen-deficient crystalline catalysts (crystal phase 2). The phase transition between them is associated with reduction–oxidation (redox) chemistry. The mechanism also includes the case in which the reaction proceeds with lattice oxygen transfer through the oxygen vacancies formed during the redox process over the entire crystal without changing the starting crystal phase. Structural elasticity of bulky crystal seems a fundamental requisite for this mechanism. In both cases, most of the lattice oxygen in the metal oxide crystal is participated in the catalytic reaction and has equivalent chemical potential to the oxygen molecules in the gas phase. In this chapter, the case in which lattice oxygen of the entire crystal is

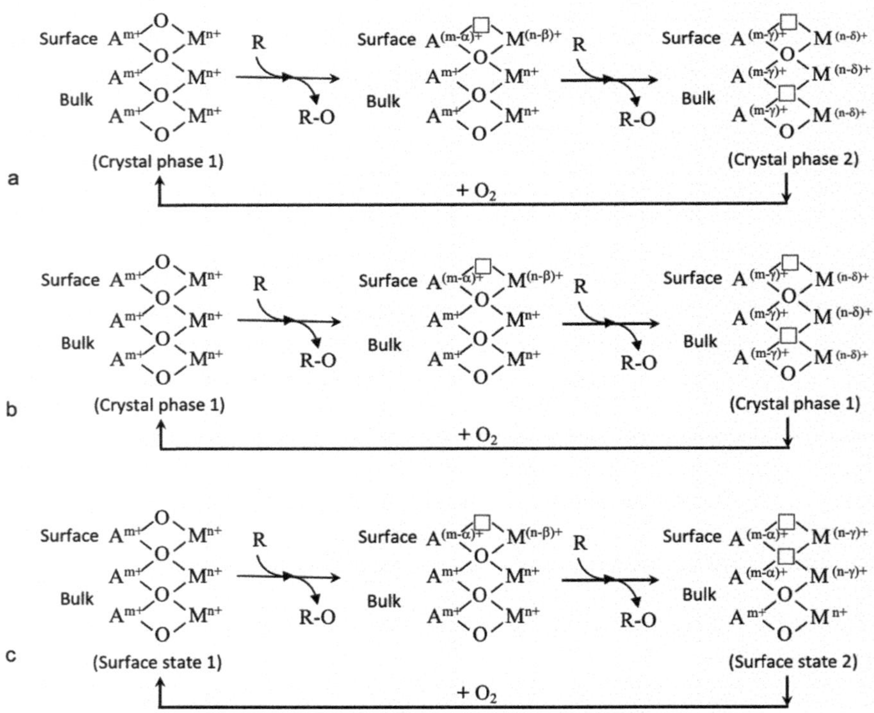

Fig. 2.1 Bulk-type MvK mechanism with crystal phase transformation (**a**), without the transformation (**b**), and localized MvK mechanism (**c**)

involved in the catalytic reaction will be referred to as the bulk-type MvK mechanism (Fig. 2.1a and b).

There is another MvK mechanism different from the above bulk-type one: Catalytic reaction, in which the local structure site or surface structure of the solid catalyst is only considered to undergo microscopic dynamic reversible displacement under the steady catalytic cycle. This can be categorized in the MvK mechanisms since the dynamic phenomenon at the locally limited region of solid catalyst takes place without accompanying a clear crystal phase change. The reaction is here called the localized MvK mechanism (Fig. 2.1c). Even when an oxygen-deficient site is formed on the surface of a crystalline oxide catalyst and acts as a strong adsorption site, small surface structure displacement inevitably occurs through the dynamic formation of oxygen vacancy and also during adsorption and catalysis, which means that the local structure site or surface structure of the solid catalyst has also a character of dynamic reversible displacement definitely similar to the case of the bulk-type MvK mechanism. This mechanism, therefore, can even apply to polyoxometalate catalysts with discrete metal oxide clusters [12]. Not only redox-type reactions but also catalytic hydrogenation of CO_2 over various complex metal oxide catalysts is also an example of this category because strong adsorption of CO_2 adsorption on oxygen vacancy and leaving oxygen in catalysts with electron transfer are involved, those of which strongly force deformation in the local structure on the surface of solid catalyst [13–15]. For this type of reaction, the Langmuir–Hinshelwood mechanism and the Rideal-Eley mechanism, which are static reaction mechanisms taking adsorption into main consideration, have been used to explain the reaction so far, but from the oxide catalyst state point of view, the localized MvK mechanism is more realistic [16, 17] and gives a better understanding of catalysis. Thus, the dynamic reversibility of solid metal oxide catalysts under steady-state catalytic reactions is important in many catalytic reactions in relation to the crystal structure of the metal oxide catalysts. In this chapter, crystalline complex metal oxide catalysts are classified into two types according to the bulk-type MvK mechanism and the localized MvK mechanism.

2.1.3 Defect Structure and Ion-Conduction

In the bulk-type MvK mechanism involving redox of whole crystalline complex metal oxide catalysts, the formation of oxygen defect structures and ionic conductivity within the solid catalyst are essential. The structural transformation is always accompanied by a valence change of the constituent elements and a defect structure. This kind of transformation is accompanied to some extent even in the bulk-type MvK mechanism of the case without exhibiting a clear crystalline structural transition. In fact, many crystalline oxide catalysts are known to adopt the bulk-type MvK mechanism without a clear crystalline structural transition, and perovskite-type oxides are typical examples (see also Chap. 10). They have been widely applied as oxidation catalysts, NOx removal, solid electrolytes, electrocatalysts, etc. This is because their bulk properties such as anion and cation defect structure formation, valence control, anomalous valence, mixed valence, ionic conduction, and absorption of ions and molecules based on their stable crystal structures are clearly reflected in various catalytic functions. Interestingly, the bulk properties brought about by the crystalline structure can accurately control the solid catalyst properties. These facts mean that the bulk crystal structure is continuous to the surface to some extent, a metastable state formed on the surface structure is controlled by the bulk structure, and the chemical state of the bulk can be linked to the surface structure. To use this tendency, crystalline metal oxides have an advantage, on the other hand, it is difficult to efficiently manifest these properties by controlling amorphous solid materials.

What is the view of defect structures and ionic conductivity of crystalline complex metal oxide catalysts in the case of a localized MvK mechanism working? It is reasonable to assume that defect structures exist as a normal state in metastable surface structures, but the dynamic displacement of the surface structure under steady-state conditions of catalytic reactions is difficult to understand quantitatively because the structure is highly dependent on the defect nature. The mechanism is particularly complicated when ionic conduction occurs at the surface structure. If fast ionic conduction, such as proton transfer with a time constant smaller than the localized MvK mechanism, occurs during the catalytic reaction, the localized MvK mechanism accompanying dynamic local structure deformation is unlikely to work majorly. On the other hand, reactions involving the transfer of oxygen ions and desorption processes of strongly adsorbed reaction products on oxide surfaces may well obey the localized MvK mechanism.

Comparison between the bulk-type MvK mechanism and the localized MvK mechanism reveals something very important, which is the effect only in the bulk-type MvK mechanism for catalytic oxidation. Molecular oxygen receives an electron at the catalyst surface, quickly forms two lattice oxygen ions, migrates through the defect structure, and finally reappears at the surface as a suitable electronic state of the lattice oxygen ions depending on the properties of the adjacent component metal ions. This phenomenon can only happen in the bulk-type MvK mechanism. On the other hand, in the localized MvK mechanism the extent of electron transfer taking place to adsorbed oxygen molecules at local structural sites will be limited if electrons cannot be supplied sufficiently from the oxide catalysts. This results in a different bonding state between the metal cation and oxygen ions from that in the original structure, and thus in a different catalytic activity. Therefore, the optimal reaction differs between the bulk MvK mechanism and the localized MvK mechanism.

2.1.4 Valence Control and Mixed Valence

As mentioned above, the defect structure is very important when considering the function of the crystalline complex metal oxide catalyst. A good example is the above-mentioned perovskite type oxide catalyst. If the ionic radius and charge conditions are satisfied for maintaining the basic perovskite structure, the catalyst can be composed of a variety of elements, which makes it possible to control the valence of metal ions and to introduce oxygen deficiency. Taking $La_{1-x}Sr_xBO_3$ (B: B-site ion) as an example, when a part of La^{3+} is replaced by Sr^{2+}, the valence of the B-site ion is forced to change to balance the total charge, or the formation of oxygen vacancies (Vo) is accompanied. If the B-site ion is Al^{3+} of which valence does not change, oxygen vacancies are formed (Eq. 2.1). If the B-site ion can take a valence of 4+ like Mn, the net charge is compensated by the valence change and the mixed valence (Eq. 2.2). If the B-site ion can take a valence of 4+ but the valence is not stable, for example, Co and Fe, the valence change of the B-site ion occurs giving mixed valence states and thus forming oxygen vacancies at the same time (Eq. 2.3).

$$(1 - x)LaAlO_3 + xSr^{2+} \rightarrow La_{1-x}Sr_xAlO_{3-x/2}Vo_{(x/2)} \quad (2.1)$$

$$(1 - x)LaMnO_3 + xSr^{2+} \rightarrow La_{1-x}Sr_xMn^{3+}_{1-x}Mn^{4+}_xO_3 \quad (2.2)$$

$$(1 - x)LaCoO_3 + xSr^{2+} \rightarrow La_{1-x}Sr_xCo^{3+}_{1-x+y}Co^{4+}_{x-y}O_{3-y/2}Vo_{(y/2)} \; (0 \leqq y \leqq x) \quad (2.3)$$

These three formulas can be clearly classified as solid-state chemistry, but at the same time, there is a clear relationship between solid-state chemistry and catalysis. In Eqs. 2.1 and 2.3, oxygen vacancies (Vo) are formed, so that oxygen ion conductivity appears, and the material becomes a good solid electrolyte. In Eq. 2.3, electron conductivity is added at the same time, so that the mixed conductivity is obtained. The reaction of Eq. 2.3 causes a change in the valence state of the B-site ion, which is mainly responsible for catalytic activity, and facilitates the transfer of lattice oxygen throughout the solid with the formation of oxygen vacancies (Vo), thus bringing about functions as a solid catalyst. Furthermore, the value of y in the solid obtained by Eq. 2.3 changes depending on the oxygen partial pressure of the environment, so that the valence state of the metal ion changes, which in turn causes a change in the catalytic function. It can be said that the solid obtained by Eq. 2.3 is a typical catalytic crystalline material where the bulk type MvK mechanism works in the catalytic reaction.

On the other hand, the solid obtained by Eq. 2.2 is stable in the high valence state, and almost no oxygen vacancies are created within the crystal, so that the catalytic function associated with oxygen ion-conduction in the solid does not occur, but only the state of oxygen ions bound to high valence metal ions on the solid surface and oxygen vacancies on the surface are involved in catalysis. Catalytic reaction over this type of catalyst should follow the localized MvK mechanism under the consideration of the mechanism definition. Thus, it is always necessary to keep in mind that solid-state chemistry and crystal synthesis based on it even affect the mechanism of catalytic reactions. This is a phenomenon unique to crystalline composite oxide catalysts.

Such solid-state chemistry methodologies are also being developed for electrocatalysts in oxygen reduction reaction (ORR) and oxygen evolution reaction (OER), where catalytic performance is being improved by the formation of defect structures as well as the resulting coordination structure of B-site ions, electron occupancy of e_g orbitals, and O_{2p} band structure (see descriptors in Table 1.7).

2.1.5 Introducing Coordinatively Unsaturated State

The introduction of a coordinatively unsaturated environment around the metal elements constituting the solid catalyst is extremely important for the generation of a catalytic function, and the crystalline complex metal oxide catalyst is no exception. Introducing a coordinatively unsaturated state into metal ions while preserving the crystal structure is an important key to catalyst design. Since the termination of an elemental sequence of a crystal structure is the truncation at the surface of a normal metal oxide, a coordinatively unsaturated state is inevitably introduced into the surface metal element depending on the exposed crystal plane and the kind of coordination polyhedral. However, as far as which exposed surface has a coordinatively unsaturated state is not clear, it is hard to selectively expose the specific crystal plane without doing a trial-and-error material synthesis. After all, constructing of a coordinatively unsaturated state directly by complexing elements while maintaining a specific crystal structure is realistic. The following is a list of some of the methodologies for introducing a coordinatively unsaturated environment.

1. A solid-state chemical methodology for controlling valences, in which the complexation of metal elements with different ionic radii and electronegativity is used. Doping of cation is also applicable.
2. A method of using crystals with non-stoichiometric oxides, such as crystals with a fluorite structure.
3. Hydrogen reduction to remove lattice oxygen by keeping the structure under the condition that the defective structure cannot be eliminated by forming a shear structure.
4. Formation of metal oxide crystal composed of polyhedral with a coordination number less than that is optimal for the radius of the constituent metal ions.
5. The metal oxyhydroxide is heated to desorb water and form oxygen vacancies.
6. High-temperature heating under vacuum to release oxygen from oxide bulk.
7. Introduction of coordinatively unsaturated state due to removal of organic ligand by heat treatment from synthesized metal oxide crystals containing strongly coordinating organic ligand.
8. The coordinatively unsaturated surface is selectively exposed by controlling the crystal habit by element complexation.
9. Make crystals nano-size.
10. Coordinatively unsaturated surface exposure due to strong support effect of crystalline carrier on metal oxide.

The reason why there are various methods as briefly listed above is that the introduction of a coordinatively unsaturated environment is obviously an important issue in establishing a solid catalyst, and therefore, various synthetic approaches have been made and developed for that purpose. From a catalytic chemistry point of view, it is sufficient to construct a catalytically active local site that contains the minimum coordinatively unsaturated environment. At the same time, it is desirable to maximize the exposure extent of the surface having a coordinatively unsaturated environment. In other words, the ideal crystalline material is one which is homogeneous and three-dimensionally constructed with the connection of catalytically active local structure units, like oxide clusters, that should retain a coordinatively unsaturated state around constituting elements, and more preferably is porous enough to allow reactant molecules to freely access all the active coordinatively unsaturated sites.

Efforts toward the realization of this ideal catalyst state have a long history, beginning with crystalline layered compounds, followed by zeolites with oxygen tetrahedra as the basic building block, and most recently, the appearance of all-inorganic porous crystals with oxygen octahedra as the basic building block. At the same time, there has been a remarkable development in the direction of MOFs, which are porous crystals made from inorganic–organic compounds, although MOF materials have still difficulties in introducing coordinatively unsaturated states around building units to reveal catalytic activity. Coordinatively unsaturated state is not necessarily ensured in the stage of solid-state synthesis but should be generated with its own structural characteristics under pretreatment conditions or catalytic reactions.

In layered compounds, metal ions are introduced between layers by ion exchange, and the introduced metal ions are held in the interlayer spacing of the negatively charged metal oxide layer, so that the coordinatively unsaturated state of the metal ions attains easily under catalytic reaction conditions. Similarly, metal ions in zeolites can be introduced into the cavity or the inner walls of the pore structure by ion exchange, resulting in a coordinatively unsaturated state. A coordinatively unsaturated state is also created when the tetrahedrally coordinated metal ions in the zeolite framework are replaced by other metal ions, resulting in a change in the coordination number. Recently, all-inorganic porous crystals with the oxygen octahedron as the basic building block are synthesized as crystalline complex metal oxide catalysts [Fig. 1.12b, c], in which not only the coordinatively unsaturated state is produced around metal ion introduced by ion-exchange in the pore structure in the same way as zeolite, but also the metal-oxygen polyhedral themselves, which are necessary to constitute the framework here, can have a coordinatively unsaturated state due to crystal structure regulation. It should be emphasized that since the complex metal oxide itself is a porous crystal, all the constituent elements can involve in the catalytic reaction. In this situation, this structure type of solid catalyst is regarded as the metastable state material. Furthermore, the material is the ultimate solid catalyst in which a localized MvK mechanism is established for all crystal constituents.

2.1.6 Catalytic Collaboration among Constituting Elements and Interplay of Different Structural Phases

In solid catalysis chemistry, complexation and diversification of the constituent elements have very often produced positive effects, leading to the realization of many practical catalysts. Such complexation effects are not necessarily limited to crystalline solid catalysts. However, as seen in the history of the development of crystalline complex metal oxide catalysts described in Chap. 1, introduction of multiple elements in the crystalline state has been successful with a certain bonding pattern and electronic state, as a result of the pursuit of improved catalytic performance. In this trial, establishing a certain spatial arrangement of multiple elements in bulk crystal structure or local site structure on the surface is indispensably important for the evolution of catalytic function. In other words, structure, especially local structure, is essential for the generation of the complexation effect. It should be kept in mind that the manifestation of the effect is first brought about by a change in the individual constituent elements themselves, which in turn causes a change in the catalytic activity. The factors governing these will be summarized later as descriptors in Table 2.1. The individual effects are illustrated by specific examples in the contents of the related chapters.

Here, we focus on multi-elemental cooperation due to the proximity effects of elements, and edge shared polyhedral dimer is taken as the example. First, for those classified as localized MvK mechanisms, their characteristic structural sites can be roughly divided into catalytic functions as shown in Fig. 2.2. In all cases, different metal elements face each other via metal–oxygen bonds and can work together in a close environment without losing the characteristics of each element. Furthermore, the electron density of the intervening lattice oxygen ions varies greatly depending on the type of surrounding atoms, bonding distance, bond angle, and number of bonds, which in turn greatly affects the catalytic activity. When metal elements have different charges, protons are located on oxygen to maintain charge balance (Case 1). This proton can play as acid, resulting in the material possessing acid catalytic property. If protons are removed as water by reacting with lattice oxygen, the oxygen site in the metal–oxygen bond becomes empty and forms a coordinatively unsaturated state on metal cation which can work as a Lewis acid site (Case 2). Owing to the strong covalent bonding character between oxygen and metal in Cases 1 and 2, the oxygen defects cannot form easily, do not readily migrate, and are not replaced with others, which means that the state is rather static. However, there are still a few changes in the metal–oxygen bonding situation occurring under dynamic catalytic reaction circumstances. In Case 3, a redox active element and non-redox element collaboratively work to accept proton and electron concertedly (coupled) from organic reactant to activate the C–H bond. If electron transfer becomes easier, the metal–oxygen bond weakens, allowing retaining of the oxygen deficiency, resulting in a coordinatively unsaturated state (Cases 3, 4, 5, 6). The vacancy can accept oxygen and hydrogen from reactant molecules. In Cases 4 and 5, if two unpaired electrons are created when the oxygen vacancy is formed and are located around oxygen vacancy,

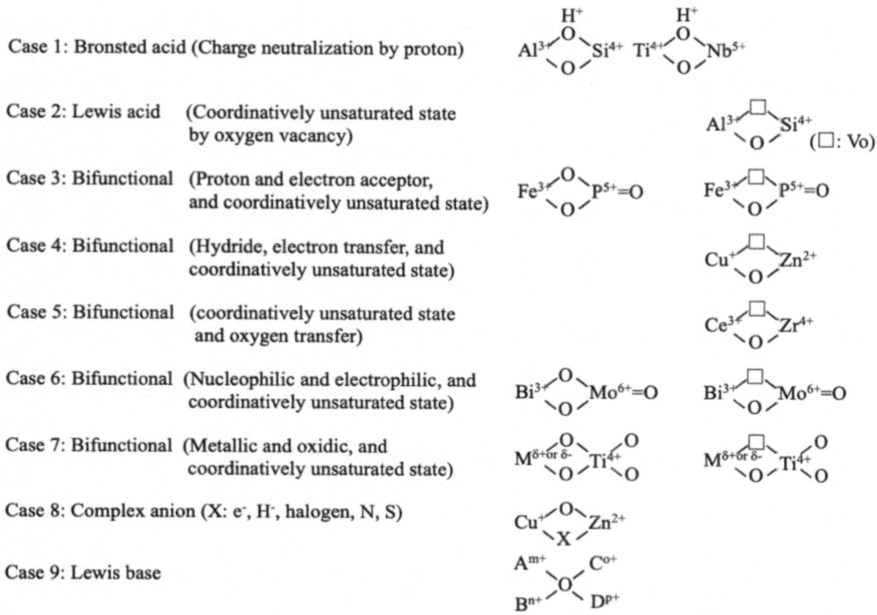

Fig. 2.2 Proximity effects of two different elements in edge shared polyhedral dimers on catalytic functions

electrons in this situation play as Lewis base, adsorbing acidic molecules like CO_2, assisting C–O breaking, and accepting O atom in oxygen vacancy [18] (more details in Chap. 12). In Case 6, there are two types of lattice oxygen, one connected with Bi shows nucleophilicity and the other connected with Mo shows electrophilicity, thus separately catalyzing hydrogen abstraction and oxygen insertion. Case 7 is rather unique, but electron transfer from a metallic element to an adjacent non-redox element or vice versa to create a catalytically active site. Case 8 is another unique one in which anionic species like electron, hydride, halogen anions, N ion, and S ion get the position of lattice oxygen [19], causing various catalytic functions (related examples in Chap. 7), like photocatalysts (more details in Chap. 9). The final case is Case 9, where the coordination number around one lattice oxygen is high and forms multiple bonding with metal elements. This bonding situation creates lattice oxygen with high nucleophilicity that can promote the breaking of a strong C–H bond (an example in Chap. 10). As the variety of constituent elements increases and hence the complexity of the local structural sites increases, several of the above classified cooperative effects will operate simultaneously depending on constituting elements. At present, a depth understanding of these effects has not well progressed, but more multi-elemental catalytic systems with control of oxygen vacancy [20] will be advanced further [21].

Next, regarding the establishment of different element collaboration and functional separation based on the crystalline composite oxide catalyst in which the bulk-type MvK mechanism is established, the multi-component Mo–Bi catalyst, which is well known as the olefin oxidation catalyst, is taken as an example. The multi-component Mo–Bi oxide catalyst for propene selective oxidation is composed of Mo as the main component and with nearly ten kinds of elements like Bi, Fe, Co, Ni, K, and P [22] which is the most complicated catalyst among the industrial catalysts. The oxidation activity of the multi-component Mo–Bi oxide catalysts has dozens of times of that of simple crystalline Mo–Bi–O oxide catalyst (Fig. 1.20) which is known as a basic active crystalline catalyst. The catalyst shows activity even at 100 °C lower reaction temperature and high selectivity for the acrolein formation. In reflection of the complicated component, the multi-component Mo-Bi oxide catalysts consist of various metal oxide phases and complex metal oxide phases either in crystal or in amorphous states. When $Mo_{12}Bi_1Fe_3Co_8Ox$ is taken as an example (Fig. 2.3) [23, 24], Co and Fe are mainly located in the bulk of the molybdate crystal particle, and on the other hand Bi condensates on the molybdate particle surface because the Bi has larger ionic radius than Co and Fe. As a result, a Mo–Bi–O phase forms fine crystalline particles on the surface region of the crystal particle of Co(+Fe) molybdate solid-solution (Fig. 1.21). This whole structure quite resembles RuO_2/TiO_2 catalyst shown in Fig. 1.8.

The oxidation of propene over the catalysts proceeds through the dissociative activation of oxygen molecules, the abstraction of allyl hydrogen of propene, the oxygenation of allyl intermediate, the electron transfer in the catalyst through the lattice defect, and a migration of the active oxygen in the bulk of the oxide catalysts. The oxidation can appreciably take place over the Bi–Mo–O catalyst at a relatively high temperature, indicating that the catalytic cycle is not optimal for each element to work effectively in this catalyst. The problem is due to excessive roles imposed on

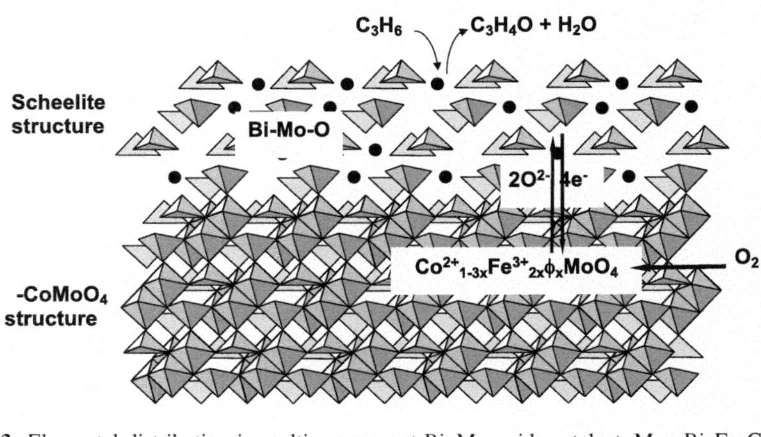

Fig. 2.3 Elemental distribution in multi-component Bi–Mo oxide catalyst, $Mo_{12}Bi_1Fe_3Co_{11}O_x$, active for propene oxidation to acrolein and catalytic function separation based on crystal structure and phase cooperation [23, 24]

Bi which must directly participate in the whole reaction cycle. Therefore, the mean number of the Bi site involved in the rate-determining step and in oxygen activation could be limited at the dynamic reaction cycle condition. This is considered to be the reason for the low catalytic activity of the simple Bi–Mo–O catalyst. On the other hand, in the multi-component Mo–Bi oxide catalysts, tri-valent cation Fe takes part in the oxygen molecule activation in place of Bi, and activated oxygen can migrate through the lattice defect in Co(+Fe) molybdate to the surface where Bi locates (Fig. 2.3). Then Bi can devote itself to the process of hydrogen abstraction from propene predominantly, resulting in high oxidation activity. Furthermore, a useless oxidation reaction with oxygen species formed by partial reduction of molecular oxygen causing lower selectivity will be diminished when the rapid migration of lattice oxygen and electron transfer. This situation acts profitably in many respects that reduction of oxygen with the Fe site occurs at lower temperature and sustains full reversibility of the local catalytic active site under dynamic micro-kinetics conditions, thus resulting in a remarkable increase in the catalytic activity. This cannot be attained by a simple phase mixture, although almost all elements are practically exposed on the surface, emphasizing the importance of the ordered arrangement of constituent elements in the crystal structure. This is a representative example that the bulk-type MvK mechanism works most effectively and makes a clear function separation and elemental collaboration supported by the catalyst structure and multi-element complexation possible. Element complexation effect based on particular crystal structure formation is derived neither from the summation of catalytic functions of each element nor from simple multiplication of the function but must be based on block chain structure where electrons and lattice oxygens rapidly move to exchange the state information of elements.

2.1.7 *Crystal Facet Dependence of Catalytic Function*

It is often seen that the catalytic activity of crystalline single metal oxide or complex metal oxide catalysts differs greatly depending on the exposed crystal plane. This dependency has been demonstrated by methods such as crystal morphological change, selective poisoning of exposed surfaces, grinding effect, and so on. Although the structure and elemental composition of the crystal surface are not always clarified, such dependency does not occur unless the bulk crystal structure affects the structural termination at the surface. Therefore, the crystal facet dependence of catalytic activity is important.

The example here is the crystalline divanadyl pyrophosphate [25, 26] catalyst which is active in the selective oxidation of n-butane to maleic anhydride. The characteristic point of this catalyst is that the twin octahedral vanadyl cluster is structurally isolated by the six surrounding tetrahedral phosphate anions (Fig. 2.4). Only crystallized divanadyl pyrophosphate particles showed exclusive activity and selectivity for the reaction to other V-P-O crystalline compounds, indicating that the particular crystalline phase formation is indispensable for the catalysis in this case. Before the

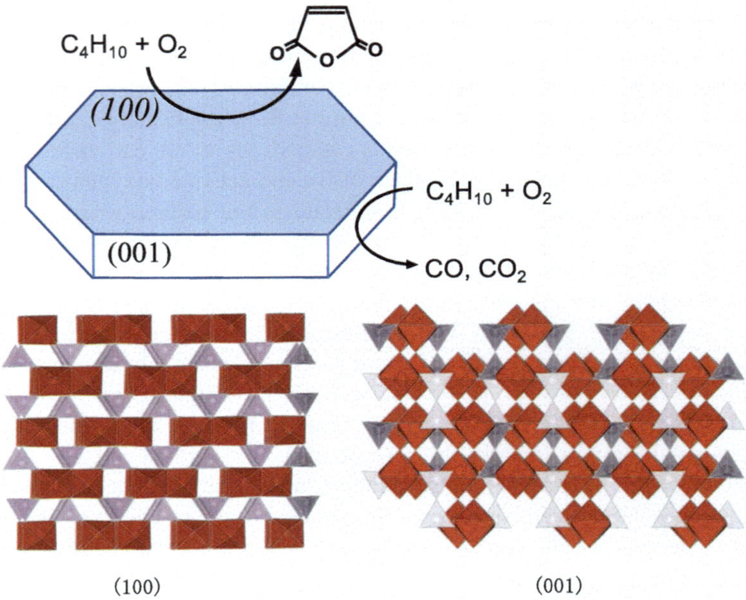

Fig. 2.4 Facet dependency in the case of butane oxidation over $(VO)_2P_2O_7$ catalyst and site-isolation structure on (100), which shows the isolated V octahedra dimer by six tetrahedral PO_4 surround [25, 26]

discovery of the crystallized divanadyl pyrophosphate catalyst, an amorphous type of V–P–O catalyst has been known for 1-butene oxidation to malic anhydride. However, the amorphous type of V–P–O was not as good as the divanadyl pyrophosphate catalyst for the n-butane oxidation, suggesting that a particular crystal structure is needed for forming oxygen species active enough for n-butane oxidation than for 1-butene oxidation which has much weaker C–H bond than n-butane has. Furthermore, the entire catalyst material must be uniform, and if not, various undesirable reactions will occur, causing lower product selectivity. In fact, under the progressive development of V–P–O catalysts, the introduction of crystal structure has been successfully done and turned out to be a highly active catalyst for the n-butane oxidation to maleic anhydride.

It is believed that the twin vanadyl sites surrounded by six tetrahedral phosphate anions can provide active oxygen species through dynamic structure deformation (MvK mechanism) during the catalysis and also transfer 14 electrons to complete the reaction of n-butane to maleic anhydride (site isolation concept [10, 11]). This twin structure responsible for the catalysis appears on the basal plane only of the plate-type crystal of divanadyl pyrophosphate. Therefore, the surface dependence of this catalyst is clear [25, 27, 28], and the structural specificity of the catalytically active surface also supports this phenomenon. The same facet dependency was also demonstrated in crystalline $Mo_3VO_{11.2}$ catalyst for ethane oxidative dehydrogenation as introduced in Chap. 4.

It is natural that the catalytic activity is significantly different between a surface having a simple element arrangement structure in terms of crystal structure and a surface having a complicated and discrete element arrangement structure. In other words, it is not natural that surface dependence does not occur in such characteristic crystal particles. An important conclusion comes here, that is, the local elemental arrangement in structure has a great significance for catalytic chemistry. Even a simple elemental arrangement on the surface has a catalytic function, but the surface of a complicated but stable local structure is mostly catalytically superior. Therefore, crystal particles having a particular surface predominantly exposed result in a successful catalyst consequently. It is not the conclusion that the simple surface is not doing anything as a catalyst. Rather, a simple structural surface is responsible for other necessary reaction steps like the uptake of oxygen molecules while the complex surface is responsible for the selective main reaction, as exemplified using the above-mentioned multidimensional Mo–Bi composite oxide catalyst where the bulk-type MvK mechanism is working. Although fundamental studies have yet progressed from that point of view, the separation of catalytic functions and the collaboration of constituting elements and facet dependency are rather a natural phenomenon when considering the dynamic behavior in the crystal structure under catalytic reaction conditions. As a matter of fact, crystalline complex metal oxides are industrially used as an excellent catalyst and this fact is strong evidence supporting that catalytic function is derived from or based on crystal structures. However, there are still a lot of problems. Even if an attempt is made to introduce an element in anticipation of the complexation effect of the element on catalysis in order to further improve the catalytic function, the crystal structure does not accept to accommodate alien elements in itself because the crystal structure always tries to maintain the original structural characteristics. This might be the main reason why further advancement of V–P–O catalyst performance has almost ceased in the last two decades. Exactly the same situation seems going on in the MoVTe(Sb)NbO catalysts for alkane oxidation. The catalyst was invented more than 20 years ago, but even today the basic component of this catalyst is the same although many attempts have been made to improve its catalytic performance. The obvious reason here is again the mandatory requisite of specific crystal formation. This trend seems characteristic in crystalline solid catalysts and thus a new approach based on crystals, crystal engineering, is required to find out post-V–P–O catalyst and post-MoVTe(Sb)NbO catalysts.

2.1.8 Catalysis based on Molecularity of Discrete Structure

The above-mentioned crystal plane dependence makes it clear that the complicated but discretely arranged structure, oxide cluster, is extremely important for creating catalytic function. Here is another important factor in addition to this. Take a consideration about the skeleton of the discretely arranged structure based on the oxide cluster structure with the same elements. The oxide cluster referred to here is, for example, a dimer in which oxygen octahedrons share corners, edges, and faces. Not only the cluster structure itself has a catalytic function, but also the metal element that needs to connect the oxide clusters during crystallization and another element that is arranged in the void between the clusters have catalytic functions too. Constructed structural environment in such a way can provide a local catalysis field where constituent element cooperates with each other and ultimately generates an excellent catalytic function, like $(VO)_2P_2O_7$ catalysts described already in Sect. 2.1.7.

Such a discrete structure environment may appear when a high index crystal plane is preferentially exposed even in a dense crystalline complex metal oxide having a simple crystal structure, but this approach is not designable. Otherwise, the method of forming oxide clusters on a surface of two-dimensional metal oxide crystal support gives rise to the same structural environment. Actually, this support-assisted method is now widely advanced (Figs. 1.7 and 1.8) and makes it possible to design a discretely element-arranged structure by the chemical interaction between the surface of the support and the cluster.

Based on a discrete structure (cluster) composed of polyhedral, a three-dimensional crystal structure having the discrete structure environment can be formed as well. For instance, an anionic oxide cluster together with another metal cation forms an ionic crystal (Figs. 1.4, 1.5 and 1.14). Polyoxometallate cluster is a typical example and is condensed each other to form a framework-type oxide structure (Fig. 1.12). It becomes possible to transform into a crystal having micropore properties at the same time as the oxide clusters are three-dimensionally connected with added metal cations. Note that this idea of catalyst structure design is naturally derived from the requirement of a discrete elemental arrangement structure for catalysis. It has already been known that there are a variety of oxide clusters as illustrated in Fig. 1.14 and are formed in many cases in the intermediate stages of oxide solid material formation, as shown in Fig. 1.26, so there is a high probability that these molecular oxide clusters formed in the intermediate stages are further condensed and join with other metallic elements to form a structure that eventually develops into a three-dimensional crystal. The formation of discrete elemental structures seems a very natural phenomenon. Therefore, efforts to synthesize metal oxide clusters in zeolite-type pore structures (Fig. 1.11) have recently become widespread, and there have been many reports of successful development of specific catalytic functions because there are many kinds of possible discrete oxide clusters as shown in Fig. 1.14.

2.1.9 Descriptors

As described above, the characteristics of crystalline composite oxide catalysts are summarized in terms of crystal structure, and in some cases local structures are especially focused. However, there are not many cases where the actual local structure responsible for catalysis has been clarified. This situation should be regarded as

Table 2.1 Various descriptors for catalytic functions[a] of crystalline metal oxides

State	Descriptor					
	Metal cation		Metal-(μ-oxygen)-metal or Metal-(*t*-oxygen)		Oxygen	
	Structure	Electronic	Structure	Electronic	Structure	Electronic
Bulk	Ionic radius Coordination number Crystal field Metal–metal distance	Electronegativity Electrophilicity Electron density Charge density Oxidation state *e* and *t* energy level Spin state	Geometry Bond strength Bond length Bond angle Bond order Bond symmetry Multiple bonding Cation defect Oxygen defect Dislocation Deformation Madelung energy	Electronic structure Covalency Ionic character Band structure Band-gap energy Density of state Ligand-to-metal charge transfer Charge localization-delocalization Inter valence charge transfer Bader charge transfer Super-exchange interaction Band overlapping Electron and hole Hybridization	Share number Occluded oxygen Interstitial oxygen Oxygen migration	Electron density Charge density *e*-filling

(continued)

Table 2.1 (continued)

State	Descriptor					
	Metal cation		Metal-(μ-oxygen)-metal or Metal-(*t*-oxygen)		Oxygen	
	Structure	Electronic	Structure	Electronic	Structure	Electronic
Surface	Coordinatively unsaturated state Segregation Migration	Electron density Oxidation state *e* and *t* energy level Spin state	Cation defect (vacancy) Oxygen defect (vacancy) Lattice plane (Facet) Edge, corner, face share M=O bond O–H bond (protonation) Proton transfer Dangling bond Dynamics Reconstruction Displacement Surface contaminant	Surface potential Fermi surface Electron transfer	Coordination number Share number Oxygen species Adsorbed oxygen species O–O bonding Surface migration	Electron density Radical state Surface oxygen nnvacancy formation energy

[a]Catalytic functions are oxidation, redox property, dehydrogenation, acid–base, electrocatalysis, photocatalysis, and so on

common but not good for understanding and designing the catalytic activity in a true sense. Although reliable local structure information is essential for these purposes, the clear and strong effect of bulk crystal structure on the catalytic property of crystalline composite oxide catalysts can be utilized in one way. As an alternative, we can understand and explain the catalytic activity with descriptors focusing on metal cation, lattice oxygen, and metal–oxygen bond, which are the smallest units of oxide catalysts but are based on crystal structure information. Relevant descriptors as many as possible are listed in Table 2.1. Many efforts have been made to explain catalytic functions using a variety of descriptors in many studies [29] (see more in Chaps. 12 and 13). However, it must be reminded that no single descriptor is sufficient, and here again, a complex approach is needed. The problems lying behind the complexity are the following.

2.2 Catalysis Chemistry in Complexity

2.2.1 Structure Difference Between Bulk and Surface of Complex Metal Oxides

This has been a long-standing controversy and may be an eternal issue peculiar to the catalysis chemistry field. Even with the same crystal catalyst, the understanding of the surface may be completely different depending on the analysis method and analysis conditions. Some researchers argue that the surface is completely different from the bulk, and some emphasize the peculiarities of the surface while reflecting the bulk structure. Some believe that the surface is just the end of a continuous bulk structure but not exactly in the same environment as bulk, and thus support that the catalytic function can be understood as reflecting the bulk crystal structure. Recently, research efforts are being focused on knowing what the surface of a crystal catalyst looks like in an actual catalytic field [1, 8]. It is really the right direction, but it seems like a long way off to see the results in a clear manner. The reason is that modern analytical methods are not yet fully capable of identifying true catalytic local surface structures. Nevertheless, this approach should be continued as it may result in a deeper understanding of the state and function of catalysts. However, nowadays we are not allowed to stop making efforts to introduce new catalysts required by modern society until the progress of analytical technology is made and the surface state is fully clarified. A more realistic and reliable interpretation is needed. Below are the facts that can positively emphasize the importance of the bulk crystal structure.

1. The catalytic function can be controlled in a correlative manner by applying a methodology for controlling bulk physical properties (defect structure, valence control, band structure, DOS, etc.; see Table 2.1) [7].
2. Catalytic performance may differ significantly between catalytic materials with the same constituent elements but different crystal structures.

3. Pore structural crystals such as zeolite have already lost their surface uncertainty described above.
4. Catalytic function of the three-dimensional crystal catalyst composed of a discrete metal oxide cluster is determined by the cluster.
5. The transfer of lattice oxygen, protons, electrons, etc. in the crystal depends on the crystal structure, which in turn greatly affects the catalytic reaction.
6. A better catalyst is established by well and uniform crystallization of the constituent elements by appropriate synthetic methods.
7. Catalyst life can be improved by increasing the resilience of the crystal catalyst by adding an element that does not become a catalytically active element.
8. Catalytic activity is produced only when multiple constituent elements are arranged at fixed positions in the crystal structure.
9. The catalytic performance is maximized at extremely narrow or simple constituent element ratios, that is, crystal formation.
10. Even if element segregation occurs on the crystal surface, the element often forms a structure (epitaxial, etc.) while maintaining the arrangement of the underlying elements, and the entire crystal structure is often reflected.
11. The intrinsic catalyst performance is basically independent of the particle size of the crystal catalyst.
12. Crystal catalysts are often in the form of small crystal particles, which inevitably causes phenomena such as element segregation, impurity phase formation, and condensation at grain boundaries, but such formation is suppressed at the catalyst preparation stage. Thus, the catalyst performance is improved by adopting the method.
13. Polymorphic problems. If polymorphs exist in crystals, polymorphs slightly thermodynamically unstable can form unavoidably near the surface, but this also falls into the category of crystal catalysts in the end.
14. There is an optimum calcination temperature. Calcination just before the structure changes or collapses brings good catalytic performance in many cases.
15. The proportion of surface-exposed metal ions varies depending on whether they are in the oxygen octahedron or oxygen tetrahedron in the crystal structure. This is because the stability on the coordinatively unsaturated surface differs depending on the polyhedron, and the exposure of oxygen octahedron has priority [8].

Crystalline complex metal oxides are representative in the field of ceramics and ionics, and new materials have been actively developed. As a result, new materials and new physical/chemical properties have been discovered and put into practical use. Looking at such circumstances and movements in the field of ceramics and ionics, it is unfortunate that only the field of catalysts sticks to the surface and neglects bold catalyst development. Efforts to put a scientific scalpel into understanding the true surface state should be greatly promoted, but at the same time, exploration of new catalytic materials, especially those based on crystallinity, should be actively promoted.

2.2.2 Difficulty in Determination of Active Site Number on Metal Oxide Surface

For better catalytic activity comparison, it is desirable to determine the reaction rate (or formation rate) per active site, which is indispensable for performance comparison between catalysts. As described above, the catalytic functions are based on a local active site structure in which the catalytic constituent elements are accumulated or clustered depending on the bulk crystal structure. If the local active site is a single metal, it is possible to define the active site number by using the surface-exposed metal element, and then it might be possible to measure the amount of surface-exposed metal. However, there must be exposed surface dependency and effect of defects in the case of the crystalline metal oxide crystal catalyst; it is difficult to evaluate the catalytic activity correctly unless the amount of true active surface metal under working conditions of the catalytic reaction is obtained. At present, the amount of active surface sites must be determined by ensuring conditions that can be reasonably deduced or extrapolated. If the active site is formed by a local structural unit composed of various elements, a simple adsorption method is not always useful to determine the number of true active site numbers, unless there is a strong proof to support the correlation between adsorption amount and active site number. It is, therefore, necessary to devise a method for quantifying the active site in connection with the catalytic reaction. Ideally, a truly working active site can be specified only when the structural state and constituent element state of the catalytically active site can be identified, and then the number can be accurately determined. In recent years, the constituent elements of metal catalysts have tended to be extremely complicated, making it difficult to accurately compare functions between catalysts, and as a result, the role of constituent elements has become increasingly uncertain. The uncertainties associated with such complications prevent the essential function of the catalyst from being clarified.

On the other hand, what about crystalline metal oxide catalysts? Since the catalysts classified as single atom in Table 1.2 are placed in the same situation as the above-mentioned metal catalysts, the active site number seems to be strictly defined. However, again, since not all of them contribute to catalysis as a single atom site does, it is still necessary to quantify by some method under various restrictions. The discrete molecular oxide cluster-type catalyst categorized into the crystalline metal oxide catalyst (Table 1.2) ensures a solid structure and homogeneity. The number of catalytically active sites can be defined by the amount of substance of the oxide cluster used, if it is recovered in the same state after use in the reaction. This is in the same classification as the organometallic complex catalyst, which is the opposite of the solid metal catalyst.

In the surface layer or interlayer compounds (two-dimensional) in the second classification of Table 1.2, it is extremely difficult to identify the active site and its structural mode, and it is almost impossible to determine the number of active sites. This category poses a problem for catalytic chemistry because the locally layered surface structure is a frequent form in solid catalyst materials and is easily prepared. The problem will be discussed in detail in the next section.

How about a metal oxide catalyst with a more advanced dimension? Needless to say, a simple structural oxide catalyst has the same level of complexity as the two-dimensional material described above and has similar difficulty in the determination of active site number. However, even if the long-period order of elements is not established, it is possible to determine the number of active site structure by applying some assumptions that the catalyst has high structure homogeneity in the element distribution and chemical bond mode as a whole catalyst material. No matter what the crystal particle states, the particle size distribution of the solid, and the aggregated state of any particles are, the local parts that make up the catalytic active site structure are nearly homogeneous. It is permissible to make a comparative evaluation of catalytic activity based on the outer surface area as the first step. However, if the constituent elements become complicated, the relative arrangement of the constituent elements can change in various ways in crystal structures, so that the homogeneity is greatly reduced. In this respect, the trend of the increasing complexity of solid catalysts is hindering the development of catalytic chemistry.

The densely crystallized oxide catalyst is generally uniform in the bulk structure, but there is inevitably a limited number of non-uniformities on the surface, which is the catalytic field. This is because there are always different crystal exposed surfaces. And since the constituent elements are exposed in various states on each surface, it makes virtually no sense to specify the number of active sites if the catalytic reaction proceeds with different performances on all exposed surfaces. However, as described above, since a crystalline metal oxide catalyst having activities and high selectivity has a strong crystal plane dependence, in other words, the stronger crystal plane dependence results in a better catalyst; it is possible to derive the active site number if the active plane can be specified with an active structural characteristic. However, strictly speaking, if defects are present and is involved in the catalytic reaction, it is necessary to measure the number of defects by some method. However, at present, there is no effective method because the defect state is dynamic under catalytic conditions. The possible way is to evaluate from chemisorption or from the reactivity using the pulse reaction and temperature programmed surface reaction (TPSR) [30], but still further advancement of this methodology is necessary.

Apparently, there are various limits to the determination of the number of active sites. Nevertheless, under such limitations tremendous catalysis research has achieved and actually developed various new solid-state catalysts. It started with crystalline layered materials, and with the advent of microporous composite oxide crystals, in which the above problems could be eliminated. In the example of a crystalline layered material, protons are located between layers, the number of which is determined from the valence of the constituent elements, and the interlayer spacing freely functions as a catalytic reaction field. Therefore, the number of protons is directly equal to the active site number. When the metal oxide layer itself exhibits catalytic activity, the homogeneity of the surface is guaranteed, so that the calculation can be done with the assumption of the active site based on the crystal structure and the value of the outer surface area.

The crystalline catalyst that can specify the number of active sites more strictly is the recently introduced microporous crystalline complex metal oxide crystal catalyst. Since they are crystalline substances and also porous, the characteristic of this material is that the environment of the constituent elements in the crystal is uniform, and that all the constituent elements uniformly contact the reactant molecule and work as a catalyst. Therefore, the active site can be easily defined, and the number can be calculated based on the elemental composition. Indeed, the identification of active site structures and the determination of the number become possible by the realization of a completely new crystalline complex metal oxide. This approach has been difficult so far. In the context of catalytic active site number, it can be recognized here again how important crystalline complex metal oxides having various structural characteristics is.

2.2.3 *Micro-Structure in Amorphous Complex Metal Oxides*

The amorphous state is not necessarily inferior to the crystalline catalyst, and it is often seen that amorphous oxide has an ordered structure to some extent when viewed microscopically, and it can be sometimes regarded as a quasi-crystal locally. This state is clearly exemplified in the case of $Mo_3VO_{11.2}$ oxide catalysts (Fig. 1.17). Fig. 2.5 shows HAADF-STEM images of two samples having the same elemental composition [31]; one is a well-crystallized solid and the other is XRD amorphous, which are synthesized by changing pH conditions under the identical hydrothermal condition. The characteristic structural motif of the so-called pentagonal unit can be observed clearly in both cases, but the ordering of the pentagonal unit is different. The crystal specimen showed clear order while the amorphous specimen showed a random arrangement of the motif, which is typical in an amorphous state. However, if we take a closer look at the images, one can see the same local structure arrangement (highlighted in Fig. 2.5) in both cases. Apparently, the number of this local structure is lower in the amorphous material than in the corresponding crystal, and in fact the catalytic activity is superior to the crystal than the amorphous one. Historically, the Mo–V–O type catalyst developed for acrolein oxidation in the gas phase was the amorphous one because of the conventional metal oxide preparation process. This catalyst state has been long used industrially after elemental complexation because the resulting catalyst shows reasonably high oxidation activity and extremely high product selectivity, but recently even simple crystalline $Mo_3VO_{11.2}$ catalysts show much superior catalytic performance by keeping the same high selectivity, emphasizing the importance of the introduction of crystalline state.

Considering the atomic level region of the active sites, short-range order of elements should be enough in principle for these states fully exerting catalytic activity. However, if a locally well-ordered atomic arrangement is essential for generating catalytic activity and if this arrangement is possible with crystalline catalysts, the active site density of crystalline catalysts must be higher than that of amorphous catalysts and accordingly, the catalytic activity becomes higher per unit surface area.

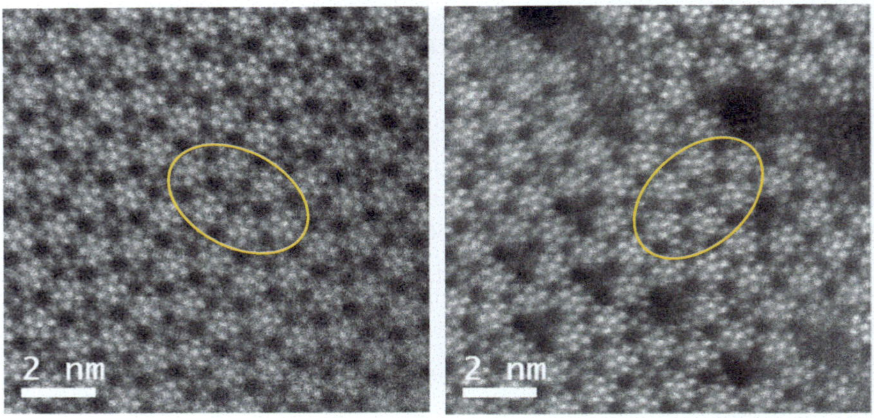

Fig. 2.5 HAADF-STEM images of **a** orthorhombic Mo_3VO_x, and **b** amorphous Mo_3VO_x. The highlight represents the basic structural motif that appeared in both the materials [31]

Furthermore, even if the atomic arrangement around the catalytic active site is satisfied to some extent in the amorphous oxides, subtle differences such as bond angle, metal–oxygen bond distance, and metal-to-metal distance will inevitably occur, so that the intrinsic activity of the active site between the crystalline catalyst and the amorphous state are different. In other words, both can never be the same.

On the other hand, crystallinity seems to be more advantageous in many cases in terms of selectivity. Crystallinity induces high selectivity because the homogeneity of the catalytic active structure almost always promotes the reaction in the specified direction, but the increased inhomogeneity creates the possibility of different reactions occurring, which leads to lowing of reaction selectivity. Inhomogeneity in the catalytic active structure is derived from non-identical elemental mutual position and deformed metal–oxygen bond formation. This situation is more prominent in the case of amorphous solids even with the same constituting elements and in the case that constituting elements become more multiple and complicated. This is another type of problem since catalyst development goes in the direction of further element complexation which causes hardly controlling of each constituting element's position in the structure, resulting in selectivity problems. Here, it should be reminded that higher complexation is still high potential in catalyst development, particularly for catalytic reactions with no need to take selectivity into consideration, for example, total oxidation. Further, the appearance of completely different local structural units is unavoidable in the amorphous form. The inhomogeneity pointed out here cannot be eliminated even with a crystalline catalyst, for example, because the exposed surface appears in various ways and forms a local structure depending on each. These have a negative effect on selectivity. Nevertheless, most of the catalysts that have successfully appeared as catalysts with excellent selectivity have strong exposure surface dependence of catalytic performance as described above, so such a possibility can be minimized. Otherwise, excellent selectivity cannot be created. Then, a highly ordered structural mode and characteristic bonding mode are inevitably formed on that surface, which are not formed on other surfaces. In other words, it is a good solid catalyst that a unique local structure is formed in the crystalline material.

The above is ideal, but it is rare to reach a crystalline substance at the beginning of the real catalyst development process, as introduced in Chap. 1. Oxide catalyst is often established in the supported form or the amorphous state. Eventually, the crystallized solid catalyst appeared by the sophistication of the catalyst synthesis method. Not all catalysts have gone through this process to reach complexation, but this process seems inevitable. However, it is not always a good idea to pursue crystalline material synthesis from the beginning. This is because most of the synthetic methodologies for giving crystalline materials have employed processes such as high-temperature calcination and resulted in a thermodynamically stable crystalline phase. As mentioned earlier, the catalytically active structure is not always in an extremely stable structural state but mostly metastable, so that catalytically active crystalline solid cannot be reached without careful consideration of the synthesis methodology. Therefore, a way to develop crystalline catalysts is not a simple pursuing of crystal, but under prioritizing catalytic activity development process should be followed by identifying the active ingredients of the catalyst while sacrificing some selectivity, embodying them in the form of carrier or simple composite, and then proceeding to the process of purifying the active structure. What is interesting is that even in the process of such development, there are many similarities with the final ideal crystal catalyst, for example, the optimal elemental composition is not so different. This is not the case with the supported catalyst type, but if the bonding surface between the carrier and the catalytic substance can be accurately separated, it can probably become a new catalytic substance that exists in nature. The fact that a certain degree of selectivity is obtained even with a mixed-type catalyst means that the catalytically active site is quite limited in percentage, but the formation is still dependent on the overall constituent element ratio. Since the abundance of such active sites cannot be expected to be large, active structure purification should proceed through crystal formation in the process of catalyst development later, and then the high activity will be achieved. Again, crystallization is not prioritized in catalyst development at the beginning, but more development of mixed or amorphous forms of solid catalysts proceeds because the methodology of complexing elements in this case is easily adopted in order to improve catalyst performance. The reason is that mixed or amorphous solid catalysts can accept any number of composites of elements. The complexing effect obtained here will be utilized later in the crystal catalyst evolution.

2.2.4 *Structural Surface Material Formation*

One of the major advances in the long history of solid catalyst development is that the catalytic material is immobilized on the surface of the solid carrier. The supported catalytic materials work not only by themselves, but also work with strong assistance from the solid carrier through electronic and structural interaction. This has greatly changed and advanced the view of solid catalysts. Take a system that a metal oxide catalyst is supported on an oxide carrier as an example. The formation of catalytic materials in a monolayer state with the electronic effect from the support carrier and with the influence on the catalyst structure (particle size, morphology, etc.) was the first concern. For example, the chemical bond with the oxide carrier has been recognized as an important factor in understanding the catalytic property, which in turn regulates the structure of the catalytic materials on the carrier (Fig. 2.6) [32]. This has recently led to the understanding of the formation of a molecular cluster-type surface substance. At this point, it can be considered that a highly ordered structure in the discrete range is locally formed between the crystal structure of the carrier and the surface oxide materials like oxide clusters, beyond the dimension of the supported catalyst. This is regarded as a structural surface substance. Since this material form seems unstable but relatively homogeneous on the carrier surface, it may not be impossible to synthesize the materials as an isolated single crystal structure. If the analysis advances the firm identification of such structural surface materials, it will be possible to extract and synthesize them as crystals by making full use of synthetic catalytic chemistry.

A system in which a metal catalyst is supported on an oxide carrier follows similar advancement. The ultimate example is a supported gold catalyst. The gold particles themselves show little catalytic activity, but when properly immobilized on a metal oxide carrier such as TiO_2, a completely new catalytic state is created and it exhibits significant catalytic activity for CO oxidation. The origin is now believed to be the junction boundary region between the gold particles and the carrier, that is, the perimeter region. The gold atom is incorporated into the titanium oxide matrix at the atomic level, and more locally, the Au_n–O–Ti bond can be established (Fig. 1.9), and in such a state, Au is no longer a metal, maybe in a positive or negative charge. This state is maintained only when titanium and gold particles are joined, and it is not possible to take out the gold in the state of Au_n–O–Ti. Au_n–O–Ti might be also the

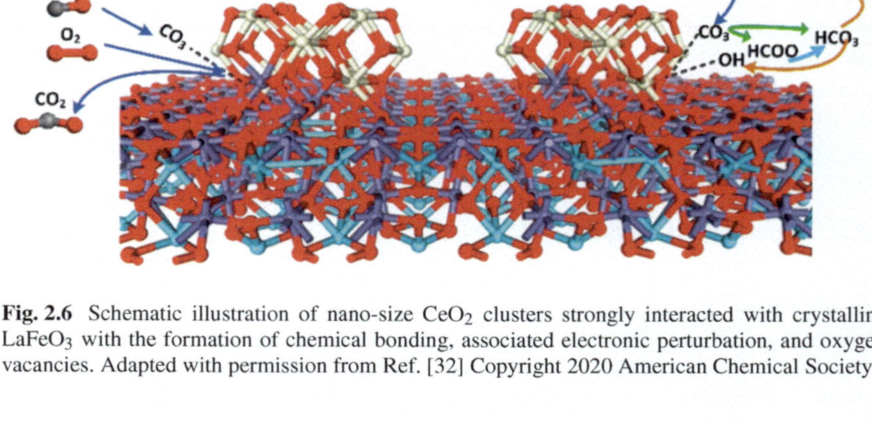

Fig. 2.6 Schematic illustration of nano-size CeO_2 clusters strongly interacted with crystalline $LaFeO_3$ with the formation of chemical bonding, associated electronic perturbation, and oxygen vacancies. Adapted with permission from Ref. [32] Copyright 2020 American Chemical Society

state of a structural surface material generated during the development of catalysts, but if the state of Au_n–O–Ti can be extracted and synthesized in a form of crystal structure in the same way as the above-mentioned oxide system, much different catalytic material understanding will be advanced greatly. Although this approach is not progressing yet, extraction will become more realistic if the chemical theory of catalyst improvement can be developed, for example, the elemental complexation effect can be developed to further stabilize the structural surface material. The completely opposite system is a metal oxide catalyst supported on metal particles (Fig. 1.7). The basic view of the catalyst structure is the same, only the structure is reversed, and in fact many examples have recently been found. Here as well, the identification of structural surface substances will be further advanced in the future, and it will be possible to take them out as crystal structures.

As described above, a clear local structure can be seen over the traditional supported solid catalyst when closely looking at the catalytically active site and it plays a major role in the catalytic reaction. The catalytic elements are not randomly distributed, and the electronic states are not scattered, and thus homogeneously formed active site naturally shows high catalytic performance. Now the catalytic activities can be understood both from the structure of the crystalline composite oxide catalyst and from the structural surface material on the carrier are on the same level. Therefore, there is no difference between them from the viewpoint of catalytic activity even if they are not a clear and pure crystalline solid form in a solid-state catalyst. In other words, more new crystalline solid catalysts will be created, and more new catalytic functions such as micropores, structural coordinatively unsaturated state, ordered elemental defects, and anomalous valence content will be added. Then, new catalytic chemistry can be developed with many crystalline catalysts. At this stage, there is no need to be particularly aware of the difference between surface and bulk as discussed earlier. This is the way that scientific progress should be made, and breaking the old catalytic chemistry. At the same time, we will be able to escape from our current strong dependency on a precious metal element as solid catalysts by developing crystalline metal oxide catalysts, thanks to their superiority of compositional and structural diversity and flexibility and to elemental abundancy and environmental friendless.

References

1. Trunschke A, Noack J, Trojanov S, Girgsdies F, Lunkenbein T, Pfeifer V, Hävecker M, Kube P, Sprung C, Rosowski F, Schlögl R. The impact of the bulk structure on surface dynamics of complex Mo-V-based oxide catalysts. ACS Catal. 2017;7:3061–71.
2. Zhang S, Shan J, Zhu Y, Nguyen L, Huang W, Yoshida H, Takeda S, Tao F. Restructuring transition metal oxide nanorods for 100% selectivity in reduction of nitric oxide with carbon monoxide. Nano Lett. 2013;13:3310–4.
3. Bai L, Polo-Garzon F, Bao Z, Luo S, Moskowitz BM, Tian H, Wu Z. Impact of surface composition of $SrTiO_3$ catalysts for oxidative coupling of methane. ChemCatChem. 2019;11:2107–17.

4. Guo K, Wang Y, Huang J, Lu M, Li H, Peng Y, Xi P, Zhang H, Huang J, Lu S, Xu C. *In Situ* activated $Co_{3-x}Ni_xO_4$ as a highly active and ultrastable electrocatalyst for hydrogen generation. ACS Catal. 2021;11:8174–82.
5. Polo-Garzon F, Bao Z, Zhang X, Huang W, Wu Z. Surface reconstructions of metal oxides and the consequences on catalytic chemistry. ACS Catal. 2019;9:5692–707.
6. Polo-Garzon F, Wu Z. Acid-base catalysis over perovskites: a review. J Mater Chem A. 2018;6:2877–94.
7. Foo GS, Polo-Garzon F, Fung V, Jiang D, Overbury SH, Wu Z. Acid-base reactivity of perovskite catalysts probed via conversion of 2-propanol over titanates and zirconates. ACS Catal. 2017;7:4423–34.
8. Wach IE, Routray K. Catalysis science of bulk mixed oxides. ACS Catal. 2012;2:1235–46.
9. Vos YD, Jacobs M, Voort PVD, Driessche IV, Snijkers F, Verberckmoes A. Development of stable oxygen carrier materials for chemical looping processes—a review. Catalysts. 2020;10:926.
10. Grasselli RK. Site isolation and phase cooperation: two important concepts in selective oxidation catalysis: a retrospective. Catal Today. 2014;238:10–27.
11. Mars P, Krevelen DWV. Oxidations carried out by means of vanadium oxide catalysts. Chem Eng Sci. 1954;3:41–57.
12. Efremenko I, Neumann R. Computational insight into the initial steps of the mars–van krevelen mechanism: electron transfer and surface defects in the reduction of polyoxometalates. J Am Chem Soc. 2012;134:20669–80.
13. Wang J, Zhang G, Zhu J, Zhang X, Ding F, Zhang A, Guo X, Song C. CO_2 hydrogenation to methanol over In_2O_3-based catalysts: from mechanism to catalyst development. ACS Catal. 2021;11:1406–23.
14. Jiang X, Nie X, Guo X, Song C, Chen JG. Recent advances in carbon dioxide hydrogenation to methanol via heterogeneous catalysis. Chem Rev. 2020;120:7984–8034.
15. Li K, Chen JG. CO_2 hydrogenation to methanol over ZrO_2-containing catalysts: insights into ZrO_2 induced synergy. ACS Catal. 2019;9:7840–61.
16. Kattel S, Ramírez PJ, Chen JG, Rodriguez JA, Liu P. Active sites for CO_2 hydrogenation to methanol on Cu/ZnO catalysts. Science. 2017;355:1296–9.
17. Zhang Y, Zhao Y, Otroshchenko T, Lund H, Pohl M, Rodemerck U, Linke D, Jiao H, Jiang G, Kondratenko EV. Control of coordinatively unsaturated Zr sites in ZrO_2 for efficient C-H bond activation. Nat Commun. 2018;9:3794.
18. McFarland EW, Metiu H. Catalysis by doped oxides. Chem Rev. 2013;113(6):4391–427.
19. Liu Y, Wang W, Xu X, Veder JM, Shao Z. Recent advances in anion-doped metal oxides for catalytic applications. J Mater Chem A. 2019;7:7280–300.
20. Kotiuga M, Zhang Z, Li J, Rodolakis F, Zhou H, Sutarto R, He F, Wang Q, Sun Y, Wang Y, Aghamiri NA, Hancock SB, Rokhinson LP, Landau DP, Abate Y, Freeland JW, Comin R, Ramanathan S, Rabe KM. Carrier localization in perovskite nickelates from oxygen vacancies. PNAS. 2019;116(44):21992–7.
21. Wexler RB, Gautam GS, Stechel EB, Carter EA. Factors governing oxygen vacancy formation in oxide perovskites. J Am Chem Soc. 2021;143:13212–27.
22. Sprenger P, Kleist W, Grunwaldt J. Recent advances in selective propylene oxidation over bismuth molybdate based catalysts: synthetic, spectroscopic, and theoretical approaches. ACS Catal. 2017;7:5628–42.
23. Moro-oka Y, Ueda W. Selective oxidation and ammoxidation of propane: catalysts and processes. Catalysis. 1994;11(6):223–43.
24. Moro-oka Y, Ueda W, He D. Surface reaction controlled by the bulk migration of oxide ion in multicomponent metal oxide system. In: Tamaru K, editor. Dynamic processes on solid surfaces. New York: Plenum Publishing Co. Ltd. 1993, 11, p. 288–306.
25. Inumaru K, Okuhara T, Misono M. Active crystal face of vanadyl pyrophosphate for catalytic oxidation of n-butane to maleic anhydride. Chem Lett. 1992;21:1955–8.
26. Ziolkowski J, Bordes E, Cortine P. Dynamic description of the oxidation of n-butane on various faces of $(VO)_2P_2O_7$ in terms of the crystallochemical model of active sites. J Mol Catal. 1993;84:307–26.

27. Védrine JC. Revisiting active sites in heterogeneous catalysis: their structure and their dynamic behavior. Appl Catal A-Gen. 2014;474:40–50.
28. Hiyoshi N, Yamamoto N, Ryumon N, Kamiya Y, Okuhara T. Selective oxidation of n-butane in the presence of vanadyl pyrophosphates synthesized by intercalation-exfoliation-reduction of layered $VOPO_4 \cdot 2H_2O$ in 2-butanol. J Catal. 2004;221:225–33.
29. Getsoian AB, Zhai Z, Bell AT. Band-gap energy as a descriptor of catalytic activity for propene oxidation over mixed metal oxide catalysts. J Am Chem Soc. 2014;136:13684–97.
30. Wachs IE, Jehng J, Ueda W. Determination of the chemical nature of active surface sites present on bulk mixed metal oxide catalysts. J Phys Chem B. 2005;109:2275–84.
31. Konya T, Murayama T, Kato T, Sadakane M, Ishikawa S, Battrey D, Ueda W. An orthorhombic Mo_3VOx catalyst most active for oxidative dehydrogenation of ethane among related complex metal oxides. Catal Sci Technol. 2013;3:380–7.
32. Zheng Y, Xiao H, Li K, Wang Y, Li Y, Wei Y, Zhu X, Li H, Matsumura D, Guo B, He F, Chen X, Wang H. Ultra-fine CeO_2 particles triggered strong interaction with $LaFeO_3$ framework for total and preferential CO oxidation. ACS Appl Mater Interfaces. 2020;12:42274–84.

Chapter 3
Polyoxometalate Unit Assembling for Crystal Catalysts

Haruka E. Ooyama and Masahiro Sadakane

3.1 Introduction

3.1.1 Polyoxometalates

Early transition metals such as tungsten (W) and molybdenum (Mo) form anionic multi-metal oxide molecules, so-called "polyoxometalates (POMs)" [1–3]. The metal oxides such as MO_3 (W and Mo) are dissolved by adding bases such as NaOH or NH_3 to form Na_2MO_4 or $(NH_4)_2MO_4$, respectively. The $[MO_4]^{2-}$ is a monomeric molecule. The addition of a sufficient amount of acid or full removal of the base reversibly produces polymeric metal oxides, n MO_3 by dehydrative condensation. On the other hand, if the amount of acid or removal of the base is controlled, dimeric to polymeric metal oxide molecules "polyoxometalate" are formed (Fig. 3.1).

Polyoxometalates with only one early transition metal are called as "isopolyoxometalate", and those with more than two cationic elements in the molecule framework are called as "heteropolyoxometalate". It is possible to make a variety of structures by changing synthesis conditions (kinds and amounts of additional elements, solvents, pH, redox states, temperature, and so on) [1–3].

Polyoxometalates show acid–base properties [4, 5], redox properties [5–9], and photochemical properties [10, 11] which are important for catalysis and utilized not only in academic research but also in industries [12].

H. E. Ooyama · M. Sadakane (✉)
Department of Applied Chemistry, Graduate School of Advanced Science and Engineering, Hiroshima University, 1-4-1, Kagamiyama, Higashi-Hiroshima 739-8527, Japan
e-mail: sadakane09@hiroshima-u.ac.jp

H. E. Ooyama
e-mail: ooyama20@hiroshima-u.ac.jp

W. Ueda (ed.), *Crystalline Metal Oxide Catalysts*,
https://doi.org/10.1007/978-981-19-5013-1_3

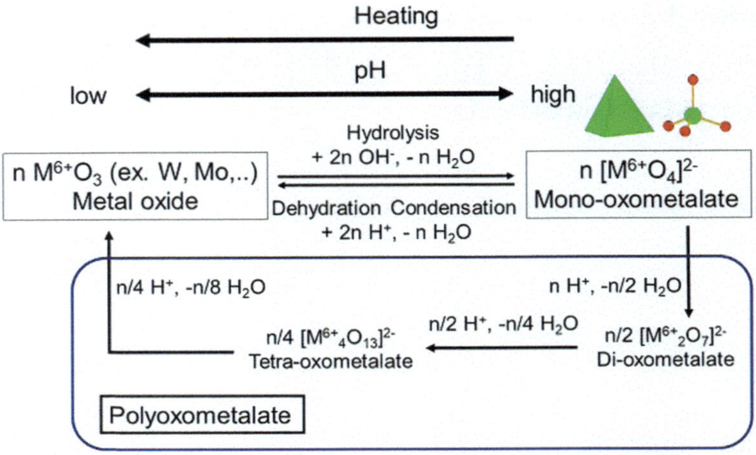

Fig. 3.1 Polyoxometalate formation mechanism

3.1.2 Structure of Polyoxometalates

The structure of the polyoxometalates is classified into three types, (1) molecule structure, (2) crystal structure together with counter cation, and (3) aggregation of crystal (Fig. 3.2).

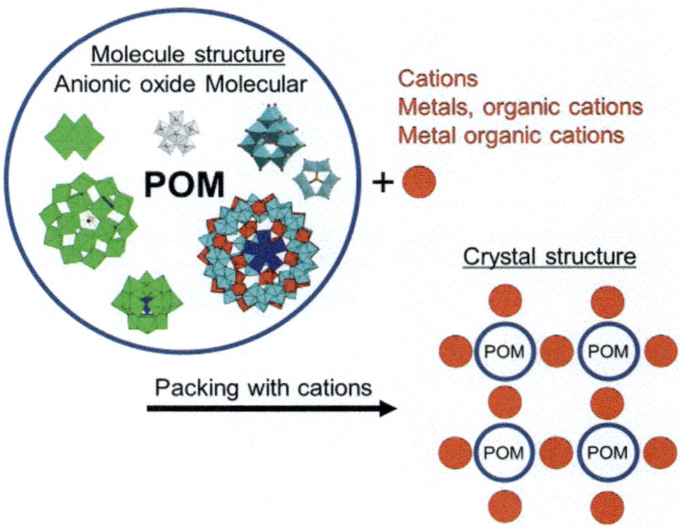

Fig. 3.2 Polyoxometalate Crystal Structure

The primary structure is molecule structure. Variety of molecule structure and anionic charge is possible with a different number of component elements, and several important properties such as acidities, redox properties, and photochemical properties of the molecule are tunable by changing structures. Therefore, control of the primary structure is the most important for catalyst design both in homogeneous and heterogeneous systems.

Anionic molecules interact with a variety of cationic species such as proton, alkaline and alkaline earth cations, transition metals, lanthanide metals, organic cations, organometallic cations, and organic compounds [1–3, 13], and ordered packing of the anionic polyoxometalates and cationic components forms crystals, whereas unordered packing forms amorphous materials.

Aggregation of polyoxometalate crystals forms micropores (pore size less than 2 nm) between the crystals where small molecules such as N_2, O_2, CO_2, CH_4, and CH_3CH_3 can enter [5, 14].

3.1.3 Scope and Limitations

If we search papers with the keywords of "polyoxometalate", "crystal", and "catalysis", a lot of papers are listed because polyoxometalates are important key species for catalysis. Papers are divided into two categories; polyoxometalates crystals are utilized as catalysts and polyoxometalates are loaded into the crystalline supports. In recent years, many papers include crystalline materials of MOFs (metal-organic frameworks) in which polyoxometalate molecules or nanoparticles are loaded. Catalytic applications of polyoxometalate/MOF crystals [15–17] are excluded. Table 3.1 summarizes recent (later than 2021) research on polyoxometalate crystals that can be utilized for catalysis.

In this chapter, we describe polyoxometalate crystal materials where polyoxometalate units have bonds with linkers to form crystals. After the introduction of the molecule structure of polyoxometalates, the connection of polyoxometalates with cationic linkers to form crystalline materials is presented.

3.2 Design of Polyoxometalate Crystals

3.2.1 Molecule Structure

3.2.1.1 Isopolyoxometalates

In the basic condition, monooxometalte, $[MO_4]^{2-}$ (M = W or Mo), is formed. The addition of proton or removal of base promotes dehydrative condensation to produce a variety of structures depending on the formation condition [1, 35]. An example

Table 3.1 Summary of recent (later than 2021) reported polyoxometalate-based crystals applicable to catalysts

No.	Crystal	POM	Application
1. Liu et al. [18]	$Co_3(4\text{-}H_2dpye)[MnMo_9O_{32}]$ $Co_3(4\text{-}H_2dpyb)[MnMo_9O_{32}]$ 4-H_2dpye: N,N′-bis(4-pyrimidinecarboxamido)-1,2-ethane 4-H_2dpyb: N,N′-bis(4-pyrimidinecarboxamido)-1,4-butane	Waugh-type $[MnMo_9O_{32}]^{8-}$	Oxidation of sulfide
2. Chen et al. [19]	$[\varepsilon\text{-}PMo_{12}O_{37}(OH)_3][Zn_2(bipy)(H_2O)_2]_2$ bipy: 4,4'-bipyridine	ε-Keggin-type $[\varepsilon\text{-}PMo_{12}O_{37}(OH)_3]^{8-}$	Oxidation of sulfide
3. Zhang et al. [20]	$Cu(en)_2Cu(pdc)(en)Cu(en)_2[BW_{12}O_{40}]$ $Cu_5(pz)_6[BW_{12}O_{40}]$ en: ethylenediamine pz: pyrazine	α-Keggin-type $[BW_{12}O_{40}]^{5-}$	Photocatalysis (for degradation of dye)
4. Li et al. [21]	$[(DPNDIH)(Ni_6(OH)_3(H_2O)_9SiW_9O_{34})]_2$ DPNDI: N,N′-di(4-pyridyl)-1,4,5,8-naphthalenediimide	Ni6-substituted trivacant Keggin-type $[Ni_6(OH)_3(H_2O)_9SiW_9O_{34}]^{1-}$	Photochemical C-N bond formation
5. Wang et al. [22]	$HFe_4O_2(pydc)_3[PW_{12}O_{40}]$ Pydc: pyridine-2,5-dicarboxylic acid	α-Keggin-type $[PMo_{12}O_{40}]^{3-}$	Oxidation of alkylbenzene
6. Xu et al. [23]	$Cu^{I}{}_{12}Cl_2(trz)_8[HPW_{12}O_{40}]$ trz = 1,2,4-triazole	α-Keggin-type $[PMo_{12}O_{40}]^{3-}$	Oxidation of alkylbenzen
7. Wang et al. [24]	$Cu_8(dpyh)_4[\alpha\text{-}\gamma\text{-}Mo_8O_{26}][\beta\text{-}Mo_8O_{26}]$ H_2dpyh = N,N-bis(3-pyrazolamide)-1,2-hexahydrobenzene	α-γ $[Mo_8O_{26}]^{4-}$	Oxidation of sulfide
8. Zhao et al. [25]	$Co_{2.5}(LOH)(LO)_2[PW_{12}O_{39}]$ $Zn_{1.5}(LOH)_3[PMo_{12}O_{40}]$ $Cd_{1.5}(LOH)_3[PW_{12}O_{40}]$ $Mn(LOH)_2[PW_{12}O_{40}]$ LOH = 2,6-Bis(2′-pyridyl)-4-hydroxypyridine	α-Keggin-type $[PMo_{12}O_{40}]^{3-}$ $[PW_{12}O_{40}]^{3-}$	Cycloaddition reaction of CO_2 Oxidation of sulfide

(continued)

Table 3.1 (continued)

No.	Crystal	POM	Application
9. Zhang et al. [26]	$Co_2(L)_2[CrMo_6(OH)_5O_{19}]$ $Ni_2(L)_2[AlMo_6(OH)_5O_{19}]$ $Co_3(L)_2[TeMo_6O_{24}]$ $Ni_3(L)_2[TeMo_6O_{24}]$ L = N,N-bis(3-pyrazolamide)-1,2-ethane	Anderson-type $[CrMo_6(OH)_5O_{19}]^{4-}$ $[TeMo_6O_{24}]^{6-}$	Electrochemical sensor Electrochemical oxidation of dopamine
10. Li et al. [27]	$Ni_3L_2(CH_3OH)_6][PMo_{12}O_{40}]_2$ $Ni_3L_2(CH_3OH\ )_6[PW_{12}O_{40}]_2$ $Ag_3L[PMo_{12}O_{40}]$ L = tetra-[5-(mercapto)-1-methyltetrazole]-thiacalix [4]arene	α-Keggin-type $[PMo_{12}O_{40}]^{3-}$	Oxidation of sulfide
11. Chang et al. [28]	$K_3[Cu(tza)_2]_2[Cu(Htza)_2][BW_{12}O_{40}]$ $H_3K_3[Cu(Htza)_2]_3[SiW_{12}O_{44}]$ Htza: tetrazol-1-ylacetic acid	α-Keggin-type $[BW_{12}O_{40}]^{5-}$ $[SiW_{12}O_{44}]^{12-}$	Oxidation of alkylbenzene
12. Li et al. [29]	$Ni_3(4\text{-}NH_2\text{-}trz)_6][V_6O_{18}]$ $[Co_3(4\text{-}NH_2\text{-}trz)_6][V_6O_{18}]$ $[Cu_3OH(4\text{-}NH_2\text{-}trz)_3H_2O][VO_3]_5$ $4\text{-}NH_2\text{-}trz$ = 4-amino-1,2,4-triazole	Ring-type $[V_6O_{18}]^{6-}$ Open ring-type $[V_5O_{15}]^{5-}$	Knoevenagel condensation Oxidation of aromatic alcohol
13. Zhang et al. [30]	$(NH_4)_{3.7}Bi_2[\varepsilon\text{-}VMo_{9.4}V_{2.6}O_{40}]$	ε-Keggin-type $[\varepsilon\text{-}VMo_{9.4}V_{2.6}O_{40}]^{9.7-}$	Epoxidation of olefine
14. Yin et al. [31]	ε-Keggin-based FeMo oxide	ε-Keggin-type $[\varepsilon\text{-}FeMo_{12}O_{40}]^{n-}$	Oxidation of ciprofloxacin
15. Zhu et al. [32]	ε-Keggin-based FeMo oxide	ε-Keggin-type $[\varepsilon\text{-}FeMo_{12}O_{40}]^{n-}$	Alcohol oxidation
16. Tao et al. [33]	Cubane-based TiVW oxide and ZrVW oxide	Cubane-type $[W_4O_{16}]^{8-}$	NH_3-SCR
17. Tao et al. [34]	Cubane-based MoWV oxide	Cubane-type $[Mo_xW_{4-x}O_{16}]^{8-}$	NH_3-SCR

is shown in Fig. 3.3, the reaction of ammonia with MoO_3 produces ammonium monomolybdate, $(NH_4)_2MoO_4$. Heating in the solid state releases NH_3 and H_2O, and $(NH_4)_2MoO_4$ is transformed into a variety of polyoxomolybdates such as ammonium heptamolybdate, $(NH_4)_6Mo_7O_{24}$; ammonium dodecamolybdate, $(NH_4)_8Mo_{10}O_{34}$; ammonium tetramolybdate, $(NH_4)_2Mo_4O_{13}$; and then molybdenum oxide, MoO_3 [36]. On the other hand, control of pH in the solution also produces a variety of polyoxomolybdates. A lot of different structures have been produced depending on synthesis conditions [1]. One of the most common isopolyoxomolybdate species is heptamolybdate, $[Mo_7O_{24}]^{6-}$, which is produced both in the solid state and in solution. The structure is controlled by the synthesis conditions such as counter cations, concentrations, pH, solvents, and the presence of reducing reagents.

The reduction of Mo(6+) to Mo(5+) or Mo(4+) produces different structures, and ball-type, wheel-type, and cage-type polyoxomolybdates have also been reported (Fig. 3.4) [37]. A huge cage-type polyoxomolybdate containing 368 Mo is also reported.

In the case of isopolytungstates, heptatungstate, $[W_7O_{24}]^{6-}$; paratungstate, $[H_2W_{12}O_{42}]^{10-}$ [38]; metatungstate, $[H_2W_{12}O_{40}]^{6-}$ [39]; and decatungstate, $[W_{10}O_{32}]^{4-}$ are most common species. However, many other structures have been reported [1].

In the case of isopolyoxovanadate, isopolyoxoniobate, and isopolytantalate, hexaoxometalate, $[M_6O_{19}]^{8-}$, is the most common species (Fig. 3.5) [1].

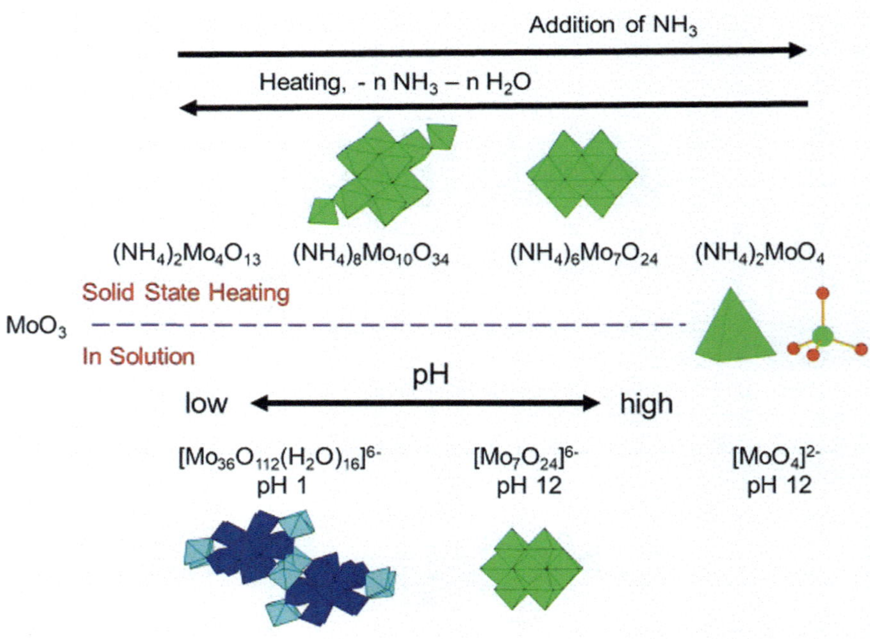

Fig. 3.3 Formation of isopolyoxomolybdate in (upper) solid-state heating and (lower) solution

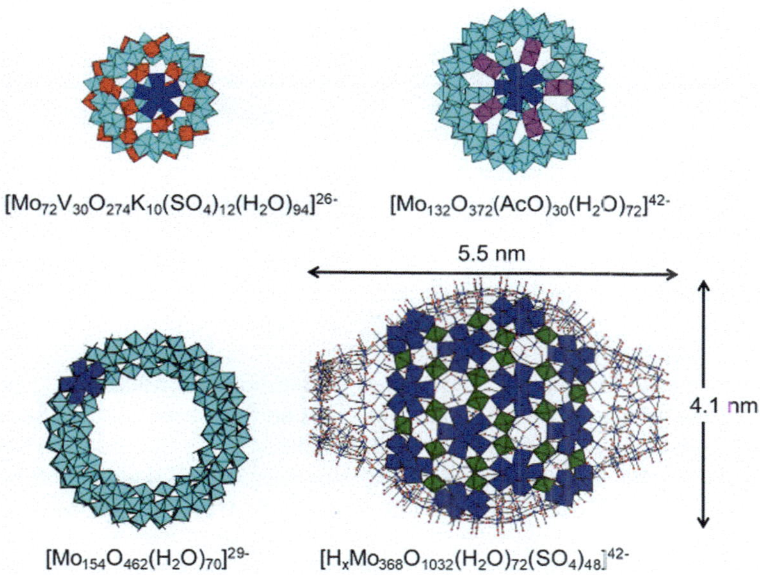

Fig. 3.4 Examples of reduced Isopolyoxomolybdates

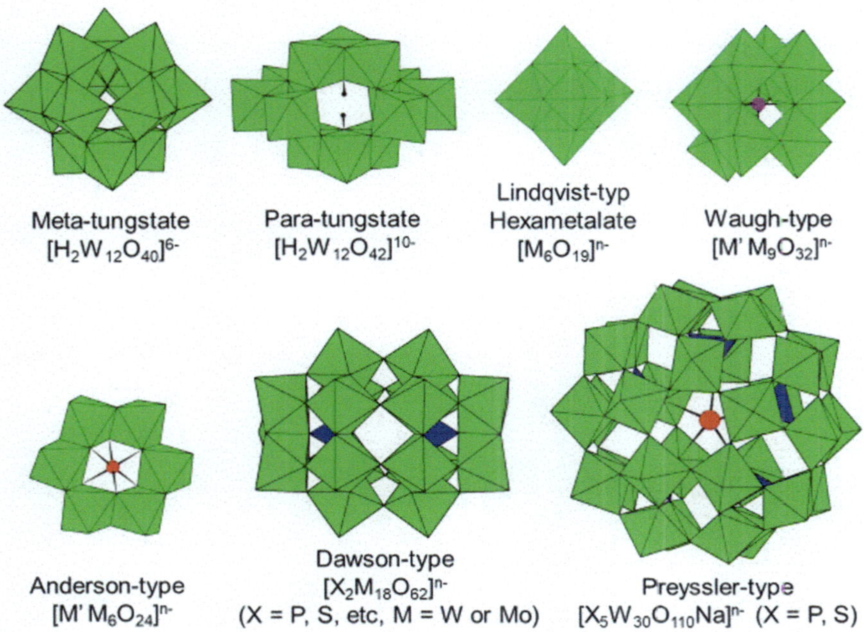

Fig. 3.5 Structure of polyoxometalates

3.2.1.2 Heteropolyoxometalate

Polyoxometalates containing other cationic elements in the molecule framework are called as "heteropolyoxometalate". The most known example is a Keggin-type structure with one tetrahedral XO_4 (X = P, Si, S, Co, B, etc.) surrounded by 12 octahedral MO_6 to form a ball-type molecule (Fig. 3.6). Four edge-sharing M_3O_{12} units are connected to the central X. According to the rotation of the M_3O_{12} unit, there are α-, β-, γ-, δ-, and ε-isomers. The α-Keggin-type phosphotungstate, $[PW_{12}O_{40}]^{3-}$, is formed in an acidic solution containing phosphate and tungstate. The same combination also produces other molecules, and the two most often observed molecules are Dawson-type, $[P_2W_{18}O_{62}]^{6-}$, and Preyssler-type phosphotungstates, $[P_5W_{30}O_{110}Na]^{14-}$ (Fig. 3.5).

It is possible to incorporate other metals in the molecule frameworks. Several [M = O] units in the polyoxometalate molecules are removable to form so-called "lacunary" species. Figure 3.7 shows examples of the lacunary species of the Keggin-type polyoxometalates. Removals of one, two, or three [M = O] units form mono-, di-, or tri-lacunary Keggin-type complex, respectively (Fig. 3.7). Removal of [M = O] produces new terminal oxygens in the lacunary sites which can react with cationic species such as transition metals, lanthanide metals, organometallic cations, and organic molecules. The reaction of mono-lacunary species with transition metals forms mono-transition metal substituted species, where the incorporated transition metals occupy the removed metal position. In the case of di-lacunary species, two metals incorporated species and sandwiched species are produced. In the case of

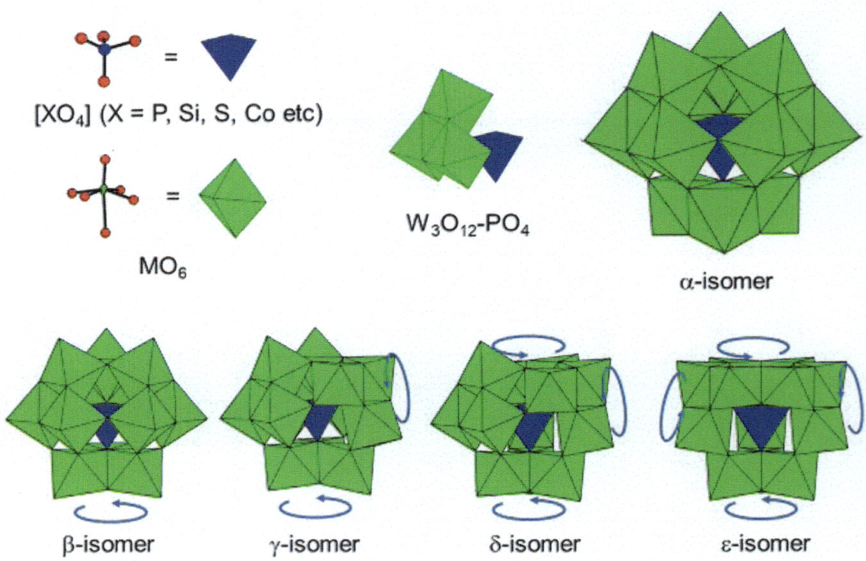

Fig. 3.6 Structures of Keggin-type polyoxometalates

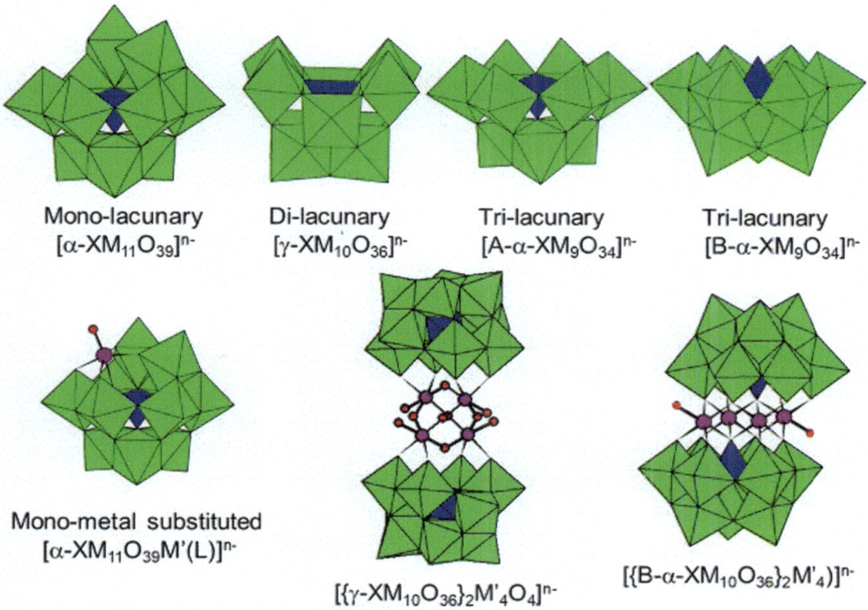

Fig. 3.7 Structures of lacunary Keggin-type polyoxometalates and other metal-containing complexes

tri-lacunary species, sandwiched complexes where multi-metals are sandwiched by the two tri-lacunary species are produced (Fig. 3.7).

The structure of other polyoxometalates such as Dawson-type, Preyssler-type, Anderson-type, and Lindqvist-type (Fig. 3.5) can also be tuned by removal of molecule and incorporation of other metals. All 12 metal sites of α-Keggin-type polyoxometalates are equivalent, and therefore, mono-metal substituted α-Keggin-type polyoxometalates have only one isomer. On the other hand, there are two kinds of metal sites in the case of Lindqvist-type and Dawson-type, and there are two isomers of mono-metal substituted Lindqvist-type and Dawson-type polyoxometalates (Fig. 3.8). It is also possible to substitute more than two metals in the polyoxometalates, and there are increased number of isomers in the multi-metal substituted polyoxometalates. In the case of Anderson-type polyoxometalates where transition metals are surrounded by six corner-sharing octahedra of Mo or W, it is possible to change the central transition metals.

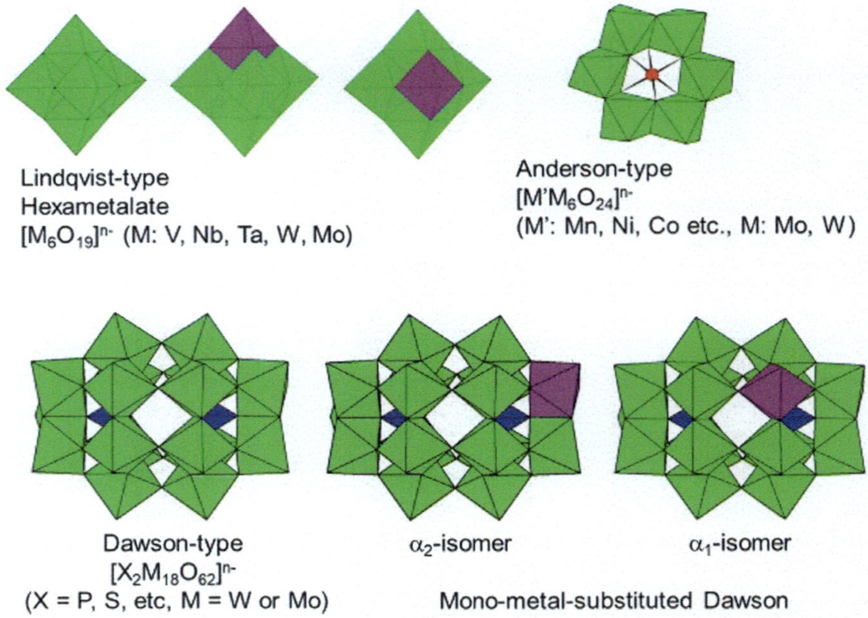

Fig. 3.8 Substitution of molecular metals in polyoxometalates

3.2.2 Polyoxometalate Unit-Based Crystal

3.2.2.1 Pentagonal Polyoxometalate Unit

The first example of polyoxometalate unit to form metal oxide crystals is pentagonal polyoxometalate, $[M_6O_{21}]^{6-}$ (Fig. 3.9). This molecule has not been isolated, but is has been found as a building unit in many polyoxometalates such as $[Mo_{36}O_{112}(H_2O)]^{6-}$, ball-type, wheel-type, and cage-type polyoxometalate (Figs. 3.3 and 3.4) [37]. The pentagonal unit has one pentagonal bipyramidal MO_7 surrounded by five octahedral MO_6. Two equatorial terminal oxygens of neighboring MO_6 on the molecular plane bind to other cations (red arrows in Fig. 3.9) and bind with octahedral linkers of Mo and V to form a rectangle sheet or diamond-shape sheet. Stackings of the rectangle and diamond-shaped sheets through axial Mo–O–Mo bonds vertical to the sheets produce orthorhombic and trigonal Mo–V oxide [40, 41], respectively. One unit cell of orthorhombic oxide contains four pentagonal units, four six-membered rings of MO_6, and four seven-membered rings of MO_6. On the other hand, one unit cell of trigonal oxide contains three pentagonal units, three six-membered rings, and three seven-membered rings. The stacking of the six-membered and seven-membered rings produces two kinds of channels. The diameter of the seven-membered ring channel is ca. 0.4 nm [42], into which molecules smaller than ethane can enter. Both oxides are redox active, and the pore diameter is tunable by redox treatments [43]. In the case

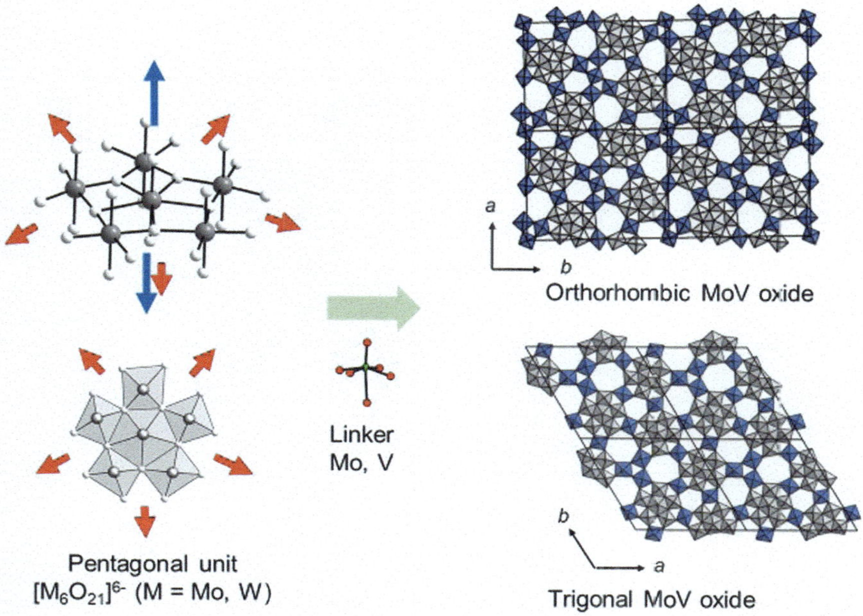

Fig. 3.9 Pentagonal polyoxometalate unit assembly to orthorhombic and trigonal Mo–V oxides

of Mo-V-based oxides, pentagonal units contain exclusively Mo, and the linkers are composed of Mo and V octahedral [44–46]. The catalytic activities of these crystalline metal oxides are summarized in Chap. 4.

3.2.2.2 ε-Keggin-Type Polyoxometalate Unit

ε-Keggin-type polyoxometalates has T_d symmetry, and there are four face directing tetrahedral fashion (Fig. 3.10). There are three oxygen atoms (red in Fig. 3.10) that can bind to metal atoms. Transition metals such as Bi, Mn, Zn, Co, and Fe can bind to the three oxygens in an octahedral fashion. Linkage of two ε-Keggin units by one metal forms metal oxides with 3D diamond-like frameworks (Fig. 3.10) [47]. There are all inorganic ε-Keggin-type polyoxometalate-based oxides where ε-Keggin-type polyoxometalates are linked by metal cations and organic-ε-Keggin-type polyoxometalate-based hybrid materials where ε-Keggin-type polyoxcmetalate are linked by organic linkers. In the case of the first all inorganic ε-Keggin-type polyoxometalate-based oxide, Bi-V-Mo oxide. ε-Keggin-type $[VMo_{9.4}V_{2.6}O_{40}]^{9.7-}$ where central tetrahedral VO_4 is surrounded by 12 octahedral MO_6 (M: Mo and V) is linked by Bi(3+). In the case of M'–Mo oxide (M': Zn, Mn, Co, Fe), ε-Keggin-type

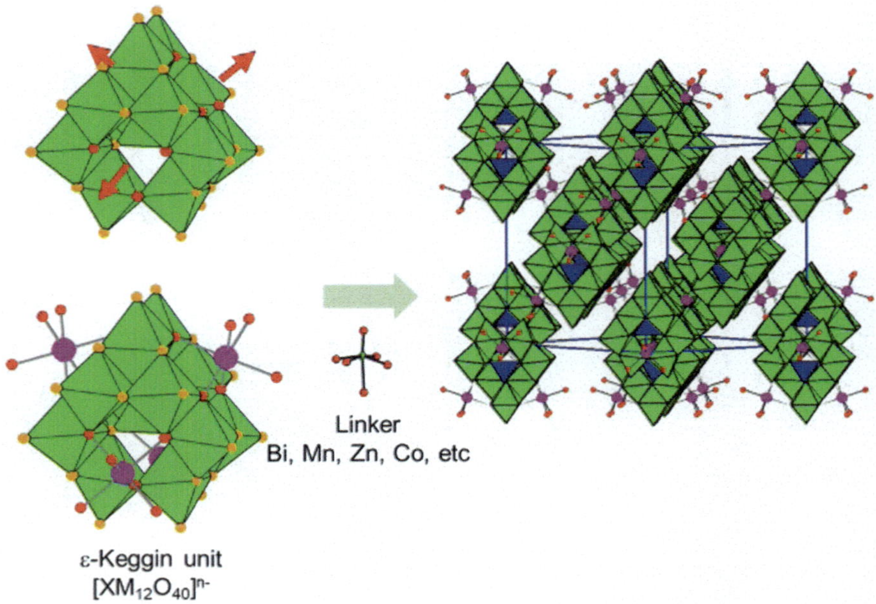

Fig. 3.10 ε-Keggin-type polyoxometalate unit assembly to metal oxides with diamond-like frameworks

$[M'Mo_{12}O_{40}]^{n-}$ where tetrahedral $M'O_4$ is surrounded by 12 MoO_6 is linked by octahedral M'. In the case of organic-ε-Keggin-type polyoxometalate-based hybrid materials, ε-Keggin-type polyoxometalates containing four transition metals (Fig. 3.10, pink balls) bound on the four sets of three oxygens (Fig. 3.10, red oxygens) are linked by organic linkers. All inorganic one is more thermally stable and water molecules in between polyoxometalates can be removed without collapse of the framework, and 3D micropores in the oxide can be available.

The catalytic activity of the ε-Keggin-type polyoxometalate crystals has been reported. In 2021, epoxidation reaction catalyzed by Bi–V–Mo oxide (No. 13 in Table 3.1) [30] and oxidation of ciprofloxacin [31] and alcohol oxidation [32] (No. 14 and 15 in Table 3.1) catalyzed by FeMo oxide have been reported. The synthesis and catalytic activities of these crystalline metal oxides are summarized in Chap. 5. There are many examples where organic ligands link ε-Keggin-type polyoxometalates [48, 49]. Crystalline organo-polyoxometalate material where ε-Keggin-type $[\varepsilon\text{-}PMo_{12}O_{37}(OH)_3]^{8-}$ with four-surface four Zn(2+), $[\varepsilon\text{-}PMo_{12}O_{37}(OH)_3]Zn_4$ is linked by 4,4'-bipyridine shows catalytic activity for oxidation of sulfide (No. 2 in Table 3.1) [19].

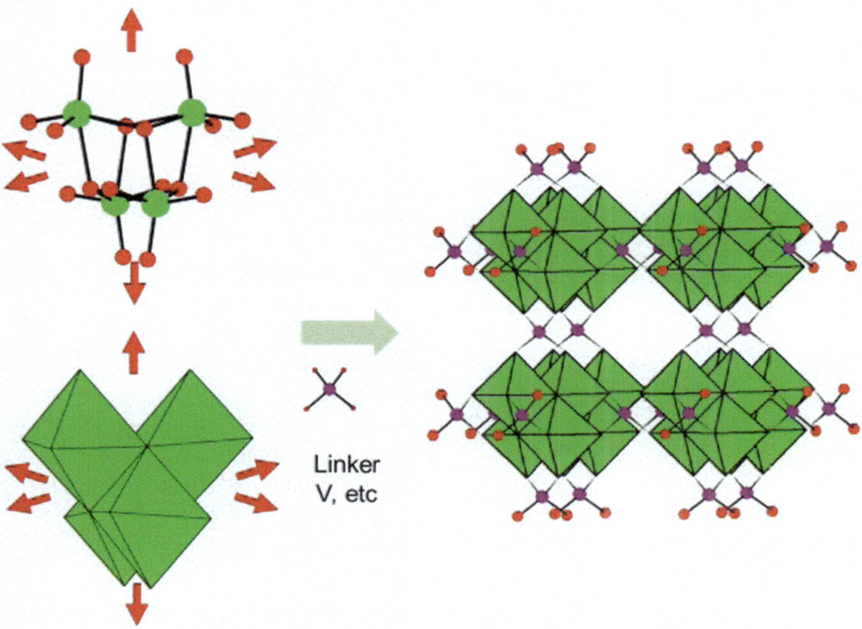

Fig. 3.11 ε-Keggin-type polyoxometalate unit assembly to metal oxides with diamond-like frameworks

3.2.2.3 Cubane-Type Polyoxometalate Unit

It is known to form a reaction of $[MoO_4]^{2-}$ with Cp^*IrCl_2 (Cp*: pentamethylcyclopentadienyl) form cubane $[Mo_4O_{16}]^{8-}$ polyoxomolybdate (Fig. 3.11) unit where four Cp*Ir moieties are attached on the surface [50]. Octahedra are bound by edge-sharing and each metal in the cubane $[M_4O_{16}]^{8-}$ polyoxometalate has three terminal oxygens, and it is possible to bind metal cations. It is possible to link the cubane $[W_4O_{16}]^{8-}$ polyoxotungstate with tetrahedral V to form cubic cubane-type polyoxometalate-based oxide which shows high porosity. Furthermore, a variety of linkage modes promotes diversity of the zeolite-like structure [51]. Substitution of W with Mo and V with Zr or Ti is possible, and catalytic activity for NH_3–SCR (selective catalytic reduction of NOx with NH_3) can be tuned (No. 16 and 17 in Table 3.1) [33, 34]. The synthesis of the zeolite like tungsten oxide and catalytic activities are summarized in Chap. 5.

3.2.2.4 Anderson-Type Polyoxometalate Unit

In some polyoxometalates, it is possible to substitute three molecule oxygens with tri-methanol-methane derivatives. Then, trismethanolmethane moiety can

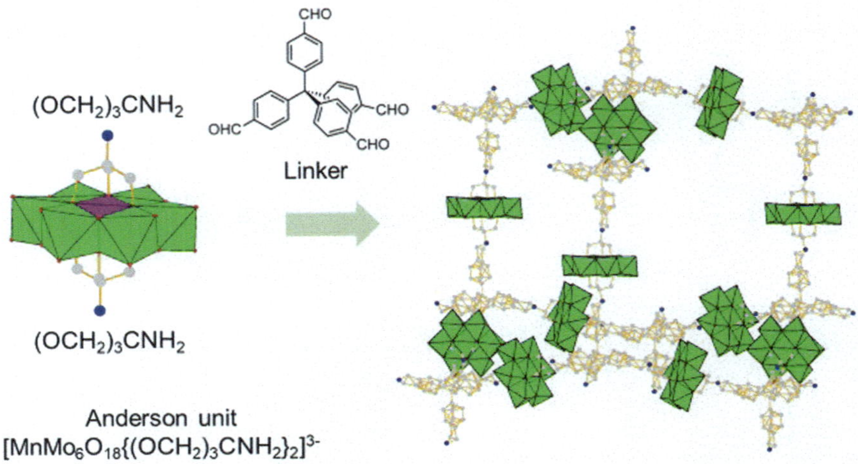

Fig. 3.12 Anderson-type polyoxometalate unit linkage with the organic linker to organo-polyoxometalate crystal with adamantane-like frameworks

be bound on the surface of polyoxometalates. Anderson-type polyoxometalate, $[M'M_6O_{24}]^{n-}$ (M' = Ni, Mn, etc., M = W or Mo) can bind two trismethanol-methane on both sides (Fig. 3.12, left). Anderson-type, $[MnMo_6O_{24}]^{10-}$, with two trimethanolaminomethane (Tris) react with tetrahedral aldehyde, tetrakis(4-formylphenyl)methane to form a 3D framework crystal with adamantane-like (Fig. 3.12) [52].

3.2.2.5 Polyoxometalate-Based Ionic Crystals

There are many reports about polyoxometalate ionic crystals where anionic polyoxometalate molecules interact with cationic species by the Coulomb force. The most common example is ammonium or cesium salt of Keggin-type polyoxometalates which forms cubic crystals (Fig. 3.13) [5]. Increasing the size of counter cations increases unit cell parameters and forms porosity in the crystals. Large organometallic cation-polyoxometalate materials which show catalytic activity are intensively investigated [53, 54].

The presence of metals and organic ligands forms a variety of crystals with different polyoxometalate structures. A variety of Keggin-type complexes are utilized to form crystals showing catalytic activities (No. 3, 4, 5, 6, 8, 10, 11 in Table 3.1) [20–23, 25, 27, 28]. Other structures such as Waugh-type (No. 1 in Table 3.1) [18], octamolybdate (No. 7 in Table 3.1) [24], Anderson-type (No. 9 in Table 3.1) [26], and cyclic and non-closed vanadate (No. 12 in Table 3.1) [29] are packed with organic ligands to make crystals showing catalytic activities.

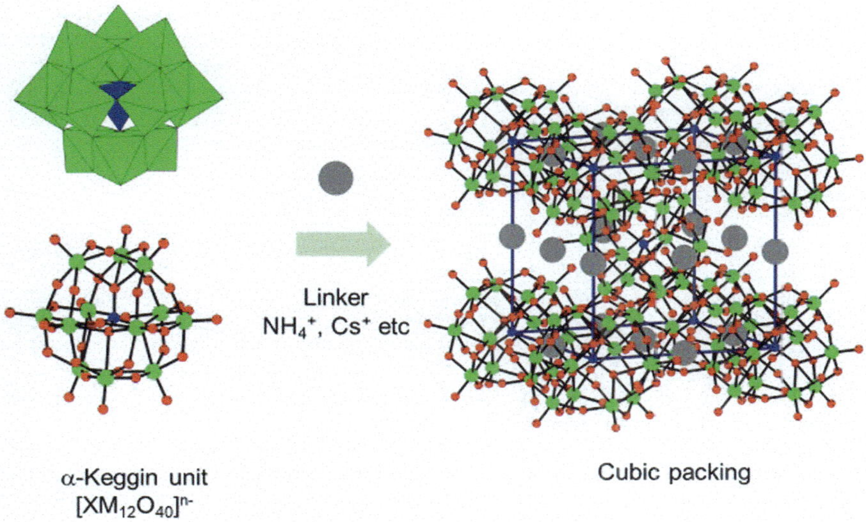

Fig. 3.13 Packing of α-Keggin-type polyoxometalates with mono-valent cations such as NH_4^+ or Cs^+ to form cubic packing

3.3 Perspective

Polyoxometalates, anionic metal oxide molecules with a variety of components and structures, are important components for metal oxide crystalline catalysts. Molecule properties such as acidity, redox potentials, and photochemical property are tunable by changing structure and component elements. Furthermore, there are a variety of strategies to bind anionic polyoxometalates with cationic metals, organic cations, organometallic cations, and organic compounds. Here, we presented examples of the structure and component of polyoxometalate molecules and crystals together with recent examples. We hope that this article helps further development of polyoxometalate-based crystalline catalysts.

References

1. Pope MT. Heteropoly and isopoly oxometalates. Berlin: Springer-Verlag; 1983.
2. Hill CL. Chem Rev. 1998;98:1–380.
3. Cronin L. Chem Soc Rev. 2012;41:7325–648.
4. Kozhenvikov IV. Catalysis by heteropoly acids and multicomponent polyoxometalates in liquid-phase reactions. Chem Rev. 1998;98:171–99.
5. Mizuno N, Misono M. Heterogeneous catalysis. Chem Rev. 1998;98:199–218.
6. Najari S, Saeidi S, Concepcion P, Dionysiou DD, Bhargava SK, Lee AF, Wilson K. Oxidative dehydrogenation of ethane: catalytic and mechanistic aspects and future trends. Chem Soc Rev. 2021;50:4564–605.

7. Horn MR, Singh A, Alomari S, Goberna-Ferron S, Benages-Vilau R, Chodankar N, Motta N, Ostrikov K, MacLeod J, Sonar P, Gomez-Romero P, Dubal D. Polyoxometalates (POMs): from electroactive clusters to energy materials. Energy Environ Sci. 2021;14:1652–700.
8. Stregiou AD, Symes MD. Organic transformations using electro-generated polyoxometalate. Catal Today. 2022;384–386:146–55.
9. Sadakane M, Steckhan E. Electrochemical properties of polyoxometalates as electrocatalysts. Chem Rev. 1998;98:217–38.
10. Han Q, Ding Y. Recent advances in the field of light-driven water oxidation catalytzed by transition-metal substituted polyoxometalates. Dalton Trans. 2018;47:8180–8.
11. Yamase T. Photo- and electrochromism of polyoxometalates and related materials. Chem Rev. 1998;98:307–26.
12. Katsoulis DE. A survey of applications of polyoxometalates. Chem Rev. 1998;98:359–88.
13. Sanchez C, Soler-Illia GJAA, Ribot F, Lalot T, Mayer CR, Cabuil V. Designed hybrid organic-inorganic nanocomposites from functional nanobuilding blocks. Chem Mater. 2001;13:3061–83.
14. Inumaru K. Roles of interfaces in nanostructured composites: nanocrystals, sponge crystals and thin films. J Ceram Soc Jpn. 2016;124:1110–5.
15. Mialane P, Mellot-Draznieks C, Gairola P, Duguet M, Benseghir Y, Oms O, Dolbecq A. Heterogenisation of polyoxometalates and other metal-based complexes in metal-organic frameworks: from synthesis to characterization and applicatio0n in catalysis. Chem Soc Rev. 2021;50:6152–220.
16. Zhang M, Li H, Zhang J, Lv H, Yang G-Y. Research advances of light-driven hydrogen evolution using polyoxometalate-based catalysts. Chin J Catal. 2021;42:855–71.
17. Du Z-Y, Yu Y-Z, Li N-F, Xue Y-S, Xu L-X, Mei H, Xu Y. Polyoxometalate-induced "cage-within-cage" metal-organic frameworks with high efficiency towards CO_2 photoreduction. Sus Energy Fuels. 2021;5:3876–83.
18. Liu Q, Lin H, Wang X-L, Wang X, Xu N, Tian Y, Yang L, Li X, Sun J. Two novel polyoxometalate-based metal-organic complexes with chiral waugh-type $[MnMo_9O_{32}]^{6-}$ anions as high-efficiency catalytic oxidative desulfurization catalysts. Cryst Growth Des. 2021;21(12):7015–22.
19. Chen Y, Chang S, An H, Li Y, Zhu Q, Luo H, Huang Y. Two polymorphic polyoxometalate-based metal-organic frameworks for the efficient synthesis of functionalized sulfoxides and detoxification of mustard gas simulants. ACS Sustain Chem Eng. 2021;9(46):15683–93.
20. Zhang W, Gong L, Du N, Wang C, Yu K, Wang C, Zhou B. $BW_{12}O_{40}$ hybrids modified by in situ synthesized rigid ligand with supercapacitance and photocatalytic properties. Inorg Chem. 2021;60(21):16357–69.
21. Li J, Jiao J, Chang J, Li M, Han Q. Visible-light-driven C-N bond formation by a hexanickel cluster substituted polyoxometalate-based photocatalyst. Inorg Chem. 2021;60(13):10022–9.
22. Wang Q, Xu B, Wang Y, Wang H, Hu X, Ma P, Niu J, Wang J. Polyoxometalate-incorporated framework as a heterogeneous catalyst for selective oxidation of C-H bonds of alkylbenzenes. Inorg Chem. 2021;60(11):7753–61.
23. Xu B, Xu Q, Wang Q, Liu Z, Zhao R, Li D, Ma P, Wang J, Niu J. A copper-containing polyoxometalate-based metal-organic framework as an efficient catalyst for selective catalytic oxidation of alkylbenzenes. Inorg Chem. 2021;60(7):4792–9.
24. Wang XL, Zhang JY, Chang ZH, Zhang Z, Wang X, Lin HY, Cui ZW. Alpha-gamma-type $[Mo_8O_{26}]$(4-)-containing metal-organic complex possessing efficient catalytic activity toward the oxidation of thioether derivatives. Inorg Chem. 2021;60(5):3331–7.
25. Zhao Y-Q, Liu Y-Y, Ma J-F. Polyoxometalate-based organic-inorganic hybrids as heterogeneous catalysts for cycloaddition of CO2 with epoxides and oxidative desulfurization reactions. Crystal Growth Des. 2021;21(2):1019–27.
26. Zhang J-Y, Chang Z-H, Wang X-L, Wang X, Lin H-Y. Different Anderson-type polyoxometalate-based metal–organic complexes exhibiting –OH group-directed structures and electrochemical sensing performance. New J Chem. 2021;45(6):3328–34.

27. Li J, Du P, Liu YY, Ma JF. Assembly of polyoxometalate-thiacalix[4]arene-based inorganic-organic hybrids as efficient catalytic oxidation desulfurization catalysts. Dalton Trans. 2021;50(4):1349–56.
28. Chang S, Chen Y, An H, Zhu Q, Luo H, Huang Y. Polyoxometalate-based supramolecular porous frameworks with dual-active centers towards highly efficient synthesis of functionalized p-benzoquinones. Green Chem. 2021;23(21):8591–603.
29. Li J, Wei C, Han Y, Mei Y, Cheng X, Huang X, Hu C. Triazole-directed fabrication of polyoxovanadate-based metal-organic frameworks as efficient multifunctional heterogeneous catalysts for the Knoevenagel condensation and oxidation of alcohols. Dalton Trans. 2021;50(29):10082–91.
30. Zhang Z, Li D, Zhu Q, Hara M, Li Y, Ueda W. Preparation of zeolitic bismuth vanadomolybdate using a ball-shaped giant polyoxometalate for olefin epoxidation. New J Chem. 2021;45(46):21624–30.
31. Yin S, Wang J, Tong Q, Jiang X, Lu P, Zhu Q, Zhang Q, Zhang Z, Ueda W. Degradation of ciprofloxacin with hydrogen peroxide catalyzed by ironmolybdate-based zeolitic octahedral metal oxide. Appl Catal A Gen. 2021;626.
32. Zhu Q, Yin S, Zhou M, Wang J, Chen C, Hu P, Jiang X, Zhang Z, Li Y, Ueda W. Aerobic alcohol oxidation by a zeolitic octahedral metal oxide based on iron vanadomolybdates under mild conditions. ChemCatChem. 2021;13(7):1763–71.
33. Tao M, Ishikawa S, Zhang Z, Murayama T, Inomata Y, Kamiyama A, Nakaima I, Jing Y, Mine S, Shimoda K, Toyao T, Shimizu K-I, Ueda W. Synthesis of zeolitic Ti, Zr-substituted vanadotungstates and investigation of their catalytic activities for low temperature NH_3-SCR. ACS Catal. 2021;11(22):14016–25.
34. Tao M, Ishikawa S, Murayama T, Inomata Y, Kamiyama A, Ueda W. Synthesis of zeolitic Mo-doped vanadotungstates and their catalytic activity for low-temperature NH_3-SCR. Inorg Chem. 2021;60(7):5081–6.
35. Gumerova NI, Rompel A. Polyoxometalates in solution: speciation under spotlight. Chem Soc Rev. 2020;49:7568–601.
36. Kovács TN, Hunyadi D, de Lucena ALA, Szilágyi IM. Thermal decomposition of ammonium molybdates. J Therm Anal Calorim. 2016;124(2):1013–21.
37. Muller A, Gouzerh P. From linking of metal-oxide building blocks in a dynamic library to giant clusters with unique properties and towards adaptive chemistry. Chem Soc Rev. 2012;41(22):7431–63.
38. Jassal AK, Mudsainiyan RK, Shankar R. A rational assembly of paradodecatungstate anions from clusters to morphology-controlled nanomaterials. Mater Chem Front. 2021;5:1090–125.
39. Sukmana NC, Sadakane SM. Structure and thermal transformations of methylammonium tungstate. Z anorg allgem Chem. 2021;647:1930–37.
40. Sadakane M, Watanabe N, Katou T, Nodasaka Y, Ueda W. Crystalline Mo_3VO_x mixed-metal-oxide catalyst with trigonal symmetry. Angew Chem Int Ed. 2007;46(9):1493–6.
41. Sadakane M, Endo K, Kodato K, Ishikawa S, Murayama T, Ueda W. Assembly of a pentagonal polyoxomolybdate building block, $[Mo_6O_{21}]^{6-}$, into crystalline MoV oxides. Euro J Inorg Chem. 2013;2013(10–11):1731–6.
42. Sadakane M, Kodato K, Kuranishi T, Nodasaka Y, Sugawara K, Sakaguchi N, Nagai T, Matsui Y, Ueda W. Molybdenum-vanadium-based molecular sieves with microchannels of seven-membered rings of corner-sharing metal oxide octahadra. Angew Chem Int Ed. 2008;47:2493–6.
43. Sadakane M, Ohmura S, Kodato K, Fujisawa T, Kato K, Shimizu K-I, Murayama T, Ueda W. Redox tunable reversible molecular sieves: orthorhombic molybdenum vanadium oxide. Chem Commun. 2011;47:10812–4.
44. Sadakane M, Yamagata K, Kodato K, Endo K, Toriumi K, Ozawa Y, Ozeki T, Nagai T, Matsui Y, Sakaguchi N, Pyrz WD, Buttrey DJ, Blom DA, Vogt T, Ueda W. Synthesis of orthorhombic Mo-V-Sb oxide species by assembly of pentagonal Mo_6O_{21} polyoxometalate building blocks. Angew Chem Int Ed. 2009;48:3782–6.

45. Ishikawa S, Kobayashi D, Konya T, Ohmura S, Murayama T, Yasuda N, Sadakane M, Ueda W. Redox treatment of orthorhombic $Mo_{29}V_{11}O_{112}$ and relationships between crystal structure, microporosity and catalytic performance for selective oxidation of ethane. J Phys Chem C. 2015;119:7195–206.
46. Sadakane M, Kodato K, Yasuda N, Ishikawa S, Ueda W. Thermal behavior, crystal structure, and solid-state transformation of orthorhombic Mo-V oxide under nitrogen flow or in air. ACS Omega. 2019;4:13165–71.
47. Zhang Z, Sadakane M, Murayama T, Izumi S, Yasuda N, Sakaguchi N, Ueda W. Tetrahedral connection of ε-keggin-type polyoxometalates to form an all-inorganic octahedral molecular sieve with an intrinsic 3D pore system. Inorg Chem. 2014;53(2):903–11.
48. Dolbecq A, Mialane P, Sécheresse F, Keita B, Nadjo L. Functionalized polyoxometalates with covalently linked bisphosphonate, N-donor or caboxylate ligands: from electrocatalytic to optical properties. Chem Commun. 2012;48:8299–316.
49. Rousseau G, Rodriguez-Albelo LM, Salomon W, Mialane P, Marrot J, Doungmene F, Mbomekakke I, de Oliviera P, Dolbecq A. Tuning the dimensionality of polyoxometalate-based materials by using a mixture of ligands. Cryst Growth Des. 2015;15:449–56.
50. Hayashi Y, Toriumi K, Isobe K. Novel triple-cubane type organometallic oxide clusters: $[MCp^*MoO_4]_4.nH_2O$ ($M = Rh$ and Ir; $Cp^* = C_5Me_5$; $n = 2$ for Rh and 0 for Ir). J Am Chem Soc 1988;110:3666–8.
51. Zhang Z, Zhu Q, Sadakane M, Murayama T, Hiyoshi N, Yamamoto A, Hata S, Yoshida H, Ishikawa S, Hara M, Ueda W. A zeolitic vanadotungstate family with structural diversity and ultrahigh porosity for catalysis. Nat Commun. 2018;9:3789.
52. Xu W, Pei X, Diercks CS, Lyu H, Ji Z, Yaghi OM. A metal-organic framework of organic vertices and polyoxometalate linkers as a solid-state electrolyte. J Am Chem Soc. 2019;141(44):17522–6.
53. Shimoyama Y, Uchida S. Structure-function relationships of porous ionic crystals (PICs) based on polyoxometalate anions and oxo-centered trinuclear metal carboxylates as counter cations. Chem Lett. 2021;50:21–30.
54. Ogiwara N, Iwamoto T, Ito T, Uchida S. Proton conduction in ionic crystals based on polyoxometalates. Coord Chem Rev. 2022;462: 214524.

Chapter 4
Mo-V-Based Crystalline Complex Metal Oxide Catalysts

Satoshi Ishikawa and Wataru Ueda

4.1 Introduction

In recent years, there has been a growing interest in a carbon–neutral society to realize a sustainable future [1]. Particularly, there has been a lot of interest in technology developments related to energy and chemical productions, such as fuel cells, energy carriers, and conversion of various resources into valuable chemicals [1–3]. Catalysis is one of the key technologies to accelerate these developments, and heterogeneous catalytic processes using solid-state catalysts can play a central role to realize the practical application of these technologies [2, 3].

Complex metal oxides are the most important classes of solid-state catalysts due to their excellent performance and durability under catalytic reaction conditions. In addition, the materials have a designability in their crystal structures and elemental compositions, allowing the optimization of catalytic functions to suit catalytic reactions. Much effort has been devoted to the development of new complex metal oxide catalysts. The selection of elements in the development is crucially important in order to control the electronic state of solid catalysts since the electronic state dominates the catalytic properties [4]. Therefore, metal oxide catalysts are synthesized using multiple elements to achieve the appropriate electronic state. Crystal structure constructed by the interaction of cation (metal) and anion (oxygen) is also important to control the electronic state of metal oxide catalysts.

This trend toward complexation and structurization is pronounced in a selective oxidation catalyst [5] because oxidation reaction normally requires multiple catalytic sites to activate not only the substrate but also the oxygen source at the same time.

S. Ishikawa (✉) · W. Ueda
Department of Material Chemistry and Life, Kanagawa University, 3-27-1, Rokkakubashi, Kanagawa-ku, Yokohama 221-8686, Japan
e-mail: sishikawa@kanagawa-u.ac.jp

W. Ueda
e-mail: uedaw@kanagawa-u.ac.jp

W. Ueda (ed.), *Crystalline Metal Oxide Catalysts*,
https://doi.org/10.1007/978-981-19-5013-1_4

Table 4.1 Examples of complex metal oxide catalysts used for gas-phase oxidation reactions

Reaction type	Reaction example	Catalyst
Oxyhydration	Ethylene/O_2 → Acetic acid	Pd/Si–W–O[a]
Allyl oxidation	Propylene/O_2 → Acrolein	K–Bi–Mo–Fe–Co–O
Ammoxidation	Propane/O_2/NH_3 → Acrylonitrile Propylene/O_2/NH_3 → Acrylonitrile	Mo–V–Nb–Sb–O/SiO_2 K–Bi–Mo–Fe–Co–O
Alcohol oxidation	Methanol/O_2 → Formaldehyde	Mo–Fe–O
Aldehyde oxidation	Methacrolein/O_2 → Methacrylic acid Acrolein/O_2 → Acrylic acid	Cs–P–Mo–V–Cu–Sb–O Pd/Si–W–O Mo–V–W–Cu–O
Alkane oxidation	Ethane/O_2 → Acetic acid Propane/O_2 → Acrylic acid Butane/O_2 → Maleic anhydride	Pd–Mo–V–Nb–O/TiO_2 Mo–V–Nb–Te–O/SiO_2 V–P–O
Aromatic oxidation	Benzene/O_2 → Maleic anhydride o-xylene/O_2 → Phthalic anhydride	Mo–V–P–O V–Ti–Si–O
Oxidative dehydrogenation	Ethane/O_2 → Ethylene Propane/O_2 → Propylene	Mo–V–Nb–Te–O V–Mg–O

[a] Heteropoly acid

Necessary multiple catalytic functions are such as adsorption/desorption properties of substrates/products and activation properties of substrates/molecular oxygen. Accordingly, the catalysts employed in the oxidation processes have been developed in the direction of multi-component systems. Table 4.1 lists the representative examples of the multi-component complex metal oxide catalysts for gas-phase selective oxidations.

There are many important structural features desired for high catalytic oxidation performance in metal oxide catalysts. The most important one seems the site isolation concept proposed by Grasselli [6, 7]. The idea of this concept is that limiting the number of active oxygens available for the reaction can prevent the undesired sequential reactions, and thus, the selectivity to partial oxidation products can improve. This concept can be realized by the crystal structure organization where catalytically active sites consisting of a necessary number of elements polyhedra isolate each other by being surrounded by other element polyhedra. The construction of a site-isolated structure locally but periodically in a bulk crystal must be profitable to create highly selective oxidation catalysts. For this purpose, clarification of the catalytically active structure responsible for catalysis beforehand is the foremost task toward a rational catalyst design. However, the proposal of crystalline metal oxide catalysts on the basis of the concept has scarcely been reported so far. One of the reasons is that it is almost impossible to deduce a desired crystal structure of metal oxide catalysts from a given catalytic reaction. Another reason is the lack of rational synthetic methodologies for oxide catalysts. These difficulties are obviously derived from the complexity of solid catalysts.

Historically, long-year efforts have successfully realized various highly efficient metal oxide catalysts for oxidations and made it possible to use them as industrial processes. For example, Mo-V-based complex metal oxides were developed as highly active catalysts for gas-phase oxidations of acrolein and alkanes. Recently, we have synthesized a crystalline Mo_3VOx catalyst extremely active for these oxidations for the first time by using the hydrothermal method. Interestingly, this crystalline oxide catalyst has a complicated local structure that perfectly satisfies the site-isolation requisite.

This chapter focuses on the crystalline Mo_3VOx catalyst which has a unique crystal structure responsible for oxidation catalysis and shows how the catalytic functions are generated and tuned in terms of crystal structure.

4.2 Mo-V-Based Oxide Catalysts for Selective Oxidations

4.2.1 Development of Industrial Oxidation Catalysts

Mo-V-based complex metal oxides show catalytic activity for a variety of oxidations of light alkanes and unsaturated aldehydes. Several reactions have been industrialized using these catalysts. Here, we briefly summarize the development of the catalysts in the selective oxidation of acrolein, propane, and ethane from their discovery to the present (Fig. 4.1).

Mo-V-based oxide catalysts have a long history as industrial catalysts in the selective oxidation of acrolein to produce acrylic acid. An excellent oxidation property of these catalysts is firstly discovered by Toyo Soda in 1963 [8]. Since this discovery, much effort has been made for the development of these catalysts, and the basic elemental compositions were determined to be Mo–V–W–Cu–O by 1974 [9]. Acrylic acid yields of more than 95% have been currently achieved since the discovery and industrial research are still ongoing to improve their catalytic performance further.

In 1995, Mitsubishi Kasei discovered the crystalline Mo–V–Te–Nb oxide as an extremely active catalyst for the ammoxidation of propane [10]. This catalyst showed far superior catalytic activity to V–Sb–W–O and achieved the acrylonitrile yield close to the industrially applicable level [11]. The reaction over this catalyst was operatable at a lower reaction temperature below 450 °C, making this crystalline catalyst further attractive. 20 years later from the discovery, a propane ammoxidation process has been successfully industrialized by Asahi-Kasei in 2013 in Thailand using Mo–V–Sb–Nb–O having an identical crystal structure to Mo–V–Te–Nb–O [12]. Acrylonitrile yield is reported approximately 60% in this process.

The crystalline catalyst exhibited excellent catalytic activity for selective oxidation of propane to acrylic acid as well as ammoxidation. This reaction was also discovered by Mitsubishi Kasei [13]. The process has not yet been industrialized because the yield of acrylic acid in a recent report [14] is still not high enough for the industrialized levels.

At the beginning of the discovery of Mo–V–Te–Nb–O and Mo–V–Sb–Nb–O catalysts, it has already been recognized that the formation of a crystalline state is

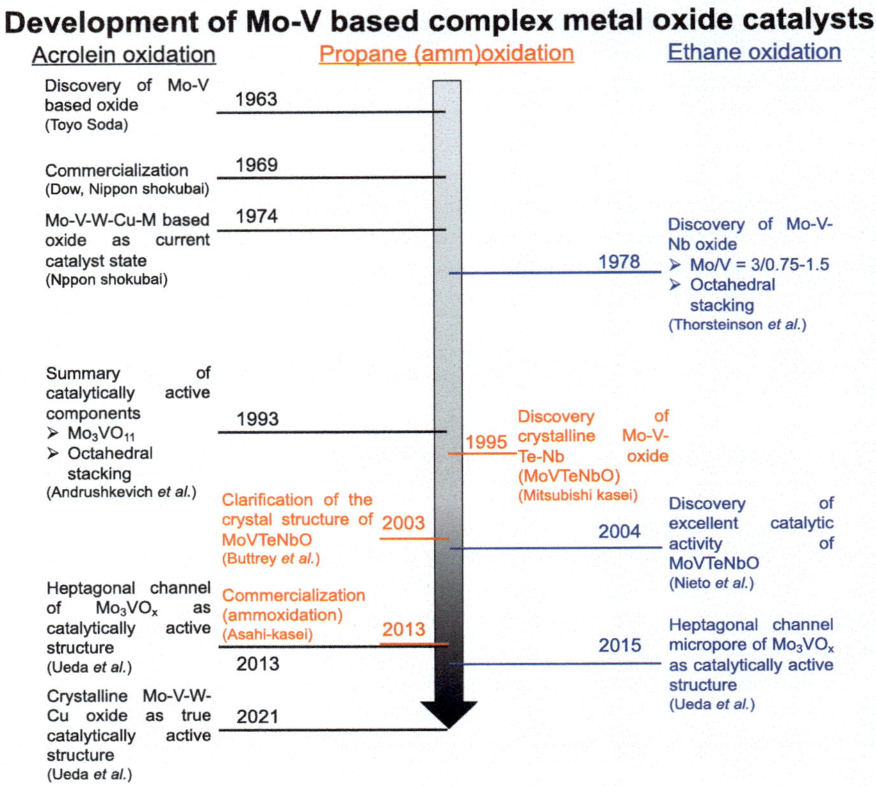

Fig. 4.1 Development of Mo-V-based complex metal oxide catalysts in the field of acrolein oxidation (black), propane (amm)oxidation (red), and ethane oxidation (blue)

crucial for generating the catalytic property and also that the catalytic preparation had to be carefully done. Soon the crystal structure was determined in the company and a hydrothermal method was demonstrated as a better reproducible preparation of the catalysts [15–18]. Since then, fundamental research on this crystalline catalyst, including detailed structural analysis, chemical states, and catalytic roles of constituting elements, were extensively conducted. The most striking point in the catalysts is that the crystal structure is highly complicated [19]. The structure was based on the arrangement of pentagonal units composed of Mo and Nb, which were connected by VO_6 octahedral to form the basal plane constituting hexagonal and heptagonal textures. The basal planes stacked each other along the *c*-direction, forming hexagonal and heptagonal channels in the crystal structure. Te (or Sb) was located mainly at the hexagonal channel and partially at the heptagonal channel in the framework structure. The elemental composition of Mo–V–Te–Nb–O was determined with some deviation depending on the reports but was in the range of Mo/V/Te/Nb = 3/0.6–1.2/0.2–0.6/0.3–0.7, and the group of Mitsubishi Kasei adopted the elemental composition of Mo/V/Te/Nb = 3/1.2/0.3/0.6 as the most active and selective catalyst in their initial studies [10, 19–23].

This crystalline catalyst also showed prominent catalytic properties for oxidative dehydrogenation of ethane to form ethylene. Thosteinson et al. discovered that Mo–V–Nb–O promoted the selective oxidation of ethane to form ethylene and further to acetic acid depending on the reaction conditions [24]. They mentioned that the active catalysts for the ethane conversion are constructed by the stacking of octahedral with the Mo/V ratio of 3/0.75 ~ 1.5. Later, Nieto et al. reported the catalytic performance of Mo–V–Te–Nb–O in the oxidative dehydrogenation of ethane [25, 26]. This discovery again highlighted the importance of crystal structure in the selective oxidations over Mo-V-based complex metal oxide catalysts. The yield of ethylene over these catalysts reported to date is around 75% [27]. This catalytic performance over Mo–V–Te–Nb–O affords the ethylene yield almost at the commercialization levels. The practical application of this process has been a central topic in the selective oxidation since the shale gas revolution in the US due to the huge demand for ethylene which is the most produced organic compound in the world [5]. Attempts to further improve the ethylene yield are ongoing. In addition to this issue, attention has focused on preventing the sublimation of Te for practical applications. Replacement of Te with Sb or introduction of Bi into Mo–V–Te–Nb–O is investigated [28–30]. As this catalyst has a specific crystal structure, much work has been done to elucidate the relationship between the crystal structure and catalytic activity to obtain fundamental information about their catalytic function. However, a consensus has not yet been reached due to the complexity of their catalyst structure and elemental composition [31–34].

Although the Mo-V-based complex metal oxide catalysts have been developed differently depending on the reactions, there are common characteristics in the catalysts. Andruskevich et al. summarized the characteristics of the catalytically active components in the Mo-V-based oxide catalysts for acrolein oxidation [35]. They concluded that catalytically active components are constructed by the stacking of octahedral with the elemental composition of Mo_3VO_{11}. Vogel et al. also confirmed the optimum Mo/V ratio of approximately 3 [36, 37]. The octahedral stacking and compositional (Mo/V ratio = 3) characteristics are also valid for the ethane oxidation and the propane (amm)oxidation catalysts.

Furthermore, catalytic activity over these catalysts was strongly dependent on the crystal structure. Mo–V–O having a tetragonal Mo_5O_{14} structure was once considered as the catalytically active phase. Although this phase can satisfy the Mo/V ratio and other characteristics, the catalytic activity of pure tetragonal crystal was negligible. The complicated crystal structure for the Mo–V–Te–Nb–O catalyst was found inherently active for the reactions and only specific structures exposed on the cross-section of the rod-shaped crystal played crucial roles in these reactions [15, 38]. This discovery implies that the formation of a local catalyst structure is responsible for the oxidation catalysis.

4.2.2 Evolution of Crystalline Mo_3VO_x Catalysts

Minimization of constituting elements of metal oxide catalysts with keeping the intrinsic catalytic property, that is, with keeping the original crystal structure, is an indispensable approach for understanding the fundamental catalytic function and their origins. We approached the above introduced crystalline Mo-V-based complex metal oxide catalyst in this context and we successfully synthesized crystalline orthorhombic Mo_3VO_x catalyst (Orth-MoVO) by hydrothermal synthesis [39]. This catalyst has an identical crystal structure with Mo–V–Te–Nb–O and is composed of the structural arrangement of $\{Mo_6O_{21}\}^{6-}$ pentagonal unit and MO_6 (M = Mo, V) octahedral. There are no elements inside the hexagonal and heptagonal channels different from Mo–V–Te–Nb–O. The empty heptagonal channel works as a micropore with a diameter of approximately 0.40 nm, which adsorb small molecules like light alkanes [40]. Interestingly, Orth-MoVO satisfied all the structural and compositional characteristics reported by Andruskevich et al. as active components of the Mo-V-based oxide catalysts for the selective oxidation of acrolein. Different from the Mo_5O_{14}-type tetragonal crystal structure, Orth-MoVO exhibited excellent catalytic activity for acrolein oxidation, and the activity was even superior to that of the industrial catalysts [41]. These facts suggest that this catalyst contains a catalytically active component for this reaction. In the (amm)oxidation of propane, Orth-MoVO showed comparable or even higher catalytic activity for the conversion of propane with that of Mo–V–(Te,Sb)–(Nb)–O, while the selectivity toward acrylonitrile or acrylic acid was significantly poorer than those of these catalysts [42, 43]. This observation suggests that the Mo–V–O texture is involved in the activation of propane and that additional elements are needed to obtain high selectivity for partial (amm)oxidation products. In the selective oxidation of ethane, Orth-MoVO showed excellent catalytic activity for the ethane oxidation to ethylene, achieving comparable ethane conversion at a reaction temperature lower than 100 °C for Mo–V–Te–Nb–O under similar reaction conditions [44].

We then further synthesized the crystalline Mo_3VO_x catalysts with the same structural units (pentagonal unit and octahedral), the same crystal morphology (rod-shaped crystal), and the same elemental composition, but with the different structural arrangement in the cross-section of the rod [44–47]. These are Orth-MoVO, trigonal Mo_3VO_x (Tri-MoVO), tetragonal Mo_3VO_x (Tet-MoVO), and amorphous Mo_3VO_x (Amor-MoVO) (Fig. 4.2a). Among them, Tet-MoVO was the one proposed as a catalytically active component in the selective oxidation of acrolein once. Orth-MoVO, Tri-MoVO, and Amor-MoVO constitute the pentagonal, hexagonal, and heptagonal channels in the crystal structure, while there is no heptagonal channel in Tet-MoVO. Since all these catalysts satisfied the structural and compositional characteristics, the catalytic property of these catalysts was investigated in the selective oxidation of acrolein and ethane and compared to each other [46]. In both the reactions, Orth-MoVO, Tri-MoVO, and Amor-MoVO having the heptagonal channel in the crystal structure exhibited the catalytic activities (Fig. 4.2). The catalytic activity of Tet-MoVO was almost negligible. The order of the catalytic activity was Tri-MoVO > Orth-MoVO > Amor-MoVO ≫ Tet-MoVO (almost inactive) in the acrolein oxidation, while Orth-MoVO > Tri-MoVO > Amor-MoVO ≫ Tet-MoVO (almost inactive) in the ethane oxidation. The former was consistent with the number of the heptagonal channel on the catalyst surface. The latter corresponded to the micropore volume of the heptagonal channel rather than the number of the heptagonal channel. These results indicate that the local catalyst structure around the heptagonal channel is primarily responsible for the catalytic oxidations and that the reaction field where the oxidation takes place depends on the reactions, acrolein oxidation over the surface heptagonal structure and ethane oxidation inside the heptagonal channel.

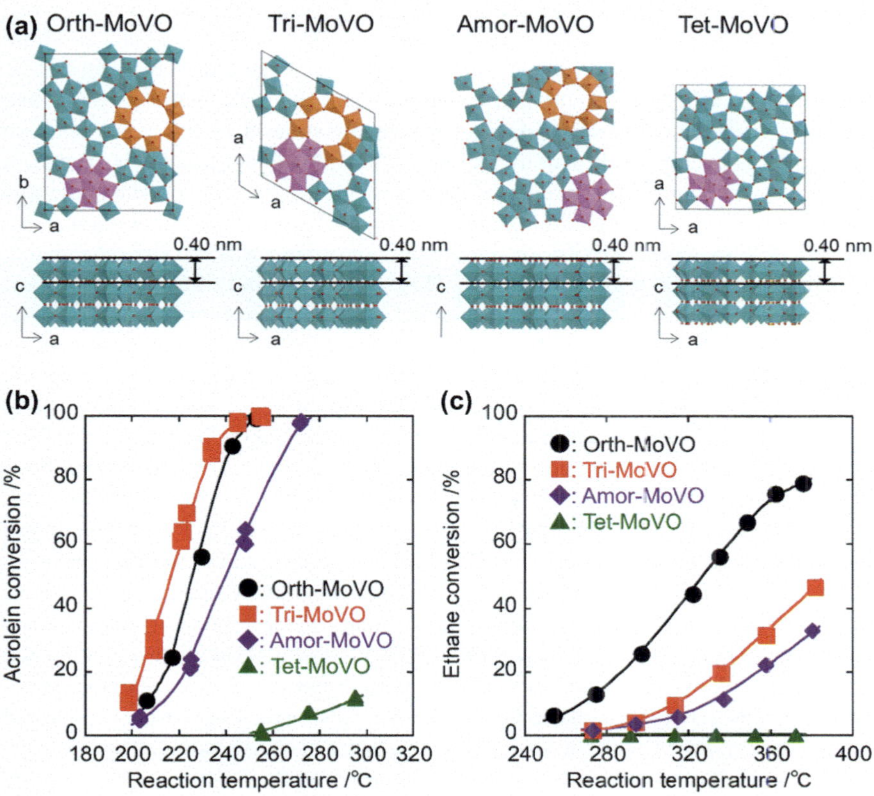

Fig. 4.2 **a** Crystal structures of four distinct crystalline Mo_3VO_x. **b** Acrolein conversion as a function of reaction temperature over the four distinct crystalline Mo_3VO_x catalysts. Reaction conditions: reaction gas, acrolein/O_2/(N_2 + He)/H_2O = 1.3/4.0/31.8/13.0 mL min^{-1} (total: 50.1 mL min^{-1}); catalyst amount, 0.15 g. **c** Ethane conversion as a function of reaction temperature over the four distinct crystalline Mo_3VO_x catalysts. Reaction conditions: reaction gas, ethane/O_2/N_2 = 5/5/40 mL min^{-1} (total: 50 mL min^{-1}); catalyst amount, 0.50 g

4.3 Catalytic Property of Crystalline Mo_3VO_x

4.3.1 Catalysis Field based on Crystal Structure

As briefly introduced above, the heptagonal channel formed in the crystalline Mo_3VO_x catalysts is proved to play essential roles in selective oxidations. To gain insight into the crystal structure-catalytic activity relationship, it is necessary to elucidate the field where the oxidation catalysis takes place [48, 49]. Orth-MoVO with the different external surface areas but with the same micropore volumes were synthesized by changing the hydrothermal synthesis conditions and tested for selective oxidation of ethane and acrolein (Fig. 4.3 a–d). Figure 4.3e shows the relationship between the external surface area and the conversions of ethane and acrolein. The acrolein conversion increased with the increase of the external surface area. This is normal when the heptagonal channel site on the surface acts as the active site. In fact, Orth-MoVO without grind treatments showed much lesser catalytic activity since the cross-section of the rod, where the heptagonal channel is located, is hardly exposed. Similar dependency of the catalytic activity on the crystal plane was observed in (amm)oxidation of propane over Mo–V–Te–Nb–O, as mentioned above [15, 38]. The catalytic activity for the ethane conversion was, on the other hand, independent of its external surface area. This observation indicates that the ethane conversion takes place inside the heptagonal channel micropore. Concomitantly, Orth-MoVO without the grind treatments exhibited catalytic activity comparable to that of the ground catalyst. Recently, it was reported that the ethane conversion also takes place inside the heptagonal channel micropore of Mo–V–Te–Nb–O catalyst, although their heptagonal channel is partially occupied by Te, retarding the gas diffusion [33]. Ethane conversion inside the heptagonal channel micropore was also confirmed in Tri-MoVO [50]. This catalyst was amenable to introducing the additional elements without altering the basic crystal structure, and Tri-MoVO containing Fe inside the heptagonal channel (Tri-MoVFeO) and that containing W in the pentagonal units (Tri-MoVWO) were successfully synthesized by hydrothermal synthesis. Tri-MoVO and Tri-MoVWO containing no elements inside the heptagonal channels showed almost the same catalytic activity for selective oxidation of ethane, while the catalytic activity over Tri-MoVFeO was almost negligible at the same reaction conditions. Fe inside the heptagonal channel made the diffusion of ethane in the heptagonal channel micropore difficult, resulting in a significant decrease in the ethane conversion. The difference in the catalysis field depending on the reactions is simply due to the different molecular sizes of reactants and products, because molecules larger than the micropore diameter cannot react inside the channel but easily access the catalytically active heptagonal area. The other reason besides molecular size is functional groups in molecules as a polarity derived from these groups would strongly interact with polar sites over the catalyst surface.

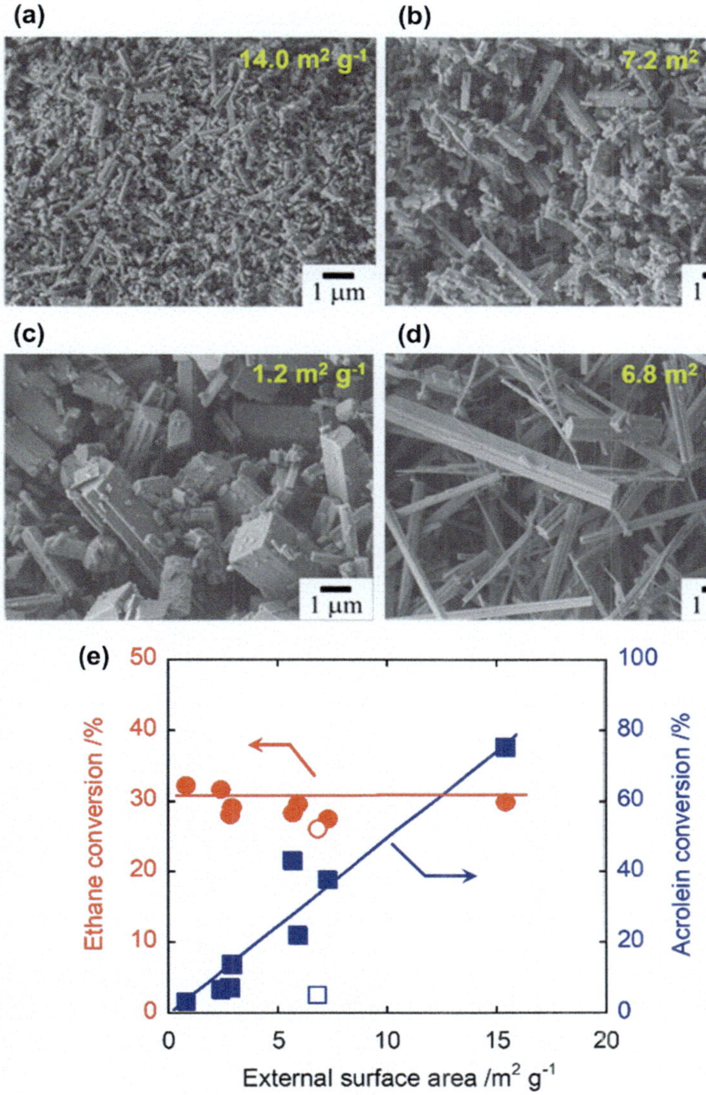

Fig. 4.3 **a–d** SEM images of Orth-MoVO with controlled crystal size. External surface areas obtained by N_2 adsorption measurements are shown in the images. **d** is the same material as **b** but is not ground. **e** Ethane (left) and acrolein (right) conversions as a function of external surface areas of Orth-MoVO. Open symbols represent the results of the unground catalyst (SEM image is shown in (**d**)). Reaction conditions in acrolein oxidation: reaction gas, acrolein/O_2/(N_2 + He)/H_2O = 2.5/8.0/70.0/27.0 mL min^{-1} (total: 107.5 mL min^{-1}); catalyst amount, 0.125 g. Reaction conditions in ethane oxidation: reaction gas, ethane/O_2/N_2 = 5/5/40 mL min^{-1} (total: 50 mL min^{-1}); catalyst amount, 0.50 g

4.3.2 Oxidation Catalysis for Ethane Oxidation

Since the bulk properties of metal oxide catalysts can be evaluated by various characterization methods, further details of the relationship between crystal structure and catalytic activity are investigated in depth. Among the inorganic microporous materials reported so far, Orth-MoVO is a particularly important class of material in terms of the constituent elements [51–54]. Different from the typical inorganic microporous materials such as zeolites, Orth-MoVO is composed of transition metals that majorly adopt octahedral configuration. Since the size of the octahedral can be tuned by controlling the oxidation state of metal elements, the size of the heptagonal channel micropore was tunable by controlling the oxidation state of constituent elements [55–57].

The effect of reduction degree on the microporosity and the catalytic property for ethane oxidation has been investigated in Orth-MoVO [58]. The degree of reduction of Orth-MoVO was controlled by controlling the reduction treatment conditions and was expressed as the number of lattice oxygens removed from the crystal structure. Orth-MoVO with the lattice oxygens vacancy is abbreviated as MoVO (δ), where δ indicates the number of lattice oxygens vacancy formed from the Orth-MoVO unit cell ($Mo_{29}V_{11}O_{112-\delta}$). Ethane adsorption measurements were conducted on MoVO (δ) and the amounts of ethane adsorbed in MoVO (δ) are plotted as a function of δ (Fig. 4.4a). Orth-MoVO can adsorb ethane up to the theoretically calculated microporous volume of the heptagonal channel (23.4×10^{-3} cm^3 g^{-1}) when the material was calcined in the air (MoVO (0)). This capacity was not significantly changed despite the removal of lattice oxygens up to MoVO (4.2). However, the adsorbed amount of ethane drastically decreased by the reduction above MoVO (5.4). Structural analysis revealed that the pentagonal unit started to expand by the reduction above MoVO (5.4) which led to the shrinkage of the heptagonal channel. The drastic decrease in the micropore adsorption amounts was the result of this shrinkage. The ethane conversions at 10 min from the start of the reaction (initial catalytic activity) over MoVO (δ) are also plotted in Fig. 4.4a. Since the ethane oxidation takes place inside the heptagonal channel micropore, the ethane conversion drastically decreased by the reduction above MoVO (5.4). It is noteworthy that the ethane conversion increased from 13.7 to 33.0% by the reduction from MoVO (0) to MoVO (4.2), although the adsorption capacity of ethane was comparable. Since Orth-MoVO catalyzes the ethane oxidation inside the heptagonal channel micropore, possible changes in the local structure around the heptagonal channel might be related to this observation. The drastic changes in the microporosity and the catalytic activity upon the reduction at the threshold of MoVO (5.4) may also be related to the local structural changes.

When MoVO (6.8) was calcined in air at 400 °C (MoVO (6.8)-AC), the ethane conversion drastically increased. Interestingly, the ethane conversion of MoVO (6.8)-AC was approximately three times higher than that of MoVO (0) and was similar to that of MoVO (4.2). Despite the air-calcined catalysts at the same conditions, the ethane conversion over MoVO (0) and MoVO (6.8)-AC was significantly different. Through the investigations of the local catalyst structure around the heptagonal

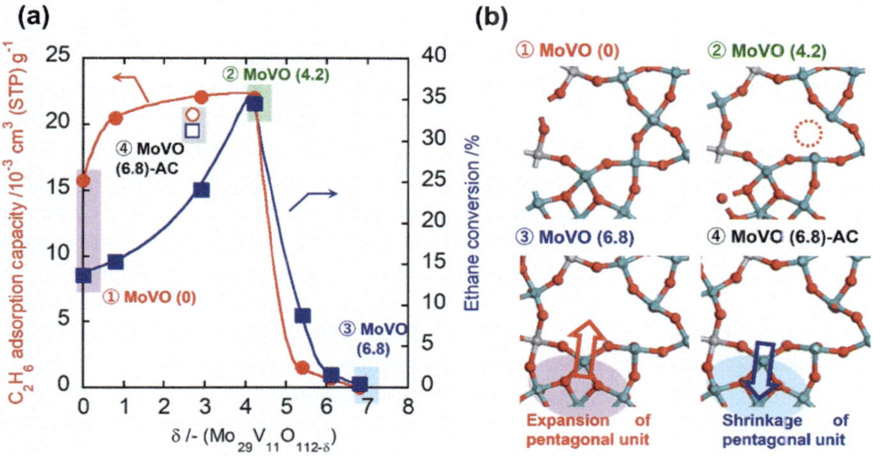

Fig. 4.4 **a** Micropore volumes of MoVO (δ) measured by ethane adsorption (left) and ethane conversions at 10 min from the start of the reaction (right) as a function of d ($Mo_{29}V_{11}O_{112-\delta}$). Reaction conditions: reaction gas, ethane/O_2/N_2 = 5/5/40 mL min^{-1} (total: 50 mL min^{-1}); catalyst amount, 0.50 g; reaction temperature, 300 °C. **b** Structural images around the heptagonal channels of MoVO (δ)-(AC). α-oxygen defect site is represented as a dotted circle in MoVO (4.2)

channel, it was found that there were two types of lattice oxygens in Orth-MoVO. One was the lattice oxygen removed in the early stage of the reduction and hardly come back to the structure by the oxidation treatment. This lattice oxygen is almost completely removed by the reduction up to MoVO (4.2). The other was the lattice oxygen removed continuously by the reduction and reversibly come back to the structure by the reoxidation. The former and the later lattice oxygens are abbreviated as α-oxygen and β-oxygen, respectively. α-oxygen is identified as the lattice oxygen in the pentamer unit which is isolated in the framework structure based on the pentagonal unit network, and β-oxygen is identified as the lattice oxygen connecting the *a-b* plane along the *c*-direction. The physicochemical properties of MoVO (0) and MoVO (6.8)-AC were similar, but the occupancy of α-oxygen was significantly different since α-oxygen is not able to come back to the structure despite the oxidation treatment. After removing α-oxygen, the micropore size expanded as confirmed by various adsorption experiments. This observation indicates that α-oxygen defect site is concentrated at the pentamer unit facing the heptagonal channel. Based on the fact that the catalytic activity drastically increased by the removal of α-oxygen, it is speculated that active oxygen species form at the α-oxygen defect site. These species efficiently promote the ethane oxidation to form ethylene and water inside the heptagonal channel micropore. The similar catalytic activity between MoVO (4.2) and MoVO (6.8)-AC is also derived from the occupancy of α-oxygen since α-oxygen is almost completely removed from the structure by the reduction up to MoVO (4.2). As mentioned above, the microporosity and the catalytic activity dropped drastically with the reduction above MoVO (5.4) due to the expansion of the pentagonal unit, while no such changes were observed up to MoVO (4.2). Removal of α-oxygen is a possible trigger to initiate the expansion of the pentagonal unit. Local structural changes around the heptagonal channel of Orth-MoVO by the redox treatments are summarized in Fig. 4.4b. Details have not yet been understood why the oxidation treatment did not refill α-oxygen defect. We are tentatively considering the electron localization. Upon removing α-oxygen from the structure, the pentamer unit gains four electrons since this unit contains two α-oxygens (see Fig. 4.5). If these electrons are localized at the specific sites in the pentamer unit, particularly on the metals adjacent to the oxygen defect site, four-electron oxygen reduction is not likely to refill the lattice oxygen ($O_2 + 4e^- \rightarrow O^{2-}$). Instead, a limited number of electron transfers might occur to molecular oxygen during the oxidation catalysis, resulting in partially reduced oxygen species like superoxide (O_2^-) or peroxide (O_2^{2-}). These species might work as catalytically active oxygen species.

Based on the results obtained above, the oxidation catalysis over Orth-MoVO for the ethane oxidation is proposed as follows: (1) both ethane and molecular oxygen are captured inside the heptagonal channel; (2) molecular oxygen is activated at the oxygen defect site at the pentamer unit facing the heptagonal channel; (3) activated molecular oxygen reacts with ethane to form ethylene and water. During the oxidation catalysis, the pentamer unit constituting the α-oxygen defect is allowed to be kept under the reaction conditions since this unit is stabilized in the stable framework structure based on the pentagonal unit network. The role of the crystal structure for ethane oxidation is summarized in Fig. 4.5. Although further works are necessary to clarify how the reaction cycle closes and what types of active oxygen species are indeed generated, it is noteworthy that the highly organized local catalyst structure constituting an isolated structural site within the framework is responsible for the oxidation catalysis.

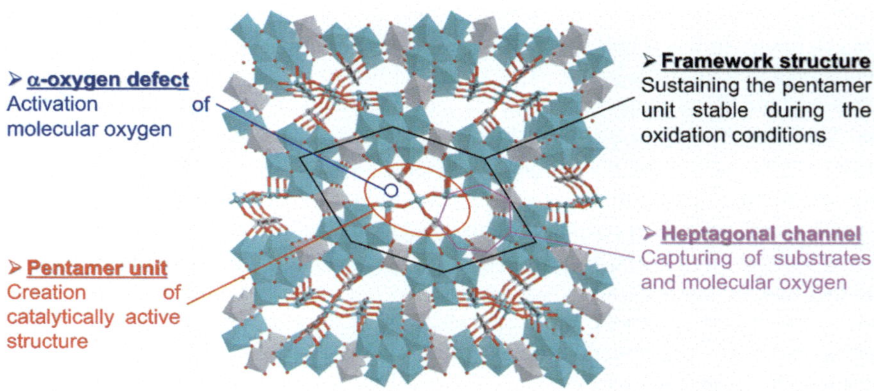

Fig. 4.5 Role of the crystal structure of Orth-MoVO for oxidation catalysis in selective oxidation of ethane

4.3.3 Catalytically Active Structure for Acrolein Oxidation

In the case of acrolein oxidation, as mentioned above, the local catalyst structure around the heptagonal channel of the crystalline Mo_3VO_x works as the catalytically active structure. In consideration of the fact that these catalysts satisfied the structural and compositional characteristics as catalytically active components, it seems that the local catalyst structure around the heptagonal channel is the catalytically active structure for this reaction. In the typical industrial catalysts for acrolein oxidation, on the other hand, Mo-V-based complex metal oxide catalysts contain various promoter elements [9, 59–61]. Cu and W are used in most of the cases as the promoter elements in the industrial catalysts, although their catalytic roles have not yet been clarified.

Cu can be introduced into the crystal structure of Orth-MoVO by optimizing the hydrothermal conditions (Orth-MoVCuO) [62]. Cu is placed at the interstitial spaces between the lattice oxygens in the pentamer units facing the heptagonal channel, the interstitial spaces between α-oxygens (Fig. 4.6). The catalytic activity for acrolein oxidation was almost halved with the introduction of Cu. However, the selectivity to acrylic acid was improved from 92 to 98% by the introduction of Cu, irrespective of the acrolein conversion (Fig. 4.6). Since Cu is located between α-oxygens, Cu is considered to affect the number of electrons transferred to molecular oxygen. In this regard, it was reported that the heat of adsorption of molecular oxygen decreased when Cu is introduced into the Mo-V-based catalysts [35, 63]. The introduction of Cu seems to decrease the activity of active oxygen species due to the change in the number of electrons transferred to molecular oxygen. This active oxygen state may not cause undesired oxidations like C–C scission to generate CO_x, resulting in the improvement of acrylic acid selectivity.

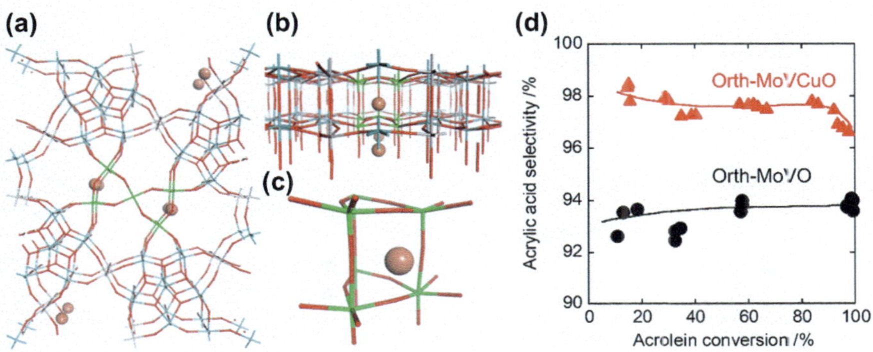

Fig. 4.6 **a** Structural images of Orth-MoVCuO. Mo, light blue; V, gray; O, red; mixture of Mo and V, light green; Cu, dark red. The crystal structure in the c-direction and the local structure around Cu are shown in (**b**) and (**c**). **d** Acrylic acid selectivity change as a function of acrolein conversion over Orth-MoVO (circle) and Orth-MoVCuO (triangle). Reaction conditions: catalyst amount, 0.02–0.25 g; reaction temperature, 250 °C; reaction gas, acrolein/O_2/(N_2 + He)/H_2O = 1.3/4.0/31.8/13.0 mL min^{-1} (total: 50.1 mL min^{-1}). Adapted with permission from Ref. [62]. Copyright 2019 American Chemical Society

The catalytic role of W for acrolein oxidation can be understood based on the crystal structure [64]. Although W was hardly introduced into Orth-MoVO, Tri-MoVO containing W was successfully obtained by optimizing the hydrothermal synthesis conditions. W was placed in the $\{Mo_6O_{21}\}^{6-}$ pentagonal units in Tri-MoVO. The introduction of W improves the tolerance of the acrolein conversion upon the decrease in the water pressure, although the introduction did not alter the acrolein conversion or selectivity to acrylic acid. The water tolerance of the acrolein conversion was further improved by increasing the amount of W introduced, which can be seen in the long-term reactions (Fig. 4.7). Under the low water pressure region, with partial pressure of H_2O less than 5%, the acrolein conversion over Tri-MoVO gradually decreased with reaction time, while the catalytic activity over Tri-MoVWO was stable. In the case of Tri-MoVO, the decrease of the acrolein conversion in the low water pressure region is mainly caused by the strong adsorption of reaction intermediates such as acrylate species, which is known to be firmly bound to the catalyst surface [37]. The introduction of W promoted the dissociative adsorption of water under the reaction conditions and consequently, desorption of the reaction intermediates is promoted by the replacement of the hydroxyl group, leading to the formation of acrylic acid. A similar catalytic role of W was also reported in W-V-based complex metal oxide catalyst for selective catalytic reduction of NO_x with NH_3 (NH_3–SCR) [65, 66].

Tri-MoVWO can accommodate Cu inside the structure. The resulting catalyst (Tri-MoVWCuO) contains Cu cations at the interstitial spaces between the lattice oxygens in the trimer unit, which is isolated by the framework structure like the pentamer unit in Orth-MoVO. Although the acrolein conversion decreased with the introduction of Cu, Tri-MoVWCuO showed stable acrolein conversion even under

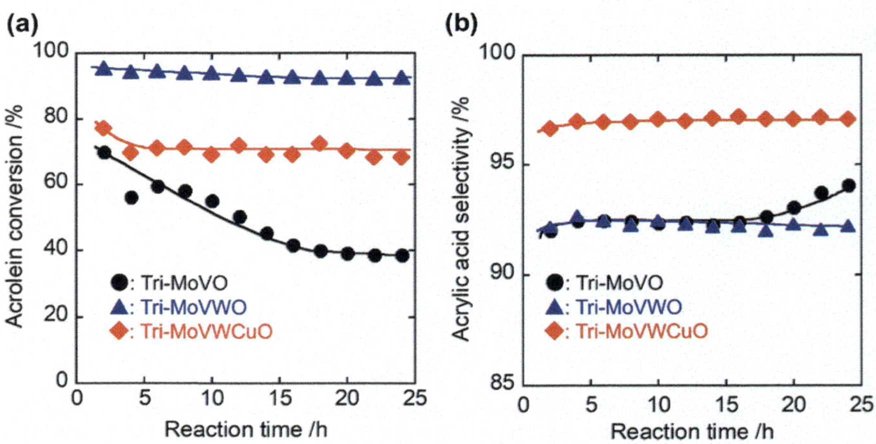

Fig. 4.7 **a** Acrolein conversion and **b** acrylic acid selectivity changes as a function of reaction time over Tri-MoV(W)(Cu)O. Reaction conditions: catalyst amount, 0.15 g; reaction temperature, 250 °C; reaction gas, acrolein/O_2/(N_2 + He)/H_2O = 1.3/4.0/55.0/3.0 mL min^{-1} (total: 63.3 mL min^{-1})

the low water pressure condition in a manner similar to Tri-MoVWO (Fig. 4.7). In addition to the improvement of the tolerance of the acrolein conversion under the low water pressure region, the selectivity to acrylic acid was substantially increased by the introduction of Cu as was also observed in Orth-MoVCuO. As a result of the structural arrangement to accommodate multiple elements in each structural site, the constituent elements can work cooperatively to achieve multiple catalytic functions in Tri-MoVWCuO, including catalytic activity, water tolerance, and excellent acrylic acid selectivity.

The physicochemical properties and catalytic functions of Tri-MoVWCuO were compared with the industrial Mo–V–W–Cu–Sb–O catalyst (Ind-MoVWCuSbO) synthesized according to the patent [60]. These catalysts satisfied the structural and compositional characteristics reported by Andruskevich et al. as catalytically active components for acrolein oxidation. Both the catalysts indeed showed catalytic activity for acrolein conversion as can be expected from their structural and compositional characteristics, while the activity per catalyst weight over Tri-MoVWCuO was approximately twice of Ind-MoVWCuSbO. HAADF-STEM measurements of Ind-MoVWCuSbO unraveled their local catalyst structure (Fig. 4.8). This catalyst consisted of the pentagonal unit and octahedral like the crystalline Mo_3VO_x catalysts, and the local catalyst structure was identical to Tri-MoVWCuO. The observations strongly suggest that the local catalyst structure around the heptagonal channel of Tri-MoVWCuO is the true catalytically active structure for the selective oxidation of acrolein over industrial Mo-V-based complex metal oxide catalyst (Fig. 4.8).

Mo-V-based complex metal oxides have been used for various oxidation reactions. Although the catalysts were optimized to suit each catalytic reaction, the developed catalysts resulted in common structural and compositional characteristics. The appearance of the crystalline Mo_3VO_x catalysts successfully unraveled the reason for these common observations. The catalytically active component of Mo-V-based complex metal oxides visualized in the crystalline Mo_3VO_x was a highly organized local structure consisting of micropores. The history of the developments of Mo-V-based complex metal oxides is summarized in Fig. 4.9 based on their crystal structure.

4.4 Conclusions and Future Outlook

The oxidation catalysts have been developed based on complex metal oxides for many years and are extremely advanced by introducing multi-elements and crystal structuring. The chapter highlighted the most prominent example, which is Mo-V-based complex metal oxides and more specifically, crystalline Mo_3VO_x catalysts. Thanks to the crystalline state of the catalysts and to the clear structure effect on catalytic functions, depth understanding of catalysis and the catalytic roles of each constituent become possible on the basis of a clear picture of crystal structure and local structure. This chapter visualized the local catalyst structure responsible for oxidation catalysis and provided fundamental insight for rational catalyst design.

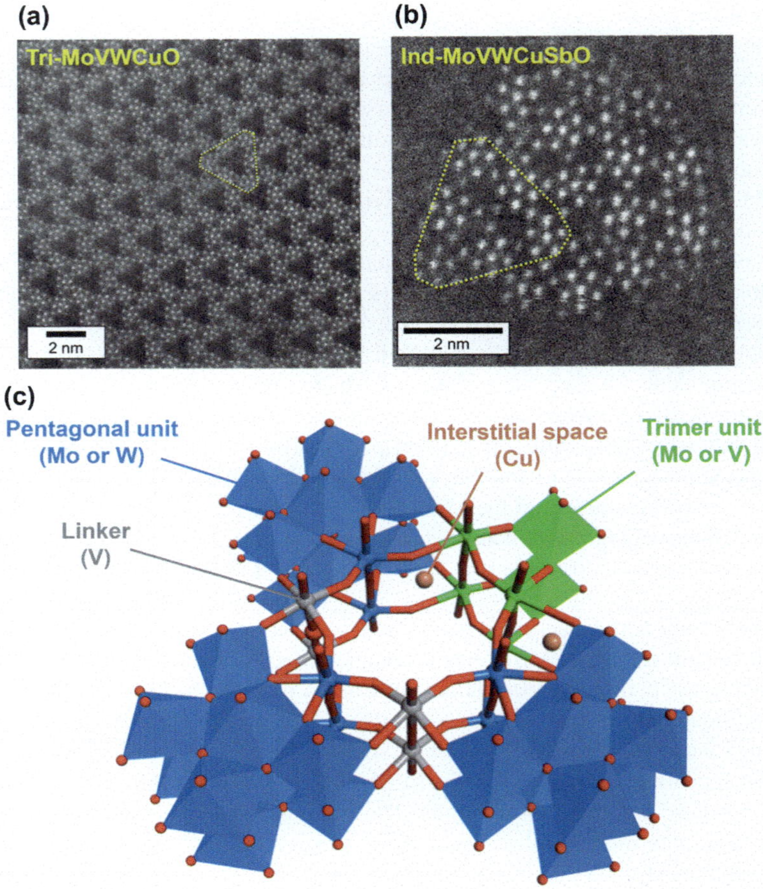

Fig. 4.8 HAADF-STEM images of **a** Tri-MoVWCuO and **b** Ind-MoVWCuSbO. **c** Image of the true catalytically active structure in industrial Mo-V-based complex metal oxide catalysts for the selective oxidation of acrolein. Blue, mixture of Mo and W; gray, V; light green, mixture of Mo and V; dark red, Cu; red, O. Atoms constituting the heptagonal channel micropore are represented in ball-and-stick manner. Adapted with permission from Ref. [64]. Copyright 2021 American Chemical Society

Crystal structure clearly gives us the advantage of correctly accessing the real catalysis field. This situation may not only be valid for metal oxide catalysts but also for other material's state of catalysts. Therefore, it should be highly valuable to pick up crystalline solids as catalysts because there are a huge number of crystalline solids which are yet used as catalysts.

There is no doubt that technology development toward a carbon-neutral society will accelerate in the near future. It may be necessary to replace current industrial

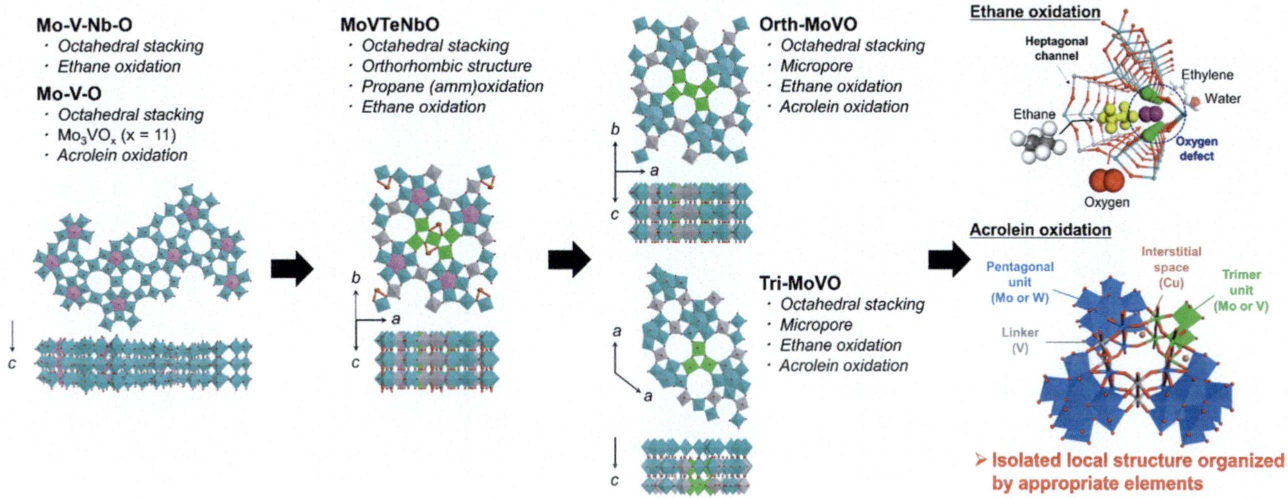

Fig. 4.9 Development of Mo-V-based complex metal oxide catalysts from the past to the present summarized based on the catalyst structure

processes with the ones realizing extremely high selectivity toward desired products with low environmental impacts. For this realization, the development of solid catalysts will play central roles.

References

1. Zhao Q. A review of pathways to carbon neutrality from renewable energy and carbon capture. E3S Web Conf. 2021;245.
2. Védrine JC. Metal oxides in heterogeneous oxidation catalysis: state of the art and challenges for a more sustainable world. Chemsuschem. 2019;12:577.
3. Védrine JC. Recent developments and prospectives of acid-base and redox catalytic processes by metal oxides. Appl Catal A Gen. 2019;575:170.
4. Toyao T, Maeno Z, Takakusagi S, Kamachi T, Takigawa I, Shimizu K-I. Machine learning for catalysis informatics: recent applications and prospects. ACS Catal. 2020;10:2260.
5. Grant JT, Venegas JM, McDermott WP, Hermans I. Aerobic oxidations of light alkanes over solid metal oxide catalysts. Chem Rev. 2018;118:2769.
6. Grasselli RK. Genesis of site isolation and phase cooperation in selective oxidation catalysis. Top Catal. 2001;15:93.
7. Grasselli RK. Fundamental principles of selective heterogeneous oxidation catalysis. Top Catal. 2002;21:79.
8. Izawa S, et al. Japan Patent 1966001775 (1966); assigned to Toyo Soda.
9. Wada M, et al. Japan Patent 1974117419 (1974); assigned to Nippon Shokubai.
10. Ushikubo T, et al. US Patent 5,472,925 (1995); assigned to Mitsubishi Chemical Corporation.
11. Ushikubo T, Oshima K, Kayou A, Vaarkamp M, Hatano M. Ammoxidation of propane over catalysts comprising mixed oxides of Mo and V. J Catal. 1997;169:394.
12. Tateno E, et al. EP Patent 3,513,876 B1 (2021); assigned to Asahi Kasei.
13. Ushikubo T, et al. US Patent 5,380,933 (1995); assigned to Mitsubishi Kasei Corporation.
14. Tu X, Niwa M, Arano A, Kimata Y, Okazaki E, Nomura S. Controlled silylation of MoVTeNb mixed oxide catalyst for the selective oxidation of propane to acrylic acid. Appl Catal A Gen. 2018;549:152.
15. Ueda W, Oshihara K. Selective oxidation of light alkanes over hydrothermally synthesized Mo-V-M-O (M=Al, Ga, Bi, Sb, and Te) oxide catalysts. Appl Catal A Gen. 2000;200:135.
16. Watanabe H, Koyasu Y. New synthesis route for Mo-V-Nb-Te mixed oxides catalyst for propane ammoxidation. Appl Catal A Gen. 2000;194–195:479.
17. Tsuji H, Koyasu Y. Synthesis of MoVNbTe(Sb)Ox composite oxide catalysts via reduction of polyoxometalates in an aqueous medium. J Am Chem Soc. 2002;124:5608.
18. Celaya Sanfiz A, Hansen TW, Girgsdies F, Timpe O, Rödel E, Ressler T, Trunschke A, Schlögl R. Preparation of phase-pure M1 MoVTeNb oxide catalysts by hydrothermal synthesis—influence of reaction parameters on structure and morphology. Top Catal. 2008;50:19.
19. DeSanto P, Buttrey DJ, Grasselli RK, Lugmair CG, Volpe AF, Toby BH, Vogt T. Structural characterization of the orthorhombic phase M1 in MoVNbTeO propane ammoxidation catalyst. Top Catal. 2003;23:23.
20. Murayama H, Vitry D, Ueda W, Fuchs G, Anne M, Dubois JL. Structure characterization of orthorhombic phase in MoVTeNbO catalyst by powder X-ray diffraction and XANES. Appl Catal A Gen. 2007;318:137.
21. Deniau B, Bergeret G, Jouguet B, Dubois JL, Millet JMM. Preparation of single M1 phase MoVTe(Sb)NbO catalyst: study of the effect of M2 phase dissolution on the structure and catalytic properties. Top Catal. 2008;50:33.
22. Amakawa K, Kolen'ko YV, Villa A, Schuster ME, Csepei L-I, Weinberg G, Wrabetz S, Naumann d'Alnoncourt R, Girgsdies F, Prati L, Schlögl R, Trunschke A. Multifunctionality of crystalline MoV(TeNb) M1 oxide catalysts in selective oxidation of propane and benzyl alcohol. ACS Catal. 2013;3:1103.

23. Melzer D, Xu P, Hartmann D, Zhu Y, Browning ND, Sanchez-Sanchez M, Lercher JA. Atomic-scale determination of active facets on the MoVTeNb Oxide M1 phase and their intrinsic catalytic activity for ethane oxidative dehydrogenation. Angew Chem Int Ed. 2016;55:8873.
24. Thorsteinson EM, Wilson TP, Young FG, Kasai PH. The oxidative dehydrogenation of ethane over catalysts containing mixed oxides of molybdenum and vanadium. J Catal. 1978;52:116.
25. Nieto JML, Botella P, Vázquez MI, Dejoz A. The selective oxidative dehydrogenation of ethane over hydrothermally synthesised MoVTeNb catalysts. Chem Commun. 2002;1906.
26. Botella P, García-González E, Dejoz A, Nieto JML, Vázquez MI, González-Calbet J. Selective oxidative dehydrogenation of ethane on MoVTeNbO mixed metal oxide catalysts. *J.* Catal. 2004;225:428.
27. Najari S, Saeidi S, Concepcion P, Dionysiou DD, Bhargava SK, Lee AF, Wilson K. Oxidative dehydrogenation of ethane: catalytic and mechanistic aspects and future trends. Chem Soc Rev. 2021;50:4564.
28. Zenkovets GA, Shutilov AA, Bondareva VM, Sobolev VI, Marchuk AS, Tsybulya SV, Prosvirin IP, Ishchenko AV, Gavrilov VY. New multicomponent $MoVSbNbCeO_x/SiO_2$ catalyst with enhanced catalytic activity for oxidative dehydrogenation of ethane to ethylene. ChemCatChem. 2020;12:4149.
29. Svintsitskiy DA, Kardash TY, Lazareva EV, Saraev AA, Derevyannikova EA, Vorokhta M, Šmíd B, Bondareva VM. NAP-XPS and in situ XRD study of the stability of Bi-modified MoVNbTeO catalysts for oxidative dehydrogenation of ethane. Appl Catal A Gen. 2019;579:141.
30. Lazareva EV, Bondareva VM, Svintsitskii DA, Kardash TY. Preparing MoVTeNbBiO catalysts for the selective oxidative conversion of light alkanes. Catal Ind. 2020;12:39.
31. Cheng MJ, Goddard WA. In silico design of highly selective Mo-V-Te-Nb-O mixed metal oxide catalysts for ammoxidation and oxidative dehydrogenation of propane and ethane. J Am Chem Soc. 2015;137:13224.
32. Zhu Y, Sushko PV, Melzer D, Jensen E, Kovarik L, Ophus C, Sanchez-Sanchez M, Lercher JA, Browning ND. Formation of oxygen radical sites on MoVNbTeOx by cooperative electron redistribution. J Am Chem Soc. 2017;139:12342.
33. Annamalai L, Liu Y, Ezenwa S, Dang Y, Suib SL, Deshlahra P. Influence of tight confinement on selective oxidative dehydrogenation of ethane on MoVTeNb mixed oxides. ACS Catal. 2018;8:7051.
34. Liu Y, Annamalai L, Deshlahra P. Effects of Lattice O atom coordination and pore confinement on selectivity limitations for ethane oxidative dehydrogenation catalyzed by vanadium-oxo species. J Phy Chem C. 2019;123:28168.
35. Andrushkevich TV. Heterogeneous catalytic oxidation of acrolein to acrylic acid: mechanism and catalysts. Catal Rev. 1993;35:213.
36. Kampe P, Giebeler L, Samuelis D, Kunert J, Drochner A, Haass F, Adams AH, Ott J, Endres S, Schimanke G, Buhrmester T, Martin M, Fuess H, Vogel H. Heterogeneously catalysed partial oxidation of acrolein to acrylic acid-structure, function and dynamics of the V-Mo-W mixed oxides. PhysChemChemPhys. 2007;9:3577.
37. Drochner A, Ohlig D, Knoche S, Gora N, Heid M, Menning N, Petzold T, Vogel H. Mechanistic studies on the transition metal oxide catalysed partial oxidation of (Meth)acrolein to the corresponding carboxylic acids. Top Catal. 2016;59:1518.
38. Shiju NR, Liang X, Weimer AW, Liang C, Dai S, Guliants VV. The role of surface basal planes of layered mixed metal oxides in selective transformation of lower alkanes: propane ammoxidation over surface ab planes of Mo-V-Te-Nb-O M1 phase. J Am Chem Soc. 2008;130:5850.
39. Katou T, Vitry D, Ueda W. Hydrothermal synthesis of a new Mo-V-O complex metal oxide and its catalytic activity for the oxidation of propane. Chem Lett. 2003;32:1028.
40. Sadakane M, Kodato K, Kuranishi T, Nodasaka Y, Sugawara K, Sakaguchi N, Nagai T, Matsui Y, Ueda W. Molybdenum-vanadium-based molecular sieves with microchannels of seven-membered rings of corner-sharing metal oxide octahedra. Angew Chem Int Ed. 2008;47:2493.
41. Chen C, Kosuke N, Murayama T, Ueda W. Single-crystalline-phase Mo_3VO_x: an efficient catalyst for the partial oxidation of acrolein to acrylic acid. ChemCatChem. 2013;5:2869.

42. Watanabe N, Ueda W. Comparative study on the catalytic performance of single-phase Mo−V−O-based metal oxide catalysts in propane ammoxidation to acrylonitrile. Ind Eng Chem Res. 2006;45:607.
43. Trunschke A, Noack J, Trojanov S, Girgsdies F, Lunkenbein T, Pfeifer V, Hävecker M, Kube P, Sprung C, Rosowski F, Schlögl R. The impact of the bulk structure on surface dynamics of complex Mo-V-based oxide catalysts. ACS Catal. 2017;7:3061.
44. Konya T, Katou T, Murayama T, Ishikawa S, Sadakane M, Buttrey D, Ueda W. An orthorhombic Mo_3VO_x catalyst most active for oxidative dehydrogenation of ethane among related complex metal oxides. Catal Sci Technol. 2013;3:380–7.
45. Sadakane M, Watanabe N, Katou T, Nodasaka Y, Ueda W. Crystalline Mo_3VO_x mixed-metal-oxide catalyst with trigonal symmetry. Angew Chem Int Ed. 2007;46:1493.
46. Ishikawa S, Ueda W. Microporous crystalline Mo-V mixed oxides for selective oxidations. Catal Sci Technol. 2016;6:617.
47. Sadakane M, Kodato K, Yasuda N, Ishikawa S, Ueda W. Thermal behavior, crystal structure, and solid-state transformation of orthorhombic Mo–V oxide under nitrogen flow or in air. ACS Omega. 2019;4:13165.
48. Ishikawa S, Yi X, Murayama T, Ueda W. Heptagonal channel micropore of orthorhombic Mo_3VO_x as catalysis field for the selective oxidation of ethane. Appl Catal A Gen. 2014;474:10.
49. Ishikawa S, Yi X, Murayama T, Ueda W. Catalysis field in orthorhombic Mo_3VO_x oxide catalyst for the selective oxidation of ethane, propane and acrolein. Catal Today. 2014;238:35.
50. Ishikawa S, Murayama T, Kumaki M, Tashiro M, Zhang Z, Yoshida A, Ueda W. Synthesis of trigonal Mo–V–M3rd–O (M3rd = Fe, W) catalysts by using structure-directing agent and catalytic performances for selective oxidation of ethane. Top Catal. 2016;59:1477.
51. Haushalter RC, Mundi LA. Reduced Molybdenum phosphates: octahedral-tetrahedral framework solids with tunnels, cages, and micropores. Chem Mater. 1992;4:31.
52. Maspoch D, Ruiz-Molina D, Veciana J. Old Materials with new tricks: multifunctional open-framework materials. Chem Soc Rev. 2007;36:770.
53. Nakazawa N, Ikeda T, Hiyoshi N, Yoshida Y, Han Q, Inagaki S, Kubota Y. A microporous aluminosilicate with 12-, 12-, and 8-ring pores and isolated 8-ring channels. J Am Chem Soc. 2017;139:7989.
54. Ishikawa S, Zhang Z, Ueda W. Unit synthesis approach for creating high dimensionally structured complex metal oxides as catalysts for selective oxidations. ACS Catal. 2018;8:2935.
55. Bhuvanesh NSP, Gopalakrishnan J. Solid-state chemistry of early transition-metal oxides containing d0 and d1 cations. J Mater Chem. 1997;7:2297.
56. Ok KM, Halasyamani PS, Casanova D, Llunell M, Alemany P, Alvarez S. Distortions in octahedrally coordinated d0 transition metal oxides: a continuous symmetry measures approach. Chem Mater. 2006;18:3176.
57. Lunkenbein T, Girgsdies F, Wernbacher A, Noack J, Auffermann G, Yasuhara A, Klein-Hoffmann A, Ueda W, Eichelbaum M, Trunschke A, Schlogl R, Willinger MG. Direct imaging of octahedral distortion in a complex molybdenum vanadium mixed oxide. Angew Chem Int Ed. 2015;54:6828.
58. Ishikawa S, Kobayashi D, Konya T, Ohmura S, Murayama T, Yasuda N, Sadakane M, Ueda W. Redox treatment of orthorhombic $Mo_{29}V_{11}O_{112}$ and relationships between crystal structure, microporosity and catalytic performance for selective oxidation of ethane. J Phys Chem C. 2015;119:7195.
59. Hibst H, et al. US Patent 5,807,531 (1998); assigned to BASF.
60. Sugi H, et al. US Patent 5,959,143 (1999); assigned to Nippon Kayaku.
61. Kawajiri T, et al. EP Patent 0792866 A1 (1997); assigned to Nippon Shokubai.
62. Ishikawa S, Yamada Y, Qiu C, Kawahara Y, Hiyoshi N, Yoshida A, Ueda W. Synthesis of a crystalline orthorhombic Mo-V-Cu oxide for selective oxidation of acrolein. Chem Mater. 2019;31:1408.
63. Tichý J. Oxidation of acrolein to acrylic acid over vanadium-molybdenum oxide catalysts. Appl Catal A Gen. 1997;157:363.

64. Ishikawa S, Yamada Y, Kashio N, Noda N, Shimoda K, Hayashi M, Murayama T, Ueda W. True catalytically active structure in Mo-V-based mixed oxide catalysts for selective oxidation of acrolein. ACS Catal. 2021;11:10294.
65. Broclawik E, Góra A, Najbar M. The role of tungsten in formation of active sites for no SCR on the V-W-O catalyst surface-quantum chemical modeling (DFT). J Mol Catal A Chem. 2001;166:31.
66. Inomata Y, Kubota H, Hata S, Kiyonaga E, Morita K, Yoshida K, Sakaguchi N, Toyao T, Shimizu K-I, Ishikawa S, Ueda W, Haruta M, Murayama T. Bulk tungsten-substituted vanadium oxide for low-temperature NO_x removal in the presence of water. Nat Commun. 2021;12:557.

Chapter 5
All-Inorganic Zeolitic Octahedral Metal Oxides

Zhenxin Zhang and Wataru Ueda

5.1 General Introduction of Zeolitic Octahedral Metal Oxides

Zeolites are crystalline aluminosilicates that have porous structures. The frameworks of zeolites are composed of metal–oxygen tetrahedral [1–4]. Zeolites have been applied to many fields, such as adsorption, ion exchange, separation, and catalysis. The chemical composition of zeolite is based on main group elements majorly, which limits the application fields of the materials. Therefore, expanding the chemical composition of zeolites from main group elements to transition metal elements that show unique properties, such as multi-electron transfer property, electrochemical property, photochemical property, and magnetic property, is desired. There are many attempts for incorporating transition metal elements in zeolites. Zeolitic aluminates are the typical example for this purpose [5–7], where transition metal ions can be incorporated. However, frameworks of the zeolitic aluminates are always not stable, which leads to that the micropores of the materials cannot be opened. The frameworks would be damaged when the structure directing agents inside the micropore are removed. Therefore, the applications of the materials are limited. The limited content of the transition metal elements in the material and the weak frameworks are the main shortcomings that are needed to be overcome.

Synthesis of zeolitic inorganic networks based on mainly transition metal oxide is difficult, because the coordination number of transition metal element is always

Z. Zhang (✉)
School of Materials Science and Chemical Engineering, Ningbo University, Fenghua Road 818, Ningbo, Zhejiang 315211, P. R. China
e-mail: zhangzhenxin@nbu.edu.cn

W. Ueda
Faculty of Engineering, Kanagawa University, Kanagawa-ku, Yokohama-shi, Rokkakubashi 221-8686, Japan
e-mail: uedaw@kanagawa-u.ac.jp

W. Ueda (ed.), *Crystalline Metal Oxide Catalysts*,
https://doi.org/10.1007/978-981-19-5013-1_5

higher than four, majorly six or more. However, the assembly of transition metal–oxygen octahedra to form molecular clusters is easy, and the resulting metal–oxygen clusters are called polyoxometalates (POMs) [8–10]. Introducing organic linkers with connecting metal–oxygen clusters produces porous materials based on transition metal–oxygen octahedra, which are called metal–organic frameworks (MOFs) [11–14]. MOFs show good structural and elemental diversity. MOF materials have been successfully applied to the fields, such as adsorption and separation. The success in the synthesis of MOFs demonstrates that metal–oxygen clusters are good building blocks for porous materials based on transition metal elements.

Furthermore, POM-based porous materials are desired to be synthesized for many years. In the early stage, Cs salt of POM is found to be microporous, which can adsorb small molecules [15]. The micropore of the material is not intrinsic and only derived from particle aggregation. There are other examples of POM-based porous frameworks. However, the pores of the materials always cannot be opened and the materials do not show adsorption properties [16–24].

Assembly of POM units with inorganic ions forms the crystalline microporous transition metal oxide oxides, and the materials are composed of metal–oxygen octahedra, which are called zeolitic octahedral metal oxides (ZOMOs). Different from other porous materials based on POMs, ZOMOs show a fully inorganic composition of the frameworks as well as ordered microporosity. Furthermore, the frameworks of the materials are stable enough for the removal of the guest molecules of the materials to open the micropores. Combined with the microporosity and the redox property, ZOMOs show good application potential in many fields, including adsorption, separation, ion exchange, battery, and catalysis. There are four different ZOMOs found in recent years, the pentagon unit-based vanadomolybdate, the ε-Keggin POM-based molybdates, hexagon unit-based molybdates and tungstates, and cubane unit-based vanadotungstates.

5.2 Material Synthesis

5.2.1 Conditions for Material Synthesis

ZOMOs are synthesized by a hydrothermal method in most cases, the precursor solutions of which contain simply metalates for constructing the framework, reducing agent, and cation species. The materials are formed by self-assembly of the precursor compounds under heating conditions. There are many factors that affected the formation of ZOMOs, such as temperature, time, concentration, ratio of the precursor materials, types of the precursor materials, acidity of the solution, and cation species.

Temperature, time, and concentration are important factors and show similar influences on ZOMO formation. Different ZOMOs are obtained under different temperatures. ε-Keggin-based ZOMOs are able to be obtained at the temperature range from 150 to 230 °C [25]. For hexagon unit-based ZOMOs, the materials are able to be obtained from room temperature to 175 °C [26]. Based on the current material syntheses, if the ZOMO materials can be prepared at a lower temperature the material also can be obtained at a higher temperature. Sometimes the syntheses at a higher temperature would produce impurity phases in the products [25–27]. The effect of time and concentration is similar to that of temperature. When the other conditions are suitable for one exact ZOMO material, increasing reaction time, concentration of the precursor solution, and the reaction time tends to obtain the material with high yield. Decreasing reaction time, concentration of the precursor solution, and the reaction time decreases the yield of the material. The factors of temperature, time, and concentration majorly affect the rate of the formation of the materials.

The more important factors are acidity of precursor solution, cation species, and types of starting material for obtaining ZOMOs, which not only affect the rate of formation of the materials, but also affect the crystalline phases of the materials obtained in the syntheses.

The effect of acidity of precursor solution has a great influence on the resulting ZOMOs. Generally, the ZOMO materials can be generated in acidic conditions. Different ZOMOs are obtained at slightly different pH of the precursor solutions. Increasing acidity of the precursor solutions causes speeding up condensation of metal–oxygen bond and increases the rate of formation of ZOMOs, so that it increases formation speed and yield of the products [25–27]. Furthermore, low acidity produces ZOMOs with high porosity. For zeolitic vanadomolybdates, the highly acidic condition of the precursor solution generates trigonal vanadomolybdates (t-VM) with low porosity (Table 5.1, entry 1). Decreasing the acidity of the precursor solution generates the orthorhombic vanadomolybdates (o-VM) that have the same chemical composition as t-VM. The porosity of o-VM is higher than that of t-VM (Table 5.1, entry 2) [28, 29]. Other different ZOMOs prepared at low acidic conditions, such as ε-Keggin POM-based ZOMOs (zeolitic bismuth vanadomolybdates, $(NH_4)_{2.8}H_{0.9}[\varepsilon\text{-}V^{V}{}_{1.0}Mo^{V}{}_{2.3}Mo^{VI}{}_{7.1}V^{IV}{}_{1.8}V^{V}{}_{0.8}O_{40}Bi^{III}{}_{2}]$, ε-BVM) and cubane unit-based ZOMOs (zeolitic vanadotungstate, $(NH_4)_{0.25}K_{1.5}H_{0.25}[W^{VI}{}_{4}V^{IV}{}_{3}O_{19}]$, VT-1), show high porosity compared with VMs (Table 5.1, entries 3,4). Furthermore, when only the pH of the precursor solution is changed with keeping other conditions the same, the effect of the pH of the starting reaction mixture is important for obtaining different crystalline phases. Hydrothermal reactions of reaction mixtures with pH values between 1.6 and 6.0 produces the desired ε-Keggin-based Mo oxides ($NaH_9[\varepsilon\text{-}Co^{II}Mo^{V}{}_{8}Mo^{VI}{}_{4}O_{40}Co^{II}{}_{2}]$, ε-Na-CM). The main side products are MoO_2 and $CoMoO_4{\cdot}0.75H_2O$ at lower pH values (1.6–4.2) and at higher pH values (4.8–6.0), respectively. When pH is less than 1.3, only MoO_2 is detected (Fig. 5.1) [30].

Cation species are important for preparing ZOMOs. Cation occupies micropores of ZOMOs, which affects the structure of ZOMOs. Cation species also affect the properties and applications of the materials, which will be discussed in the later part. Cation of the materials has an influence on the iso-structural zeolitic vanadotungstate formation [31], the material formed the cubic phase (VT-1) when K^+ is used as the cation, while using organic ammonium cation, trimethyl amine, only produces the

Table 5.1 Acidity and micropore volume of the materials

Entry	Material	pH of precursor	Micropore volume (cm^3/g)	References
1	t-VM	2.2	0.0048	[31]
2	o-VM	3.4	0.0122	[32]
3	ε-BVM	3.4–5	0.0202	[33]
4	VT-1	5	0.102	[34]

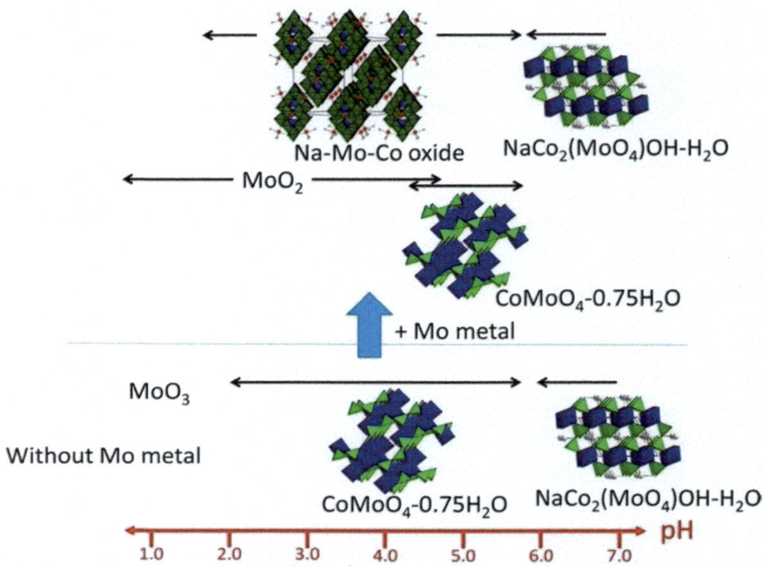

Fig. 5.1 Relationship between pH and products of hydrothermal reactions of aqueous solutions of Na_2MoO_4 and $CoCl_2$ at 170 °C for 24 h with and without Mo, Reprinted with permission from Ref. [30]. Copyright 2017 American Chemical Society

trigonal phase ($[N(CH_3)_3]_{1.4}H_2[W^{VI}{}_4V^{IV}{}_3O_{19}]$, VT-5) with almost the same conditions of other factors. Furthermore, not using proper cations make the synthesis for ZOMOs unsuccessful. In the case of o-VM, NH_4^+ is critical for obtaining the material. When Na^+ and K^+ are used for cation, no solids are obtained. Cations act as structural directing agents that help with forming different structure topologies of the materials [31].

Starting materials are important for obtaining ZOMOs. There are three factors for starting materials: valency, physical properties, and chemical properties. Valency of the starting chemicals is critical. In most cases of the ZOMO synthesis, the syntheses need a reduced state of the metal elements. As summarized in Table 5.2, the reduced state of the elements is essential to obtain the material. For ε-Keggin POM-based ZOMOs, molybdates are reduced by Mo to form ε-Keggin POM (Table 5.2). For transition metal oxide molecule wires, SeO_2 and TeO_2 with the reduced state are used.

Table 5.2 Starting materials for ZOMOs synthesis

	Metal I (valence)	Reduced Metal II (valence)	Metal III	References
t-VM	AHM (Mo^{VI})	$VOSO_4$ (V^{IV})	–	[29]
o-VM	AHM (Mo^{VI})	$VOSO_4$ (V^{IV})	–	[28]
ε-ZM	AHM (Mo^{VI})	Mo (Mo^0)	$ZnCl_2$ (Zn^{II})	[27]
h-TT	AMT (W^{VI})	TeO_2 (Te^{IV})	–	[35]
VT-1	WO_3 (W^{VI})	$VOSO_4$ (V^{IV})	–	[34]
VT-5	WO_3 (W^{VI})	$VOSO_4$ (V^{IV})	–	[34]

For cubane unit-based ZOMOs and pentagon unit-based vanadomolybdates, $VOSO_4$ is used for synthesis. The physical state of the material is also important for obtaining the material with different crystallinity. Starting materials are important for obtaining ZOMOs. Types of the starting materials not only affect the resulting ZOMOs formation, but also affect the size of the resulting materials. Using $Bi(OH)_3$ as the insoluble Bi source for ε-BVM formation produced well-crystalline ε-BVM with the size in the micrometer size. Furthermore, using soluble Bi source of $Bi(NO_3)_3$ nanocrystals of ε-BVM are able to be obtained with the size of 24 nm. The size of the material is also observed by transmission electron microscopy (TEM), showing that the size decreased from micrometer level to nanometer level [25].

5.2.2 *Design of Framework*

ZOMO materials show potential structural diversity. However, the current structure topologies of the synthesized ZOMOs are not many. Compared with MOFs and zeolites, the structural diversity of transition metal oxides is currently poor. The reason might be the highly coordinated metal ion center of transition metal oxides compared with the lowly coordinated (tetrahedrally coordinated) metal center of zeolites. The building blocks of MOFs are based on transition metal–oxygen octahedra. However, MOFs have good structural diversity, which is due to the assembly of the metal–oxygen cluster as building blocks with organic ligands as linkers. Therefore, using the assembly of POM with inorganic linkers to develop structural diversity of ZOMOs is promising, because structures of POM molecules are diverse and the connection of the POM is many and designable, which can be utilized for creating ZOMOs.

The cubane unit of $[W_4O_{16}]^{8-}$-based ZOMO materials show good structural diversity, and the framework can be extended with different connection styles of the cubane unit and linker. Ten hypothetical structures are generated based on the building unit of $[W_4O_{16}]^{8-}$, denoted as VT-1 to VT-10, respectively (Fig. 5.2). The hypothetical isomeric structures are optimized by density functional theory (DFT) calculation. The resulting energies and densities of the structures show that the energy of VT-1 is the lowest and the density of VT-1 is the highest among all the VT members, and thus

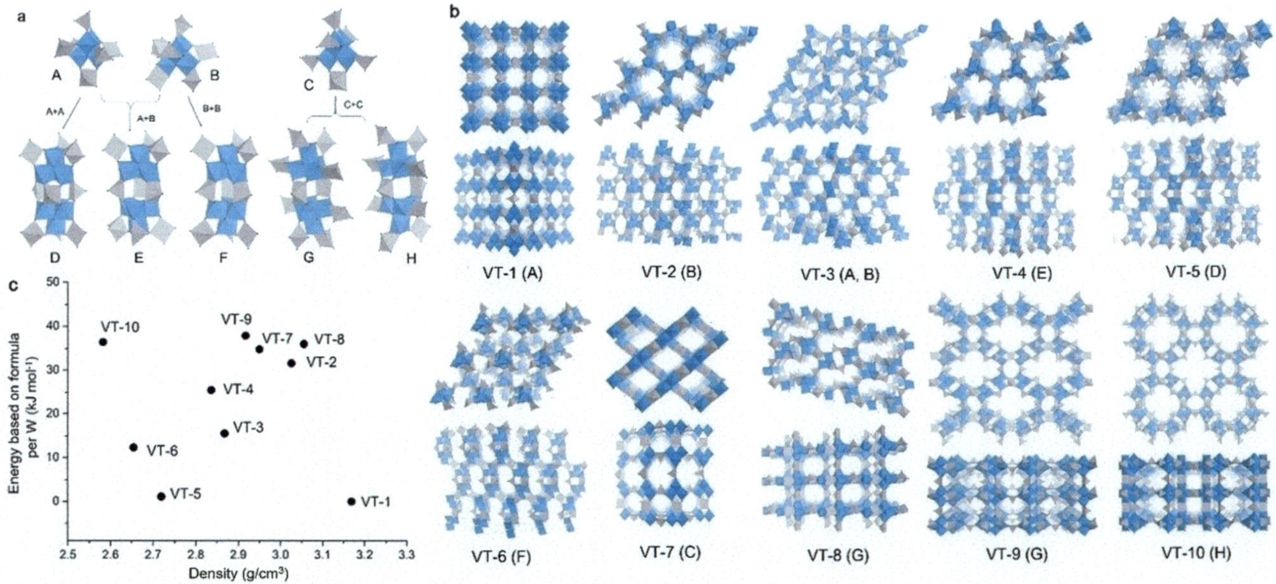

Fig. 5.2 Models of the building units and corresponding hypothetical structures. **a** Three preliminary building units (top) and five secondary building units (bottom); **b** frameworks of VT family constructed by the units from the front view (top) and side view (bottom); W (blue), V (gray), O (red); and **c** energy–density map for the proposed structures

VT-1 is the thermodynamically most stable structure among VTs (Fig. 5.2c). The energy of VT-5 is just 1.1 kJ mol^{-1} higher than that of VT-1, which demonstrated that VT-5 is the second most stable one.

5.3 Advanced Structure Determination, Step by Step

Structure determination of new crystalline materials is important because the properties and application performance of the materials are highly dependent on the structure of the materials. Furthermore, for tuning the properties and modifying the application performance of the materials, understanding the structural information of the material is also necessary. The current advanced structure determination strategies and methods are applicable to zeolitic octahedral metal oxides, but the materials show different features compared with other materials. Therefore, some slight modifications of the current structural determination are necessary for ZOMOs.

5.3.1 *Structure Determination Using Single Crystal X-ray Analysis*

Hydrothermal reaction of $(NH_4)_6Mo_7O_{24}\cdot 4H_2O$, $VOSO_4\cdot 5H_2O$, and $Bi(OH)_3$ produced crystalline ε-BVM, the size of which was less than 1 μm in one dimension, which were too small to perform single crystal structure analysis (Fig. 5.3) [32]. Large crystal of ε-BVM was prepared by the crystal growth experiments using the as-synthesized material as a seed in the reaction mixture. The repeating crystal growth produced large crystals of ε-BVM (~5 μm in one diameter), which was enough for single crystal analysis (Fig. 5.3).

The obtained crystals by the crystal growth experiments were still too small for the diffractometer in the laboratory system, and therefore, data collection was performed on a high-precision diffractometer installed in the SPring-8 BL40XU beamline. The synchrotron radiation emitted from the helical undulator was monochromated by using a Si(111) channel cut monochromator and focused with a Fresnel zone plate. A Rigaku Saturn724 CCD detector was used. The measurement was performed at 100 (2) K. An empirical absorption correction based on the Fourier series approximation was applied. The data were corrected for Lorentz and polarization effects. The structure was solved by direct methods and refined by full-matrix least-squares (SHELX-97), where the unweighted and weighted agreement factors of $R = \Sigma // Fo| - |Fc // / \Sigma |Fo|$ $(I > 2.00\sigma (I))$ and $wR = [\Sigma w(F_{o2} - F_{c2})^2/\Sigma w(F_{o2})^2]^{1/2}$, respectively, were used. Oxygen atoms of water in Mo − V − Bi oxide were refined isotropically, and other atoms were refined anisotropically. Total amounts of water and ammonium cations estimated by the elemental analysis were slightly larger than those obtained by single crystal structure analysis. This is because of the difference in the crystal

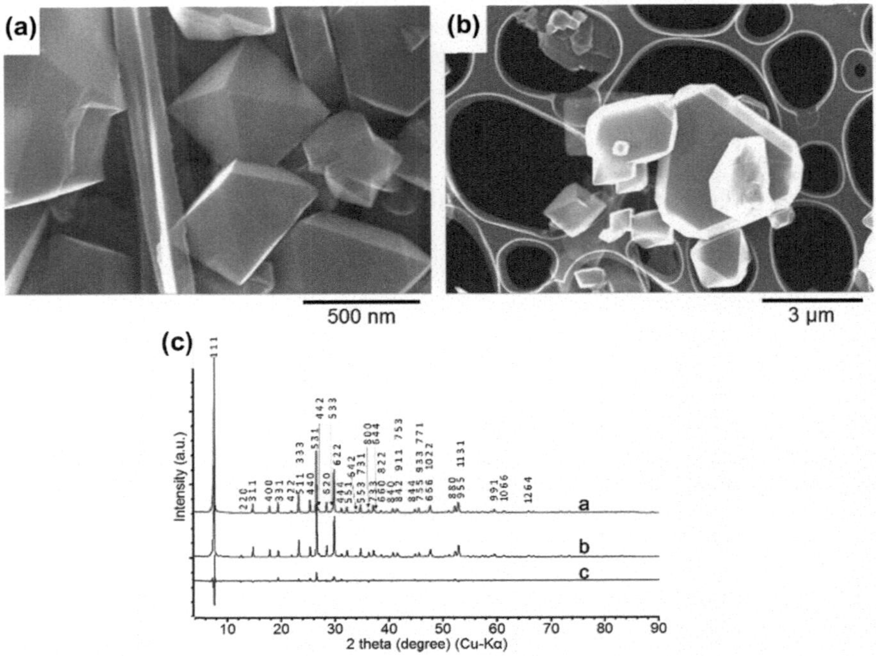

Fig. 5.3 SEM images of **a** ε-BVM, **b** large ε-BVM crystal for single crystal analysis, and **c** Comparison of a) the experimental XRD pattern with b) simulated pattern using structural data obtained by single crystal structure analysis with lattice parameter refinement (a = 19.79 Å, R_{wp} = 10.49%), and c) difference of experimental pattern and simulated pattern, Reprinted with permission from Ref. [32]. Copyright 2014 American Chemical Society

sample and bulk sample. The sample for elemental analysis may contain surface waters.

The structure of the powder sample was not completely the same as that of the material for single crystal X-ray analysis. The Rietveld refinement was used for confirming the structure of the powder sample. The initial structure of ε-BVM was refined by powder XRD Rietveld refinement. The pattern parameters and lattice parameters of the materials were refined by the Pawley method. Then, isotropic temperature factors were given for every atom in the initial structure of the material. Rietveld analysis was initiated with the initial models of the materials, and the lattice parameters and pattern parameters were from the Pawley refinement. Every atom position was refined. The occupancy of the atoms in the framework was fixed without further refinement, and the occupancies of the atoms in water and the cations were refined with consideration of the elemental analysis results. Finally, the pattern parameters were refined again to obtain the lowest R_{wp} value (Fig. 5.3). The powder X-ray diffraction pattern of ε-BVM was similar to the simulated pattern obtained by using crystal data from single crystal structure analysis. Furthermore, there were no additional peaks in the experimental data, indicating that the powder sample of ε-BVM was pure.

For different materials, the crystal growth method would be different. For $(NH_4)_2[Te^{IV}W^{VI}{}_6O_{21}]$ (h-TT), the large single crystal was obtained by low-temperature crystallization [33]. h-TT synthesized at low temperature is larger than that synthesized by the regular hydrothermal method (Fig. 5.4). XRD and FTIR confirmed that the basic structure of the material at low temperature was the same as that of the material obtained from hydrothermal synthesis. Single crystal analysis was performed using the same process of ε-BVM. Furthermore, the structure of the powder sample was confirmed by the Rietveld refinement, which showed that simulated patterns of the materials were similar to those of the experimental data, indicating that the proposed structures were correct (Fig. 5.4).

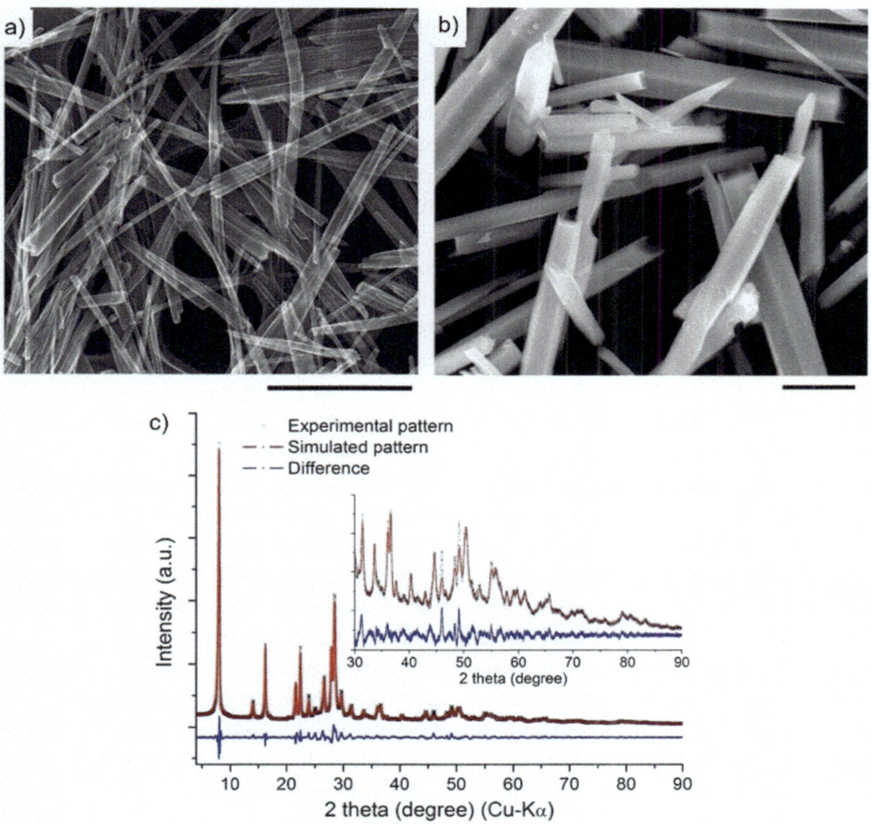

Fig. 5.4 SEM images, **a** h-TM, scale bar = 3 μm, **b** h-TM synthesized under refrigeration, scale bar = 5 μm, and **c** comparison of the experimental XRD patterns with the simulated XRD patterns using the Rietveld method

5.3.2 Structure Determination Using Powder X-ray Diffraction Analysis

The crystals of ZOMOs were always small by hydrothermal reaction, which is not suitable for single crystal analysis. The crystal growth process is not applicable to all the ZOMOs we synthesized, and thus other methods for structure determination are applied for structure determination of ZOMOs.

Crystal structure determination using powder X-ray analysis is good for solving the structure of ZOMO, particularly the material with high symmetry. The initial structure of $Na_{1.5}H_{11.4}[\varepsilon\text{-}Zn^{II}Mo^{V}{}_{10.9}Mo^{VI}{}_{1.1}O_{40}\{Zn^{II}\}_2]$·(ε-Na-ZM) was determined by powder XRD [27]. The powder XRD patterns for structural analysis were obtained on a laboratory powder X-ray diffraction apparatus. First, the powder XRD pattern was indexed by the DICVOL06 [34] and X-cell programs [35]. After performing Pawley refinement, the most reasonable space group was obtained. Then, the Le Bail method [36] was applied for intensity extraction with the EdPCR program. The initial structure was solved by a charge-flipping algorithm [37]. The positions and types of atoms were obtained by analyzing the generated electron density maps (Fig. 5.5). The framework oxygen atoms and cations that could not be found by the charge-flipping algorithm were added logically. For ε-Na-ZM, the results obtained by using the charge-flipping algorithm reveal the three most intensive peaks of the electron density map with the intensity order of surrounding metal sites > central metal site ~ linking metal site (Fig. 5.5 a-c). Elemental analysis of ε-Na-ZM reveals that the ratio of Mo: Zn is 12: 3. These results indicate that Zn is present in the central and linking metal sites and that Mo is present in the surrounding metal sites. Other sites are assigned to be oxygen atoms of the Keggin-unit, counter cations, and oxygen atoms of water. The initial structure of ε-Na-ZM was refined by Rietveld refinement, which showed that the simulated pattern was the same as the experimental one (Fig. 5.5).

For $(NH_4)_{0.4}[CoII_{1.3}(OH)Se^{IV}W^{VI}{}_6O_{21}]$ (h-CST), the structure was determined from powder XRD [38]. The powder diffraction peaks of h-CST were indexed, and the hexagonal system with *P*6 space group was obtained. The charge flipping algorithm indicated the arrangement of the heavy metal ions in the building blocks of the material. There were three sites in the material with high electron densities with the ratio of the linker site: the surrounding site: the central site = 3: 6: 1 in the unit cell (Fig. 5.6). The initial structure was refined by the Rietveld refinement. The simulated XRD pattern of the material matched the experimental pattern. No additional peaks were observed from the experimental data when compared with the simulated pattern, indicating that the powder sample of h-CST was pure (Fig. 5.6).

The XRD pattern of VT-1 was indexed to a cubic cell with lattice parameters of 17.1101 Å and a space group of PA-3 (Fig. 5.7). The initial structure was solved by the charge-flipping algorithm. The arrangement of W formed $[W_4O_{16}]^{8-}$ cubane cluster. The $[W_4O_{16}]^{8-}$ units were proposed to be connected by VO^{2+} linkers to form

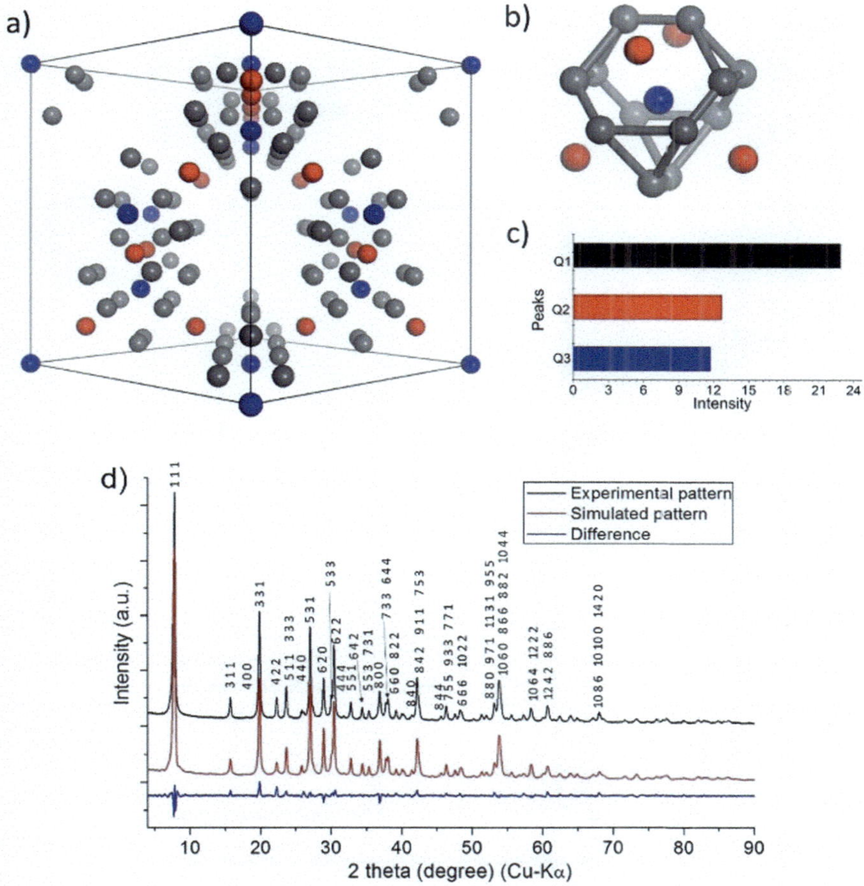

Fig. 5.5 Schematic representations of charge-flipping algorithm results: **a** electron density map from the charge-flipping method showing positions of the intensive peaks (surrounding metal sites (Q1; gray), linking metal sites (Q2; red), and central metal sites (Q3; blue)) in a unit cell, **b** an ε-Keggin unit with 4 linking metal sites, c) intensity difference of the peaks of ε-Na-ZM, and d) comparison of experimental XRD patterns with simulated XRD patterns using the Rietveld method, Mo–Zn oxide, $R_{wp} = 7.10\%$, Reprinted with permission from Ref. [27]. Copyright 2014 American Chemical Society

a 3D framework. The simulated pattern fit well with the experimental pattern after Rietveld refinement, indicating that the proposed structure was correct and that there were no obvious crystalline impurities (Fig. 5.7).

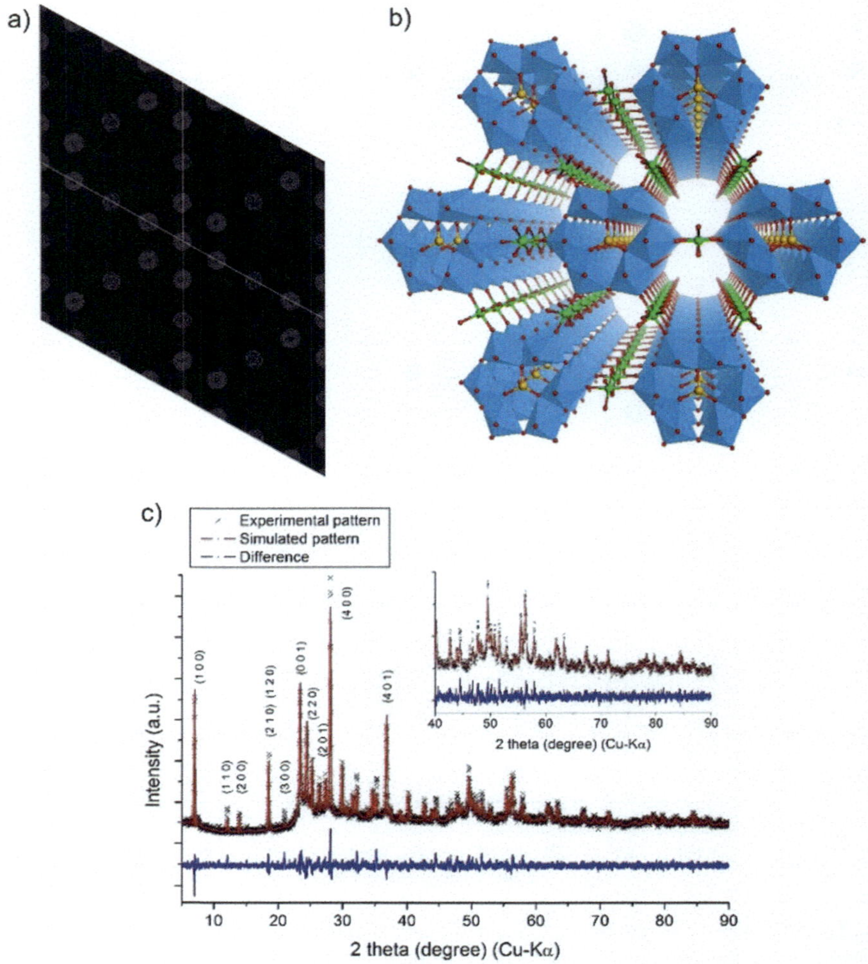

Fig. 5.6 **a** Electron density map from charge-flipping algorithm, **b** structure of h-CST, and **c** comparison of experimental XRD patterns with simulated XRD patterns using the Rietveld method

5.3.3 Structure Confirmation Using Atomic Resolution Electron Microscopy

The obtained initial structures of the materials were confirmed by atomic resolution electron microscopy. The High-angle annular dark-field scanning transmission electron microscopy (HAADF-STEM) images were obtained using an ARM-200F electron microscope (JEOL, Japan) operated at 200 kV with a CEOS probe aberration corrector. The probe convergence semi-angle was 14 mrad and the collection

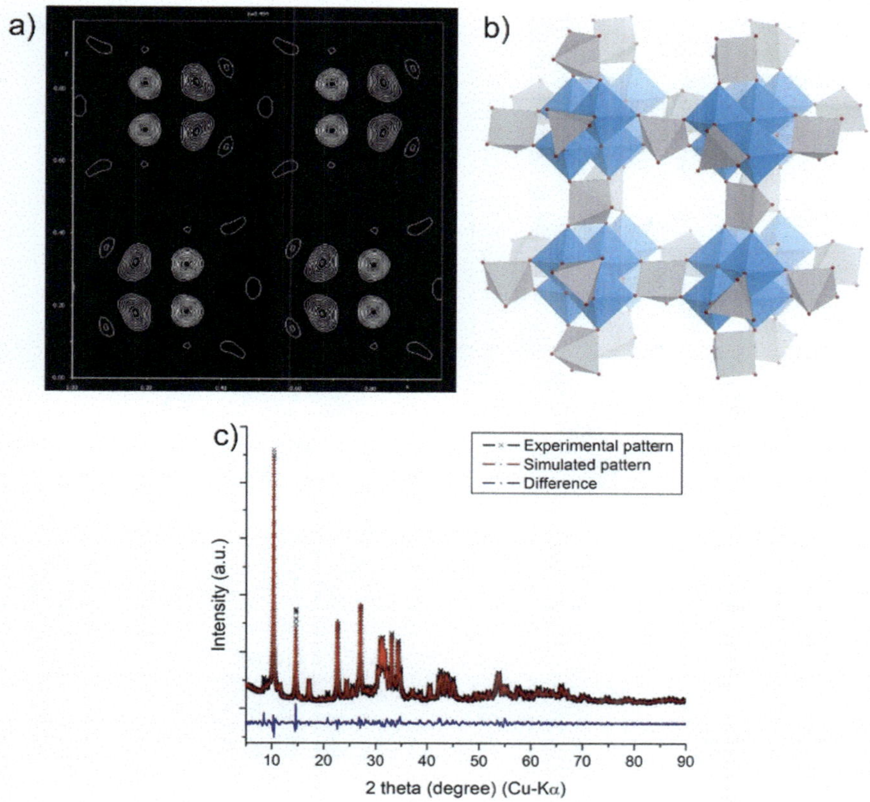

Fig. 5.7 **a** Electron density map from charge-flipping algorithm, **b** structure of VT-1, and **c** comparison of experimental XRD patterns with simulated XRD patterns using the Rietveld method

angle of the HAADF detector was 54–175 mrad. Obtained images were treated with Local 2D Wiener Filter in the HREM-Filters Pro software (HREM Research Inc., Japan) for noise removal. STEM samples for side views of molecular wire were prepared through the deposition of the wire on carbon-coated cupper grids from the aqueous suspensions. The wire was dispersed in water for a short time (~5 min). Cross-sectional STEM observation was performed using thin sections of the resin-embedded molecular wires sliced with an ultramicrotome.

The arrangement of the $[W_4O_{16}]^{8-}$ cubane units in VT-1 was observed by HAADF-STEM (Fig. 5.8). The lattice parameters determined from HAADF-STEM by directly measuring the distance were the same as the lattice parameters obtained from powder XRD. The ordered arrangement of 4 intense spots was observed in the (100) plane, attributing to the $[W_4O_{16}]^{8-}$ units. The periodic packing of the $[W_4O_{16}]^{8-}$ units in the (100) and (110) planes was identical to the proposed structure (Fig. 5.8).

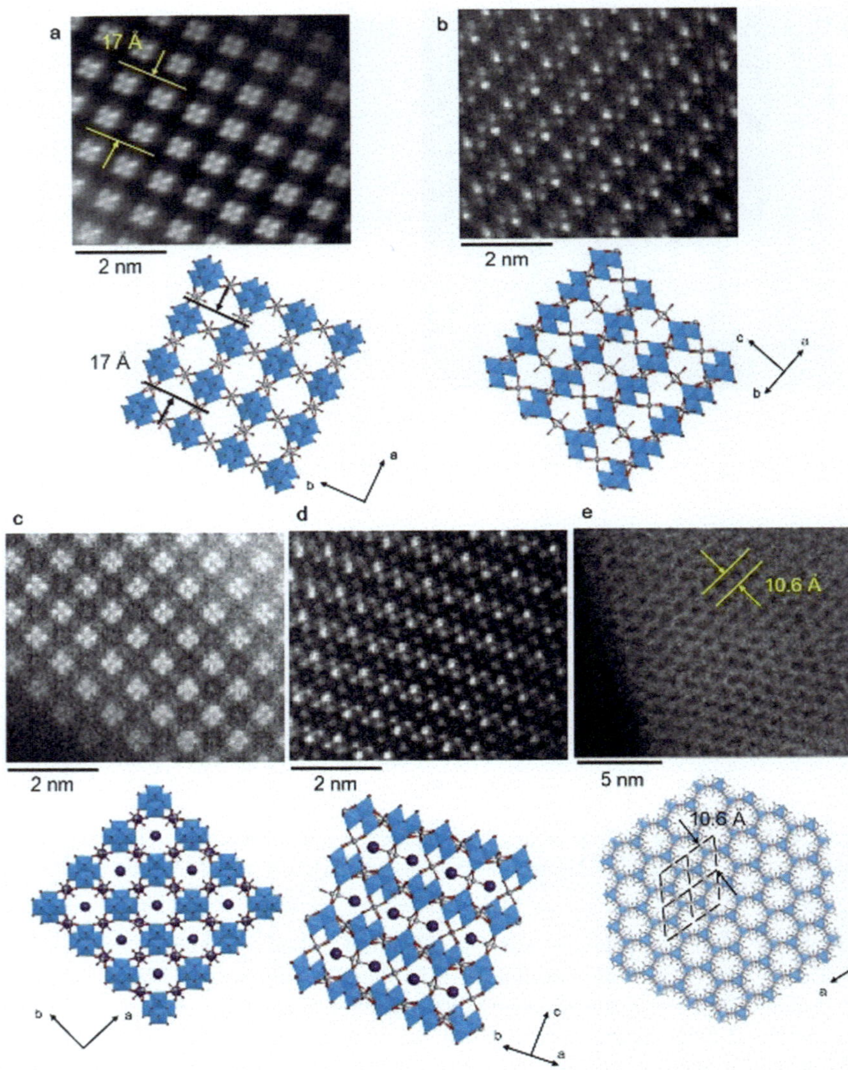

Fig. 5.8 HAADF-STEM images (top) and corresponding structure models (bottom). **a VT-1** in the (1 0 0) plane, **b** VT-1 in the (1 1 0) plane, **c** VT-5 in the (0 0 1) plane, scale bar: 5 nm, **d** Cs-VT-1 in the (1 0 0) plane, and **e** Cs-VT-1 in the (1 1 0) plane; W (blue), V (gray), O (red), and Cs (purple)

5.4 Typical Zeolitic Octahedral Metal Oxides

5.4.1 ε-Keggin POM-Based ZOMOs

The structure of ε-Keggin POM-based ZOMOs is confirmed by single crystal analysis, powder X-ray diffraction (XRD) Rietveld refinement, and high resolution high-angle annular dark-field scanning transmission electron microscopy (HAADF-STEM).

Among the iso-structural materials based on ε-Keggin POM, only the structure of $(NH_4)_{2.8}H_{0.9}[\varepsilon\text{-}V^{V}{}_{1.0}Mo^{V}{}_{2.3}Mo^{VI}{}_{7.1}V^{IV}{}_{1.8}V^{V}{}_{0.8}O_{40}Bi^{III}{}_2]$ (ε-BVM) oxide is confirmed by single crystal X-ray analysis. The crystal size of ε-BVM is the largest among the iso-structural materials based on ε-Keggin POM. The building block of ε-BVM is an ε-Keggin-type polyoxovanadomolybdate that is formed by one central VO_4 tetrahedron surrounded by twelve MO_6 (M = Mo or V) octahedra (Fig. 5.9). The tetrahedral VO_4 is surrounded by four edge-sharing M_3O_{13} (M = Mo or V) units to form the ε-Keggin POM [32].

The cavity of the material is comprised of ten ε-Keggin POM building blocks that are connected by MO_6 octahedra (Fig. 5.9) with an internal diameter of ca. 7.7 Å. The cavity is connected with each other by four pore openings with the size of ca. 3.4 Å, which constructs a periodical 3D pore system in a tetrahedral fashion. In one direction, the tunnel of the micropore is not linear but in a zigzag-like fashion.

For other iso-structural materials, the structures of the materials are confirmed by the Rietveld refinement. The simulated powder XRD patterns are in good agreement with the experimental patterns, which indicates that the structures of the materials are the same as that of ε-BVM [32]. The structures of ε-BVM and $NaH_9[\varepsilon\text{-}Co^{II}Mo^{V}{}_8Mo^{VI}{}_4O_{40}Co^{II}{}_2]$ (ε-CM) are observed by HAADF-STEM (Fig. 5.10). The (110) plane of the material shows that the signal of the HAADF-STEM image is in good agreement with the proposed structure (Fig. 5.10). When Bi is used as the linker, the intensity of the linker site is high, whereas when Co is used as the linker, the intensity of the linker site is low. The micropore of the material is also observed in HAADF-STEM [30].

The ε-Keggin POM-based ZOMOs show high elemental diversity. The surrounding site of the materials is able to incorporate Mo, V, and W, and the linker site of the materials is able to incorporate Mn, Fe, Co, Ni, Cu, Zn, and Bi. The cation of the materials is replaced by an alkali metal ion and an alkali earth metal ion [27].

5.4.2 Hexagon Unit-Based ZOMOs

The hexagon unit-based transition metal oxide molecular wires are firstly synthesized by the hydrothermal method using $(NH_4)_6Mo_7O_{24}\cdot 4H_2O$ (AHM) and Te and Se sources [33]. The large single crystal of $(NH_4)_2[Te^{IV}Mo^{VI}{}_6O_{21}]$ (h-TM) suitable for X-ray single crystal analysis is prepared by crystallization at room temperature.

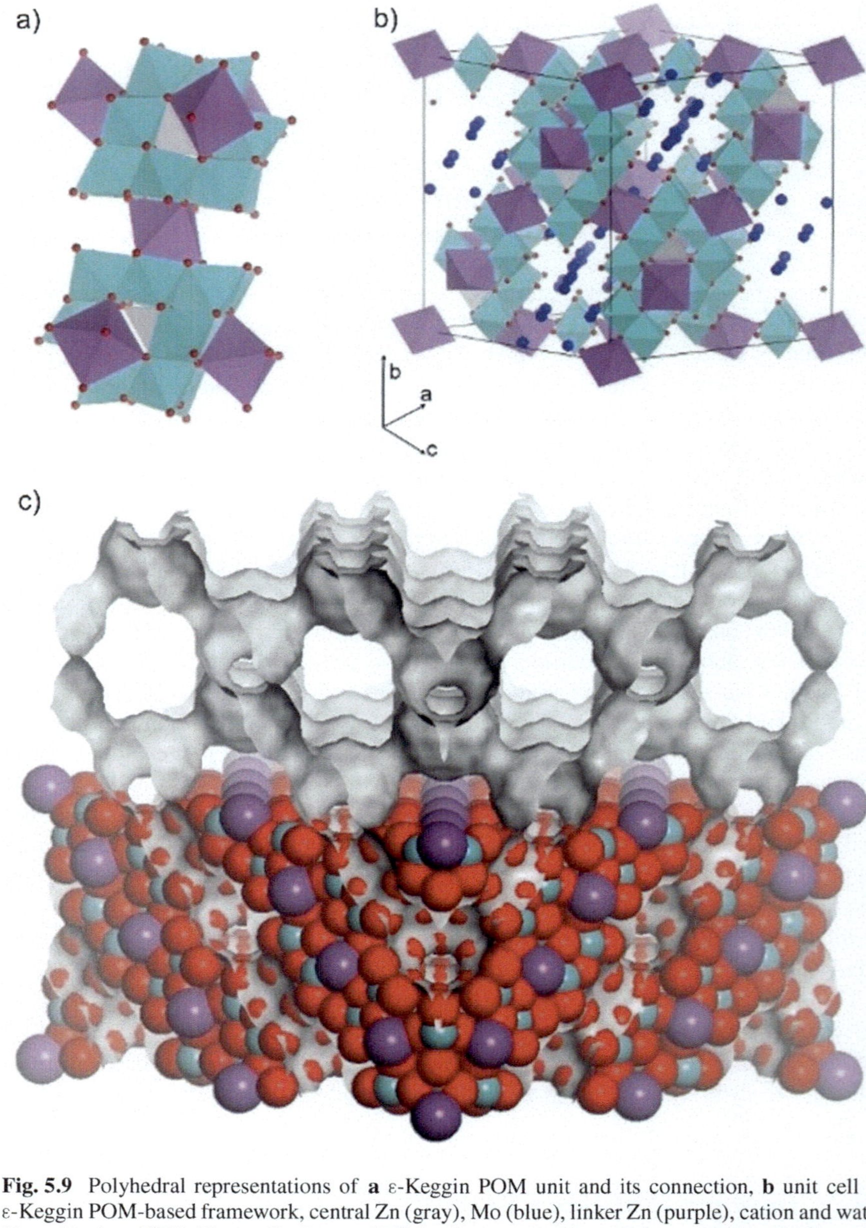

Fig. 5.9 Polyhedral representations of **a** ε-Keggin POM unit and its connection, **b** unit cell of ε-Keggin POM-based framework, central Zn (gray), Mo (blue), linker Zn (purple), cation and water (deep blue), **c** CPK (Corey, Pauling, and Koltun) representation of the material with Connelly surface (gray curved surface) that shows the micropore system of the material, Reproduced from Ref. [42] with permission from the Royal Society of Chemistry

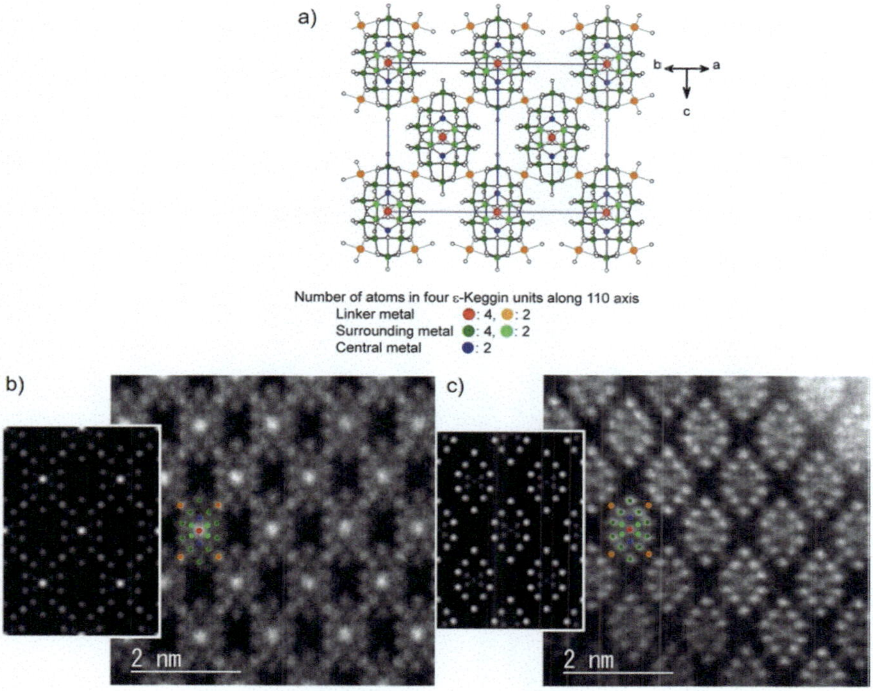

Fig. 5.10 **a** Ball-and-stick representation of ε-CM along the [110] axis, Mo (dark and light green), linker Co (red and orange), and central Co (blue). Numbers of overlapping dark green balls, light green balls, red balls, orange balls, and blue balls along the [110] axis are 4, 2, 4, 2, and 2, respectively. HAADF-STEM images of **b** ε-BVM and **c** ε-CM. (Insets) Simulated images, Reprinted with permission from Ref. [30]. Copyright 2017 American Chemical Society

The crystal structure from X-ray single crystal analysis shows that one Te ion in the central site is surrounded by six MoO_6 octahedra, forming the hexagon unit of $[TeMo_6O_{21}]^{2-}$ [33]. The Te ion in the center of $[Te^{IV}Mo^{VI}{}_6O_{21}]^{2-}$ is coordinated to three oxygen atoms. The hexagonal $[TeMo_6O_{21}]^{2-}$ units assemble along the c-axis to form the molecular wires (Fig. 5.11). The molecular wires are further packed parallel in a hexagonal fashion to form the material, in between which there are ammonium cations and water.

The material shows good elemental diversity. Mo can be successfully replaced by W [39] and V [40], and the central $[Te^{IV}O_3]$ is able to be replaced by $[Se^{IV}O_3]$, $[HP^{III}O_3]$, and $[P^{V}{}_2O_5]$ [41]. Furthermore, the local structure of the central element of the material changed with the central elements. In the case of using the reduced metal elements, such as Se^{IV}, Te^{IV}, and P^{III}, the central element is in a triangle shape. When central of the material is replaced by P^{V}, and the central of the material is a corner-sharing PO_4 tetrahedra.

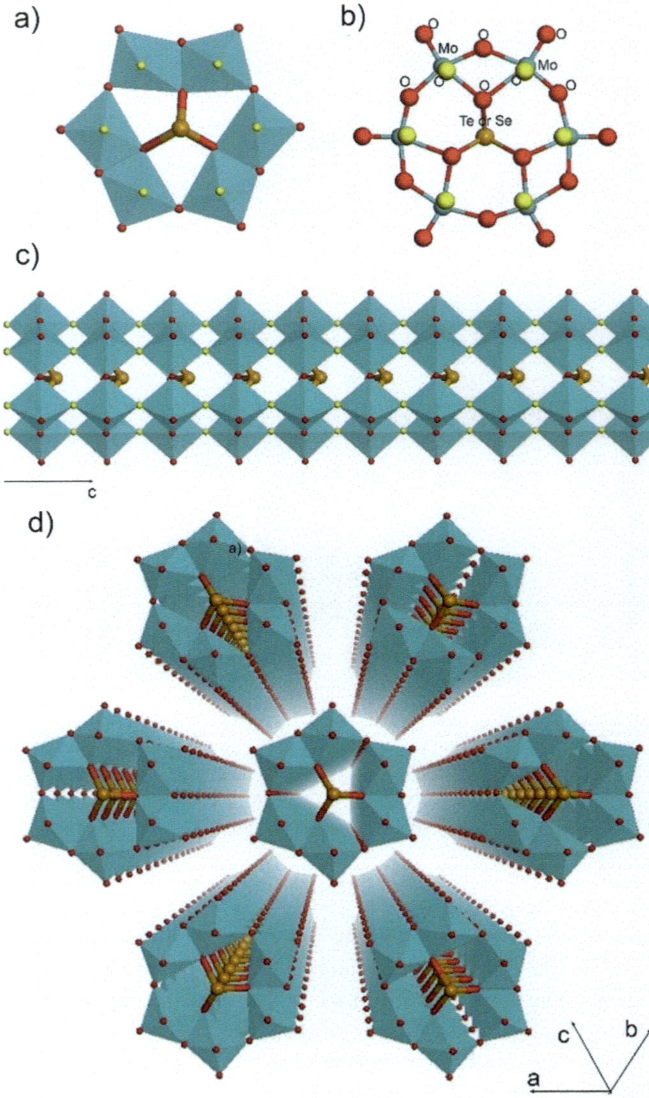

Fig. 5.11 Structural representations. **a** Polyhedral representation of a hexagonal unit of $[Te^{IV}Mo^{VI}{}_6O_{21}]^{2-}$ or $[Se^{IV}Mo^{VI}{}_6O_{21}]^{2-}$, **b** ball-and-stick representation of a hexagonal unit of $[Te^{IV}Mo^{VI}{}_6O_{21}]^{2-}$ or $[Se^{IV}Mo^{VI}{}_6O_{21}]^{2-}$ with labels, **c** a single molecular wire of Mo–Te oxide, The bridge oxygen atoms that connect the hexagonal units are highlighted in yellow, **d** assembly of single molecular wires into crystalline h-TM (or Mo–Se oxide). Mo (blue), Te (Se) (brown), O (red)

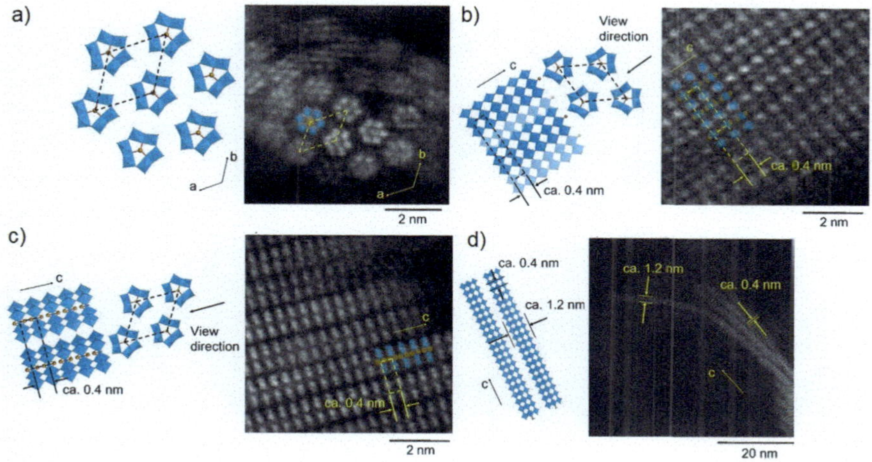

Fig. 5.12 High-resolution HAADF-STEM images (right) and proposed structures (left) of **a** h-TT in the a-b plane, **b** in the (100) plane, **c** in the (210) plane, and **d** h-ST along the c-axis, dashed line indicates the unit cell; W (blue), Te/Se (brown), O (red)

The structure of h-TT is observed by HAADF-STEM. The image observed is in good agreement with the crystal structure of the material (Fig. 5.12). The molecule wire can be isolated from the material by ultrasonication. The isolated single molecular wires are observed by HAADF-STEM. The diameter of the single molecule wire is ca. 1.2 nm. The thickness of the molecular wire is also measured by atomic force microscopy (AFM), showing the thickness of the molecule wire is ca. 1.2 nm, which is a good agreement with that from HAADF-STEM [39].

Introducing Co in the synthesis process of $(NH_4)_2[Se^{IV}W^{VI}{}_6O_{21}]$ (h-ST) produces a new porous ZOMO framework based on Co linker and the transition metal oxide molecule wire as the building unit, h-CST [38]. The framework of the material is constructed by the connection of the h-ST molecular wire with the Co ions (Fig. 5.13), which connect the terminal W = O bond of the h-ST molecular wire in the a-b plane. There are micropores surrounded by three molecular wire building blocks and Co linkers (Fig. 5.13), which are occupied by water or NH_4^+ in the as-synthesized material. The (1 0 0) plane is ascribed to the packing of the molecular wire building blocks, and the (0 0 1) plane is derived from the stacking of the hexagonal units ($[SeW_6O_{21}]^{2-}$) along the c-axis. The size of the pore opening is ca. 0.4 nm.

5.4.3 *Cubane Unit-Based ZOMOs*

The unit of VT-1 is a cubane unit of $[W_4O_{16}]^{8-}$, which is comprised of four tetrahedrally linked WO_6 octahedra. The building block of the cubane unit is connected to six VO^{2+} linkers nearby through coordination with the terminal W = O bonds of the

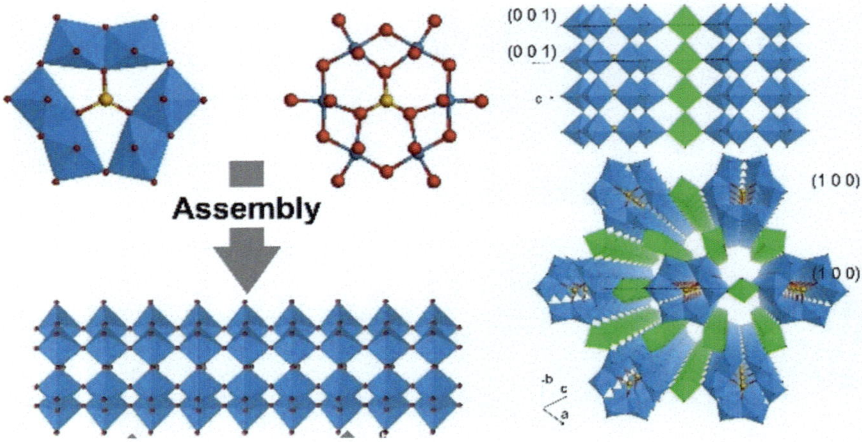

Fig. 5.13 **a** Assembly of hexagon unit to form the $\{[SeW_6O_{21}]^{2-}\}_n$ molecular wire and the 3D structure of h-CST; W (light blue), Se (yellow), Co (green), O (red), and N (deep blue)

units (Fig. 5.14a). The micropore opening is formed by an eight-membered oxygen ring with a diameter of ca. 4.3 × 4.3 Å (Fig. 5.14b). The micropore system of the material is cross-linked and unblocked, forming a 3D pore system in the material (Fig. 5.14 c). The structural topology of the material is similar to that of 4A zeolite.

In the case of VT-5, the secondary building block of the material is formed by two $[W_4O_{16}]^{8-}$ units that are firstly connected by three VO^{2+} linkers. The formed secondary building unit is in the trigonal symmetry (Fig. 5.14). The secondary building units are connected by six VO^{2+} linkers to form a porous network. The trimethyl amine occupies the micropores in the as-synthesized material. There are two different types of pore openings in the material (Fig. 5.14d). The micropore channels vertical to the a-b plane is surrounded by oxygens and is in trigonal symmetry with the size of ca. 7.4 × 7.4 Å (Fig. 5.14e). The other channel with the pore opening composed of 12 oxygen atoms is in a prolate shape with a size of ca. 3.9 × 7.9 Å, which is parallel to the a-b plane (Fig. 5.14f). Both of the pore openings are cross-linked to form a 3D microporous structure in VT-5, which is completely different from that of VT-1. The structure of VT-5 is the same as that of the zeolite IRY.

5.5 Properties and Applications

5.5.1 *Microporosity for Adsorption and Separation*

Micropores of ZOMOs are originally occupied by guest molecules. The stability of the materials is good enough to remove the guest molecules to open the micropore. The micropore of the materials can be analyzed by molecule adsorption.

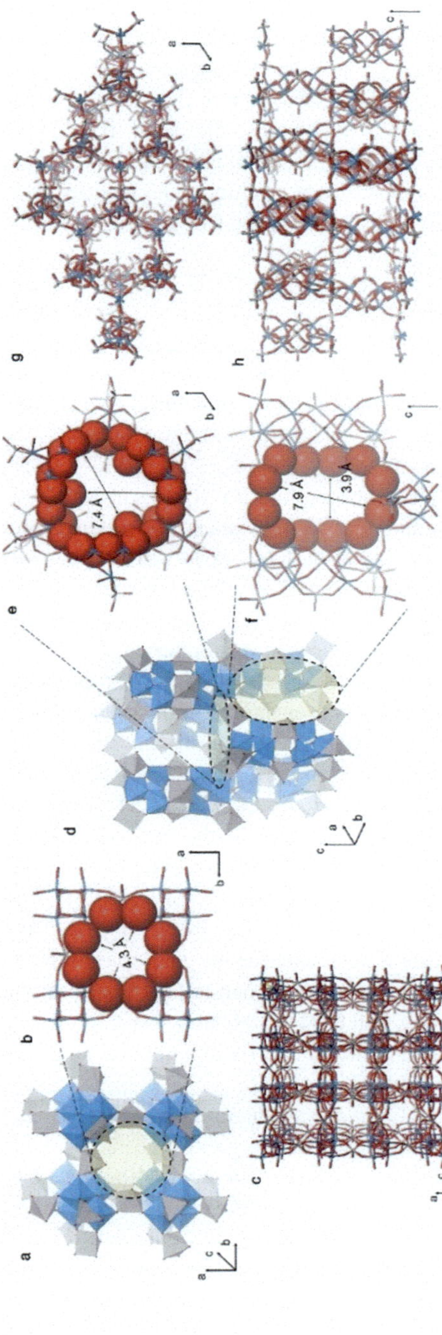

Fig. 5.14 Structural models of the synthesized VT family. **a** Polyhedral representation of the connection of $[W_4O_{16}]^{8-}$ units with VO^{2+} linkers for VT-1; **b** opening to the channel in VT-1; **c** extended framework of VT-1; **d** polyhedral representation of the connection of $[W_4O_{16}]^{8-}$ units with VO^{2+} linkers for VT-5; openings to **e** the vertical channel and **f** the parallel channel in VT-5, **g** extended framework of VT-5 in the a-b plane; and **h** extended framework of VT-5 along the c-axis; W (blue), V (gray), and O (red)

In the case of the ZOMOs based on ε-Keggin POM, there is water in the micropore of the as-synthesized materials, which is removed by heat treatment at over 200 °C. The structures of the materials do not collapse and the micropores are opened. The micropores of the materials are accessible to small molecules with proper size. The N_2 adsorption–desorption isotherm of ε-BVM shows that the material is a microporous material [32]. The surface area of this material is calculated to be ca. 60 m^2/g. The pore size distribution shows that the diameter of the micropores is ca. 5.5 Å, which is close to the average diameters of the pore openings and cavities. ε-BVM is able to adsorb other molecules, such as CO_2, CH_4, and C_2H_6, respectively, whereas a slightly larger molecule (C_3H_8) cannot be adsorbed [32].

The ε-Keggin POM-based ZOMOs can be used for the separation of gas molecules. ε-Na-ZM is used for selective adsorption of CH_4/CO_2 [42]. The mixed gas selective adsorption experiment shows that the material preferentially adsorbed CO_2 from the mixture of CO_2/CH_4 both at high- and low-pressure conditions with the selectivity of 75–52 (Fig. 5.15). Furthermore, the cation species of the material affect the selective adsorption of CO_2/CH_4. Na^+ is better than $NH_4{}^+$ for selective adsorption of CO_2. Selectivity of CO_2/CH_4 increases when Na^+ is the cation of the material. Computer-based simulation shows that CO_2 locates in the cavity with cation first, and then in the cavity without cation. There is a strong electrostatic interaction between CO_2 and the cation of the material. CH_4 locates in the cage without cation first, and then in the cavity with cation (Fig. 5.16). The reason might be that CH_4 is an electrostatic neutral molecule, which has a weak interaction with cation species, and the cavity without cation has a larger space for locating CH_4.

Furthermore, ε-Na-CM can be used as an adsorbent for C_2 molecule separation [43]. Single gas adsorption isotherms show that the adsorbed amounts of C_2H_6, C_2H_4, and C_2H_2 are different in the material, indicating that the material can separate C_2 hydrocarbons. Furthermore, the separation performance of the material can be tuned by changing the linker of the material. The IAST selectivity decreases when changing the linker from Co to other ions of the material, indicating that Co is the most suable linker for C_2 hydrocarbon separation. The breakthrough measurements of the binary mixtures of C_2H_4/C_2H_6, C_2H_2/C_2H_6, and C_2H_2/C_2H_4 with the ratio of 1:1 (v/v) for the materials show that the material can effectively separate C_2 hydrocarbon. The separation performance for gas separation increased with changing the linker of the material from Zn and Mn to Co. The breakthrough selectivities of C_2H_4/C_2H_6, C_2H_2/C_2H_6, and C_2H_2/C_2H_4 for ε-Na-CM are 12.5, 13.1, and 14.4, respectively (Fig. 5.17). The material can be regenerated and reused after the separation experiments.

The adsorption structure of the C_2 hydrocarbon molecules in the material is simulated based on the DFT calculation and system energies of the material after adsorbing different C_2 hydrocarbon molecules (Fig. 5.18). The interaction of C_2H_2 to the material is the strongest and C_2H_6 to the material is the weakest. The trend of the energy change after adsorbing the molecules based on the DFT calculation is in good agreement with the experimental adsorption enthalpy. The author found that the electrostatic interaction of the adsorbed molecule with the cation and the framework is important for good separation performance. The unsaturated bonds of C_2H_4 and

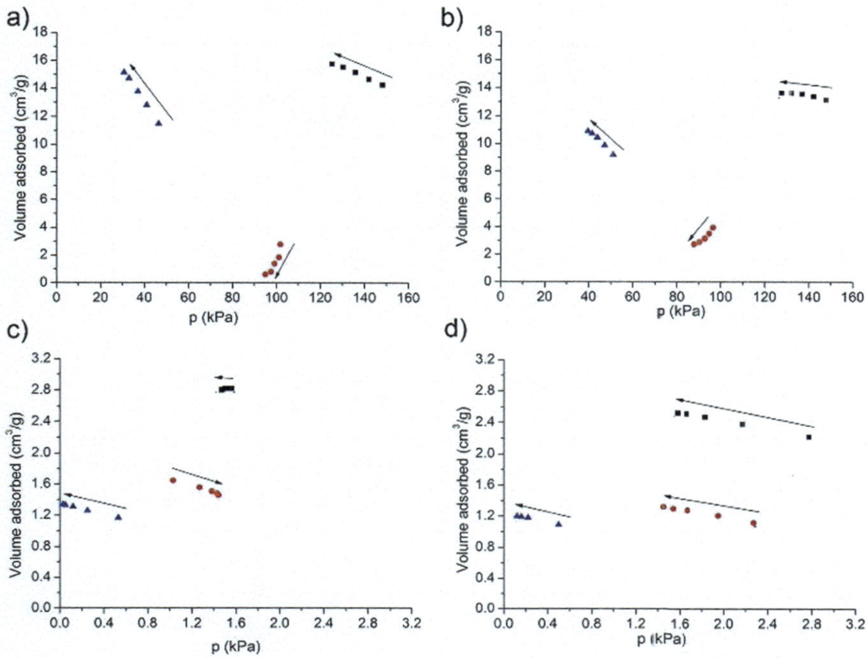

Fig. 5.15 CO_2/CH_4 co-sorption results of **a** ε-Na-ZM at high pressure, **b** ε-NH_4-ZM at high pressure, **c** ε-Na-ZM at low pressure, and **d** ε-NH_4-ZM oxide at low pressure, black square: system total pressure (x-axis) and adsorbed amount (y-axis), red cycle: CH_4 partial pressure (x-axis) and adsorbed amount (y-axis), blue triangle: CO_2 partial pressure (x-axis) and adsorbed amount (y-axis), Reproduced from Ref. [42] with permission from the Royal Society of Chemistry

C_2H_2 show the pi-Na interaction between the molecule and the material, which is stronger than C_2H_6.

The microporosity changes with the structural topology of the framework of ZOMOs. The micropores of ε-Keggin POM-based ZOMOs are composed of six oxygen atoms with a pore size of 3.4 Å, while VTs showed a larger pore size of the material.

The framework of VT-1 is a microporous framework. Water in the as-synthesized material is able to be removed by heat treatment, and the micropore of the material can be opened. N_2 adsorption–desorption measurement shows that the material is a microporous material. BET surface area of the material is ca. 310 m^2/g with a pore diameter of 0.4 nm and pore volume of ca. 0.1 cm^3/g. The isomeric structural material, VT-5, is also porous. Water in VT-5 can be removed by heating. However, the organic ammonium cation, $N(CH_3)_3$, in the material cannot be removed from the cavity by heat treatment at a low temperature (<300 °C). Therefore, the micropore of VT-5 cannot be fully opened. N_2 adsorption–desorption measurement shows that N_2 cannot be adsorbed [31].

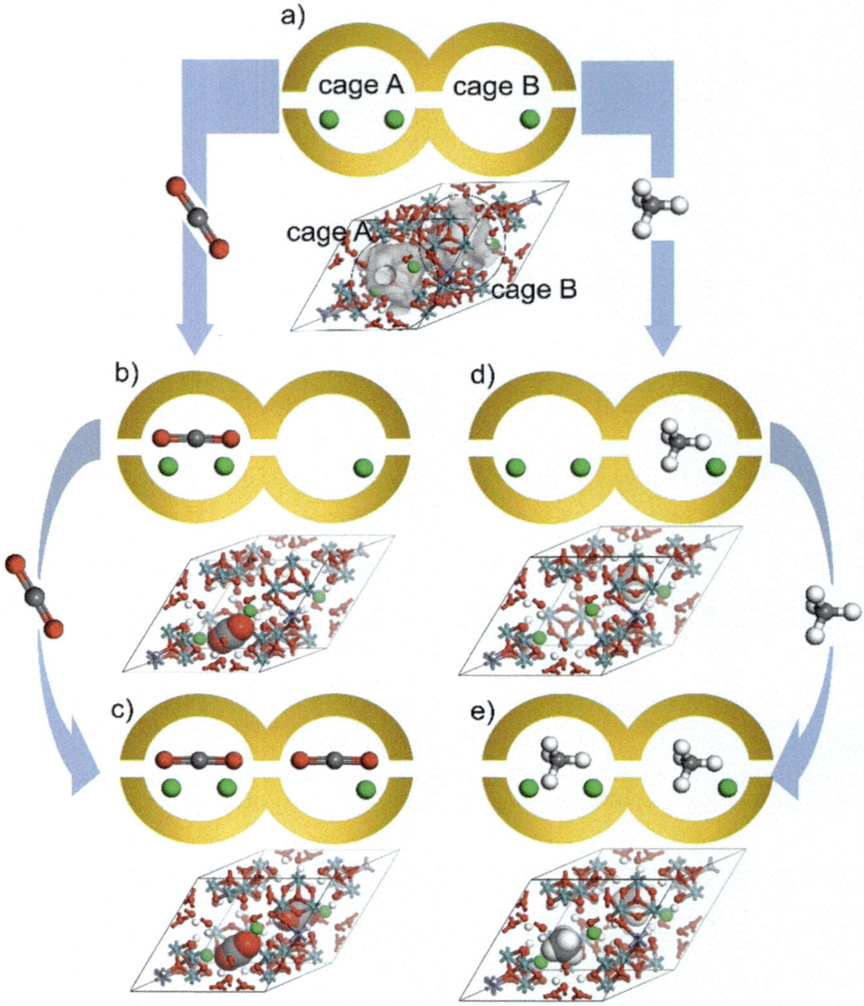

Fig. 5.16 Representations of adsorbed structures of ε-Na-ZM from Monte Carlo simulation, upper: ball-and-stick representations, lower: schematic representations. **a** primitive cell with cage A and cage B, **b** ε-Na-ZM adsorbed first CO_2, **c** ε-Na-ZM adsorbed second CO_2, **d** ε-Na-ZM adsorbed first CH_4, and **e** ε-Na-ZM adsorbed second CH_4, Mo (blue), Zn (purple), O (red), (white), C (gray), and Na (green), Reproduced from Ref. [42] with permission from the Royal Society of Chemistry

VT-1 adsorbs other small molecules, such as CO_2, CH_4, C_2H_6, C_3H_8, and nC_4H_{10}. We found that the linear molecules can be adsorbed by VT-1, and the size of the pore diameter just allowed the linear molecules to enter the cavity, such as nC_4H_{10}. The molecules with side branches, such as iC_4H_{10}, cannot be adsorbed by the material.

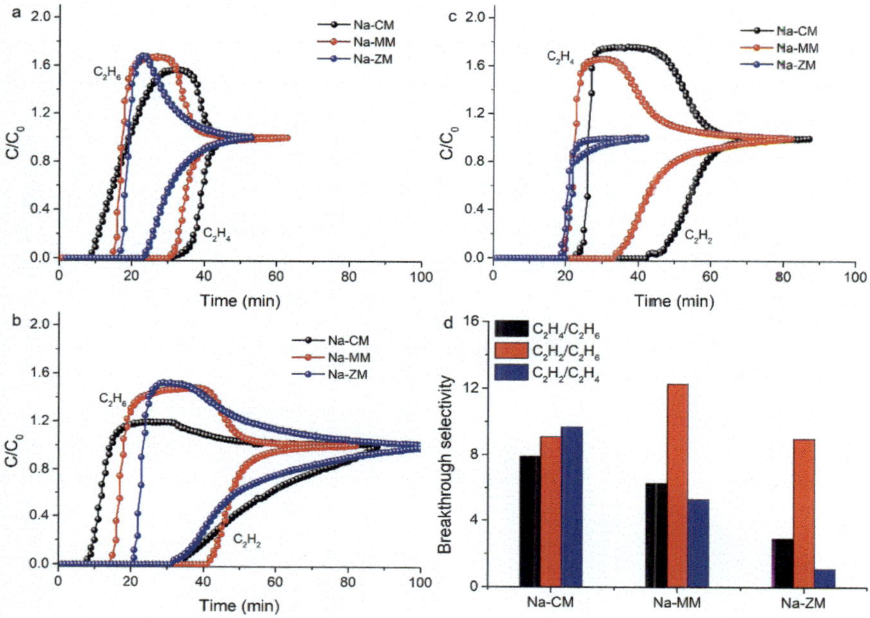

Fig. 5.17 Breakthrough measurements of the binary C2 hydrocarbons with the ratio of 1/1 (v/v) for ε-Na-CM, $Na_{0.6}H_{12}[Mn^{II}_{0.7}Mo^{VI}_2Mo^{V}_{10}O_{40}\{Mn^{II}_2\}]$ (ε-Na-MM), and ε-Na-ZM measured at 0 °C under 100 kPa. Breakthrough curves of **a** C_2H_4/C_2H_6, **b** C_2H_2/C_2H_6, **c** C_2H_2/C_2H_4, and **d** corresponding breakthrough selectivity

Based on the single gas adsorption measurements, VT-1 s can be used as the separator for nC_4H_{10}/iC_4H_{10} separation [44]. IAST selectivity for VT-1 s of nC_4H_{10}/iC_4H_{10} reached to 5100 at 21 °C under 100 kPa. The materials preferentially adsorbed nC_4H_{10} and would have high separation performance for nC_4H_{10}/iC_4H_{10}. The separation performance of the material is affected by the cation of the material. Li-VT-1 with Li^+ shows the highest selectivity, while the material with NH_4^+ showed the lowest selectivity.

Breakthrough measurements show that VT-1 effectively separates the binary gas mixture of nC_4H_{10}/iC_4H_{10} at 0 °C under 100 kPa using a ratio of 1/1 (v/v) (Fig. 5.19). The material does not adsorb iC_4H_{10}, and iC_4H_{10} is detected first while nC_4H_{10} is detected later for VT-1 s. The cation in the material has an important influence on the separation performance. VT-1 with Li^+ shows the best separation performance among VT-1 s with different cations. Li^+ would be the most suitable cation of VT-1 for nC_4H_{10}/iC_4H_{10} separation.

The transition metal oxide molecular wire is connected by Co ion to form ZOMOs with a unique soft framework, h-CST. The material is able to adsorb water at room temperature, and the structure of the material changes during the water adsorption process. There are three stages in the adsorption isotherm (Fig. 5.20). The first stage consists of a pressure range from 0 to 1.75 kPa with the adsorbed amount of ca.

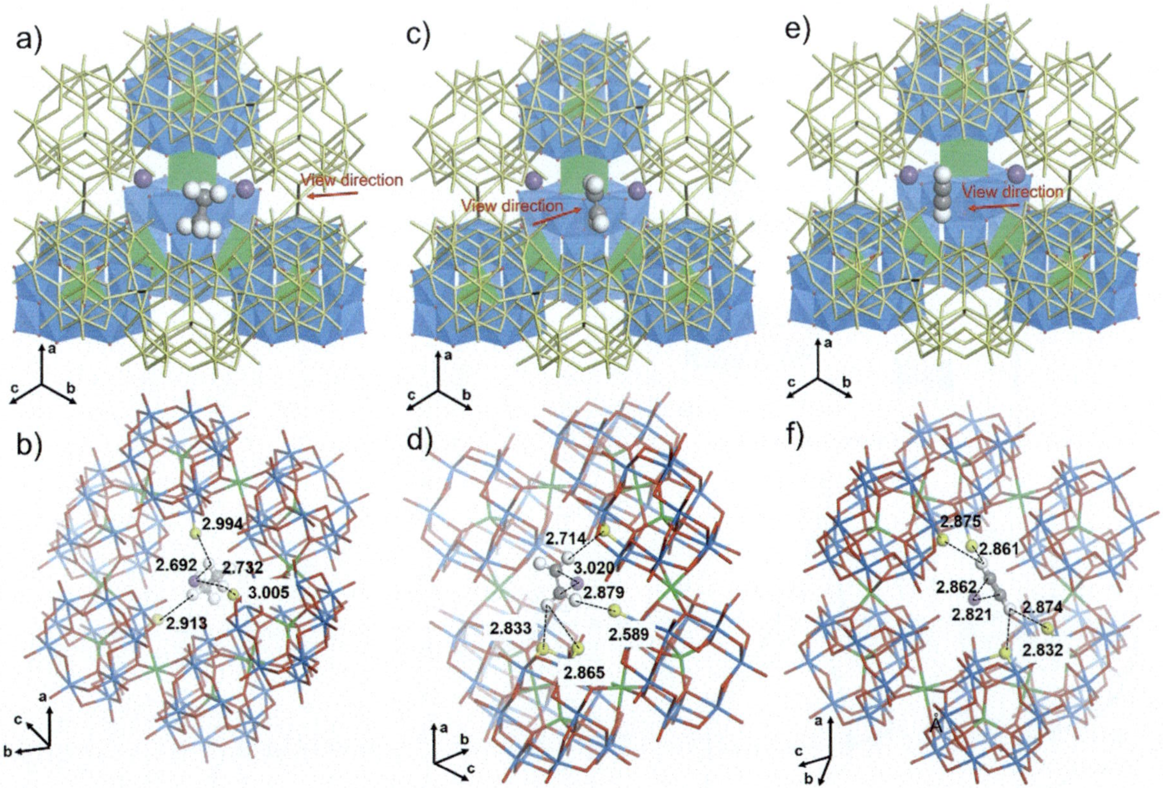

Fig. 5.18 Structural models of ε-Keggin POM-based ZOMOs. **a** C_2H_6 in the cavity, **b** interaction of C_2H_6 and the framework, **c** C_2H_4 in the cavity, **d** interaction of C_2H_4 and the framework, **e** C_2H_2 in the cavity, **f** interaction of C_2H_2 and the framework; Mo (blue), Co (green), O (red), C (gray), and H (white)

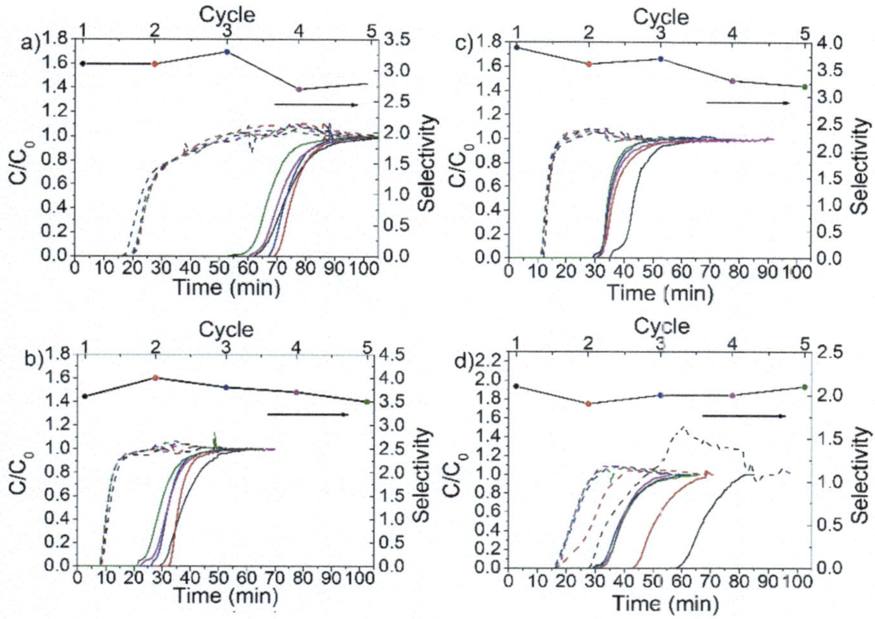

Fig. 5.19 Breakthrough curves and breakthrough selectivities of nC_4H_{10}/iC_4H_{10} with the ratio of 1/1 (v/v), **a** Li-VT-1, **b** Na-VT-1, **c** K-VT-1, **d** NH_4-VT-1, black: first cycle, red: second cycle, blue: third cycle, purple: fourth cycle, green: fifth cycle, dash line: iC_4H_{10}, full line: nC_4H_{10}, measurement conditions: total gas pressure: 100 kPa; flow rate of each gas: 0.3 mL/min, 1 g of VT-1 s, 0 °C

3 mol/mol. The second stage is from 1.75 to 2.5 kPa with the adsorbed amount of ca. 4.5 mol/mol. The final one is up to 3 kPa with the adsorbed amount of ca. 3.5 mol/mol [38].

The structure evaluation of the material during water adsorption is confirmed. The relative humidity-dependent powder XRD shows three stable phases with increasing humidity (Fig. 5.21). The molecular wire building blocks of the material are arranged in the "edge to edge" manner under a low humidity condition (17%) with a short distance, which is the first stable phase. With the increase in the humidity to 65%, the second stable phase is generated with the same packing manner with an increase in lattice parameter a. The phase with the highest humidity shows the arrangement of the $\{[Se^{IV}W^{VI}{}_6O_{21}]^{2-}\}_n$ molecular wires with Co linkers in the "corner to corner" manner.

5.5.2 *Redox Property for Catalysis and Battery*

ZOMOs have both advantages of porous materials and transition metal oxides, which show multi-electron redox properties. In the case of ε-Keggin POM-based

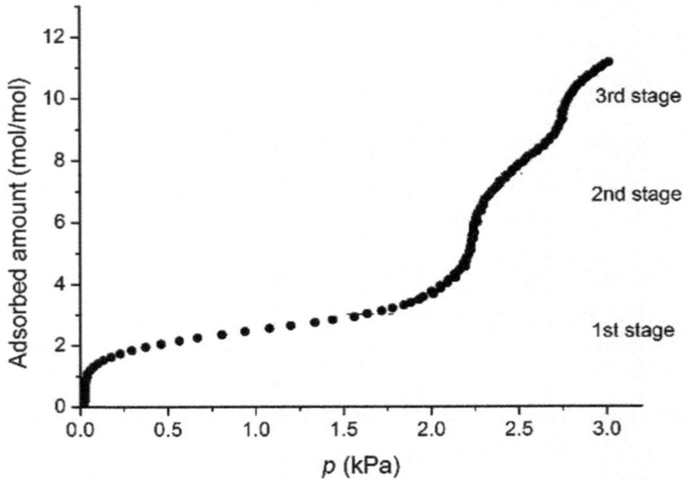

Fig. 5.20 Water vapor adsorption isotherms of h-CST

ZOMOs, the oxidation state of Mo in the material is in the reduced state of Mo^{V} [45]. The reduced Mo^{V} can be easily oxidized to Mo^{VI} in O_2. Particularly, zeolitic iron molybdate ($(NH_4)_2Fe^{II}{}_{0.6}H_{11.7}[Fe^{II}{}_{2.0}Mo^{VI}{}_{1.1}Mo^{V}{}_{10.9}O_{40}]$, ε-IM) can activate O_2 at room temperature. Further investigation shows that the material adsorbs O_2 at room temperature, which produces an irreversible adsorption isotherm. Fe is critical for the irreversible adsorption isotherm, without which the material lost the ability of O_2 adsorption at room temperature. Temperature programmed oxidation (TPO) shows that the material with Fe is active for O_2 adsorption and shows redox property at room temperature (Fig. 5.22) [45]. X-ray photoelectron spectrometry (XPS) analysis demonstrates that the irreversible adsorption isotherm of O_2 is due to the oxidation of Mo^{V} to Mo^{VI} during the oxidation.

The local structure of ε-Keggin POM of ε-IM during redox reaction slightly changed with changing the oxidation state of Mo (Fig. 5.23). The distance between neighboring Mo in the material changes. The distance of Mo--Mo in the Mo_3O_{13} trimer of the as-synthesized ε-IM is 3.410 Å. After O_2 adsorption, the distance decreases dramatically to 3.260 Å (Fig. 5.23). The distance of Mo--Mo in the Mo_2O_{10} dimer increases from 2.646 to 2.910 Å after oxidation of the material.

The redox-active ZOMO, ε-IM, shows catalytic activity for the oxidation of hydroxyl. The material can be used for the oxidation of ethyl lactate (EL) to form ethyl pyruvate (EP) with O_2 as an oxidant. The catalytic performance is highly dependent on the chemical composition of the material. The catalytic performance of ε-IM is higher than other iso-structural catalysts. Fe is critical, and after the incorporation of Fe, the inactive iso-structural materials became active [45]. Furthermore, introducing V in the material, denoted as ε-IV_xM where x is the V content, further improves the catalytic activity of the material without a change in the selectivity (Fig. 5.24). The

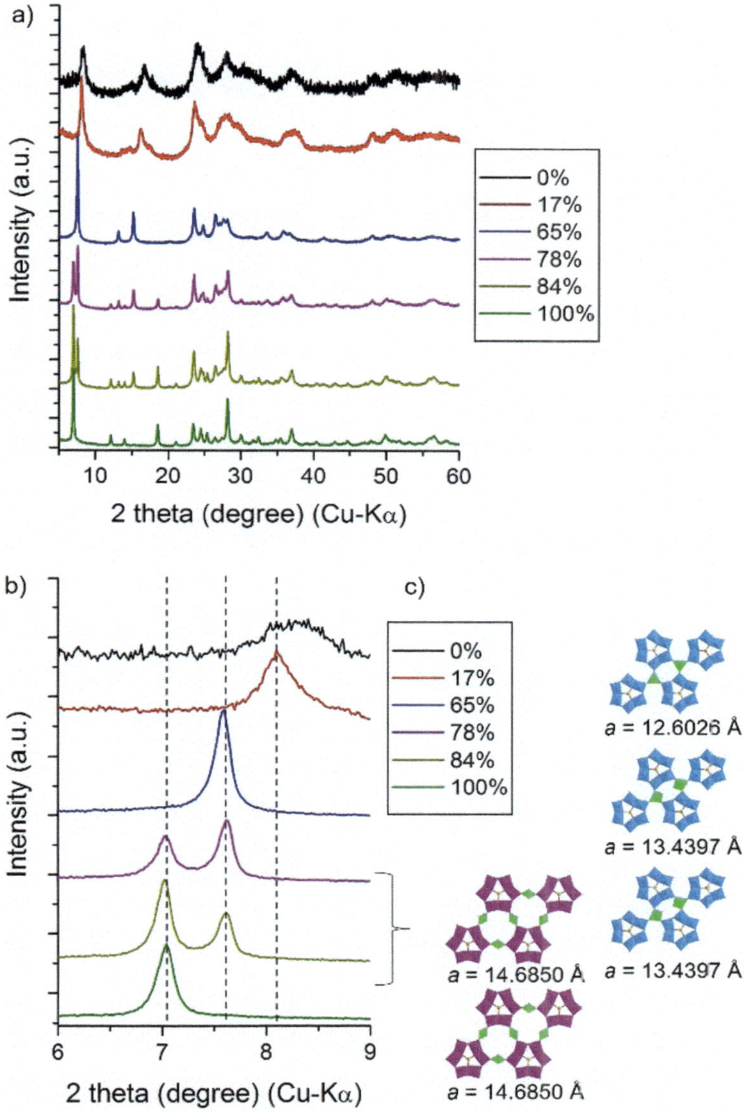

Fig. 5.21 **a** Relative humidity-dependent XRD patterns of h-CST, **b** Enlarged XRD patterns, and **c** corresponding structures of h-CST at different humidity values; the experiments were carried out at 25 °C. W (the "edge to edge" manner) (blue), W (the "corner to corner" manner) (purple), Se (yellow), Co (green), O (red)

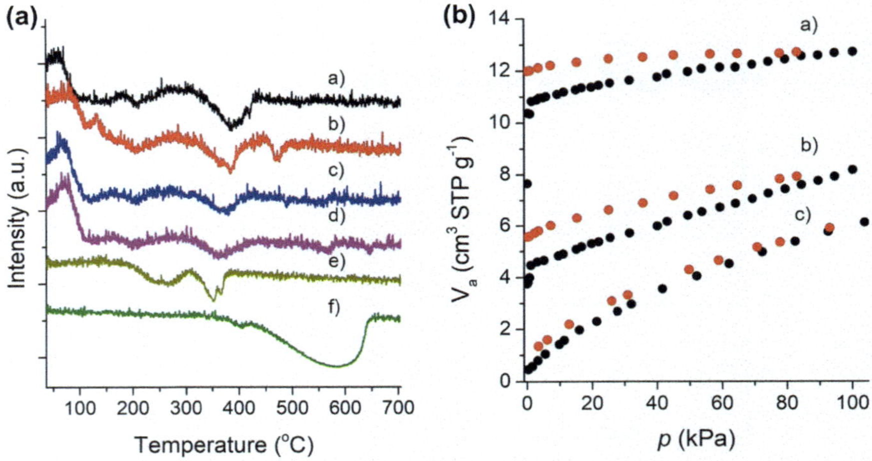

Fig. 5.22 **a** TPO profiles of a) ε-IM, b) ε-$IV_{0.07}$ M, c) ε-$IV_{0.12}$ M, d) ε-$IV_{0.18}$ M, e) MoZnO, and f) MoO_2 and **b** O_2 adsorption–desorption isotherms of a) ε-IM, b) ε-ZM, and c) Fe-ε-ZM at 25 °C, Reprinted with permission from Ref. [45]. Copyright 2019 American Chemical Society

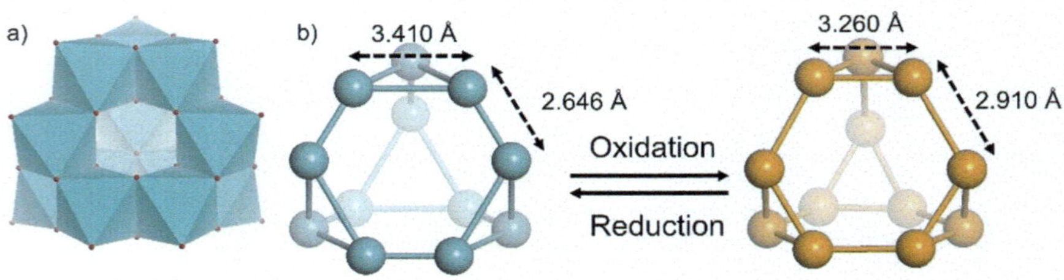

Fig. 5.23 Structural representations of the ε-Keggin unit, **a** the polyhedral representation, the ball-and-stick representation of **b** the ball representation of the material during redox, Mo (blue), Mo oxidized (orange), O (red), Reprinted with permission from Ref. [45]. Copyright 2019 American Chemical Society

materials of ε-IM and ε-IV_xM are stable during the reaction, and the structures do not change after the reaction.

Moreover, the same catalyst can activate O_2 at a lower temperature, which enables the material for oxidation of alcohols at a lower temperature. ε-IV_xM is able to catalyze the oxidation of a variety of primary aromatic alcohols, such as p-methoxybenzyl alcohol, 4-methylbenzyl alcohol, 3-methylbenzyl alcohol, benzyl alcohol, and 4-bromo benzyl alcohol, using air as the oxidant in toluene at 110 °C (Table 5.3) [46]. The catalyst is stable and is able to be reused 5 times without loss of the catalytic activity. The yield to the corresponding aldehydes is over 90% with high selectivity. The molecule structure, crystal structure, and micropore structure of the reused catalyst do not change, which demonstrates that the material is stable.

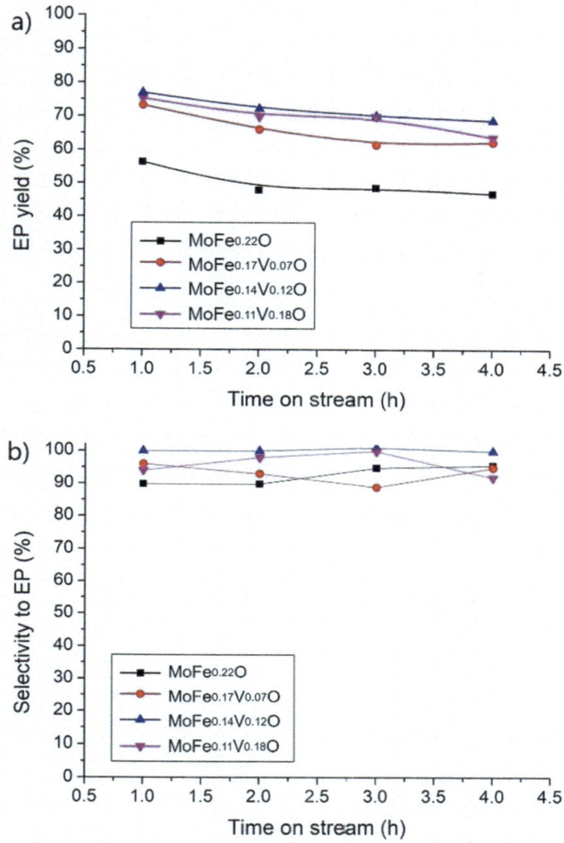

Fig. 5.24 The catalytic activity of the materials for EL oxidation to form EP. **a** the EP yield and **b** the selectivity to EP of the materials based on different V content. Reaction conditions: air 20 mL/min (N_2 15.6 mL/min, O_2 4.2 mL/min), EL 1.95 mL/min, catalyst 0.5 g, temperature 200 °C, Reprinted with permission from Ref. [45]. Copyright 2019 American Chemical Society

The transition metal oxide molecular wires based on molybdophosphite ($K_{1.7}H_{0.3}[(HP^{III}O_3)Mo^{VI}{}_6O_{18}]$, h-$MP^{III}$) and molybdophosphate ($Na_{0.2}K_{3.8}[(P^{V}{}_2O_7)Mo^{VI}{}_{12}O_{36}]$, h-$MP^{V}$) also show interesting redox property, and the local structures of the materials change during the redox reaction. $\{HP^{III}O_3\}$ and $\{P^{V}{}_2O_7\}$ are the central of the hexagon unit of the transition metal oxide molecular wires.

The building block of h-MP^{III} is based on the hexagon unit of $[(HP^{III}O_3)Mo^{VI}{}_6O_{18}]^{2-}$ with a protonated $\{HPO_3\}$ center in a "face to face" mode [41]. The hexagon unit grows along the c-axis forming the molecule wire of the material. The chemical formula of h-MP^{III} is $K_{1.7}H_{0.3}[(HP^{III}O_3)Mo^{VI}{}_6O_{18}]$. The building unit of h-MP^{V} is $[(P^{V}{}_2O_7)Mo^{VI}{}_{12}O_{36}]^{4-}$ with a corner-sharing $\{P_2O_7\}$ tetrahedron in the center. The chemical formula of h-MP^{V} is $Na_{0.2}K_{3.8}[(P^{V}{}_2O_7)Mo^{VI}{}_{12}O_{36}]$ (Fig. 5.25).

The local structure change of the temperature sensitive $\{HP^{III}O_3\}$-containing molecular wire via the solid-state structural rearrangement from

Table 5.3 Oxidation of different alcohols catalyzed by ε-IV_xM with air

Entry	Substrate	Product	Conv (%)	Y. (%)	Sel. (%)
1	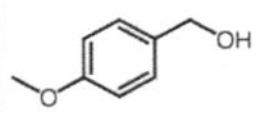	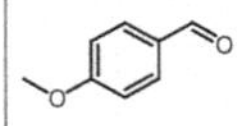	99	92	92
2	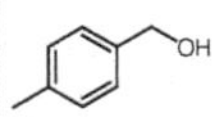	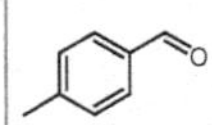	62	62	99
3	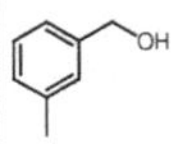	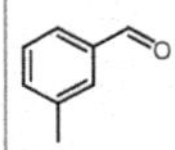	81	80	99
4	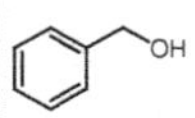	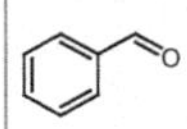	84	78	93
5	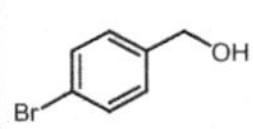	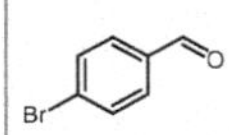	99	93	93
6	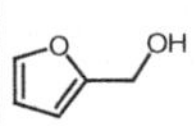	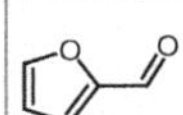	72	53	74
7	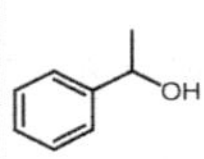	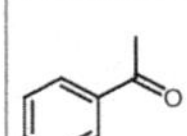	0	0	–
8	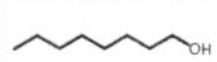	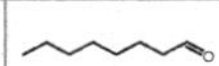	0	0	–
9	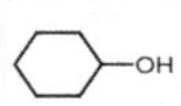	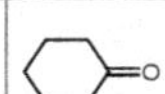	0	0	–

[a] Reaction conditions: ε-IV_xM: 0.04 g, alcohol: 0.8 mmol, TBAB: 0.01 g, TEA: 0.01 mL, toluene: 0.5 mL, decane: 0.25 mmol, temperature: 110 °C, time: 24 h

$\{[(HP^{III}O_3)Mo^{VI}{}_6O_{18}]^{2-}\}_n$ to $\{[(P^{V}{}_2O_7)Mo^{VI}{}_{12}O_{36}]^{4-}\}_n$ undergoes an unusual electron transfer and oxygen transfer (ET-OT) process. The packing manner of P changes from the "face to face" mode to the "face to end" mode at 200 °C. The structure changes from $\{[(HP^{III}O_3)Mo^{VI}{}_6O_{18}]^{2-}\}_n$ to $\{[(P^{V}{}_2O_7)Mo^{VI}{}_{12}O_{36}]^{4-}\}_n$ via blue intermediates of $\{[(P^{V}{}_2O_7)_x(P^{III}{}_2O_6)_{1\text{-}x}(Mo^{VI}{}_{12\text{-}4x}Mo^{V}{}_{4x}O_{35\text{-}x})]^{4-}\}_n$ at a temperature over 250 °C. The heat-triggered ET-OT process causes the triangle-shaped $\{HP^{III}O_3\}$ unit to transfer to the corner-sharing $\{P^{V}{}_2O_7\}$ unit. Meanwhile the $\{Mo^{VI}{}_6O_{18}\}_n$ shell is reduced to form the defective $\{(Mo^{VI}{}_{12\text{-}4x}Mo^{V}{}_{4x}O_{35\text{-}x})\}_n$ shell with oxygen defects, which are compensated by O from O_2 to obtain the fully oxidized h-MP^{V}.

Furthermore, V can be incorporated in h-MP^{III} which replaces MoO_6 in the hexagon unit forming the V-substituted hexagon unit of $[(HP^{III}O_3)(Mo^{VI}{}_5O_{15})(V^{V}O_3)]^{3-}$ (h-MVP^{III}) (Fig. 5.26) [40]. Incorporating V in the material changes the heat-triggered intramolecular redox property via the ET-OT process and the structural evaluation process. Oxygen transfers preferentially from $\{V^{V}O_6\}$ to $\{HP^{III}O_3\}$, while electron transfers from $\{HP^{III}O_3\}$ to $\{V^{V}O_6\}$

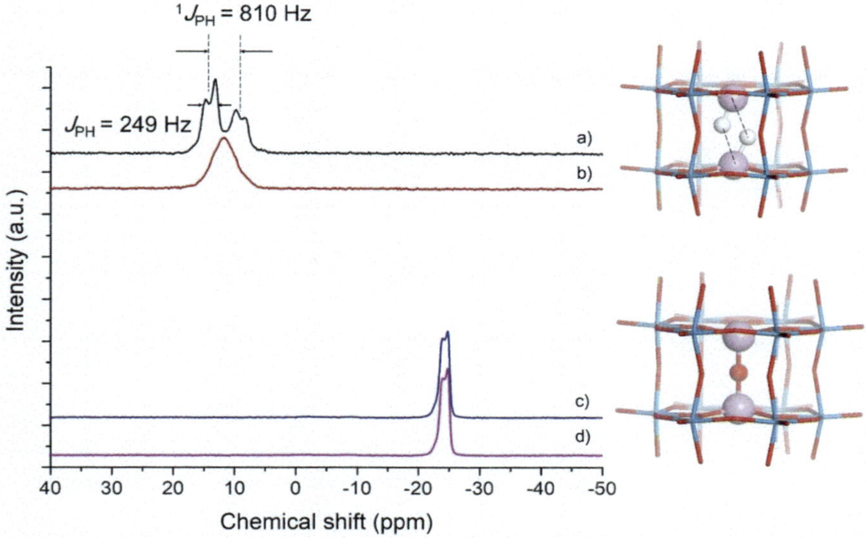

Fig. 5.25 Solid state ^{31}P MAS NMR profiles of **a** h-MP^{III} with H-coupling mode, **b** h-MP^{III} with H-decoupling mode, **c** h-MP^{V} with H-coupling mode, and **d** h-MP^{V} with H-decoupling mode, inserted images: structural models of a hexagonal unit of h-MP^{III} and h-MP^{V}, Mo (blue), P (purple), O (red), H (white), Reprinted with permission from Ref. [41]. Copyright 2019 American Chemical Society

compared with $\{Mo^{VI}O_6\}$. The redox property of the V incorporated molecule wire (h-MVP^{III}) is more active, and the structure of the material is more stable compared with h-MP^{III}.

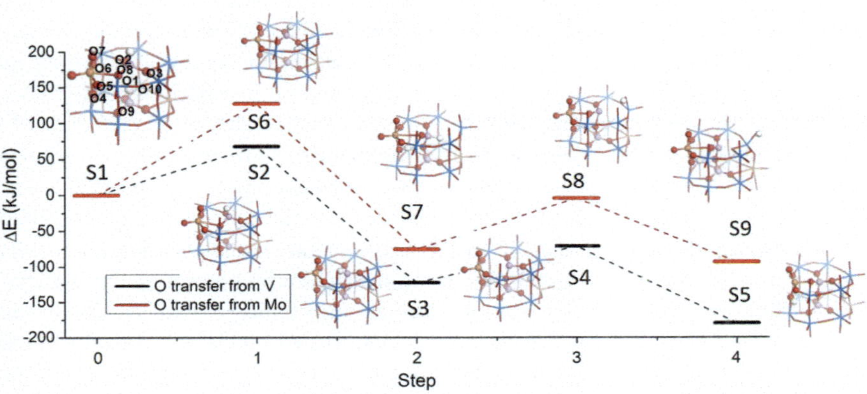

Fig. 5.26 The proposed structures change and the corresponding energy by DFT calculation in the intramolecular redox reaction W (blue), V (orange), O (red), P (purple), and H (white), Reprinted with permission from Ref. [40]. Copyright 2020 American Chemical Society

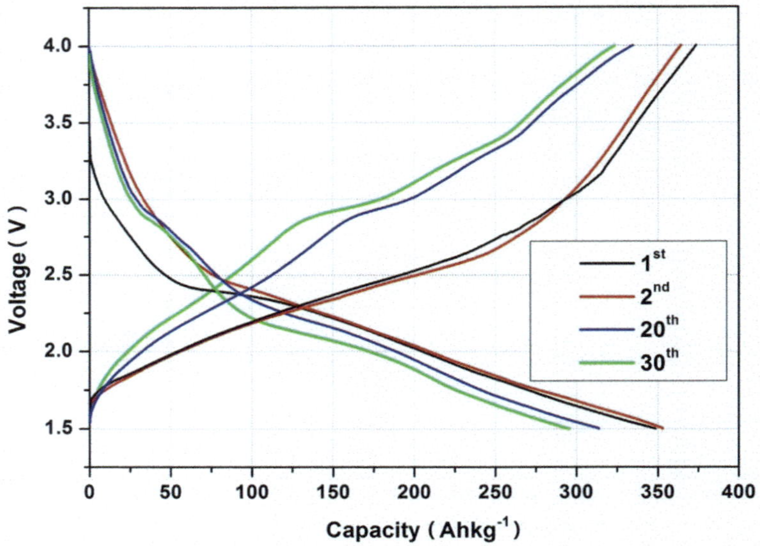

Fig. 5.27 Charging/discharging curves of o-Bi-VM

ZOMOs show multi-electron transfer properties as well as microporosity, which are good for the cathode of the battery. The bismuth incorporated orthorhombic zeolitic vanadomolybdates (o-Bi-VM) is used as a cathode-active material for lithium batteries (LBs). The first discharging capacity is ca. 380 Ah/kg (Fig. 5.27) [47]. The cycle performance of o-Bi-VM is good, and the battery only lost ca. 8% of the initial capacity after 20 times' cycles. o-Bi-VM is a stable and active cathode material for LBs.

The valence change of Mo during the charging/discharging process is monitored by operando X-ray absorption near edge structure (XANES). The initial Mo valence at open circuit voltage is ca. 6.0, which remains constant during the 1st charging. In the 1st discharging, the oxidation state of Mo decreases from 6.0 to 3.8 (3.0–1.5 V). The 2nd charging increases the valence from 3.8 to 6.0 in the range of 1.5–3.0 V, demonstrating that the Mo^{IV} ions are fully re-oxidized to Mo^{VI} reversibly. V valence is ca. 4.0 in the as-synthesized material. In the 1st discharging from 4.0 to 2.5 V, the V valence also reversibly changes from 3.35 to 4.0 V in the voltage range from 4.0 to 2.5 V, indicating that most of the V^{4+} ions in o-Bi-VM are reduced to V^{3+} reversibly.

The extended X-ray absorption fine structure (EXAFS) spectra of the material after discharging show that the bond length increases from 1.64 to 1.88 Å during the discharging (Fig. 5.28a). The increase of the Mo–O bonds is attributed to the intercalation of Li^+ with O of the material in battery reactions, which increases the Mo–O bond when Li inserts into the structure. Furthermore, the peak at 2.2 Å does not change but enhances, which might be caused by a shortening of the Mo-Mo distance due to the reduction of Mo^{VI} to Mo^{IV} during the discharging process.

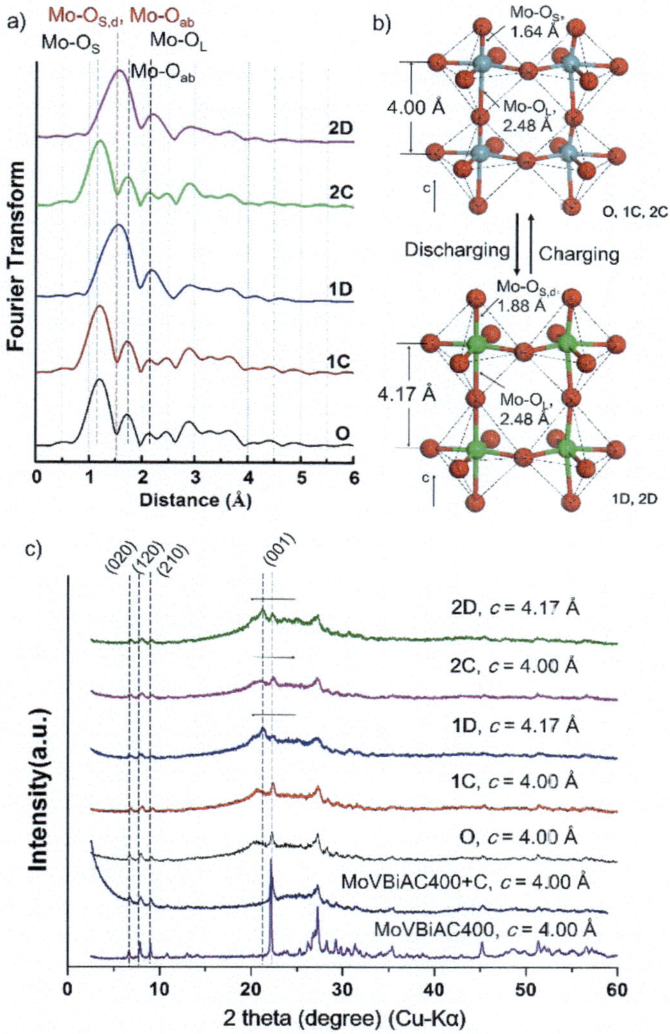

Figure 5.28 **a** Fourier transforms of the Mo K-edge EXAFS spectra, **b** local structure change of the material based on EXAFS, Mo^{6+} (blue), Mo^{4+} (green), O (red), and c) ex situ XRD patterns for O, 1C, 1D, 2C, and 2D, Reprinted with permission from Ref. [47]. Copyright 2017 American Chemical Society

The ex situ XRD further confirms the crystal structure of the material during the battery reaction. After discharging the peak for the (001) plane shifts to a lower angle (Fig. 5.28c), demonstrating the increase in the lattice parameter c (4.00–4.17 Å) and the distance of Mo–O along the c direction increases correspondingly. The in situ data shows that during the charging/discharging process Li removal and inserting from and in the structure reversibly, which does not cause the structure change of the material.

Furthermore, because of the microporosity of ZOMOs, the materials also allow the entrance of large cations, such as Na^+, into the structure and it can be used for cathode material of Na battery [48]. Four different types of isomeric vanadomolybdates are used to test the battery performance (Fig. 5.29). The trigonal vanadomolybdates (t-VM) show the best performance as the cathode for Na battery. Different from Li battery, the Na battery process causes the structure change of the material. During discharging, Na^+ inserts into the structure, due to the limited space inside the micropore of the material, the migrating Na^+ would disorder the crystalline a-b plane to form the amorphous a-b plane without changing the pentagon units, which would make more accessible sites for Na^+ location (Fig. 5.29). The material with the amorphous a-b plane is stable during the following battery cycling tests, and the local structure and valence of Mo and V change reversibly.

5.5.3 Acidity for Catalysis

The materials of ZOMO are composed of POMs as building blocks. POMs are transition metal– oxygen clusters, which show acidity [39]. Therefore, ZOMOs are expected to show acidity and have potential applications in acid catalysis.

The molecular wires based on tungstotellurate, h-TT, and tungstoselenate, h-ST, are acidic, and the materials can be used for the hydrolysis of cellulose. The cation of the as-synthesized material is removed by heat treatment at 300–350 °C, which forms the acid sites in the material. The acid amount of the material is estimated by temperature programmed desorption and mass spectrometry (TPD-MS), which shows that ca. 50% of NH_3 transfers to acid sites by the heat treatment.

Cellobiose is hydrolyzed by h-TT at 130 °C for 4 h, yielding glucose as the main product with a high yield (90.5%) (Table 5.4). h-TT exhibits high activity for hydrolysis of other polysaccharides, such as sucrose and starch. Cellulose can be converted to hexoses (glucose and mannose) at 175 °C for 2 with some by-products of 5-hydroxymethylfurfural (HMF), levulinic acid, and formic acid. h-TT is more active for cellulose conversion than other solid acids based on the acid amount, even though it is more active compared to homogeneous catalysts [39]. The catalyst is stable and can be reused, the recovered catalyst shows the same activity compared with the as-synthesized catalysts.

5.5.4 Ion-Exchange Property for Ion Removal

Currently, the frameworks of all the prepared ZOMOs are anionic. Therefore, cation species are necessary to neutralize the anionic frameworks. Furthermore, the cations of the materials can be changed without changing the crystal structures of the

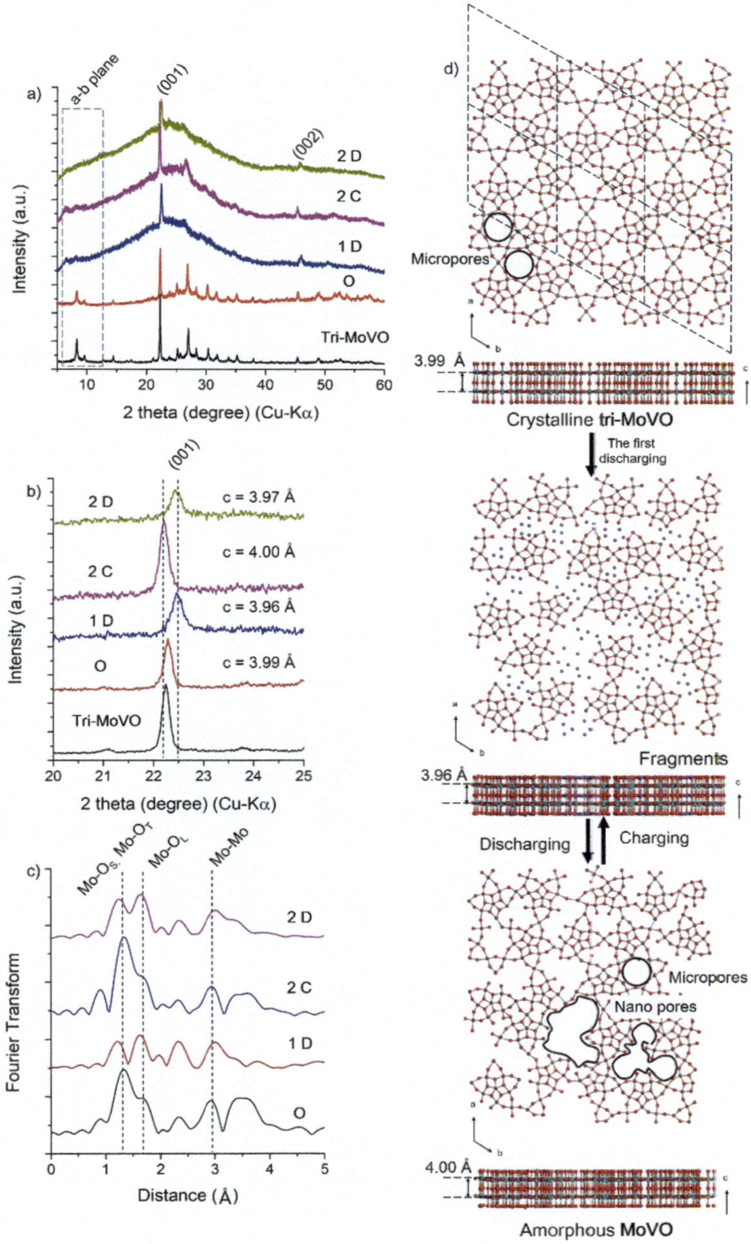

Fig. 5.29 Ex situ XRD patterns of **a** full patterns of t-VM for Li battery, **b** enlarged patterns of t-VM from 5–15 degree for Li battery, **c** operando EXAFS spectra of t-VM, and **d** structure models of t-VM before and after charging/discharging for Na battery, Mo (blue), Na (purple), O (red), Reprinted with permission from Ref. [48]. Copyright 2020 American Chemical Society

Table 5.4 Biomass hydrolysis by h-TT under different conditions[a]

Entry	Biomass	t (h)	T (°C)	Conv (%)	Yield based on carbon (%)					Total yield of organic
					Glucose	Mannose	Formic acid	Levulinic acid	HMF	(%)
1	Cellobiose	4	130	93.1	90.5	0	0.1	0.1	0.7	91.4
2	Sucrose	4	130	99	43.9	1.9	3.3	15.1	7.8	72.0
3	Starch	4	130	–	78.4	0.2	0.5	1.0	1.2	81.3
4	Microcrystalline cellulose[b]	2	175	–	8.8	2.9	0.5	1.1	0.7	14.0
5	Ball-milled cellulose	2	175	–	19.0	7.2	0.7	2.5	1.2	30.6
6	Ball-milled cellulose[c]	2	175	–	25.8	7.8	3.0	8.5	2.0	47.1
7	Microcrystalline cellulose[d]	2	175	–	0	0	0	0	0	0

[a] Reaction condition: biomass: 0.308 mmol based on glucose unit, h-TT: 0.05 g, water 0.5 mL, carbon balance was 86.7% (see detail in supporting information)
[c] Cellulose: 0.185 mmol
[d] Without catalyst

Table 5.5 Chemical formulas of ZM and MM with alkaline metal ions, Reprinted with permission from Ref. [27]. Copyright 2014 American Chemical Society

	Formula
ε-ZM	$Na_{1.5}H_{11.4}[\varepsilon\text{-}ZnMo_{12}O_{40}\{Zn\}_2]\cdot 5H_2O$
ε-H-ZM	$Na_{0.6}H_{12.3}[\varepsilon\text{-}ZnMo_{12}O_{40}\{Zn\}_2]\cdot 5H_2O$
ε-Li-ZM	$Li_{0.7}Na_{0.8}H_{11.4}[\varepsilon\text{-}ZnMo_{12}O_{40}\{Zn\}_2]\cdot 5H_2O$
ε-NH4-ZM	$(NH_4)_{1.4}Na_{0.1}H_{11.4}[\varepsilon\text{-}ZnMo_{12}O_{40}\{Zn\}_2]\cdot 5H_2O$
ε-K-ZM	$K_{1.4}Na_{0.1}H_{11.4}[\varepsilon\text{-}ZnMo_{12}O_{40}\{Zn\}_2]\cdot 5H_2O$
ε-Rb-ZM	$Rb_{1.3}Na_{0.2}H_{11.4}[\varepsilon\text{-}ZnMo_{12}O_{40}\{Zn\}_2]\cdot 5H_2O$
ε-Cs-ZM	$Cs_{1.5}H_{11.4}[\varepsilon\text{-}ZnMo_{12}O_{40}\{Zn\}_2]\cdot 5H_2O$
ε-MM	$(NH_4)_{2.1}H_{7.5}[\varepsilon\text{-}Mn_{0.2}Mo_{12}O_{40}\{Mn\}_2]\cdot 4H_2O$
ε-H-MM	$(NH_4)_{1.7}H_{7.9}[\varepsilon\text{-}Mn_{0.2}Mo_{12}O_{40}\{Mn\}_2]\cdot 3H_2O$
ε-Li-MM	$Li_{0.1}(NH_4)_{2.0}H_{7.5}[\varepsilon\text{-}Mn_{0.2}Mo_{12}O_{40}\{Mn\}_2]\cdot 3H_2O$
ε-Na-MM	$Na_{0.4}(NH_4)_{1.7}H_{7.5}[\varepsilon\text{-}Mn_{0.2}Mo_{12}O_{40}\{Mn\}_2]\cdot 3H_2O$
ε-K-MM	$K_{1.4}(NH_4)_{0.7}H_{7.5}[\varepsilon\text{-}Mn_{0.2}Mo_{12}O_{40}\{Mn\}_2]\cdot 3H_2O$
ε-Rb-MM	$Rb_{1.5}(NH_4)_{0.6}H_{7.5}[\varepsilon\text{-}Mn_{0.2}Mo_{12}O_{40}\{Mn\}_2]\cdot 3H_2O$
ε-Cs-MM	$Cs_{1.4}(NH_4)_{0.7}H_{7.5}[\varepsilon\text{-}Mn_{0.2}Mo_{12}O_{40}\{Mn\}_2]\cdot 4H_2O$

materials. The cations can be easily exchanged by ion exchange in an aqueous solution.

The ε-Keggin POM-based ZOMO shows ion-exchange properties similar to that of zeolites [27]. The cations, NH_4^+ in ε-MM and Na^+ in ε-ZM, are exchangeable with other cations. Various alkali metal ions, including H^+, Li^+, Na^+, K^+, Rb^+, and Cs^+, are tested for ion exchange with the materials. The basic structures of the materials are unchanged after the ion-exchange process with only a slight change in the lattice parameter. Elemental analysis confirms that the cations are exchanged with NH_4^+ or Na^+ and introduced into the materials. Large cations, including K^+, Rb^+, and Cs^+, show high ion-exchange capacity for both ε-ZM and ε-MM. Small ions, H^+, Li^+, and Na^+, are not as efficient as the large ions to replace NH_4^+ or Na^+ in as-synthesized materials of ε-ZM and ε-MM. Furthermore, alkaline earth metal ions are also able to be introduced into the materials by ion exchange without changing the basic structures of the materials. Elemental analysis shows that alkaline earth metal ions can exchange not only with cation species (Na^+ or NH_4^+), but also with protons in the materials (Table 5.6).

Selective removal of Cs in the alkaline metal ion solution is important. The transition metal oxide molecular wire also shows the ion-exchange property, in which the original NH_4^+ is exchanged with other cations. With this property, the material can be used for the removal of Cs^+, the isotope of which is the main radiation pollutant in nuclear waste, which has a long life-time and produces gamma and beta particles [26].

Table 5.6 Chemical formulas of ε-ZM and ε-MM with alkaline earth metal ions, Reprinted with permission from Ref. [27]. Copyright 2014 American Chemical Society

	Formula
ε-ZM	$Na_{1.5}H_{11.4}[\varepsilon\text{-}ZnMo_{12}O_{40}\{Zn\}_2]\cdot 5H_2O$
ε-Be-ZM	$Be_{0.3}Na_{0.9}H_{11.4}[\varepsilon\text{-}ZnMo_{12}O_{40}\{Zn\}_2]\cdot 4H_2O$
ε-Mg-ZM	$Mg_{0.6}Na_{0.3}H_{11.4}[\varepsilon\text{-}ZnMo_{12}O_{40}\{Zn\}_2]\cdot 6H_2O$
ε-Ca-ZM	$CaH_{10.9}[\varepsilon\text{-}ZnMo_{12}O_{40}\{Zn\}_2]\cdot 7H_2O$
ε-Sr-ZM	$Sr_{0.8}Na_{0.1}H_{11.2}[\varepsilon\text{-}ZnMo_{12}O_{40}\{Zn\}_2]\cdot 5H_2O$
ε-Ba-ZM	$Ba_{1.4}H_{10.1}[\varepsilon\text{-}ZnMo_{12}O_{40}\{Zn\}_2]\cdot 5H_2O$
ε-MM	$(NH_4)_{2.1}H_{7.5}[\varepsilon\text{-}Mn_{0.2}Mo_{12}O_{40}\{Mn\}_2]\cdot 4H_2O$
ε-Be-MM	$Be_{0.2}(NH_4)_{1.7}H_{7.5}[\varepsilon\text{-}Mn_{0.2}Mo_{12}O_{40}\{Mn\}_2]\cdot 5H_2O$
ε-Mg-MM	$Mg_{0.6}(NH_4)_{1.2}H_{7.2}[\varepsilon\text{-}Mn_{0.2}Mo_{12}O_{40}\{Mn\}_2]\cdot 3H_2O$
ε-Ca-MM	$Ca_{0.9}(NH_4)_1H_{6.8}[\varepsilon\text{-}Mn_{0.2}Mo_{12}O_{40}\{Mn\}_2]\cdot 5H_2O$
ε-Sr-MM	$Sr_{0.8}(NH_4)_1H_7[\varepsilon\text{-}Mn_{0.2}Mo_{12}O_{40}\{Mn\}_2]\cdot 5H_2O$
ε-Ba-MM	$Ba_{1.6}H_{6.4}[\varepsilon\text{-}Mn_{0.2}Mo_{12}O_{40}\{Mn\}_2]\cdot 4H_2O$

The competitive ion-exchange experiment in an aqueous solution including the same amount of the metal chlorides shows that the material preferentially exchanges Cs^+ in the mixed solution of Li, Na, K, Rb, and Cs chlorides (Fig. 5.30a).

Cs^+ is able to be removed rapidly using h-TM as an ion-exchanger. The affinity between cation and the material is described as distribution coefficient (K_d). The K_d value of h-TM is at the level of 10^4, indicating that the material is suitable for Cs exchange. The ion-exchange performance of the material is high in a wide range of pH (1–10). The material showed a better performance of Cs exchange in the neutral condition. The ion-exchange ability does not vary when the pH value of the solution changed.

5.6 Perspective

Zeolitic octahedral metal oxides (ZOMOs) are a new family of crystalline microporous materials. Compared with the reported crystalline microporous materials, such as zeolites and metal–organic frameworks, the materials almost show rigid frameworks and ordered micropores. Different from the reported crystalline microporous materials, ZOMOs show some unique features, including structural diversity, elemental diversity, and redox property. The unique features enable the properties and application performances of the materials to be tuned by different pathways or to show different performances for applications. The chapter shows the details of the recent progresses of ZOMOs, including syntheses, structures, properties, and applications.

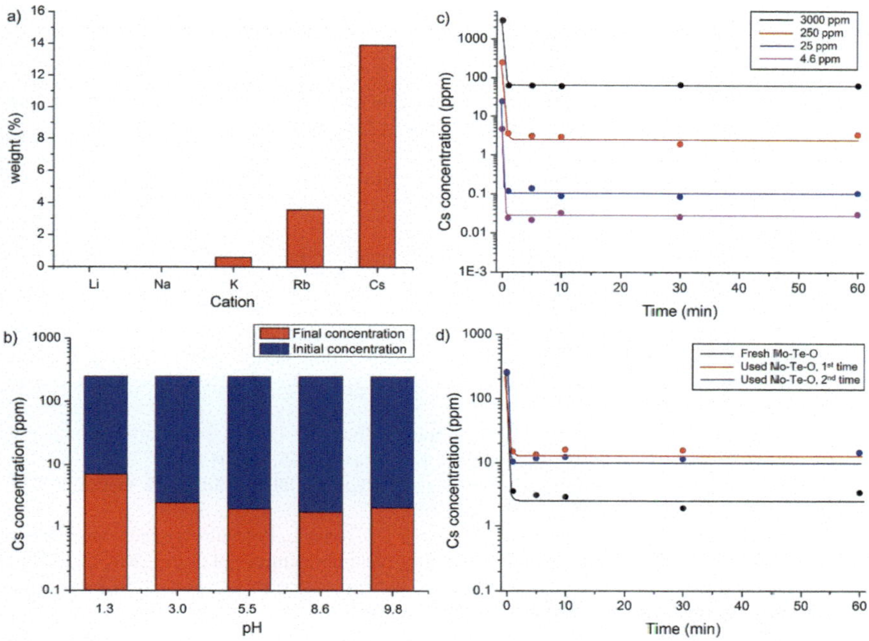

Figure 5.30 **a** competitive ion-exchange experiment of h-TM, conditions: LiCl, NaCl, KCl, RbCl, and CsCl (1.2 mmol), water 15 mL, h-TM 0.3 g, room temperature, 1 h, **b** pH dependent Cs exchange, initial Cs concentration 250 ppm h-TM 0.1 g/10 mL, room temperature, 1 h, K_d(pH 1.3) = 3425, k_d (pH 3.0) = 9944, k_d (pH = 5.5) = 12,839, k_d (pH 8.6) = 14,248, and k_d(pH 9.8) = 12,165, **c** time course Cs concentration in solution after ion-exchange with h-TM, initial Cs concentration 3000 ppm, h-TM 0.1 g/5 mL, initial Cs concentration 250 ppm, 25 ppm, 4.6 ppm, h-TM 0.1 g/10 mL, and d) time course Cs concentration in solution after ion-exchange using recovered h-TM, initial Cs concentration 250 ppm, h-TM 0.1 g/10 mL, Reproduced from Ref. [26]. With permission from the Royal Society of Chemistry

References

1. Chen L, Sun M, Wang Z, Yang W, Xie Z, Su B. Hierarchically structured zeolites: from design to application. Chem Rev. 2020;120:11194–294.
2. Corma A. From microporous to mesoporous molecular sieve materials and their use in catalysis. Chem Rev. 1997;97:2373–420.
3. Corma A. Inorganic solid acids and their use in acid-catalyzed hydrocarbon reactions. Chem Rev. 1995;95:559–614.
4. Meng X, Xiao F. Green routes for synthesis of zeolites. Chem Rev. 2014;114:1521–43.
5. Yu J, Xu R. Rational approaches toward the design and synthesis of zeolitic inorganic open-framework materials. Acc Chem Res. 2010;43:1195–204.
6. Zhou B, Yu J, Li J, Xu Y, Xu W, Qiu S, Xu R. Rational design of two-dimensional layered aluminophosphates with $[Al_3P_4O_{16}]^{3-}$ stoichiometry. Chem Mater. 1999;11:1094–9.
7. Li Y, Yu J, Jiang J, Wang Z, Zhang J, Xu R. Prediction of open-framework aluminophosphate structures using the automated assembly of secondary building units method with Lowenstein's constraints. Chem Mater. 2005;17:6086–93.

8. Hill CL (Ed). Special thematic issue on polyoxometalates. Chem Rev. 1998;98:1–390.
9. Cronin L, Müller A (Eds). Special thematic issue on polyoxometalates. Chem Soc Rev. 2012;41:7325–7648.
10. Long D-L, Tsunashima R, Cronin L. Polyoxometalates: building blocks for functional nanoscale systems. Angew Chem Int Ed. 2010;49:1736–58.
11. Férey G. Hybrid porous solids: past, present, future. Chem Soc Rev. 2008;37:191–214.
12. James SL. Metal-organic frameworks. Chem Soc Rev. 2003;32:276–88.
13. Li J, Sculley J, Zhou H. Metal À organic frameworks for separations. Chem Rev. 2012;112:869–932.
14. Meek ST, Greathouse JA, Allendorf MD. Metal-organic frameworks: a rapidly growing class of versatile nanoporous materials. Adv Mater. 2011;23:249–67.
15. Mizuno N, Misono M. Heterogeneous catalysis. Chem Rev. 1998;98:199–218.
16. Khan MI, Meyer LM, Haushalter RC, Schweitzer AL, Zubieta J, Dye JL. Giant voids in the hydrothermally synthesized microporous square pyramidal-tetrahedral framework vanadium phosphates $[HN(CH_2CH_2)_3NH]K_{1.35}[V_5O_9(PO_4)_2]xH_2O$ and $Cs_3[V_5O_9(PO_4)_2]xH_2O$. Chem Mater. 1996;8:43–53.
17. Zhou Y, Yao S, Yan J, Chen L, Wang T, Wang C, Zhang Z. Design and synthesis of purely inorganic 3D clusters and manganese linkers. Dalt Trans. 2015;44:20435–40.
18. Vilà-Nadal L, Cronin L. Design and synthesis of polyoxometalate-framework materials from cluster precursors. Nat Rev Mater. 2017;2:17054.
19. Mitchell SG, Streb C, Miras HN, Boyd T, Long D-L, Cronin L. Face-directed self-assembly of an electronically active archimedean polyoxometalate architecture. Nat Chem. 2010;2:308–12.
20. Uchida S, Mizuno N. Design and syntheses of nano-structured ionic crystals with selective sorption properties. Coord Chem Rev. 2007;251:2537–46.
21. Uchida S, Eguchi R, Nakamura S, Ogasawara Y, Kurosawa N, Mizuno N. Selective sorption of olefins by halogen-substituted macrocation-polyoxometalate porous ionic crystals. Chem Mater. 2012;24:325–30.
22. Uchida S. Frontiers and progress in cation-uptake and exchange chemistry of polyoxometalate-based compounds. Chem Sci. 2019;10:7670–9.
23. Eguchi R, Uchida S, Mizuno N. Inverse and high CO_2/C_2H_2 sorption selectivity in flexible organic-inorganic ionic crystals. Angew Chem Int Ed. 2012;51:1635–9.
24. Seino S, Kawahara R, Ogasawara Y, Mizuno N, Uchida S. Reduction-induced highly selective uptake of cesium ions by an ionic crystal based on silicododecamolybdate. Angew Chem Int Ed. 2016;55:3987–91.
25. Zhang Z, Sadakane M, Murayama T, Ueda W. Investigation of the formation process of zeolite-like 3D frameworks constructed with ε-Keggin-type polyoxovanadomolybdates with binding bismuth ions and preparation of a nano-crystal. Dalt Trans. 2014;43:13584–90.
26. Zhu Q, Zhang Z, Sadakane M, Yoshida A, Hara M, Ueda W. Synthesis of crystalline molybdenum oxides based on a 1D molecular structure and their ion-exchange properties. New J Chem. 2017;41:4503–9.
27. Zhang Z, Sadakane M, Murayama T, Sakaguchi N, Ueda W. Preparation, structural characterization, and ion-exchange properties of two new zeolite-like 3D frameworks constructed by ε-Keggin-type polyoxometalates with binding metal ions, $H_{11.4}[ZnMo_{12}O_{40}Zn_2]^{1.5-}$ and $H_{7.5}[Mn_{0.2}Mo_{12}O_{40}Mn_2]^{2.1-}$. Inorg Chem. 2014;53:7309–7318.
28. Sadakane M, Kodato K, Kuranishi T, Nodasaka Y, Sugawara K, Sakaguchi N, Nagai T, Matsui Y, Ueda W. Molybdenum-vanadium-based molecular sieves with microchannels of seven-membered rings of corner-sharing metal oxide octahedra. Angew Chem Int Ed. 2008;47:2493–6.
29. Sadakane M, Watanabe N, Katou T, Nodasaka Y, Ueda W. Crystalline Mo_3VO_x mixed-metal-oxide catalyst with trigonal symmetry. Angew Chem Int Ed. 2007;46:1493–6.
30. Igarashi T, Zhang Z, Haioka T, Iseki N, Hiyoshi N, Sakaguchi N, Kato C, Nishihara S, Inoue K, Yamamoto A, Yoshida H, Tsunoji N, Ueda W, Sano T, Sadakane M. Synthesis of E-Keggin-type cobaltomolybdate-based 3D framework material and characterization using atomic-scale HAADF-STEM and XANES. Inorg Chem. 2017;56:2042–9.

31. Zhang Z, Zhu Q, Sadakane M, Murayama T, Hiyoshi N, Yamamoto A, Hata S, Yoshida H, Ishikawa S, Hara M, Ueda W. A Zeolitic vanadotungstate family with structural diversity and ultrahigh porosity for catalysis. Nat Commun. 2018;9:3789.
32. Zhang Z, Sadakane M, Murayama T, Izumi S, Yasuda N, Sakaguchi N, Ueda W. Tetrahedral connection of ε-Keggin-type polyoxometalates to form an all-inorganic octahedral molecular sieve with an intrinsic 3D pore system. Inorg Chem. 2014;53:903–11.
33. Zhang Z, Murayama T, Sadakane M, Ariga H, Yasuda N, Sakaguchi N, Asakura K, Ueda W. Ultrathin inorganic molecular nanowire based on polyoxometalates. Nat Commun. 2015;6:7731.
34. Boultif A, Louer D. Powder pattern indexing with the dichotomy method. J Appl Crystallogr. 2004;37:724–31.
35. Neumann MA. X-cell: a novel indexing algorithm for routine tasks and difficult cases. J Appl Crystallogr. 2003;36:356–65.
36. Le Bail A. Frontiers between crystal structure prediction and determination by powder diffractometry. Powder Diffr. 2008;23:S5–12.
37. Palatinus L, Chapuis G. SUPERFLIP-a computer program for the solution of crystal structures by charge flipping in arbitrary dimensions. J Appl Crystallogr. 2007;40:786–90.
38. Zhang Z, Sadakane M, Noro S, Hiyoshi N, Yoshida A, Hara M, Ueda W. The assembly of an all-inorganic porous soft framework from metal oxide molecular nanowires. Chem Eur J. 2017;23:1972–80.
39. Zhang Z, Sadakane M, Hiyoshi N, Yoshida A, Hara M, Ueda W. Acidic ultrafine tungsten oxide molecular wires for cellulosic biomass conversion. Angew Chem Int Ed. 2016;55:10234–8.
40. Yang C, Zhu Q, Sadakane M, Zhang Z, Li Y, Ueda W. Vanadium-enhanced intramolecular redox property of a transition- metal oxide molecular wire. Inorg Chem. 2020;59:16557–66.
41. Zhang Z, Sadakane M, Hara M, Li Y, Ueda W. Intramolecular electron transfer and oxygen transfer of phosphomolybdate molecular wires. Inorg Chem. 2019;58:12272–9.
42. Zhang Z, Sadakane M, Noro S, Murayama T, Kamachi T, Yoshizawa K, Ueda W. Selective carbon dioxide adsorption of ε-Keggin-type zincomolybdate-based purely inorganic 3D frameworks. J Mater Chem A. 2015;3:746–55.
43. Wang J, Zhu Q, Zhang Z, Sadakane M, Li Y, Ueda W. Zeolitic octahedral metal oxides with ultra-small micropores for C_2 hydrocarbon separation. Angew Chem Int Ed. 2021;133:18476–82.
44. Hu P, Zhu Q, Zhang Z, Li Y, Ueda W. Zeolitic vanadotungstates with ultrahigh porosities for separation of butane and isobutane. Chem Eur J. 2021;27:13067–71.
45. Zhang Z, Ishikawa S, Zhu Q, Murayama T, Sadakane M, Hara M, Ueda W. Redox-active zeolitic transition metal oxides based on ε-Keggin units for selective oxidation. Inorg Chem. 2019;58:6283–93.
46. Zhu Q, Yin S, Zhou M, Wang J, Chen C, Hu P, Jiang X, Zhang Z, Li Y, Ueda W. Aerobic alcohol oxidation by a zeolitic octahedral metal oxide based on iron vanadomolybdates under mild conditions. ChemCatChem. 2021;13:1763–71.
47. Zhang Z, Ishikawa S, Kikuchi M, Yoshikawa H, Lian Q, Wang H, Ina T, Yoshida A, Sadakane M, Matsumoto F, Ueda W. High-performance cathode based on microporous Mo-V-Bi oxide for Li battery and investigation by operando x-ray absorption fine structure. ACS Appl Mater Interfaces. 2017;9:26052–9.
48. Zhang Z, Wang H, Yoshikawa H, Matsumura D, Hatao S, Ishikawa S, Ueda W. Zeolitic vanadomolybdates as high-performance cathode-active materials for sodium-ion batteries. ACS Appl Mater Interfaces. 2020;12:6056–6063.

Chapter 6
Position Control of Catalytic Elements in Zeolites

Ryota Osuga and Toshiyuki Yokoi

6.1 Introduction and the Recent Trend in Zeolite Catalysis

6.1.1 Zeolite Catalysts with Well-Controlled Position of Active Sites

6.1.1.1 Position Control of Heteroatoms in the Zeolite Framework

Zeolites are well-known crystalline microporous materials, which have been widely used as solid acid catalysts due to their unique and attractive characteristics: well-defined porous structure, shape selectivity to a small molecule, high surface area, controllable acid and ion-exchange properties, and high hydrothermal stability [1–11]. The acidity of zeolites is due to the bridging OH groups between the framework Si and Al atoms (Fig. 6.1 [12]). The acid strength of such Brønsted acid sites can be controlled by the isomorphous substitution of Si atoms with other trivalent heteroatoms such as B^{3+}, Fe^{3+}, and Ga^{3+}, and the order of the acid strength has been experimentally and computationally determined as $B^{3+} << Fe^{3+} < Ga^{3+} < Al^{3+}$ [13–15]. In addition, the incorporation of tetravalent heteroatoms represented by Ti^{4+} and Sn^{4+} yields Lewis acidity derived from each metal center, which functions as oxidation catalysts as well as Lewis acid catalysts [16–20]; such metal-substituted zeolites (except for Al) are called "metallosilicate". Kinds of heteroatoms, therefore, affect the acidity of zeolites. Recently, not only the type of the heteroatoms, but also the "position" of the heteroatoms have been regarded as a crucial factor

R. Osuga
Institute of Multidisciplinary Research for Advanced Materials, Tohoku University, 2-1-1 Katahira, Aoba-ku, Sendai, Miyagi 980-8577, Japan

T. Yokoi (✉)
Institute for Innovative Research, Tokyo Institute of Technology, 4259 Nagatsuta, Midori-ku, Yokohama 226-8503, Japan
e-mail: yokoi@cat.res.titech.ac.jp

W. Ueda (ed.), *Crystalline Metal Oxide Catalysts*,
https://doi.org/10.1007/978-981-19-5013-1_6

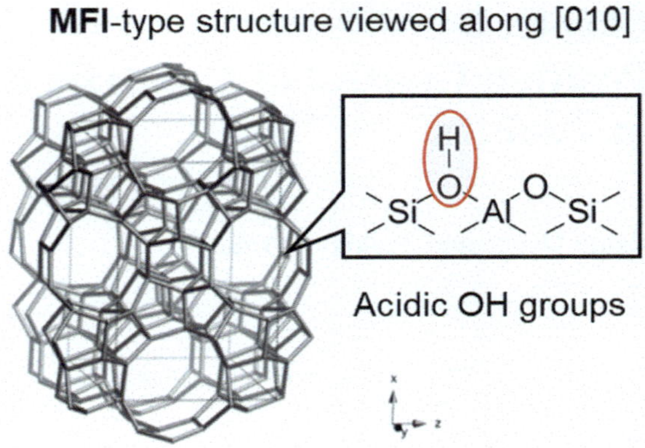

Fig. 6.1 Schematic of MFI-type topology and local structure of acidic OH groups in zeolite. Adapted with permission from the International Zeolite Association, the Structure Commission, Ref. [12]

for improving the catalytic activity and selectivity in addition to the ion-exchange property [10, 21–24] because the position of the heteroatoms is directly related to the location of both Brønsted and Lewis acid sites.

In general, the location and distribution of the framework Al atoms are controlled by the rational selection of organic/inorganic cations [21, 25–31] and starting materials [32–36] used for the zeolite synthesis. In 2014, Román-Leshkov, Davis, and their co-workers reported the control of the framework Al species in FER-type zeolite by using different organic structure-directing agents (OSDAs) that can be achieved based on the charge balance of the zeolite framework [21]. Likewise, Di Iorio and Gounder et al., have successfully synthesized CHA-type zeolites, SSZ-13, using *N,N,N*-trimethyl-1-admantylammonium hydroxide (TMAdaOH) as an OSDA with the location of the framework Al atoms controlled (Fig. 6.2 [28]); the proximity of the framework Al atoms ($Al - O(-Si - O)_x - Al$), *i.e.*, paired ($x = 1, 2$) and isolated Al atoms ($x \geq 3$), was tuned by varying $Na^+/TMAda^+$ ratios [28]. Similarly, our research group has tackled the control of framework Al atoms, which will be introduced in the next section (*See*, 6.2. Control and evaluation of the location of heteroatom in the zeolite framework).

6.1.1.2 Positional Control of Metal Species in Metal-Containing Zeolites

Metal-containing zeolites, including metal, oxide, cluster, and cation, have been developed as catalyst to realize various catalytic processes [37–47]. The position, state, and size of the metal species in zeolite are critical factors to determine the catalytic activity. These properties are oftentimes tuned by different synthetic approaches. Among them, the position of metal species, *i.e.*, intra- or extra-framework, is a hot topic in zeolite chemistry. Especially, in the case of cationic metal species, their location strongly depends on the position of the framework Al atoms (*i.e.*, ion-exchange sites), because paired Al sites can function as ion-exchange sites

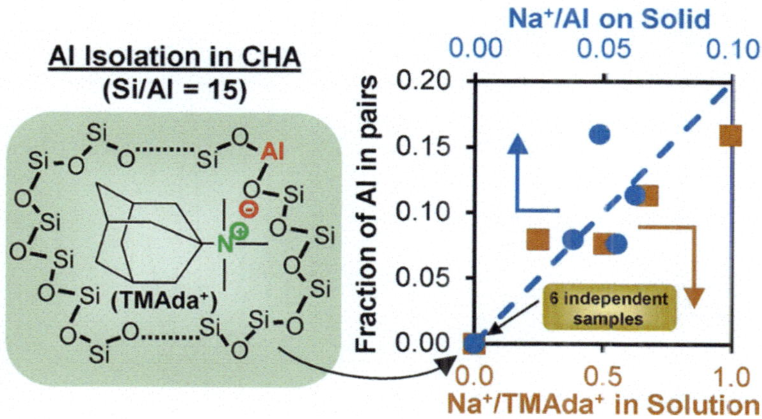

Fig. 6.2 Strategy for positional control of framework Al atoms in CHA-type structure. Adapted with permission from Ref. [28]. Copyright 2016 American Chemical Society

for the catalytically active divalent cations such as Co^{2+}, Ni^{2+}, and Cu^{2+} [37, 40, 48–50]. Hence, the synthesis of the metal-containing zeolites with the location, state and size controlled is also an attractive challenge as well as the positional control of the heteroatoms in the zeolite framework.

6.1.2 Determination of the Position of Catalytic Active Sites

6.1.2.1 Solid-State NMR Measurement

^{29}Si magic angle spinning (MAS) nuclear magnetic resonance (NMR) spectroscopy is the predominant method in order to analyze the framework Al distributions [22, 51, 52]. This method can detect $Si(OSi)_{4-n}(OAl)_n$ units (n = 0–4). Figure 6.3 shows ^{29}Si NMR spectra of Y zeolite (Si/Al = 2.4), where three peaks are observed at around −105, −100, −94, and −88 ppm attributed to the framework Si species, $Si(OSi)_4$, $Si(OSi)_3(OAl)_1$, $Si(OSi)_2(OAl)_2$, and $Si(OSi)_1(OAl)_3$, respectively. Besides, ^{27}Al MAS NMR spectroscopy is also used for determining the location of the framework Al atoms, which is a direct observation technique for Al atoms [53]. Among them, ^{27}Al–^{27}Al double-quantum single-quantum (DQ-SQ) MAS NMR measurement allows the observation of close Al atoms in the zeolite framework [54]. Moreover, ^{27}Al multi-quantum (MQ) MAS NMR spectroscopy has enabled the quantification of the framework Al atoms with different crystallographic environments combined with ^{27}Al MAS NMR measurement [22, 55, 56].

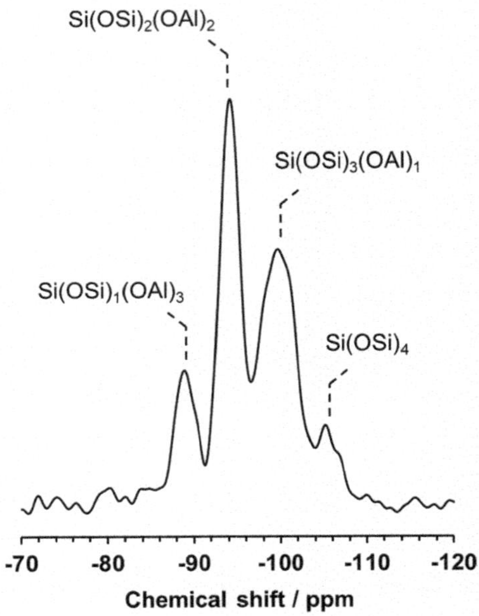

Fig. 6.3 ^{29}Si MAS NMR spectrum of FAU-type zeolite (Si/Al = 2.4)

6.1.2.2 In situ IR Spectroscopy with Probe Molecules

Infrared (IR) spectroscopy with probe molecules would be one of the most convenient and frequently used techniques for characterizing catalytic properties of zeolites [57–60]. Since the acidic OH groups are observed as the O–H stretching vibration in the IR spectra, the vibrations of the catalytic active sites (acidic OH groups) and the adsorbed molecules can be monitored simultaneously. Thus, the location of the acid sites can be distinguished based on the relationship of sizes between zeolitic pore and probe molecules. For example, pyridine cannot enter into the 8-ring pore due to size limitations, while CO molecule can access the 8-ring pore [61], indicating that we can quantify the number of the acid sites located at each size of the micropore by estimating the amount of the adsorbed probe molecules. In addition, the number of acid sites on the external surface is also quantified by using more balky probe molecules [62].

6.1.2.3 Ion-Exchange for Divalent Metal Cations

The presence of the paired Al sites can be confirmed by the ion exchange for the divalent metal cations by determining the number of the ion-exchanged cations, which represents a simple approach to quantify the paired Al sites in the zeolite framework [22]. It should be noted that the transport of the divalent cations as aqua complexes inside zeolitic micropore without any restriction is necessary. Among several divalent cations, Co^{2+} is generally accepted as the probe cations to characterize the paired

Al sites. There is an advantage in Co^{2+} ion exchange for the characterization of the framework Al distribution that the coordination of Co^{2+} can be monitored by visible absorption spectroscopy. The visible spectrum of the hexaaqua Co^{2+} complex is detectable in both solution and solid (zeolites). Besides, the perturbation of the coordinated Co^{2+} with zeolite framework is observed as a new band with a higher extinction coefficient compared to the hexaaqua Co^{2+} complex because the *d-d* transitions of the coordinated Co^{2+} with zeolites are allowed [63].

6.1.2.4 Catalytic Reaction–CI Value

Frillette and his co-workers proposed utilizing the constraint index (CI) value for the estimation of the Brønsted acid sites where located in zeolitic micropore in 1981 [64]. In this case, the CI value was defined as the reaction rate (k) ratio of the cracking of n-hexane (n-Hx) to its isomer, 3-methyl pentane (3-MP), which can be calculated by the conversion level (X) at the initial stage of both reactions under the same reaction conditions as the following equation (Eq. 6.1).

$$\mathrm{CI} = k_{\mathrm{n-Hx}}/k_{3-\mathrm{MP}} = \log\left[1 - X_{\mathrm{n-Hx}}\right]/\log\left[1 - X_{3-\mathrm{MP}}\right] \tag{6.1}$$

There are two types of reaction mechanisms in paraffin cracking [65]; monomolecular and bimolecular reactions, which proceed via the pentacoordinated carbonium ion and the carbeniumion/β-scission mechanism with a hydrogen transfer reaction, respectively. The tertiary carbon is easily reacted with the acid sites during the carbeniumion/β-scission mechanism compared to the primary carbon, meaning that the bimolecular reaction preferentially proceeds in 3-MP cracking. On the other hand, the bimolecular reaction requires a relatively wide space due to the bulky transition state of 3-MP. The higher CI, therefore, means that the acid sites are preferentially located in a relatively narrow space.

6.1.3 Recent Trend in Zeolite Catalysis

6.1.3.1 CH_4 and/or CO_2 Conversions

Methane is highly abundant and the simplest alkane that has attracted much attention as a carbon source to produce chemical feedstocks from the aspect of preventing the depletion of petroleum resources. Therefore, the catalytic conversion of methane to valuable chemicals, such as syngas, hydrocarbons, aromatics, methanol, and so on, has been extensively tackled by many research groups [9, 66–71]. The production of methanol from methane is one of the hot topics in the field of methane conversion. However, it is difficult to achieve a high methanol yield due to the high stability of methane and the reactivity of methanol. In 2017, Sushkevich and van Bokhoven et al. reported the selective anaerobic oxidation of methane to methanol over the

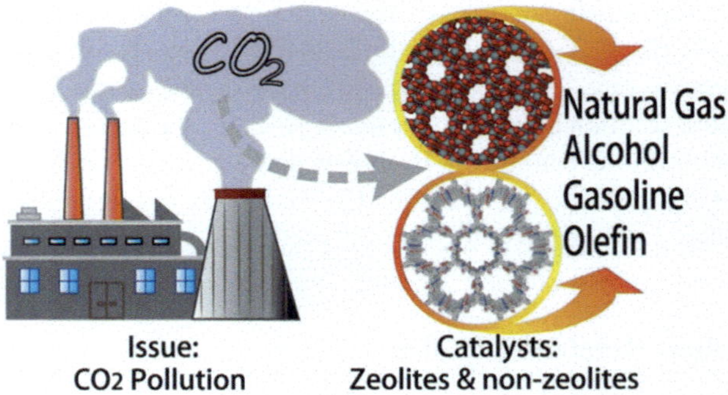

Fig. 6.4 Catalytic conversion from CO_2 to valuable chemicals. Adapted with permission from Ref. [73]. Copyright 2021 Elsevier

copper-exchanged mordenite (CuMOR) [42], where the chemical loop process was employed to improve the product selectivity. Recently, Jin and Xiao et al. realized the methane oxidation to methanol by *in situ* formed peroxide as an oxidant [72]. In this case, AuPd nanoparticles fixed within aluminosilicate zeolite crystals function as catalytic active sites for peroxide and methanol formation. It is notable that this reaction was carried out under continuous gas-flow conditions.

Carbon dioxide, an abundant greenhouse gas, is also receiving a lot of attention as a chemical resource. The development of an efficient way for converting the generated CO_2 from industry to more valuable chemicals is desired from the viewpoint of the current global warming solution [73]. Among them, zeolite-based catalysts can transform CO_2 into hydrocarbons in the presence of hydrogen. Commonly, such a reaction can be realized via two different routes: (1) Fischer–Tropsch synthesis [74] through reverse water–gas shift (RWGS) followed by hydrocarbon cracking, isomerization, and aromatization, (2) CO_2 conversion to methanol followed by methanol to hydrocarbon reaction [75–77]. In these processes, zeolite-based catalysts have been often used, where the metal-containing zeolites exhibit high catalytic activity for both reaction processes [73]. Such will be key reactions to realize sustainable chemistry (Fig. 6.4 [73]).

6.1.3.2 Catalytic Transformation of Plastics into Valuable Chemicals

The demand for the chemical conversion technology for plastic waste into fuels is going to increase [78], which can contribute to the achievement of "Sustainable Development Goals" (SDGs). The combination of pyrolysis and the catalytic reaction has been attempted to the transformation of plastics into valuable chemicals [79–81] but a high selectivity to the fuels had not been achieved. Very recently, Liu and Vlachos et al. reported the hydrocracking of the plastic wastes to fuels under mild conditions [82], where the polyolefins were converted to liquid fuels, diesel, jet, and gasoline-range hydrocarbons, with high yield using $Pt/WO_3/ZrO_2$ and HY zeolite catalysts in the presence of hydrogen at 225 °C. This is a tandem catalytic process; the polymer was primarily activated by Pt, and then the catalytic cracking occurred over WO_3/ZrO_2 and HY zeolite as well as the hydrogenation of olefin intermediates by Pt. This process can be applied to various plastic wastes including low- and high-density polyethylene, polypropylene, and polystyrene. It is expected that the achievement will be a breakthrough in this field.

6.2 Control and Evaluation of the Location of Heteroatom in the Zeolite Framework

6.2.1 Aluminosilicates

6.2.1.1 MFI-Type Zeolites

Controlling the location of the framework Al atoms in the aluminosilicate zeolites is a key aspect of tuning the catalytic activity because their position reflects the acid site location, *i.e.*, the catalytic active site. We have been tackling to control of the framework Al atoms in the aluminosilicate zeolites based on the rational selection of OSDAs, inorganic cations, and starting materials [21, 25–36]. In this section, some achievements for MFI- [25–27], MSE- [29, 83], CHA-type frameworks [33] will be introduced.

MFI-type aluminosilicates, ZSM-5, are generally synthesized by using tetrapropylammonium (TPA) hydroxide as an OSDA. The MFI-type structure has consisted of parallel and straight 10-membered ring (10R) channels intersected by sinusoidal 10R channels (Fig. 6.5 [12]). The sizes of both 10R channels are similar to that of the aromatic ring (ca. 5.5 Å). In contrast to the channels, the intersections of these 10R channels are a large spherical space (ca. 10 Å in diameter), and the TPA cations are located at the channel intersections due to the size limitation [84, 85]. Therefore, the framework Al atoms in MFI-type zeolites synthesized with TPA cations in the absence of Na cations (MFI[TPA]) should be selectively located at the channel intersections. In practice, we prepared MFI[TPA] and MFI[TPA, Na] synthesized by using TPA and Na cations, and the location of Al atoms was investigated by ^{27}Al MAS and MQMAS NMR techniques combined with the evaluation of CI value [25]. This work revealed that the framework Al atoms in the MFI-type zeolites synthesized by using TPA cation only were predominantly located at the channel intersections (Fig. 6.6 [25]).

Next, we tried the selective sitting of the framework Al atoms at the 10R channels in the MFI topology [26, 27], where various alcohols were adopted as pore-filling agents. The strategy for this study, bulky alcohol was added into the synthetic gels in the presence of Na cation. In this case, it is expected that various bulky alcohols can work as a pore-filling agent as well as the OSDA [26, 27]. Meanwhile, the pore-filling agents have no charge unlike the TPA cation; Al^{3+} atoms should be placed around

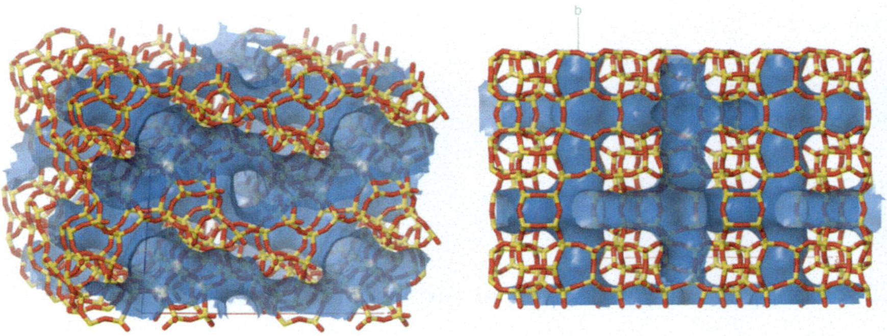

Fig. 6.5. Framework and channel system of MFI-type topology with different views. Adapted with permission from the International Zeolite Association, the Structure Commission, Ref. [12].

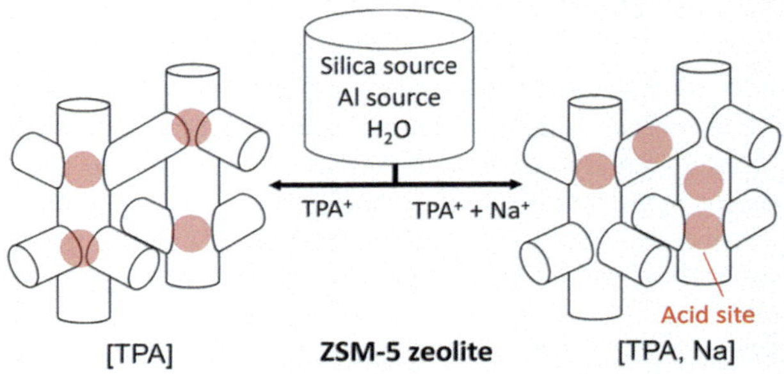

Fig. 6.6 Control of framework Al atoms in MFI-type aluminosilicates. Adapted with permission from Ref. [25]. Copyright 2015 American Chemical Society

Na^+ species, resulting in Al atoms being preferentially located at 10R channels. As a result, the thus-prepared MFI-type aluminosilicates gave remarkably higher CI values in comparison with MFI[TPA]. In particular, trimethylolethane (TME) and *tert*-butanol (TBO) were the predominant candidates (Fig. 6.7 [27]). This work manifested the importance of the kinds of OSDA and/or pore-filling agents with inorganic cations.

6.2.1.2 MSE-Type Zeolites

We have investigated the location of framework Al atoms in MSE-type zeolites not only MFI-type structure [29, 83]. The MSE-type structure has structured by interconnecting 12-10-10-ring channels and a supercage connected to a 10R channel (Fig. 6.8 [12]). MCM-68 was reported as the first example of the synthesis of MSE-type aluminosilicate by Mobil researchers [86], which was synthesized by

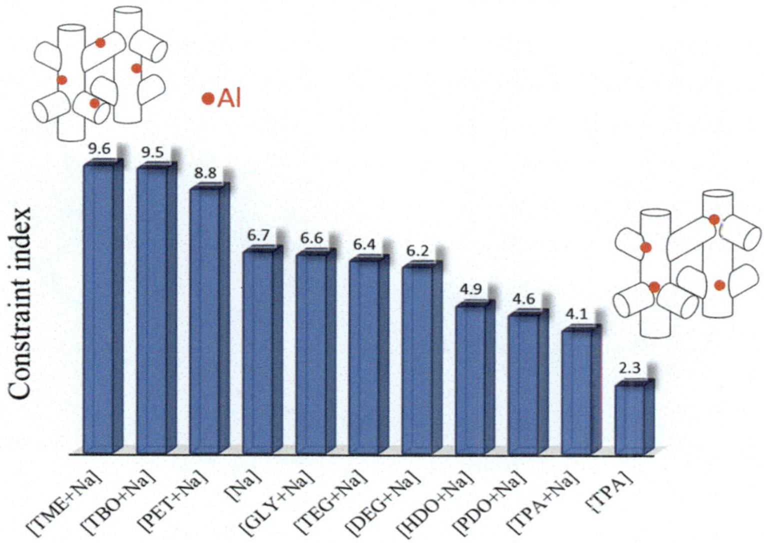

Fig. 6.7 CI values for MFI-type aluminosilicates synthesized from the synthetic gels containing various alcohols. Adapted with permission from Ref. [27]. Copyright 2017 Elsevier

using dipyrrolidinium cation, *N,N,N′,N′*-tetraethylbicyclo[2.2.2]oct-7-ene-2,3:5,6-dipyrrolidinium ($TEBOP^{2+}$, Fig. 6.9a), as an OSDA. In 2011, UZM-35, another MSE-type aluminosilicate reported by UOP LLC researchers [87], was synthesized by using dimethyldipropylammonium ($DMDPA^{+}$, Fig. 6.9b) cation. Previously, our group clarified the acid site location in MCM-68 with different Si/Al ratios by using *in situ* IR and ^{27}Al MQMAS NMR techniques [83], where the numbers of acid sites in 12R and 10R channels were quantified by *in situ* IR measurement with probe molecules, pyridine (Py) and 2,6-di-*tert*-butylpyridine (DTBPy) (Fig. 6.10 [83]), and the different distributions of the framework Al species were detected in ^{27}Al MQMAS NMR. Besides, the effect of the dealumination treatment on the acid site location and the catalytic activity for dehydration of sorbitol to isosorbide was studied. Very recently, the difference in the location of framework Al atoms between MCM-68 and UZM-35 was clarified by using the spectroscopic approaches combined with evaluating catalytic activity (Fig. 6.11 [29]). In both MSE-type aluminosilicates, 61 and 33% of Brønsted acid sites were located at the 12R channels for MCM-68 and UZM-35, respectively. Moreover, the unique catalytic activities for the hydrocarbon cracking reactions depending on the different acid site locations were observed, indicating that the kind of OSDAs for the zeolite synthesis is among the important factors for controlling the location of the framework Al atoms in the MSE-type framework.

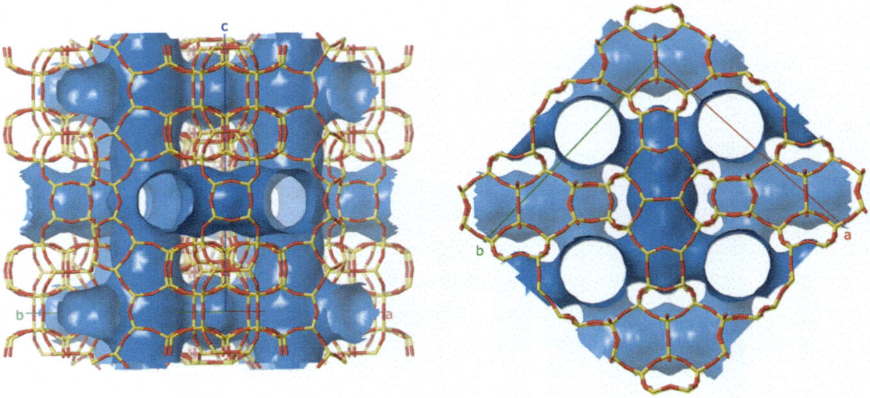

Fig. 6.8 Framework and channel system of MSE-type topology with different views. Adapted with permission from the International Zeolite Association, the Structure Commission, Ref. [12]

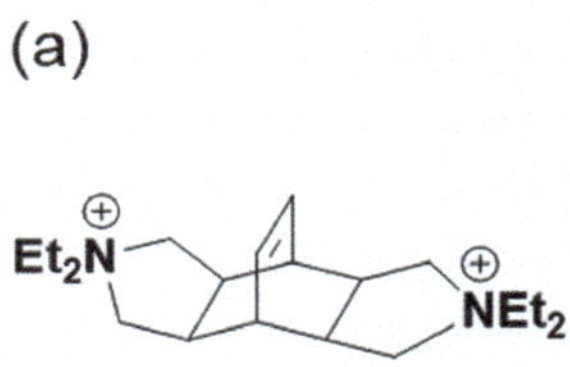

Fig. 6.9 OSDAs used for synthesis. **a** MCM-68 and **b** UZM-35

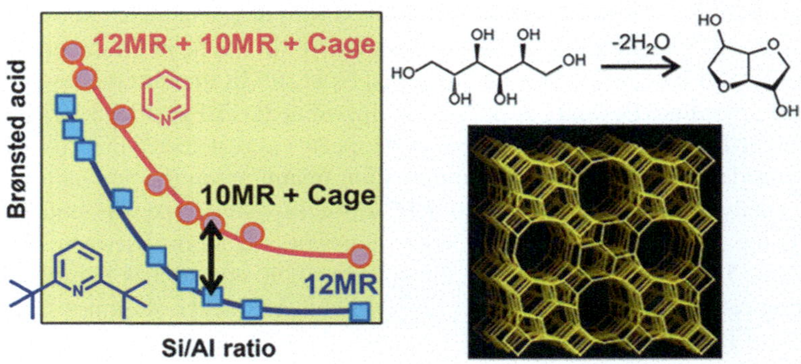

Fig. 6.10 Acid site location of MCM-68 with different Si/Al ratios. Adapted with permission from Ref. [83]. Copyright 2018 American Chemical Society

6.2.1.3 CHA-Type Zeolites

The starting materials also affect the framework Al distribution not only the kind of OSDAs. 8R zeolites have received much attention from the viewpoint of selective

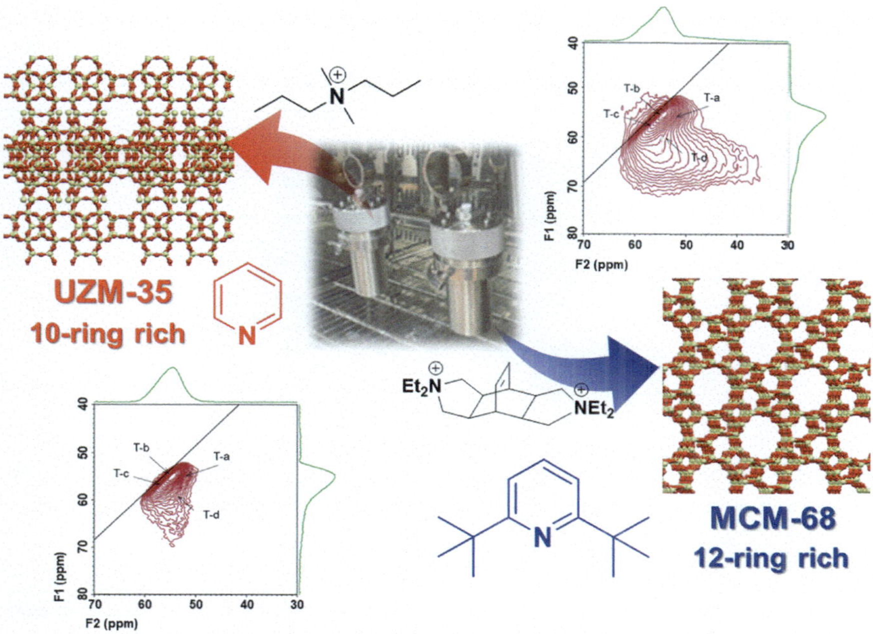

Fig. 6.11 Different acid site distributions in MSE-type aluminosilicates. Adapted with permission from Ref. [29]. Copyright 2022 The Royal Society of Chemistry

catalysis for small molecules owing to their pore size. CHA-type zeolites and cage-structured 8R zeolites (Fig. 6.12 [12]) have been extensively investigated by many groups. Among them, H^+-form and metal cation-exchanged CHA-type aluminosilicates are often used as the effective catalyst for the methanol to olefins (MTO) and the NO_x selective catalytic reduction (SCR) with ammonia reactions [7]. As already mentioned in Sect. 6.1.1.1, the control of the framework Al distribution in CHA-type topology is a crucial challenge. Our group has discovered the critical impact of the starting materials, *i.e.*, Si and Al sources, on the distribution of the framework Al atoms in CHA-type topology [33]. The CHA-type aluminosilicates can be synthesized by using amorphous silica with Al source (*e.g.*, $Al(OH)_3$, $Al(NO_3)_3$, $Al_2(SO_4)_3$, etc.) or the FAU-type zeolite as Si and Al sources [7]. In this work, CHA-type aluminosilicates were synthesized with different proportions of FAU-type zeolites in total Al source (thus-synthesized CHA-type zeolites are designated as CHA–F–*x*. *x* means the proportion of FAU-type zeolite in the total Al source for the starting gels.) (Fig. 6.13 [33]). The proportion of "Q^4(nAl)", $Si(OSi)_{4-n}(OAl)_n$ and "Q^3(nAl)", $Si(OSi)_{3-n}(OH)(OAl)_n$, in the total framework Si atoms, was estimated by ^{29}Si MAS NMR. The proportion of Q^4(2Al) was increased with increasing the proportion of the Al source derived from the FAU-type zeolite (Table 6.1). These CHA-type zeolites showed different catalytic activities for MTO reaction in addition to the hydrothermal stability. Figure 6.14 shows the changes in the methanol conversion and the selectivity for each product with the time on stream, where Q^4(1Al) rich samples yielded a longer catalytic lifetime than that for Q^4(2Al) rich samples [33].

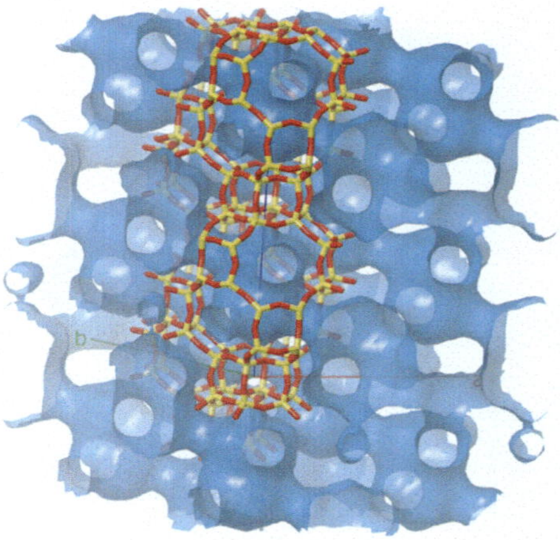

Fig. 6.12 Framework and cage system of CHA-type topology. Adapted with permission from the International Zeolite Association, the Structure Commission, Ref. [12]

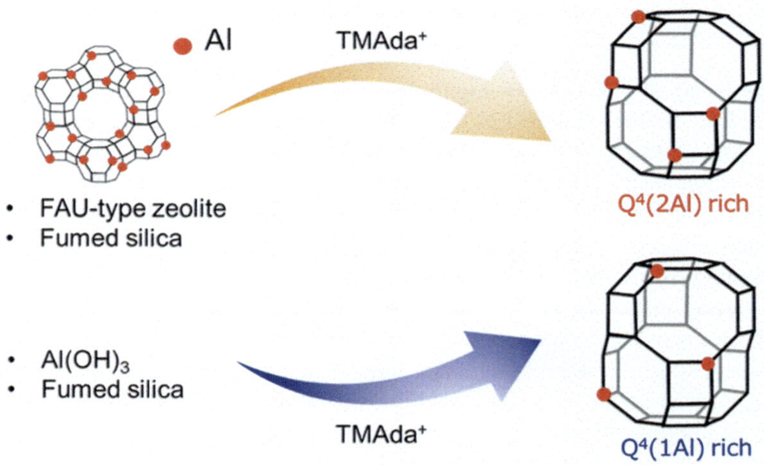

Fig. 6.13 Control of distribution of framework Al atoms in CHA-type structure. Adapted with permission from Ref. [33]. Copyright 2018 American Chemical Society

Table 6.1 Chemical composition and the proportion of Q^4(nAl) and Q^3(nAl). Acapted with permission from Ref. [33]. Copyright 2018 American Chemical Society

Samples	Si/Al ratio		Proportion of $Q^4(n\text{Al})^b$ and $Q^3(n\text{Al})^c$ (%)				
	ICP	NMR[a]	Q^4(2Al)	Q^4(2Al)	Q^3(0Al)	Q^4(1Al)	Q^4(0Al)
CHA-F-0	9	10.2	<0.1	5.5	1.2	28.1	65.2
CHA-F-0.1	10.6	11.4	<0.1	3.1	2.2	29.9	64.7
CHA-F-0.25	10.9	11.8	<0.1	4.5	5.1	24.8	65.6
CHA-F-0.5	11.3	11.0	<0.1	7.3	5	21.5	66.2
CHA-F-0.75	13.7	13.7	<0.1	5.8	4.5	17.9	71.8
CHA-F-1.0	13.5	13.0	<0.1	7.0	9	16.7	67.3
FAU[b]	2.8	2.5	14.9	47.0	<0.1	44.5	8.5

[a]Si/Al molar ratio determined by ^{29}Si MAS NMR.
[b]FAU-type zeolite (Si/Al = 2.8) used for the synthesis.

6.2.2 *Metallosilicates*

6.2.2.1 Ti-MWW Zeolites

Titanosilicates that contain tetrahedrally coordinated Ti atoms in the zeolite framework function as highly active catalysts for the liquid-phase oxidation reactions in the presence of H_2O_2 [88, 89]. A lot of titanosilicate zeolites, such as MFI- [90], *BEA- [91], MOR- [92], MSE- [93], and MWW-type [94] structures, have been developed and applied to the liquid-phase oxidation reaction so far. The positional control of the framework Ti species is becoming increasingly important like a framework Al position in the aluminosilicate. However, we have to overcome several problems to control the location of Ti species in the titanosilicates; the incorporation of Ti atoms in the zeolite framework is much more difficult than Al atoms, and the negative charge on the zeolite framework was not generated by an isomorphous substitution of Ti^{4+} with Si^{4+} [89]. In addition, no evaluation method for the distribution of the framework Ti species has been established. Consequently, we tried establishing the evaluation method for the distribution of the framework Ti species in the MWW-type framework [95]. MWW-type topology is constructed by layered structure, and there are two independent pore systems that can be accessed from 10R pores (Fig. 6.15 [12]): (1) two-dimensional sinusoidal 10R channels and (2) large supercages with 12R openings [12]. In general, Ti-MWW zeolites are synthesized by two methods [94, 96]. One is the direct synthesis (thus-synthesized sample is notated as Ti–MWW–D) and the other is the post-synthesis method (thus-synthesized sample is notated as Ti–MWW–P). In the former, the crystallization is assisted by the boron species. On the other hand, Ti–MWW–P is synthesized by the incorporation of Ti species into the deboronated MWW zeolites. In a previous study, Ti–MWW–P showed a higher catalytic activity for 1-hexene epoxidation than that for Ti–MWW–P [94, 96]. In the case of hydrocarbon cracking, the CI value can be applied to estimate the distribution

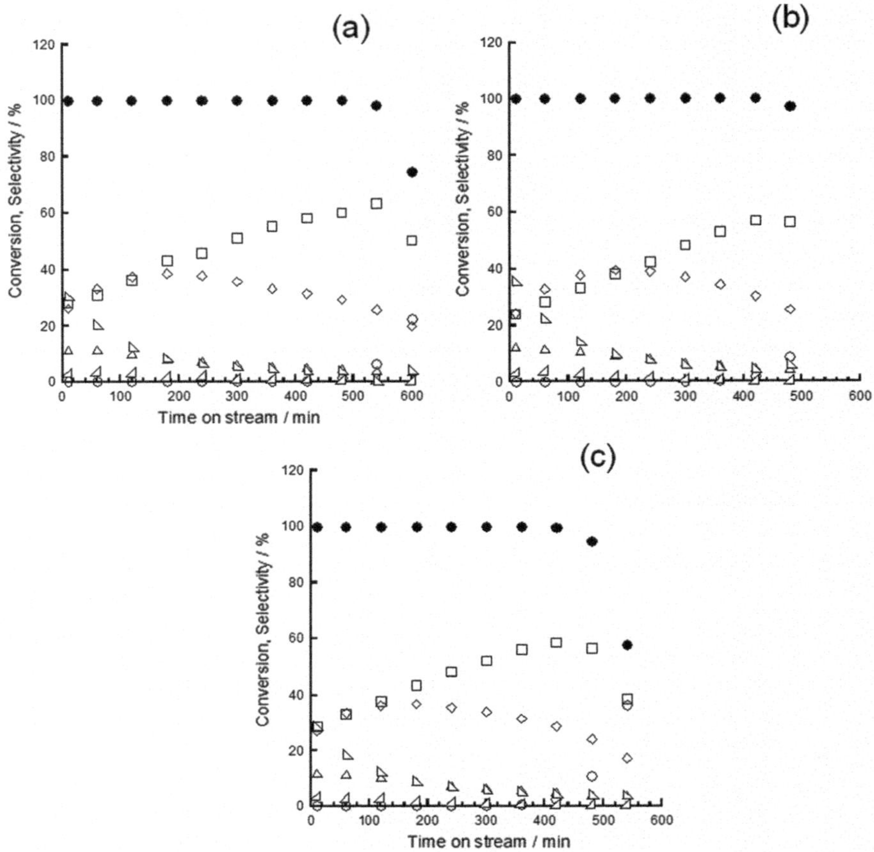

Fig. 6.14 Catalytic performance of CHA-type aluminosilicates in MTO reaction with time course over **a** CHA–F–0.1, **b** CHA–F–0.25 and **c** CHA–F–0.5. Reaction conditions: catalyst, 100 mg; P_{MeOH}, 5 kPa; $W/F = 34$ g h mol^{-1}; temperature, 623 K. ●: Conversion, ◇: propene, ◺: paraffins, □: ethene, △: C4s, ○: dimethyl ether, ◿: over C5. Adapted with permission from Ref. [33]. Copyright 2018 American Chemical Society

of the framework Al atoms. In this study, the distribution of the framework Ti species was estimated based on the catalytic activity for the epoxidation of 1-hexene (1-HX) and 2-methyl-2-pentene (2-MP). Practically, the "1-HX/2-MP index" was used for the evaluation of the distribution of the framework Ti species. This is defined as the ratio of the yields of all products in 1-HX epoxidation to those in 2-MP epoxidation similarly to the CI value. Namely, a higher 1-HX/2-MP index means a possibility for the selective siting of the framework Ti atoms into the narrow space, *i.e.*, 10R channels (Fig. 6.16). The catalytic performance and the 1-HX/2-MP index over both Ti–MWW zeolites are shown in Fig. 6.17, where clear differences were observed depending on the synthesis methods. We found that Ti atoms were selectively located

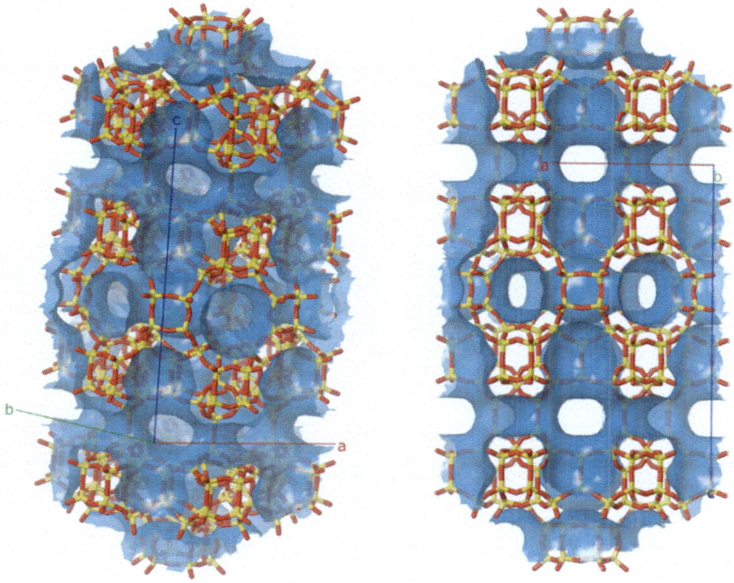

Fig. 6.15 Framework and channel system of MWW-type topology with different views. Adapted with permission from the International Zeolite Association, the Structure Commission, Ref. [12].

in narrow spaces in the Ti–MWW–P [95]. These results will contribute to the development of a new class of titanosilicates with the location of the framework Ti atoms controlled.

6.2.2.2 Sn-BEC Zeolites

Recently, Rodríguez-Fernández, Román-Leshkov, and Moliner et al. reported the selective placement of the framework Sn species in the double-4-ring (D4R) of BEC-type framework via a rationalized post-synthetic grafting method [97]. The BEC-type structure is one of the polymorphs of Beta, which possesses three types of crystallographic T-sites (Fig. 6.18 [97]). The first synthesized BEC-type zeolite, ITQ-17, is a germanosilicate, where it was suggested that Ge atoms would be preferentially placed at the right angles in the D4Rs than Si atoms. Thus, they challenged the selective incorporation of Sn atoms in the D4Rs of the BEC-type structure by taking advantage of such property of Ge-BEC. Firstly, Ge-BEC was synthesized with a high Si/Ge ratio (>150), resulting in the preferential siting of the Ge atoms in D4Rs. Then, the framework Ge species were removed, and the defect sites were generated. Finally, the Sn species were incorporated by the grafting method (Fig. 6.19 [97]). The evaluations of the location of Sn species and Lewis acidity were carried out by a solid-state NMR combination with theoretical calculations and *in situ* IR measurement with

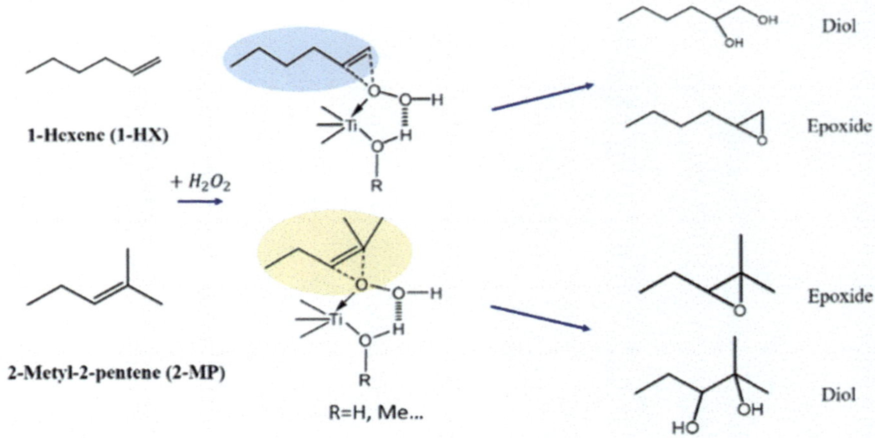

$$\text{1-HX/2-MP} = \frac{\text{Total yield of products by 1-HX oxidation}}{\text{Total yield of products by 2-MP oxidation}}$$

Fig. 6.16 Concept for 1-HX/2-MP index. Adapted with permission from Ref. [95]. Copyright 2019 The Chemical Society of Japan

CD_3CN as a probe molecule, proving the presence of a large number of open Sn sites in D4Rs. Furthermore, the impact of the location of the framework Sn atoms on the catalytic activity was investigated using the Meerwein–Ponndorf–Verley–Oppenauer (MPVO) reaction. The thus-prepared Sn-BEC exhibited a higher catalytic activity than that of the conventional Sn-BEA zeolites. This methodology presented the control of the heteroatoms with the T-sites level.

6.3 Location, State, Size, and Reactivity of Metal Species in Zeolite

6.3.1 Metal Cations in Zeolite

6.3.1.1 Fe and Cu Ion-Exchanged *BEA-Type Zeolites

We have introduced the position and the catalytic activity of the framework metal species, but the metal species included in the zeolite crystal are not only the framework atoms, but also the metal cations, clusters, and bulk species [37–47]. In this section, the location, state, size, and reactivity of the metal species will be described.

The metal cation-exchanged zeolites can work as catalysts for oxidative and reductive reactions [37–40]. The direct oxidation of benzene with H_2O_2 to phenol is among the notable oxidations because phenol is an important industrial intermediate for

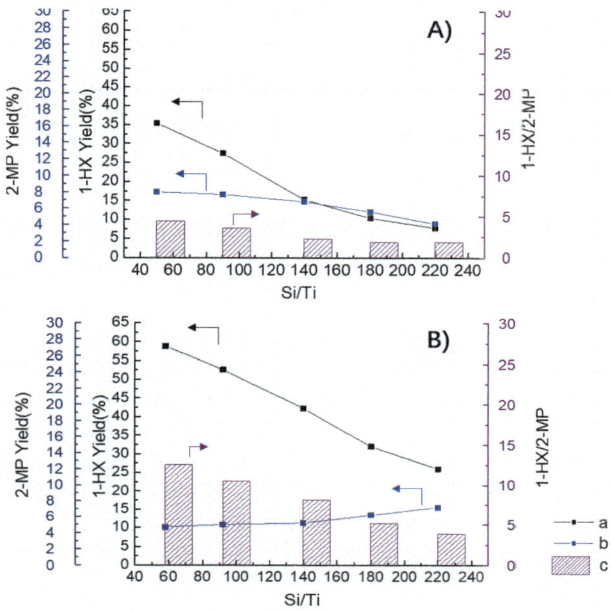

Fig. 6.17 Total yield of the epoxidation for a) 1-HX and b) 2-MP, and c) 1-HX/2-MP index over **A** Ti–MWW–D and **B** Ti–MWW–P zeolites with different Si/Ti ratios. Reaction conditions: catalyst, 50 mg in 10 mL of CH_3CN; 1-HX or 2-MP, 10 mmol; H_2O_2 (30 wt% aqueous solution), 10 mmol; temperature, 333 K, time, 2 h. Adapted with permission from Ref. [95]. Copyright 2019 The Chemical Society of Japan

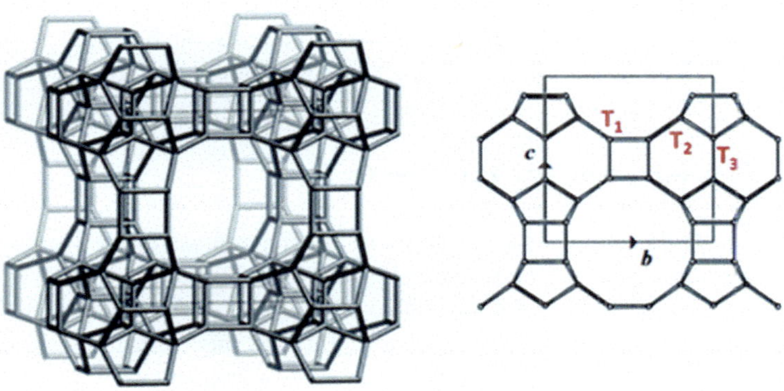

Fig. 6.18 **a** Framework structure and **b** types of crystallographic T-sites of BEC-type framework. Adapted with permission from Ref. [97]. Copyright 2020 The Royal Society of Chemistry

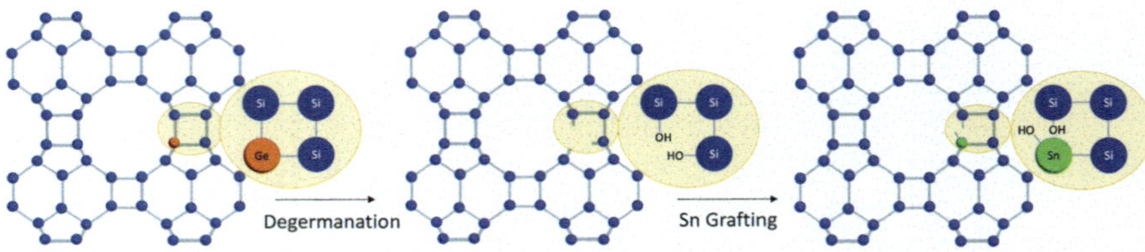

Fig. 6.19 Selectively siting Sn at the D4Rs of high-silica BEC zeolites using a multi-step synthesis procedure. Adapted with permission from Ref. [97]. Copyright 2020 The Royal Society of Chemistry

various chemicals, such as phenolic resins, bisphenol A, and adipic acid [98]. In order to catalyze this reaction, copper- and iron-based catalysts have often been used as promising candidates. Thus, our group has focused on the Cu and Fe ion-exchanged zeolite catalysts. In this research, *BEA-type zeolites were applied for the catalyst supports since reactants can enter to three-dimensional 12R channels of *BEA-type structure (Fig. 6.20 [12]), and we investigated the effect of the ion-exchange sequences on the physicochemical and catalytic properties [99]. The bimetallic ion-exchanged *BEA-type zeolites (Cu, Fe-*BEA) were prepared by ion exchange via three types of sequences: (1) Fe followed by Cu, (2) Cu followed by Fe, and (3) Fe and Cu at the same time. These catalysts exhibited different physicochemical and catalytic properties in the oxidation of benzene with H_2O_2 (Fig. 6.21 [99]). The origins of such different properties were clarified by several characterization methods, namely, *in situ* NO adsorption IR, X-ray photoelectron spectroscopy (XPS), and H_2-temperature-programmed reduction (TPR) measurements. The obtained characterization results suggested that the ion-exchange sequence influenced the chemical states of both metal species, resulting in the creation of highly active sites. Among them, the catalysts prepared by ion exchange to Cu and Fe species at the same time showed the highest phenol yield reaching 10.5%, which demonstrated the potential of bimetallic ion-exchanged zeolites as a great catalyst for the oxidative reaction.

6.3.1.2 Cu ion-Exchanged CHA-Type Zeolites

The position of the framework Al atoms is directly related to the location of the ion-exchange sites, not only Brønsted acid sites. Zhang, Liu, Meng, and Xiao et al. investigated the importance of controllable Al sites in CHA-type zeolites by crystallization pathways for NH_3–SCR reaction [100]. In this case, the location of the framework Al atoms in the CHA-type structure was controlled by using different OSDAs to improve the catalytic activity for NH_3–SCR reaction, where two OSDAs, $TMAda^+$ and *N, N*-dimethylcyclohexylammonium ($DMCHA^+$), were employed to synthesize the CHA-type aluminosilicate with the framework Al position controlled (Fig. 6.22 [100]). The thus-synthesized CHA-type zeolites were donated as CHA–T and CHA–D, respectively. Remarkably, two crystallization pathways were identified. Double-6 rings (D6Rs) were preferentially formed during the crystallization for

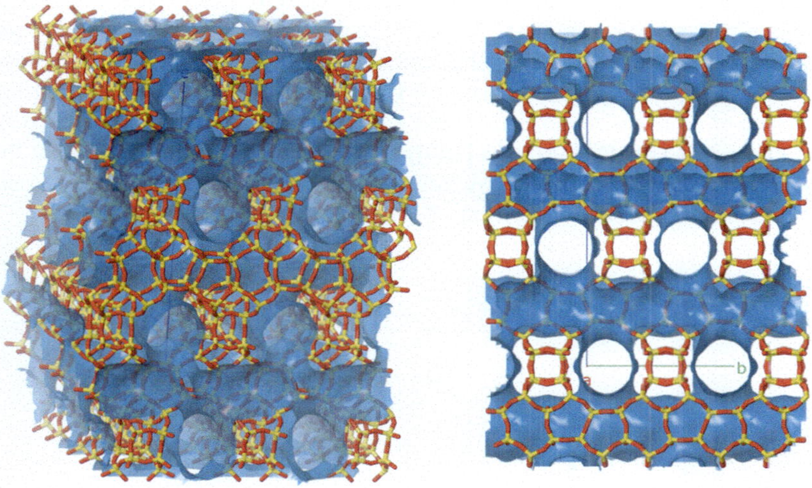

Fig. 6.20 Framework and channel system of *BEA-type topology with different views. Adapted with permission from the International Zeolite Association, the Structure Commission, Ref. [12]

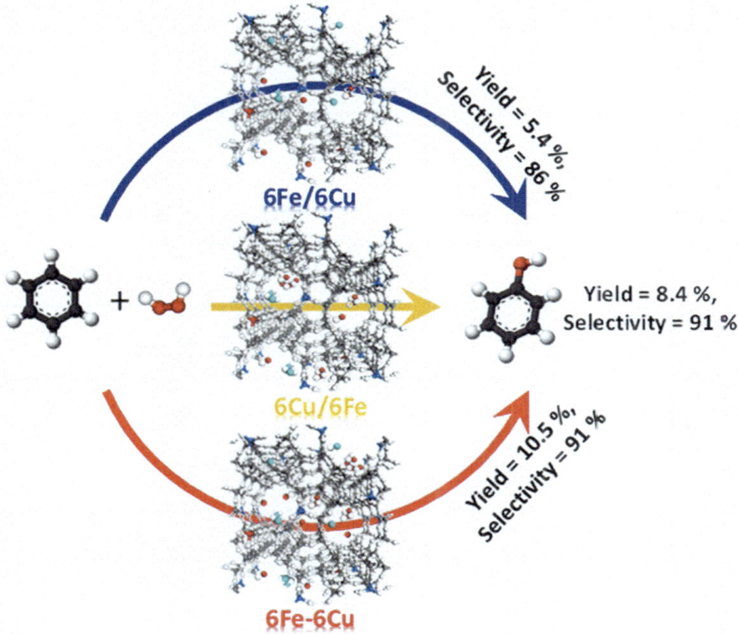

Fig. 6.21 Direct oxidation of benzene with H_2O_2 to phenol over bimetallic *BEA-type zeolites. Adapted with permission from Ref. [99]. Copyright 2020 The Royal Society of Chemistry

CHA–D, while 8Rs were favorably generated. The framework Al atoms in CHA–D were spatially closer than that for CHA–T due to different crystallization pathways. Moreover, Cu ion-exchanged CHA–D showed higher catalytic activity for NH_3–SCR reaction at 400−550 °C than that of Cu–CHA–T, which is originated from the presence of more hydrothermally stable Cu^{2+} species captured by the paired Al sites in D6Rs of CHA–D (Fig. 6.23 [100]). Their achievement suggested that the crystallization pathways controlled by rational selection of OSDAs can lead to the improvement of the catalytic activity for NH_3–SCR reaction.

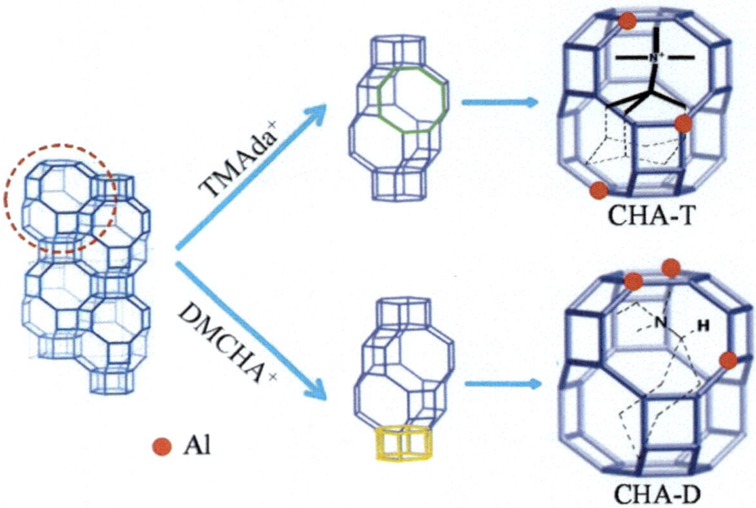

Fig. 6.22 Synthesis of CHA-type aluminosilicates with different distributions for framework Al atoms. Adapted with permission from Ref. [100]. Copyright 2020 Elsevier

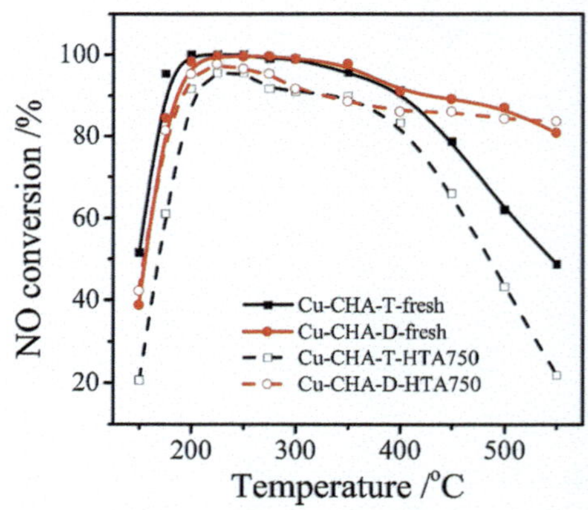

Fig. 6.23 Catalytic activity for NH_3-SCR reaction over the fresh and 750 °C aged Cu–CHA–T and Cu–CHA–D catalysts. Adapted with permission from Ref. [100]. Copyright 2020 Elsevier

6.3.2 *Metal Clusters and Oxides in/on Zeolite*

6.3.2.1 Rh-Containing MFI-type Zeolites with Framework Al Distribution Controlled

The position of the framework Al atoms can affect the chemical state of the introduced metal species via the ion-exchange method. Such a phenomenon was observed for rhodium ion-exchanged MFI-type aluminosilicate zeolites with framework Al distributions controlled [101]. In this reported study, rhodium ion-exchange for MFI[TPA] (Rh-MFI[TPA]) led to the formation of rhodium oxide clusters located at the channel intersections due to its relatively wide space, while the isolated rhodium cations were formed in MFI[TPA, Na] (Rh-MFI[TPA,Na]) induced by the relatively narrow space (Fig. 6.24 [101]). These differences were characterized by combining UV–vis and IR spectroscopies. Recently, Hou and Kobayashi et al. discovered that MOR-type zeolite-supported rhodium sub-nano clusters have a good potential as highly active catalysts for low-temperature oxidative reforming of methane [102]. Thus, we applied Rh-MFIs as catalysts to the oxidative reforming of methane. Figure 6.25 shows the conversion of methane and CO yield over Rh-MFI[TPA] and Rh-MFI[TPA,Na] with different space velocity (SV) conditions [101]. By increasing the SV values, the catalytic activities of both samples decreased, but the behavior for conversion drop differed among the catalysts. When the SV value increased from 1.2×10^6 to 4.8×10^6 ml g^{-1} h^{-1}, the CO yield over Rh-MFI[TPA, Na] evidently decreased by about 65% (from 62 to 22%). In contrast, the decreasing degree of the CO yield over Rh-MFI[TPA] was ca. 38% (from 87 to 54%). This is clear evidence that the rhodium oxide clusters in the channel intersections of MFI-type topology can work as highly active sites. These findings would contribute to discovering a new class of metal-containing zeolites with the location, state, and size of metal species controlled.

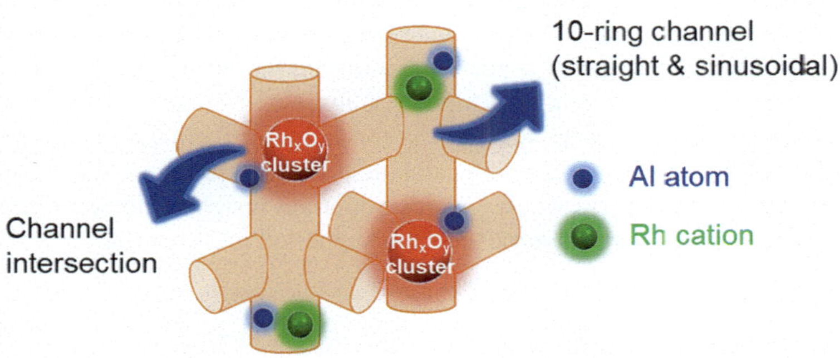

Fig. 6.24 Selective siting of rhodium species in MFI-type zeolites. Adapted with permission from Ref. [101]. Copyright 2020 The Royal Society of Chemistry

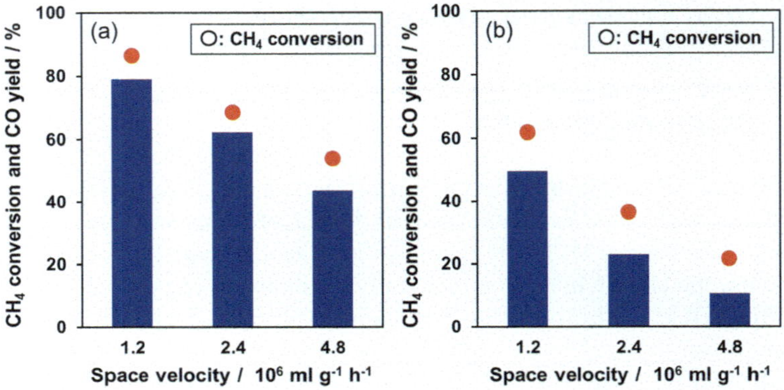

Fig. 6.25 Effect of space velocity on catalytic reforming of methane at 600 °C over **a** Rh-MFI[TPA] and **b** Rh-MFI[TPA,Na]. Reaction conditions: catalyst, 10–25 mg; partial pressure, P_{CH4}: P_{O2}: P_{Ar} = 0.0333: 0.0167: 0.95; total pressure, 0.1 MPa. Adapted with permission from Ref. [101]. Copyright 2020 The Royal Society of Chemistry

6.3.2.2 Supported Ni CHA-Type Zeolites with Different Types of Heteroatoms

The utilization of zeolite as catalyst support is an efficient way to improve the catalytic activity of the metal species [103]. The CHA-type zeolites with different types of heteroatoms, borosilicate, aluminosilicate, and gallorosilicate, were used as supports for Ni catalysts, and the effect of heteroatoms on the state of Ni species was studied (Fig. 6.26 [104]). Through the characterization in detail, we found that the types of the heteroatoms influenced the size and dispersibility of the introduced Ni species. The use of CHA-type borosilicate led to the formation of a bulk oxide on the external surface of the zeolite, while Ni cations, as well as nanoclusters, were generated in aluminosilicate and gallosilicate. Notably, the highest dispersion was achieved by using the aluminosilicate as support. In addition, the introduced Ni species on CHA-type aluminosilicate exhibited higher catalytic performance in the oxidative conversion of methane than the others. The formation of the highly dispersed and nano-sized Ni species results in a high catalytic activity to produce CO and H_2. We believe that this achievement would give us important insights into controlling the size and state of metal species introduced and supported in/on zeolites.

6.3.2.3 MOR-Type Zeolite-Supported Ultra-Small Ni Species

Finally, the catalytic property of MOR-type zeolite-supported Ni species as ultra-small particles was introduced in this section. The MOR-type framework consists of two-pore systems,12R (6.5 × 7 Å) and 8R (2.6 × 5.7 Å) (Fig. 6.27 [12]). However, most of the molecules cannot enter 8R due to the size limitation. Thus, MOR-type topology is often regarded as a one-dimensional pore system. In this study, we found

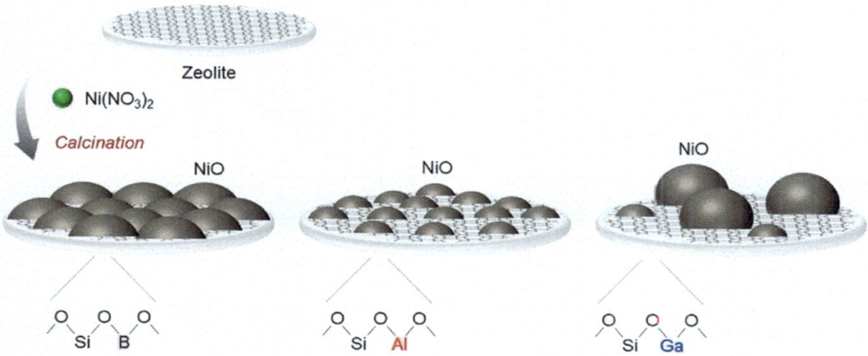

Fig. 6.26 Preparation of supported Ni on CHA-type borosilicate, aluminosilicate and gallosilicate. Adapted with permission from Ref. [104]. Copyright 2022 The Chemical Society of Japan

that MOR-type zeolites can be an excellent support for the catalytically active Ni species [105]. The development of effective catalysts for the selective oxidation of methane to syngas at low temperatures has been desired for many years [66–68]. In general, a facile synthesis is recommended from the viewpoint of the industry. The simple impregnation method, therefore, was employed to prepare the supported Ni catalysts for the oxidative reforming of methane. The formation of highly dispersed ultra-small NiO particles with 1.6 nm in size on the MOR-type zeolite resulted in excellent catalytic performance at 973 K: continuously 97–98% methane conversion, 91–92% of CO yield (Fig. 6.28 [105]). Furthermore, density functional theory (DFT) calculations demonstrated the relationship between the particle size of the supported NiO and the C-H dissociation of CH_4. In the case of the ultra-small NiO particle, the lattice oxygen of NiO actively contributed to the methane oxidation, resulting in an increase in the CO yield (Fig. 6.29 [105]). These achievements would give us a new concept of catalyst design for the industrial field.

6.4 Conclusions and Outlook

As described in this chapter, recent progress in the "position" control of catalytically active sites in the zeolites, the intra- and extra-framework metal species, was described. We have realized that it significantly influences the catalytic performance, not only selectivity, but also conversion, in various reactions. For the aluminosilicate, the appropriate selection of OSDAs and/or starting materials is a key component for the locational control of the framework Al species. In the case of metallosilicates, different synthesis methods, direct or post-synthesis, give different distributions for heteroatoms. Furthermore, the control of the location, chemical state, and size of the metal species is beginning to be achieved in the metal-containing zeolite catalysts, which are also useful for developing catalytic processes for achieving highly difficult

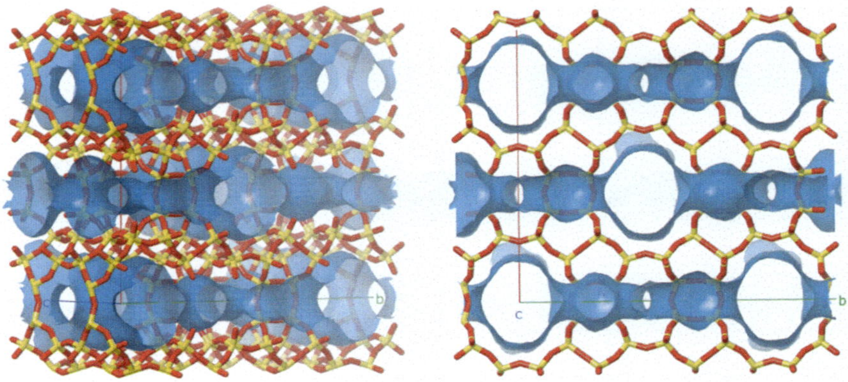

Fig. 6.27 Framework and channel system of MOR-type topology with different views. Adapted with permission from the International Zeolite Association, the Structure Commission, Ref. [12].

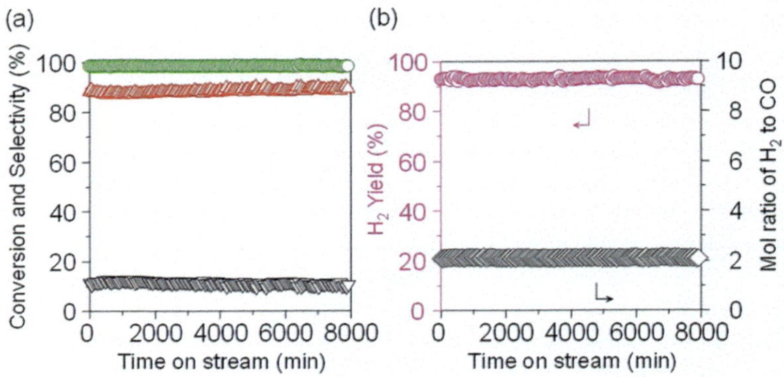

Fig. 6.28 Catalytic activity for methane oxidation over MOR-type zeolite-supported Ni catalyst in time course. **a** Conversion of methane (○) and product selectivities for CO (△) and CO_2 (▽). **b** H_2 yield (○) and molar ratio of H_2 to CO (◇). Reaction conditions: 50 mg; partial pressure, P_{CH4}: P_{O2}: P_{Ar} = 0.06: 0.03: 0.91; total pressure, 0.1 MPa. Adapted with permission from Ref. [105]. Copyright 2020 Springer Nature Limited

selective reactions and carbon neutrality. When the position, chemical state, and size of metal species are controlled, the location of the framework Al atoms is crucial in addition to the framework types. Therefore, the positional control of the catalytic active sites will become increasingly important in the future.

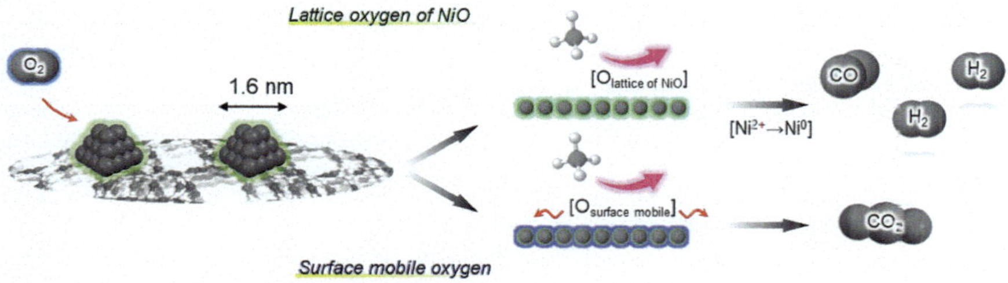

Fig. 6.29 Methane oxidation over MOR-type zeolite-supported ultra-small Ni catalyst. Adapted with permission from Ref. [105]. Copyright 2020 Springer Nature Limited

References

1. Corma A. Inorganic solid acids and their use in acid-catalyzed hydrocarbon reactions. Chem Rev. 1995;95(3):559–614.
2. Čejka J, Corma A, Zones S. Zeolites and catalysis. Wiley;2010.
3. Olsbye U, Svelle S, Bjørgen M, Beato P, Janssens TVW, Joensen F, Bordiga S, Lillerud KP. Conversion of methanol to hydrocarbons: how zeolite cavity and pore size controls product selectivity. Angew Chem Int Ed. 2012;51(24):5810–31.
4. Shi J, Wang Y, Yang W, Tang Y, Xie Z. Recent advances of pore system construction in zeolite-catalyzed chemical industry processes. Chem Soc Rev. 2015;44(24):8877–903.
5. Beale AM, Gao F, Lezcano-Gonzalez I, Peden CHF, Szanyi J. Recent advances in automotive catalysis for NOx emission control by small-pore microporous materials. Chem Soc Rev. 2015;44(20):7371–405.
6. Li Y, Li L, Yu J. Applications of zeolites in sustainable chemistry. Chem. 2017;3(6):928–49.
7. Dusselier M, Davis ME. Small-pore zeolites: synthesis and catalysis. Chem Rev. 2018;118(11):5265–329.
8. Yang M, Fan D, Wei Y, Tian P, Liu Z. Recent progress in methanol-to-olefins (MTO) catalysts. Adv Mater. 2019;31(50):1–15.
9. Zhang Q, Yu J, Corma A. Applications of zeolites to C1 chemistry: recent advances, challenges, and opportunities. Adv Mater. 2020;32(44):1–31.
10. Le TT, Chawla A, Rimer JD. Impact of acid site speciation and spatial gradients on zeolite catalysis. J Catal. 2020;391:56–68.
11. del Campo P, Martínez C, Corma A. Activation and conversion of alkanes in the confined space of zeolite-type materials. Chem Soc Rev. 2021;50(15):8511–95.
12. Baerlocher Ch, McCusker LB. Database of Zeolite Structures, http://www.iza-structure.org/databases/
13. Chu CTW, Kuehl GH, Lago RM, Chang CD. Isomorphous substitution in zeolite frameworks. II. Catalytic properties of [B]ZSM-5. J Catal. 1985;93 (2):451–8.
14. Stave MS, Nicholas JB. Density functional studies of zeolites. 2. Structure and acidity of [T]-ZSM-5 models (T = B, Al, Ga, and Fe). J Phys Chem. 1995;99 (41):15046–61.
15. Chatterjee A, Iwasaki T, Ebina T, Miyamoto A. Density functional study for estimating brønsted acid site strength in isomorphously substituted ZSM-5. Microporous Mesoporous Mater. 1998;21(4–6):421–8.
16. Tatsumi T. Metallozeolites and applications in catalysis. Curr Opin Solid State Mater Sci. 1997;2(1):76–83.
17. Corma A, Nemeth LT, Renz M, Valencia S. Sn-zeolite beta as a heterogeneous chemoselective catalyst for baeyer-villiger oxidations. Nature. 2001;412(6845):423–5.
18. Moliner M. State of the art of lewis acid-containing zeolites: lessons from fine chemistry to new biomass transformation processes. Dalton Trans. 2014;43(11):4197–208.

19. Otomo R, Kosugi R, Kamiya Y, Tatsumi T, Yokoi T. Modification of Sn-Beta zeolite: characterization of acidic/basic properties and catalytic performance in baeyer-villiger oxidation. Catal Sci Technol. 2016;6(8):2787–95.
20. Gordon CP, Engler H, Tragl AS, Plodinec M, Lunkenbein T, Berkessel A, Teles JH, Parvulescu AN, Copéret C. Efficient epoxidation over dinuclear sites in titanium silicalite-1. Nature. 2020;586(7831):708–13.
21. Román-Leshkov Y, Moliner M, Davis ME. Impact of controlling the site distribution of Al atoms on catalytic properties in ferrierite-type zeolites. J Phys Chem C. 2011;115(4):1096–102.
22. Dědeček J, Tabor E, Sklenak S. Tuning the aluminum distribution in zeolites to increase their performance in acid-catalyzed reactions. Chemsuschem. 2019;12(3):556–76.
23. Palčić A, Valtchev V. Analysis and control of acid sites in zeolites. Appl Catal A Gen. 2020;606: 117795.
24. Yabushita M, Osuga R, Muramatsu A. Control of location and distribution of heteroatoms substituted isomorphously in framework of zeolites and zeotype materials. CrystEngComm. 2021;23(36):6226–33.
25. Yokoi T, Mochizuki H, Namba S, Kondo JN, Tatsumi T. Control of the Al distribution in the framework of ZSM-5 zeolite and its evaluation by solid-state NMR technique and catalytic properties. J Phys Chem C. 2015;119(27):15303–15.
26. Yokoi T, Mochizuki H, Biligetu T, Wang Y, Tatsumi T. Unique Al distribution in the MFI framework and its impact on catalytic properties. Chem Lett. 2017;46(6):798–800.
27. Biligetu T, Wang Y, Nishitoba T, Otomo R, Park S, Mochizuki H, Kondo JN, Tatsumi T, Yokoi T. Al Distribution and catalytic performance of ZSM-5 zeolites synthesized with various alcohols. J Catal. 2017;353:1–10.
28. Di Iorio JR, Gounder R. Controlling the isolation and pairing of aluminum in chabazite zeolites using mixtures of organic and inorganic structure-directing agents. Chem Mater. 2016;28(7):2236–47.
29. Toyoda H, Osuga R, Wang Y, Park S, Yazawa K, Gies H, Gilbert CJ, Yilmaz B, Kelkar CP, Yokoi T. Clarification of acid site location in MSE-type zeolites by spectroscopic approaches combined with catalytic activity: comparison between UZM-35 and MCM-68. Phys Chem Chem Phys. 2022;24(7):4358–65.
30. Muraoka K, Chaikittisilp W, Yanaba Y, Yoshikawa T, Okubo T. Directing aluminum atoms into energetically favorable tetrahedral sites in a zeolite framework by using organic structure-directing agents. Angew Chem Int Ed. 2018;57(14):3742–6.
31. Wang Z, Chu W, Zhao Z, Liu Z, Chen H, Xiao D, Gong K, Li F, Li X, Hou G. The role of organic and inorganic structure-directing agents in selective Al substitution of zeolite. J Phys Chem Lett. 2021;12(38):9398–406.
32. Gábová V, Dědeček J, čejka J. Control of Al distribution in ZSM-5 by conditions of zeolite synthesis. Chem Commun. 2003;3(10):1196–1197.
33. Nishitoba T, Yoshida N, Kondo JN, Yokoi T. Control of Al distribution in the CHA-Type aluminosilicate zeolites and its impact on the hydrothermal stability and catalytic properties. Ind Eng Chem Res. 2018;57(11):3914–22.
34. Nishitoba T, Nozaki T, Park S, Wang Y, Kondo JN, Gies H, Yokoi T. CHA-type zeolite prepared by interzeolite conversion method using FAU and LTL-type zeolite: effect of the raw materials on the crystallization mechanism, and physicochemical and catalytic properties. Catalysts. 2020;10(10):1204.
35. Yabushita M, Kobayashi H, Neya A, Nakaya M, Maki S, Matsubara M, Kanie K, Muramatsu A. Precise control of density and strength of acid sites of MFI-type zeolite nanoparticles: via simultaneous isomorphous substitution by Al and Fe. CrystEngComm. 2020;22(44):7556–64.
36. Yabushita M, Imanishi Y, Xiao T, Osuga R, Nishitoba T, Maki S, Kanie K, Cao W, Yokoi T, Muramatsu A. Transcription-induced formation of paired Al sites in high-silica CHA-Type zeolite framework using Al-Rich amorphous aluminosilicate. Chem Commun. 2021;57(98):13301–4.
37. Sato S, Yu-u Y, Yahiro H, Mizuno N, Iwamoto M. Cu-ZSM-5 zeolite as highly active catalyst for removal of nitrogen monoxide from emission of diesel engines. Appl Catal. 1991;70(1):3–7.

38. Reddy TI, Varma RS. Rare-earth (RE) exchanged NaY zeolite promoted knoevenagel condensation. Tetrahedron Lett. 1997;38(10):1721–4.
39. Neamţu M, Zaharia C, Catrinescu C, Yediler A, Macoveanu M, Kettrup A. Fe-exchanged Y zeolite as catalyst for wet peroxide oxidation of reactive azo dye procion marine H-EXL. Appl Catal B Environ. 2004;48(4):287–94.
40. Narsimhan K, Michaelis VK, Mathies G, Gunther WR, Griffin RG, Román-Leshkov Y. Methane to acetic acid over Cu-exchanged zeolites: mechanistic insights from a site-specific carbonylation reaction. J Am Chem Soc. 2015;137(5):1825–32.
41. Wang C, Wang L, Zhang J, Wang H, Lewis JP, Xiao FS. Product selectivity controlled by zeolite crystals in biomass hydrogenation over a palladium catalyst. J Am Chem Soc. 2016;138(25):7880–3.
42. Sushkevich VL, Palagin D, Ranocchiari M, van Bokhoven JA. Selective anaerobic oxidation of methane enables direct synthesis of methanol. Science. 2017;356(6337):523–7.
43. Shan J, Li M, Allard LF, Lee S, Flytzani-Stephanopoulos M. Mild oxidation of methane to methanol or acetic acid on supported isolated rhodium catalysts. Nature. 2017;551(7682):605–8.
44. Liu L, Díaz U, Arenal R, Agostini G, Concepción P, Corma A. Generation of subnanometric platinum with high stability during transformation of a 2D zeolite into 3D. Nat Mater. 2017;16(1):132–8.
45. Zhu J, Osuga R, Ishikawa R, Shibata N, Ikuhara Y, Kondo JN, Ogura M, Yu J, Wakihara T, Liu Z, Okubo T. Ultrafast encapsulation of metal nanoclusters into mfi zeolite in the course of its crystallization: catalytic application for propane dehydrogenation. Angew Chem Int Ed. 2020;59(44):19669–74.
46. Imbao J, Van Bokhoven JA, Nachtegaal M. Optimization of a heterogeneous Pd-Cu/Zeolite y wacker catalyst for ethylene oxidation. Chem Commun. 2020;56(9):1377–80.
47. Osuga R, Yasuda S, Sawada M, Manabe R, Shima H, Tsutsuminai S, Fukuoka A, Kobayashi H, Muramatsu A, Yokoi T. Oxidative reforming of methane over Rh-containing zeolites: active species and role of zeolite framework. Ind Eng Chem Res. 2021;60(24):8696–704.
48. Nakamura K, Okuda A, Ohta K, Matsubara H, Okumura K, Yamamoto K, Itagaki R, Suganuma S, Tsuji E, Katada N. Direct methylation of benzene with methane catalyzed by Co/MFI zeolite. ChemCatChem. 2018;10(17):3806–12.
49. Hu P, Nakamura K, Matsubara H, Iyoki K, Yanaba Y, Okumura K, Okubo T, Katada N, Wakihara T. Comparative study of direct methylation of benzene with methane on cobalt-exchanged ZSM-5 and ZSM-11 zeolites. Appl Catal A Gen. 2020;601: 117661.
50. Mahyuddin MH, Shiota Y, Yoshizawa K. Methane selective oxidation to methanol by metal-exchanged zeolites: a review of active sites and their reactivity. Catal Sci Technol. 2019;9(8):1744–68.
51. Lippmaa E, Mägi M, Samoson A, Tarmak M, Engelhardt G. Investigation of the structure of zeolites by solid-state high-resolution ^{29}Si NMR spectroscopy. J Am Chem Soc. 1981;103(17):4992–6.
52. Melchior T, Vaughan DEW, Jacobson AJ. Characterization of the silicon-aluminum distribution in synthetic faujasites by high-resolution solid-state ^{29}Si NMR. J Am Chem Soc. 1982;104(18):4859–64.
53. Vjunov A, Fulton JL, Huthwelker T, Pin S, Mei D, Schenter GK, Govind N, Camaioni DM, Hu JZ, Lercher JA. Quantitatively probing the Al distribution in zeolites. J Am Chem Soc. 2014;136(23):8296–306.
54. Martineau-Corcos C, Dědeček J, Taulelle F. ^{27}Al-^{27}Al double-quantum single-quantum MAS NMR: applications to the structural characterization of microporous materials. Solid State Nucl Magn Reson. 2017;84:65–72.
55. Sklenak S, Dědeček J, Li C, Wichterlová B, Gábová V, Sierka M, Sauer J. Aluminum siting in silicon-rich zeolite frameworks: a combined high-resolution ^{27}Al NMR spectroscopy and quantum mechanics/molecular mechanics study of ZSM-5. Angew Chem Int Ed. 2007;46(38):7286–9.

56. Dědeček J, Sklenak S, Li C, Wichterlova B, Gábová V, Brus J, Sierka M, Sauer J. Effect of Al-Si-Al and Al-Si-Si-Al pairs in the ZSM-5 zeolite framework on the ^{27}Al NMR spectra. A combined high-resolution ^{27}Al NMR and DFT / MM study. J Phys Chem C. 2009;113 (4):1447–58.
57. Ryczkowski J. IR spectroscopy in catalysis. Catal Today. 2001;68(4):263–381.
58. Bordiga S, Lamberti C, Bonino F, Travert A, Thibault-Starzyk F. Probing zeolites by vibrational spectroscopies. Chem Soc Rev. 2015;44(20):7262–341.
59. Osuga R, Yokoi T, Doitomi K, Hirao H, Kondo JN. Infrared investigation of dynamic behavior of brønsted acid sites on zeolites at high temperatures. J Phys Chem C. 2017;121(45):25411–20.
60. Osuga R, Yokoi T, Kondo JN. Probing the basicity of lattice oxygen on H-form zeolites using CO_2. J Catal. 2019;371:291–7.
61. Kubota T, Osuga R, Yokoi T, Kondo JN. Consideration of acid strength of a single OH group on zeolites by isotope exchange reaction with ethane at high temperatures. Top Catal. 2017;60(19–20):1496–505.
62. Corma A, Fornés V, Forni L, Márquez F, Martínez-Triguero J, Moscotti D. 2,6-Di-Tert-Butyl-pyridine as a probe molecule to measure external acidity of zeolites. J Catal. 1998;179(2):451–8.
63. Lever ABP. Inorganic electronic spectroscopy. Elsevier;1984.
64. Frillette VJ, Haag WO, Lago RM. Catalysis by crystalline aluminosilicates: characterization of intermediate pore-size zeolites by the "constraint index." J Catal. 1981;67(1):218–22.
65. Krannila H, Haag WO, Gates BC. Monomolecular and bimolecular mechanisms of paraffin cracking: n-butane cracking catalyzed by HZSM-5. J Catal. 1992;135(1):115–24.
66. Lunsford JH. The catalytic oxidative coupling of methane. Angew Chem Int Ed. 1995;34(9):970–80.
67. Crabtree RH. Aspects of methane chemistry. Chem Rev. 1995;95(4):987–1007.
68. Lunsford JH. Catalytic conversion of methane to more useful chemicals and fuels: a challenge for the 21st century. Catal Today. 2000;63(2–4):165–74.
69. Spivey JJ, Hutchings G. Catalytic aromatization of methane. Chem Soc Rev. 2014;43(3):792–803.
70. Pakhare D, Spivey J. A review of dry (CO_2) reforming of methane over noble metal catalysts. Chem Soc Rev. 2014;43(22):7813–37.
71. Schwach P, Pan X, Bao X. Direct conversion of methane to value-added chemicals over heterogeneous catalysts: challenges and prospects. Chem Rev. 2017;117(13):8497–520.
72. Jin Z, Wang L, Zuidema E, Mondal K, Zhang M, Zhang J, Wang C, Meng X, Yang H, Mesters C, Xiao FS. Hydrophobic zeolite modification for in situ peroxide formation in methane oxidation to methanol. Science. 2020;367(6474):193–7.
73. Garba MD, Usman M, Khan S, Shehzad F, Galadima A, Ehsan MF, Ghanem AS, Humayun M. CO_2 towards fuels: a review of catalytic conversion of carbon dioxide to hydrocarbons. J Environ Chem Eng. 2021;9(2): 104756.
74. Ramirez A, Gevers L, Bavykina A, Ould-Chikh S, Gascon J. Metal organic framework-derived iron catalysts for the direct hydrogenation of CO_2 to short chain olefins. ACS Catal. 2018;8(10):9174–82.
75. Jadhav SG, Vaidya PD, Bhanage BM, Joshi JB. Catalytic carbon dioxide hydrogenation to methanol: a review of recent studies. Chem Eng Res Des. 2014;92(11):2557–67.
76. Yarulina I, De Wispelaere K, Bailleul S, Goetze J, Radersma M, Abou-Hamad E, Vollmer I, Goesten M, Mezari B, Hensen EJM, Martínez-Espín JS, Morten M, Mitchell S, Perez-Ramirez J, Olsbye U, Weckhuysen BM, Van Speybroeck V, Kapteijn F, Gascon J. Structure-performance descriptors and the role of lewis acidity in the methanol-to-propylene process. Nat Chem. 2018;10(8):804–12.
77. Yarulina I, Chowdhury AD, Meirer F, Weckhuysen BM, Gascon J. Recent trends and fundamental insights in the methanol-to-hydrocarbons process. Nat Catal. 2018;1(6):398–411.
78. Rahimi AR, Garciá JM. Chemical recycling of waste plastics for new materials production. Nat Rev Chem. 2017;1:1–11.

79. Lopez G, Artetxe M, Amutio M, Bilbao J, Olazar M. Thermochemical routes for the valorization of waste polyolefinic plastics to produce fuels and chemicals. A review. Renew Sustain Energy Rev. 2017;73:346–68.
80. Mordi RC, Fields R, Dwyer J. Thermolysis of low density polyethylene catalysed by zeolites. J Anal Appl Pyrolysis. 1994;29(1):45–55.
81. Akpanudoh NS, Gobin K, Manos G. Catalytic degradation of plastic waste to liquid fuel over commercial cracking catalysts: effect of polymer to catalyst ratio/acidity content. J Mol Catal A Chem. 2005;235(1–2):67–73.
82. Liu S, Kots PA, Vance BC, Danielson A, Vlachos DG. Plastic waste to fuels by hydrocracking at mild conditions. Sci Adv. 2021;7(17):1–10.
83. Otomo R, Nishitoba T, Osuga R, Kunitake Y, Kamiya Y, Tatsumi T, Yokoi T. Determination of acid site location in dealuminated MCM-68 by ^{27}Al MQMAS NMR and FT-IR spectroscopy with probe molecules. J Phys Chem C. 2018;122(2):1180–91.
84. Price GD, Pluth JJ, Smith JV, Bennett JM, Patton RL. Crystal structure of tetrapropylammonium fluoride containing precursor to fluoride silicalite. J Am Chem Soc. 1982;104(22):5971–7.
85. Burkett SL, Davis ME. Mechanism of structure direction in the synthesis of Si-ZSM-5: An investigation by intermolecular ^{1}H–^{29}Si CP MAS NMR. J Phys Chem. 1994;98(17):4647–53.
86. Calabro DC, Cheng JC, Crane Jr RA, Kresge CT, Dhingra SS, Steckel MA, Stern DL, Weston SC. Synthetic porous crystalline MCM-68, its synthesis and use. U.S. Patent 6,049,018, April 11, 2000.
87. Moscoso J, Jan D. UZM-35 aluminosilicate zeolite, method of preparation and processes using UZM-35. U.S. Patent 7,922,997, April 12, 2011.
88. Corma A. From microporous to mesoporous molecular sieve materials and their use in catalysis. Chem Rev. 1997;97(6):2373–419.
89. Wu P, Tatsumi T. A new generation of titanosilicate catalyst: preparation and application to liquid-phase epoxidation of alkenes. Catal Surv from Asia. 2004;8(2):137–48.
90. Taramasso M, Milanese S, Perego G, Milan NB. Preparation of porous crystalline synthetic material comprised of silicon and titanium oxides. U. S. Patent US4410501A, 1983.
91. Corma A, Camblor MA, Esteve P, Martínez A, Pérez-Pariente J. Activity of Ti-beta catalyst for the selective oxidation of alkenes and alkanes. J Catal. 1994;145(1):151–8.
92. Wu P, Komatsu T, Yashima T. Ammoximation of ketones over titanium mordenite. J Catal. 1997;168(2):400–11.
93. Sasaki M, Sato Y, Tsuboi Y, Inagaki S, Kubota Y. Ti-YNU-2: a microporous titanosilicate with enhanced catalytic performance for phenol oxidation. ACS Catal. 2014;4(8):2653–7.
94. Wu P, Tatsumi T, Komatsu T, Yashima T. A novel titanosilicate with mww structure. i. hydrothermal synthesis, elimination of extraframework titanium, and characterizations. J Phys Chem B. 2001;105 (15):2897–2905.
95. Ji X, Wang Y, Fujii T, Otomo R, Kondo JN, Yokoi T. Evaluation of Ti distribution in zeolite framework based on the catalytic activity for alkene epoxidation. Chem Lett. 2019;48(9):1130–3.
96. Wu P, Tatsumi T. Preparation of B-Free Ti-MWW through reversible structural conversion. Chem Commun. 2002;2(10):1026–7.
97. Rodríguez-Fernández A, Di Iorio JR, Paris C, Boronat M, Corma A, Román-Leshkov Y, Moliner M. Selective active site placement in lewis acid zeolites and implications for catalysis of oxygenated compounds. Chem Sci. 2020;11(37):10225–35.
98. Schmidt RJ. Industrial catalytic processes-phenol production. Appl Catal A Gen. 2005;280(1):89–103.
99. Xiao P, Osuga R, Wang Y, Kondo JN, Yokoi T. Bimetallic Fe-Cu/Beta zeolite catalysts for direct hydroxylation of benzene to phenol: effect of the sequence of ion exchange for Fe and Cu cations. Catal Sci Technol. 2020;10(20):6977–86.
100. Zhang J, Shan Y, Zhang L, Du J, He H, Han S, Lei C, Wang S, Fan W, Feng Z, Liu X, Meng X, Xiao FS. Importance of controllable Al sites in CHA framework by crystallization pathways for NH_3-SCR reaction. Appl Catal B Environ. 2020;277: 119193.

101. Osuga R, Bayarsaikhan S, Yasuda S, Manabe R, Shima H, Tsutsuminai S, Fukuoka A, Kobayashi H, Yokoi T. Metal cation-exchanged zeolites with the location, state, and size of metal species controlled. Chem Commun. 2020;56(44):5913–6.
102. Hou Y, Ogasawara S, Fukuoka A, Kobayashi H. Zeolite-supported rhodium sub-nano cluster catalyst for low-temperature selective oxidation of methane to syngas. Catal Sci Technol. 2017;7(24):6132–9.
103. Sun Q, Wang N, Yu J. Advances in catalytic applications of zeolite-supported metal catalysts. Adv Mater. 2021;33(51):1–37.
104. Yasuda S, Kunitake Y, Osuga R, Nakamura K, Matsumoto T, Sago K, Kondo JN, Yabushita M, Muramatsu A, Yokoi T. Supported nickel zeolite catalyst for oxidative conversion of methane: effect of heteroatoms in the zeolite framework on its physicochemical and catalytic properties. Chem Lett. 2022;51(1):46–9.
105. Yasuda S, Osuga R, Kunitake Y, Kato K, Fukuoka A, Kobayashi H, Gao M, Hasegawa J, Manabe R, Shima H, Tsutsuminai S, Yokoi T. Zeolite-supported ultra-small nickel as catalyst for selective oxidation of methane to syngas. Commun Chem. 2020;3(1):1–8.

Chapter 7
Crystalline Support

Masaaki Kitano and Hideo Hosono

7.1 Introduction

Supported metal catalysts have been practically used in various chemical reactions such as methanol synthesis, Fischer-Tropsch synthesis, hydrodesulfurization, steam reforming, purification of automobile exhaust gas, fuel cell, and so on [1, 2]. These reactions take place on the surface of metal sites, and therefore, metal catalysts are generally immobilized as nano-sized particles on the support materials to increase the number of the surface active sites [3]. The coordination number (CN) of the surface metal site is much lower than that of the bulk site (in most cases CN = 12), and the coordinatively unsaturated metal sites can activate various molecules. Since the CN of metal atom changes depending on the site such as edges, kinks, corners, and terraces, the structure-sensitive phenomenon is observed in various catalytic reactions. To date, tremendous efforts have been made to develop highly active supported metal catalysts with nanometer-sized metal particle such as single atom, nano-cluster, and nanoparticle [4]. The support material affects the geometric structure and the electronic structure of the supported metal nanoparticles through the metal-support interaction. In addition, the support can enhance the stability of the supported metal catalyst by preventing aggregation of metal sites. Due to their high thermal stability, metal oxides and carbon are mainly used as catalyst supports. In this chapter, we concentrate on the crystalline oxide-based support material for the catalytic reactions.

M. Kitano (✉) · H. Hosono
Materials Research Center for Element Strategy, Tokyo Institute of Technology, 4259 Nagatsuta, Midori-ku, Yokohama 226–8503, Japan
e-mail: kitano.m.aa@m.titech.ac.jp

H. Hosono
e-mail: hosono@mces.titech.ac.jp

W. Ueda (ed.), *Crystalline Metal Oxide Catalysts*,
https://doi.org/10.1007/978-981-19-5013-1_7

7.1.1 Structural Effect of the Support on Supported Metal Particles

The fraction of surface metal sites increases with decreasing the diameter of metal nanoparticles, which results in an increase in catalytic activity. For example, the particle size effect is clearly observed in CO oxidation by Au nanoparticles supported on TiO_2 and CeO_2 [5, 6]. While the turnover frequencies (TOF) of the Au catalysts are almost constant when the particle size is larger than 5 nm, the TOF drastically increases especially below 2 nm. Such a small-sized Au nanoparticle catalyst exhibits surprisingly high catalytic performance for CO oxidation even at −70 °C [7]. In the oxide-supported Au catalysts, Au clusters are preferentially immobilized at the oxygen vacancy sites on the oxide surface, which plays an important role in the stabilization of Au nanoparticles [8–10]. For maximizing the metal utilization efficiency, atomically dispersed metal catalysts, so-called single-atom catalysts (SAC), have been studied in recent years [11–13]. The SAC becomes a hot research topic in catalysis with the help of the development of advanced aberration-corrected electron microscopy. In the ideal SAC, all metal sites are present in the form of isolated atoms, meaning 100% of metal atom dispersion. Oxide supports have attracted particular attention because oxides have abundant defect sites and surface OH groups can serve as the anchoring sites for single metal atoms. FeO_x and CeO_2 are commonly investigated as SAC supports. For example, Zhang et al. prepared Pt/FeO_x SACs via the substitution of surface Fe atoms by Pt in the crystal lattice of FeO_x [11]. The Pt/FeO_x with Pt single atoms exhibits a much higher TOF than Pt clusters or nanoparticles supported on FeO_x for CO oxidation and PROX reactions. In this system, O_2 is activated at the oxygen vacancy sites near the Pt atoms in FeO_x support, and the partially vacant 5d orbitals of the positively charged Pt atoms can activate CO with a lower energy barrier for CO oxidation than Pt clusters. The scope of support materials has also been extended to non-oxides such as graphene, nitrogen-doped carbon, C_3N_4, metal-organic frameworks, MoS_2, MoC, MXene, and so on [14]. The support material not only stabilizes the metal particles but also participates directly in the catalytic reactions. Behm et al. reported that CO oxidation on Au/TiO_2 proceeds via a Mars-van Krevelen mechanism [15], in which surface oxygen sites at the Au–TiO_2 interface react with CO, resulting in the formation of oxygen vacancy. These vacancy sites are reoxidized by gas-phase O_2. The TOF values decrease with increasing Au particle size because of the decrement in the number of the Au–TiO_2 perimeter sites.

7.1.2 Electronic Effect of the Support on Supported Metal Particles

Electron transfer between the support and metal nanoparticles affects the electronic structure of the metal sites. The local charge density and distribution of the supported metal site are perturbed by exposed cations or anions of the support. This electronic metal-support interaction (EMSI) [16, 17] leads to an enhancement of catalytic performance [18]. This is in contrast to the strong metal-support interaction (SMSI), which usually results in a decrement of the catalytic activity since the surface of the supported metal sites is covered by the oxide layer of the support under reaction conditions [19, 20]. Pt-loaded CeO_2 catalysts are demonstrated to show high catalytic performance for water-gas shift reaction by the result of the EMSI effect [18]. The catalytic activity steeply increases with increasing Pt amount, reaching a maximum at a coverage of 0.2 monolayer, which corresponds to Pt particles with a diameter less than 1.7 nm. The photoemission measurements and DFT calculation reveal that the charge transfer from Pt nanoparticles to CeO_2 enhances the ability of the Pt nanoparticles to dissociate the OH bond in H_2O. Libuda and co-workers also clarified the charge transfer between CeO_2 and Pt with various particle sizes by combining synchrotron-radiation photoelectron spectroscopy, scanning tunneling microscopy, and DFT calculations [21]. The charge transfer reaches a maximum for Pt nanoparticles with 30–70 atoms, corresponding to 1.1–1.7 nm in particle size, where up to 0.11 electrons are transferred from Pt to CeO_2 to form positively charged Pt sites. At the same time, a part of Ce^{4+} ions at the surface of CeO_2 is reduced to Ce^{3+}. In contrast, the charge transfer is limited for larger or smaller Pt particles. Thus, small-sized metal nanoparticles prefer to be positively charged in the oxide-supported metal catalyst systems. On the other hand, negatively charged metal sites are formed when the metal nanoparticles are fixed on the electron donating support. Häkkinen et al. demonstrated that a negatively charged Au cluster is formed on MgO support by partial electron transfer from the oxygen vacancy F centers (2 electrons trapped at oxygen vacancy) on MgO to the adsorbed Au cluster [22]. The Au octamers bound to F centers of MgO can oxidize CO into CO_2 even at 140 K, while the same cluster on MgO without oxygen vacancies is catalytically inactive for this reaction. From these examples, it is found that the electronic and chemical properties of the supported metal site can be tuned by the interaction with simple oxide support. In the following sections, examples of metal nanoparticle catalysts fixed on unique crystalline oxide supports with heteroanions such as electron, hydride ion, and nitride ions are introduced.

7.2 $12CaO7Al_2O_3$ Electride

7.2.1 Structural Properties of $12CaO{\cdot}7Al_2O_3$ Electride

$12CaO{\cdot}7Al_2O_3$ (C12A7) is a constituent of commercially available aluminate cements and is well known as a natural mineral, mayenite. It has a cubic structure (space group $I\bar{4}3d$) with a lattice constant of 1.199 nm (Fig. 7.1) [23]. The unit cell has a positively charged framework structure composed of 12 subnanometre-sized cages with an inner diameter of 0.4 nm, and its chemical formula is expressed by $[Ca_{24}Al_{28}O_{64}]^{4+}(O^{2-})_2$. The free oxygen ions ($O^{2-}$) are incorporated in the positively

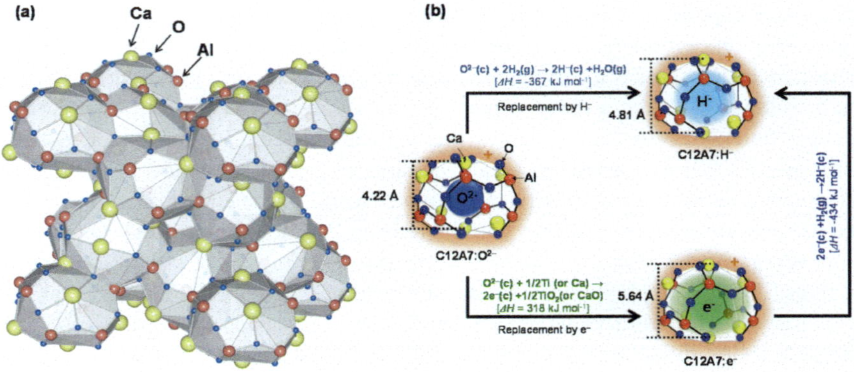

Fig. 7.1 **a** Crystal structure of $12CaO{\cdot}7Al_2O_3$ (C12A7). **b** Enthalpy changes for anion exchange between O^{2-}, H^-, and e^- in the cage of C12A7. 'g' and 'c' represent the species in the gas phase and a cage, respectively. Adapted with permission from Ref. [38]. Copyright 2015 Macmillan Publishers Limited, part of Springer Nature

charged cages to compensate for the charge balance and can be replaced by various anions such as OH^- [24], O^- [25], F^- [26], Cl^- [26], H^- [27, 28], Au^- [29], S^{2-} [30], NH^{2-} [31], and NH^{2-} [32]. Furthermore, when the free O^{2-} ions are extracted from the cages, free electrons can be injected into the cavities through the replacement of two O^{2-} ions with four electrons, leading to the formation of $[Ca_{24}Al_{28}O_{64}]^{4+}(4e^-)$ [33]. The resultant material is the first inorganic electride. The concept of "electride" was first presented in 1983 by Dye et al. and the first example was a crystalline organic compound, Cs^+(18-crown-6)$_2e^-$, in which electrons are completely isolated from the Cs^+ ions sandwiched by crown ether [34]. Hence, the electride can be regarded as a class of ionic compounds, where electrons are located at lattice cavities or channels to form individual orbitals and serve as anions individually rather than combining with other atoms [35].

The C12A7 electride (C12A7:e^-) has a high electron density ($N_e = 2.3 \times 10^{21}$ cm^{-3}), which results in high electrical conductivity (1500 S cm^{-1}) and a very low work function (2.4 eV) comparable to metal potassium [33, 36]. The C12A7:O^{2-} can be synthesized by conventional solid-phase reaction of $CaCO_3$ and Al_2O_3 with a molar ratio of 12:7 at 1300 °C. The free O^{2-} ions in the C12A7:O^{2-} can be replaced with electrons by the chemical reaction of C12A7:O^{2-} with Ca or Ti metal under vacuum heating conditions. During the heat treatment, the O^{2-} ions are diffused from the bulk to the surface by forming CaO via the reaction: $O^{2-} + Ca \rightarrow CaO + 2e^-$. The electron concentration in the C12A7 can be controlled depending on the reduction conditions. While the C12A7:O^{2-} is a typical insulator, it shows semiconducting behavior when the N_e is below 1.0×10^{21} cm^{-3}. Moreover, the C12A7:e^- with high N_e ($\geq 1.0 \times 10^{21}$ cm^{-3}) exhibits metallic conductivity. Therefore, the insulator-to-metal transition occurs at $N_e = 1.0 \times 10^{21}$ cm^{-3}, in which half of the O^{2-} ions are replaced by electrons. The anion exchange of C12A7 also leads to the structural change in the cage size (Fig. 7.1). When the encaged O^{2-} ion is replaced by electron, the cage size increases from 4.22 to 5.64 Å with a formation enthalpy of 318 kJ mol^{-1}[37]. The formation enthalpies of C12A7:H^- from C12A7:e^- (−434 kJ mol^{-1}) are larger by 67 kJ mol^{-1} than that from C12A7:O^{2-} (−367 kJ mol^{-1}) [37]. The size of the cage with H^- ion is 4.81 Å. Such structural changes caused by anion exchange have a great impact on the hydrogen absorption-desorption properties and ammonia synthesis activity [38]. Thus, these unique electronic and structural properties greatly contribute to the catalytic performance of the metal nanoparticles supported on the surface of C12A7:e^-.

7.2.2 Catalytic Applications of 12CaO·7Al_2O_3 Electride

The low work function and high electron density of C12A7:e^- can be expected to become an efficient catalyst support in ammonia synthesis reactions [35]. Generally, N_2 reduction into ammonia over transition metal (TM) catalyst is promoted by electron injection to TM sites because the rate-determining N_2 dissociation step is accelerated by electron donation from TM to the antibonding orbitals of N_2. Accordingly, the electron injection to the TM site is the key factor for high ammonia synthesis activity [39]. Among the TM catalysts, ruthenium (Ru) is known as the most active catalyst and Ru nanoparticles are supported by various materials such as carbon, Al_2O_3, MgO, CaO, CeO_2, La_2O_3, Sm_2O_3, $MgAl_2O_4$, Si_3N_4, and zeolites [40–42]. In particular, basic oxides are much more effective than acidic oxides in terms of electronic promotion effect [43]. In 2012, C12A7:e^- was demonstrated to function as a strong electronic promoting support for Ru catalyst in ammonia synthesis [44]. The Ru/C12A7:e^- exhibits high NH_3 synthesis activity comparable to Cs-Ru/MgO while the Ru particle size of the former (~20 nm) is much larger than that of the latter (<5 nm) (Fig. 7.2) [45]. It is well recognized that NH_3 synthesis on Ru catalysts is a structure-sensitive reaction, i.e., N_2 cleavage preferentially occurs on the step surface of Ru rather than on the terrace [46–48]. The active step site consists of the combination of a three-fold hollow site and a bridge site (so-called B_5-type site), which becomes optimal for particles of 1.8–2.5 nm in size [47]. Therefore, Ru/C12A7:e^- catalyst has considerably fewer active sites than Cs-Ru/MgO. As a consequence, Ru/C12A7:e^- has an order of magnitude greater turnover frequency (TOF) than those of traditional Ru catalysts. Furthermore, the apparent activation energy of Ru/C12A7:e^- (ca. 50 kJ mol^{-1}) is almost half those of other Ru catalysts (80–120 kJ mol^{-1}). As shown in Fig. 7.3, conventional CaO, Al_2O_3, and CaO·Al_2O_3 (CA) cannot promote the activity of Ru. Ru/C12A7:O^{2-} without electron in the cages shows a much lower ammonia synthesis rate and higher activation energy than Ru/C12A7:e^- even after loading the Cs-oxide promoter, indicating that the electronic promoting effect of C12A7:e^- is much stronger than the conventional alkali promoters. It should be noted that the catalytic performance of Ru/C12A7:e^- strongly depends on the electron concentration (N_e), i.e., the catalytic activity enhances 10 times at N_e of 1.0×10^{21} cm^{-3}, and the activation energy (E_a) for ammonia synthesis also reduces to about 50 kJ mol^{-1} from about 100 kJ mol^{-1} [45]. Accordingly, the catalytic performance is dominated by the metal-insulator transition of C12A7:e$^-$. Figure 7.4 shows the electronic structure of C12A7 with and without electrons. The framework conduction band (FCB) and the valence band (FVB) consist of the 4 s orbitals of Ca^{2+} ions and the 2p orbitals of O^{2-} ions in the framework, respectively [49]. In addition, the cage conduction band (CCB) derived from the empty cage in C12A7, and the energy level of free O^{2-} ion in the cage is formed at ca. 1 eV below the bottom of the FCB and at ca. 1 eV above the top of the FVB, respectively. When the O^{2-} ions in the cages are partially replaced by electrons ($N_e < 1.0 \times 10^{21}$ cm^{-3}), F$^+$-like centers are formed at the energy level of 0.4 eV below the CCB. At high N_e ($>1.0 \times 10^{21}$ cm^{-3}), the electrons become itinerant because the electrons occupy the CCB [50]. This change in the electronic structure leads to the two effects. One is the upshift of the Fermi level (E_f) and the other is the drastic contact resistance drop between C12A7:e$^-$ and loaded metal due to tunneling [51]. The color of C12A7 powder is changed from white to green, dark brown, and black with increasing N_e. The C12A7:e$^-$ with low N_e ($<1.0 \times 10^{21}$ cm^{-3}) shows semiconducting behavior, in which the intercage electron hopping takes place through the empty cages with an energy of ca. 0.4 eV since the encaged electron forms a polaron that strongly interacts with the positively charged cage framework [52]. The polaron-type electron hopping is caused by the lattice distortion in the framework (Fig. 7.1), so that C12A7:e$^-$ with low N_e ($<1.0 \times 10^{21}$ cm^{-3}) shows a semiconducting character and low electrical conductivity (100 S cm^{-1}). As a result, the electron transfer from C12A7:e$^-$ with low N_e ($<1.0 \times 10^{21}$ cm^{-3}) to Ru is not effective, which accounts for low catalytic performance in ammonia synthesis. On the other hand, the C12A7:e$^-$ with high N_e has metallic conductivity (1500 S cm^{-1}) and a very low WF (2.4 eV). The smooth electron injection occurs from C12A7:e$^-$ to Ru, which in turn facilitates N_2 activation with a low activation barrier. Metallic potassium (K) with a low WF (2.3 eV) cannot be used as an electronic promoter since it is too chemically reactive and thermally unstable under NH_3 synthesis conditions. Hence, the C12A7:e$^-$ support has unique properties combining chemical durability with low work function, which is not realized by conventional materials, resulting in high and stable ammonia synthesis activity.

Moreover, the hydrogen storage properties of C12A7:e$^-$ significantly influence the ammonia synthesis activity. Generally, the ammonia synthesis rate of Ru-based catalysts decreases with an increase in the partial pressure of H_2, where hydrogen adsorption prevents the N_2 adsorption on the Ru surface. Such a hydrogen poisoning is known as a serious problem in Ru-based ammonia synthesis catalysts [53]. However, the ammonia synthesis rate of Ru/C12A7:e$^-$ increases with H_2 pressure, resulting that the reaction order with respect to H_2 is +0.97, which is in contrast to the large negative values reported in the conventional Ru catalysts.[44] On the Ru/C12A7:e$^-$ catalyst, hydrogen adatoms on the Ru surface can be captured as H$^-$ ions in the cages through the reaction of encaged electrons with hydrogen adatom generated on Ru surface ($H^0 + e^- \rightarrow H^-$). The electrons are regenerated in the cages by releasing H atoms from the cages, H$^-$ (cage) → H^0 (release from cage) + e$^-$(cage). This reversible e$^-$–H$^-$ exchange reaction in Ru/C12A7:e$^-$ effectively prevents the hydrogen poisoning on the Ru surface. Once all electrons in (Ru-free)

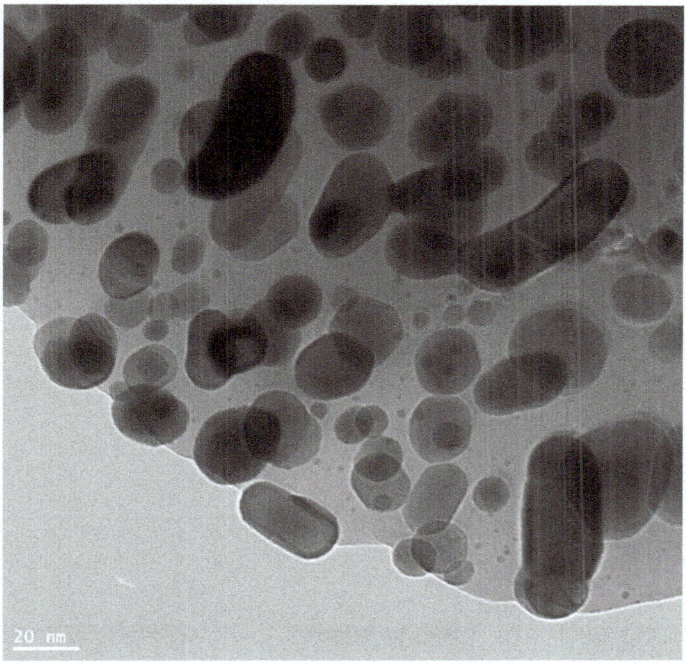

Fig. 7.2 STEM image of Ru(2wt%)/C12A7:e^- after ammonia synthesis reaction at 400 °C

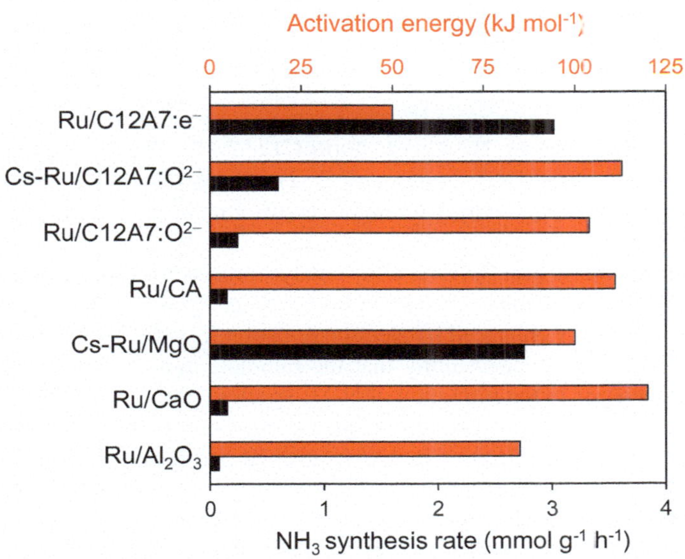

Fig. 7.3 Ammonia synthesis rates (360 °C, 0.1 MPa) and apparent activation energies of various supported Ru (2 wt%) catalysts

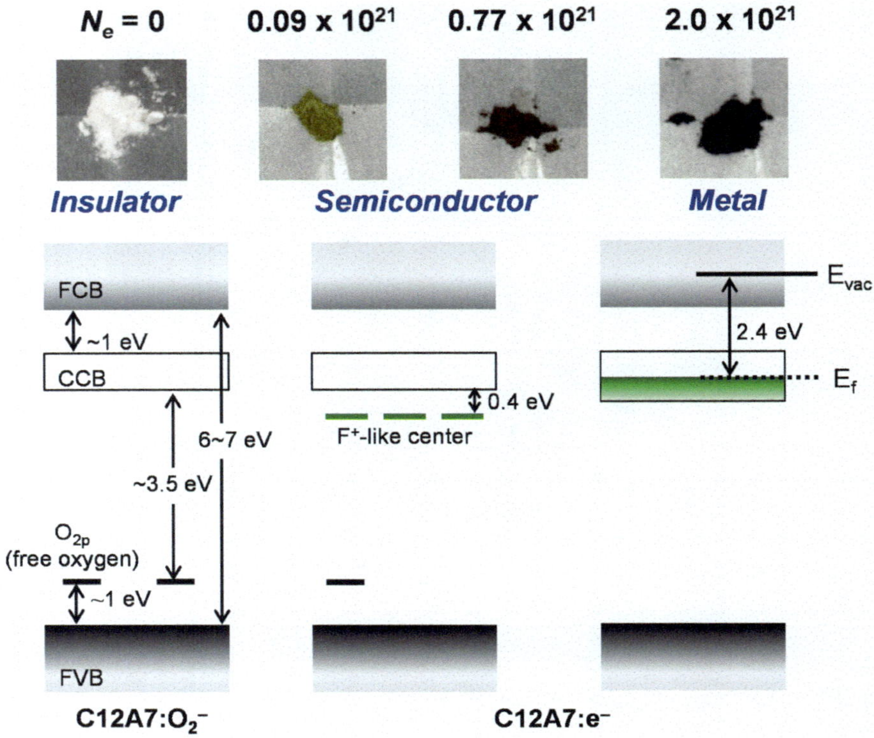

Fig. 7.4 Electronic structures and photographs of C12A7:e^- with various electron concentrations. Adapted with permission from Ref. [45]. Copyright 2015 American Chemical Society

C12A7:e^- are replaced by H^- ions to form C12A7:H^-, the reversible hydrogen storage-release reaction does not proceed.[38] Therefore, the Ru/C12A7:H^- shows negligibly low catalytic activity with higher activation energy (154 kJ mol^{-1}) than Ru/C12A7:e^-.

Ammonia synthesis over Ru/C12A7:e^- catalyst proceeds through a unique mechanism as compared with the conventional Ru-based catalysts. N_2 cleavage on the Ru surface is significantly enhanced by electron donation from C12A7:e^- support with a very low work function, resulting that the rate-determining step (RDS) shifted from N_2 dissociation to the formation of N–H_n species (Fig. 7.5) [38]. Furthermore, hydrogen adatoms on Ru are transferred into the cages of C12A7:e^- to form H^- ions through reaction with electrons, which prevents hydrogen poisoning on Ru surface. Then, the H^- ions immediately react with N species on Ru to form NH_3 with simultaneous regeneration of electrons in the cage of C12A7:e^-. This is contrary to the case of conventional Ru catalysts, where the activation of N_2 and H_2 into ammonia occurs only on the Ru surface and the RDS is the N_2 dissociation.

The catalytic properties of Ru/C12A7:e^- in ammonia synthesis have been further studied by other researchers. Kammert et al. investigated the role of H^- ion in NH_3

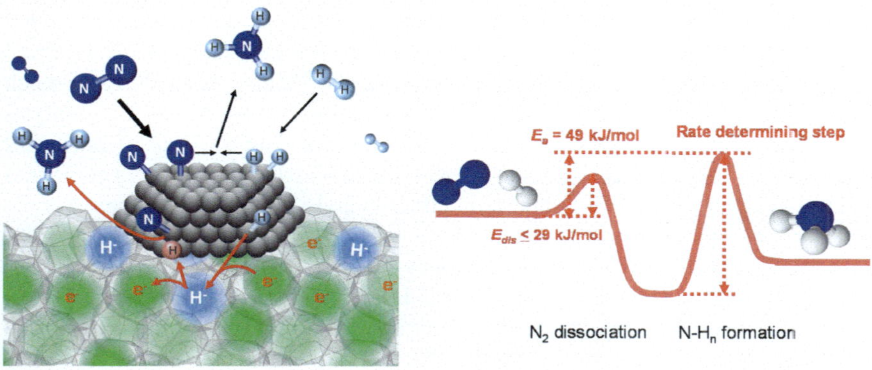

Fig. 7.5 Reaction mechanism and energy diagram for NH_3 synthesis on Ru/C12A7:e$^-$catalyst. Adapted with permission from Ref. [38]. Copyright 2015 Macmillan Publishers Limited, part of Springer Nature

synthesis over Ru/C12A7:e$^-$ catalyst by using in situ neutron-scattering techniques, DFT calculations, and steady-state isotopic transient kinetic analysis (SSITKA).[54] SSITKA experiment revealed that the total coverage of nitrogen species on the Ru/C12A7:e$^-$ reaches a maximum of 84% of the total amount of surface Ru sites, which is much higher than those of Ru/C12A7:O^{2-} (15%) and the alkali-promoted Ru/MgO ($\leq$ 14%). The extremely high coverage of nitrogen species on Ru/C12A7:e$^-$ suggests a shift away of the RDS from N_2 cleavage to the formation of the N–H bond, which is well consistent with the original work [38]. However, as demonstrated by in situ inelastic neutron-scattering spectroscopy, the encaged H^- species does not participate in NH_3 synthesis reaction because the H^- is thermally and chemically stabilized in the cage to keep electroneutrality. Furthermore, Weber et al. reported that the catalytic activity of Ru/C12A7:e$^-$ disappeared immediately above 1 MPa, which is explained by irreversible hydride formation [55]. From this result, they conclude that the application of Ru/C12A7e$^-$ catalysts at the industrial scale is limited prospects. However, in these two studies, the C12A7e$^-$ is prepared by plasma arc melting method using metal aluminum as a solid reductant and the electron concentration (N_e) of the resultant C12A7e$^-$ is obviously lower than 5.0×10^{20} cm^{-3}. As described above, the catalytic performance of Ru/C12A7:e$^-$ is drastically changed around the insulator-to-metal transition point ($N_e = 1.0 \times 10^{21}$ cm^{-3}). The deactivation caused by H^- ion incorporation occurs only on the Ru/C12A7e$^-$ with low N_e.

The C12A7:e$^-$ with high N_e can also boost the ammonia synthesis activity of a cobalt (Co) nanoparticle catalyst [56]. According to the volcano-shaped relationship between ammonia synthesis activity and nitrogen adsorption energy, Co is inferior to Ru as a catalyst for ammonia synthesis because of its lower nitrogen binding energy [57, 58]. Therefore, the Co nanoparticle (6wt%) loaded MgO catalyst shows negligibly smaller activity (31 μmol g^{-1} h^{-1}) at 400 °C and 0.1 MPa than Ru(6wt%)/MgO (1610 μmol g^{-1} h^{-1}). While Co/C12A7:O^{2-} exhibits activity comparable to Co/MgO, the catalytic activity of Co/C12A7:e$^-$ is superior to that of Ru/MgO. The operation temperature for ammonia production over Co/C12A7:e$^-$ is lower by ~200 °C than that for the Co/C12A7:O^{2-} catalyst, resulting in much lower activation energy (ca. 50 kJ mol^{-1}) than those of conventional Co catalysts (90–110 kJ mol^{-1}). The high catalytic performance of Co/C12A7:e$^-$ is attributed to the electron donation from C12A7:e$^-$ with a much lower WF (2.4 eV) compared with Co metal (5.0 eV). This promotion effect is quite different from the other Co catalysts combined with nitride or hydride such as Co_3Mo_3N [59] and Co–LiH [60], in which N_2 activation over Co is assisted by other compounds with high nitrogen adsorption energy to form nitride species as intermediates. On Co_3Mo_3N catalyst, N_2 is activated to form Mo_3N at threefold hollow Mo sites, and then ammonia is formed by the subsequent hydrogenation via the Mars-van Krevelen-type mechanism [61]. As for the Co–LiH catalysts, N_2 dissociation is accelerated at the Co-LiH interface through the intermediate Li_2NH imide formation, and subsequently, these species are hydrogenated to NH_3 [60]. This LiH-mediated mechanism also promotes the ammonia synthesis over other TM catalysts such as V, Cr, Mn, and Fe. In contrast, both N_2 and H_2 molecules are directly activated over Co nanoparticles on C12A7:e$^-$ without the aid of other additives. The Co nanoparticles on C12A7:e$^-$ maintain a metallic state without reaction with other components during NH_3 synthesis.

The C12A7:e$^-$ also has a high electronic promotion effect on the activity of supported metal catalysts in various other chemical reactions such as ammonia decomposition [62], CO oxidation, [63] and selective hydrogenation of unsaturated aldehydes [64]. In addition, C12A7 without electrons can serve as an efficient support for Pt single atom catalyst (Pt-SAC) [65]. There are a number of opened empty cavities with a positive charge on the surface of the C12A7 prepared by the hydrothermal method. These cavities can capture $[PtCl_4]^{2-}$ ions by the Coulomb interaction when the C12A7 is dispersed in $PtCl_4$/methanol solution. After the reduction of the precursor, atomically dispersed Pt is formed on the surface of the C12A7 support (Fig. 7.6). This single Pt atom on C12A7 does not aggregate even after the H_2 reduction at 600 °C. Furthermore, the Pt-SAC/C12A7 exhibits high and stable catalytic activity for the hydrogenation of 4-chloronitrobenzene to 4-chloroaniline below 60 °C. The TOF is an order of magnitude higher than that of Pt/Al_2O_3 or Pt/CaO under the same reaction conditions.

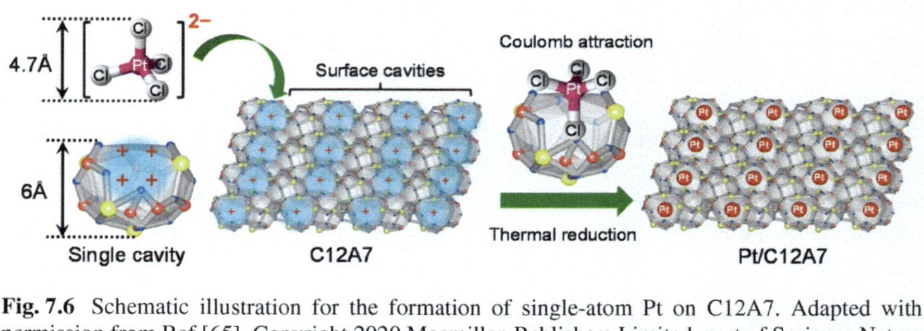

Fig. 7.6 Schematic illustration for the formation of single-atom Pt on C12A7. Adapted with permission from Ref [65]. Copyright 2020 Macmillan Publishers Limited, part of Springer Nature

7.3 Metal Oxide Bearing Heteroanion

7.3.1 *Oxyhydride*

Hydride (H^-) ion doping into oxides is a new strategy for modifying the chemical and physical properties of inorganic materials [27, 66–70]. The H^- ion consists of one proton and two electrons and has unique characteristics such as high standard redox potential (−2.2 V vs SHE), large electronic polarizability, and moderate electronegativity (2.20) [71]. In contrast to other anions, H^- ion is highly flexible in size with ionic radii of 1.3–1.5 Å, which is similar to those of O^{2-} and F^- ions. The H^- ions are formed in the solid materials by the reaction with electropositive metal elements, whereas OH^- ions with protonic hydrogen (H^+) are formed on the typical metal oxide surface. A typical metal hydride such as CaH_2 is used as a reducing agent in various chemical reactions, and recently, some oxyhydrides such as $LaFeAs(O_{1-x}H_x)$ and $La_{2-x-y}Sr_{x+y}LiH_{1-x+y}O_{3-y}$ are reported to show superconductivity and hydride-ion-conductivity [68, 70]. Thus, the oxyhydrides are unique materials that have been attracting significant attention in recent years.

The role of H^- ion in ammonia synthesis has been firstly demonstrated by Ru/C12A7:e^- and Ru/Ca_2NH catalysts [44, 72]. The Ru/Ca_2NH, in which H^- ions are stabilized at interlayer space between positively charged [Ca_2N] layers, exhibits much higher ammonia synthesis activity and lower activation energy than Ru/CaNH containing H^+. Anionic electrons are created on hydride vacancy sites at the Ru-Ca_2NH interface during the reaction, which imparts a low work function (2.3 eV) to the surface. Therefore, this material can be regarded as a quasi-electride. After this discovery, a number of hydride-based catalysts such as CaH_2 [72], LnH_2 [73], $LnH_{3-2x}O_x$ (Ln = La, Ce, Y) [74], TM-hydrides (AH_x, A = Li, Na, K, Ba, Ca) composite [60], $BaTiO_{2.5}H_{0.5}$ [75], TiH_2 [75], GdHO [76], SmHO [76], $VH_{0.39}$ [77], CaH_2–BaF_2 [78], Mo-hydride grafted SiO_2 [79], Li_4RuH_6 [80], and Ba_2RuH_6 [80] have been developed. Kageyama et al. have developed novel oxyhydrides, $BaTiO_{3-x}H_x$, with a perovskite structure by the replacement of O^{2-} sites of $BaTiO_3$ with H^- ions [69]. The $BaTiO_{2.5}H_{0.5}$ can be synthesized by heating $BaTiO_3$ with an excess amount of CaH_2 at 520–560 °C for 1 week under vacuum conditions and subsequent washing treatment with an NH_4Cl/methanol solution to remove excess CaH_2 and CaO byproduct (Fig. 7.7). The lattice H^- ions can be easily exchanged with N^{3-} by heating under N_2 or NH_3 atmosphere at low temperatures (375–550 °C), which is derived from the lability of H^- ion in $BaTiO_{2.5}H_{0.5}$ [81]. The ammonia synthesis activity of $BaTiO_{2.5}H_{0.5}$ without any transition metal catalyst is almost comparable to Cs–Ru/MgO catalyst at high temperature and high pressure conditions (400 °C, 5 MPa) [75]. In this system, Ti with a high nitrogen binding energy enhances dissociative adsorption of N_2, and the hydride partially reacts with the activated N to form nitride-hydride intermediates that are regarded as the active catalyst composition. Furthermore, the $BaTiO_{2.5}H_{0.5}$ facilitates the activity of Ru, Fe, and Co nanoparticles, which are much higher than the case of $BaTiO_3$ support [82]. When the Ba-site is substituted by other alkali-earth cations such as Sr, and

Ca, the (Ca, Sr, Ba)$TiO_{3-x}H_x$ supports exhibit much higher catalytic activity than the corresponding oxide supports. The catalytic activity increases in the order of $CaTiO_{3-x}H_x < SrTiO_{3-x}H_x < BaTiO_{3-x}H_x$. On the other hand, the catalytic activity of Ru/$BaTiO_{2.5}N_{0.2}$ without any H^- ion is negligibly low, indicating that lattice H^- ion is indispensable to enhance the catalytic activity. They also demonstrated that LnHO (Ln = Gd, Sm) with anion-ordered fluorite structure (P4/nmm)-supported Ru catalysts exhibits extremely high ammonia synthesis rate (50–168 mmol g^{-1} h^{-1}) at 400 °C and 5 MPa, although the detailed mechanism is not unveiled [76].

Lanthanide oxyhydrides such as $LaH_{3-2x}O_x$ and $CeH_{3-2x}O_x$ also have outstanding promotion effects in ammonia synthesis at low reaction temperatures (≤ 260 °C) [74]. $LaH_{3-2x}O_x$ has a fluorite-type tetragonal structure (space group: P4/nmm) and is prepared by the partial substitution of O^{2-} ion in the H^- site of LaH_3. In LaH_3, H^- ions are located in La_4 tetrahedron (T) and La_6 octahedron (Oc) sites, and the nearest neighbor of H^- site is H^- rather than La^{3+} (Fig. 7.8). The replacement of H^- site with O^{2-} leads to the creation of vacancy sites at the either T or Oc sites, and the resultant $LaH_{3-2x}O_x$ exhibits the highest H^- ion conductivity among the hydride materials [83, 84]. The $LaH_{3-2x}O_x$ can be prepared by the solid-state reaction of LaH_3 with La_2O_3, and the H^- content can be controlled by changing the mixing ratio of the LaH_3 to La_2O_3. As compared with LaH_3, diffraction peaks shifted to a lower angle side with increasing the oxygen content, which indicates the lattice expansion by the replacement of H^- ion with larger O^{2-} ion (Fig. 7.8). Figure 7.9 shows the ammonia synthesis activities of Ru catalysts supported on La-based oxide, hydride, and oxyhydride. While Ru-loaded rare earth oxides such as CeO_2 and La_2O_3 are known to show high ammonia synthesis activity, the operation temperature of Ru/$LaH_{3-2x}O_x$ is 100 °C lower than that of Ru/La_2O_3. As a result, Ru/$LaH_{3-2x}O_x$ exhibits a lower activation energy (64 kJ mol^{-1}) than Ru/La_2O_3 (81 kJ mol^{-1}). Ru/LaH_3 catalyst also works as an efficient catalyst at low reaction temperatures, but the activity gradually decreases with reaction time (Fig. 7.9). The deactivation is caused by the nitridation of the LaH_3 surface during ammonia synthesis. On the

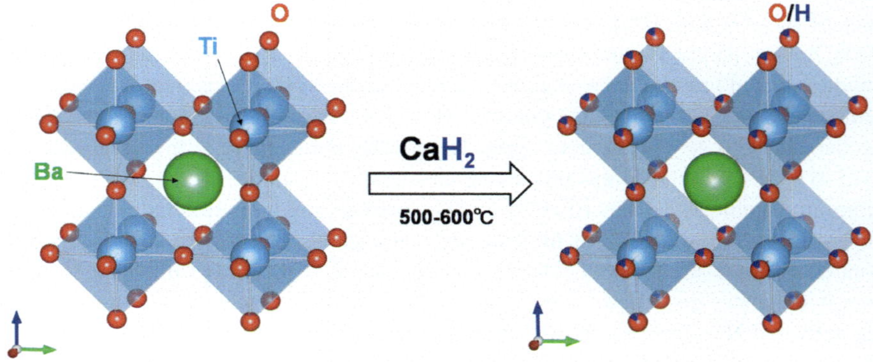

Fig. 7.7 Synthesis of $BaTiO_{3-x}Hx$ by the reaction of $BaTiO_3$ with CaH_2

other hand, Ru/$LaH_{3-2x}O_x$ has a stable catalytic activity for more than 100 h, and the total amount of produced ammonia (≈9.0 mmol) is much larger than the amount of lattice H^- ions (1.7 mmol). The surface H^- ions are preserved during the reaction because the presence of lattice oxygen prevents the transformation of $LaH_{2.5}O_{0.25}$ to LaN. However, there is no relationship between the catalytic performance and the hydrogen content ($0.25 \leq x \leq 1$) in Ru/$LaH_{3-2x}O_x$, which is contrary to the result of H^- ion conductivity of $LaH_{3-2x}O_x$ [83]. The hydride-ion conductivity of $LaH_{3-2x}O_x$ increases significantly with an increase in the H^- content, e.g., the conductivity of $LaH_{2.5}O_{0.25}$ is two orders of magnitude higher than that of LaHO. The activation energies of hydride conduction are in the range of 96.5–125.4 kJ mol^{-1}, which is much larger than that of ammonia synthesis. Accordingly, the H^- ion conduction of $LaH_{3-2x}O_x$ does not directly contribute to the catalytic performance. Nevertheless, the $LaH_{3-2x}O_x$ support shows a clear electronic promotion effect on ammonia synthesis activity since the Ru particle size and shape are the same as the La_2O_3 and LaH_3-supported Ru catalysts. Furthermore, Ru/$CeH_{3-2x}O_x$ catalysts have much higher catalytic activity than Ru/CeO_2, indicating the introduction of H^- ion in the support material is effective for low-temperature ammonia synthesis.

DFT calculation revealed that the work functions (WFs) of $LaH_3(110)$ and $LaH_{2.5}O_{0.25}(100)$ decrease from 3.5 to 1.8 eV when the H^- ion is removed from each surface. When a Ru cluster is loaded on each support, electron transfer occurs from hydride vacancy (V_H) sites to Ru cluster, leading to that the charges (Qs) of Ru on $LaH_{3-x}V_{Hx}$ and $LaH_{2.5-x}V_{Hx}O_{0.25}$ are determined to be −0.33 and −0.26, respectively. The negatively charged Ru is confirmed by XPS and FT-IR spectroscopy for CO adsorption on the catalysts. Thus, the formation of V_H at Ru-support interface is a key for the electronic promotion effect, and the electron donating ability of $LaH_{3-2x}O_x$ is comparable to that of LaH_3. The Gibbs free energy changes (ΔG) for the V_H formation on $LaH_{2.5}O_{0.25}$ is calculated to be larger than that for LaH_3.

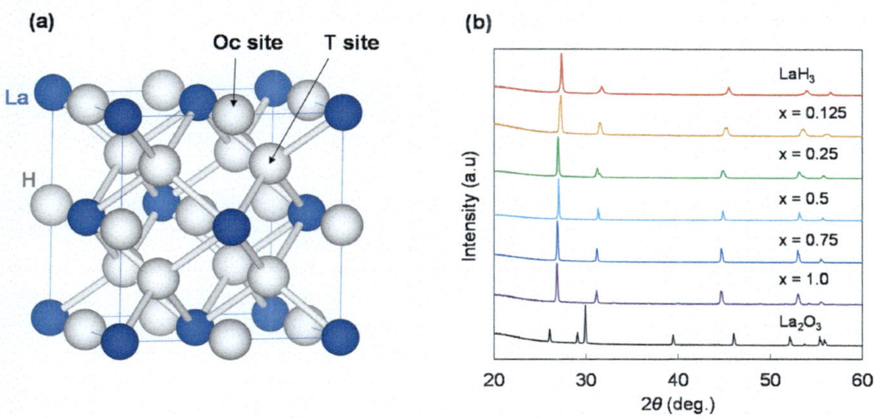

Fig. 7.8 **a** Crystal structure of LaH_3. **b** XRD patterns for $LaH_{3-2x}O_x$, LaH_3, and La_2O_3. Adapted with permission from Ref. [74]. Copyright 2020 Wiley-VCH GmbH

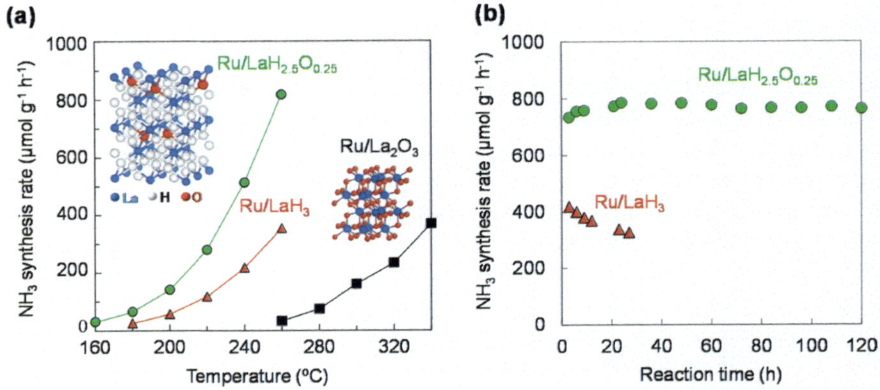

Fig. 7.9 **a** Temperature dependence of catalytic activity for various Ru (2wt%) catalysts at 0.1 MPa. **b** Reaction time profile for ammonia synthesis over Ru(5wt%)/$LaH_{2.5}O_{0.25}$ and Ru(5wt%)/LaH_3 at 0.1 MPa and 240 °C. Adapted with permission from Ref. [74]. Copyright 2020 Wiley-VCH GmbH

H_2–TPD experiments also indicated that H_2 desorption from LaH_3 takes place at a lower temperature with a lower activation energy (E_{des}: 58.4 kJ mol^{-1}) than in the case of $LaH_{2.5}O_{0.25}$ (E_{des}: 72.1 kJ mol^{-1}). The low WF surface is readily formed on LaH_3 surface rather than $LaH_{2.5}O_{0.25}$. However, the ΔG for LaN formation on LaH_3 is much more negative than that on $LaH_{2.5}O_{0.25}$, indicating that LaH_3 tends to be converted into LaN during ammonia synthesis. Indeed, the LaN formation is confirmed by XRD of Ru/LaH_3 after ammonia synthesis, which is the main reason for the deactivation of Ru/LaH_3 (Fig. 7.9). Once LaN is formed, the formation of nitride vacancy is unlikely to occur because of its larger formation energy. The LaN formation at the Ru–LaH_3 interface reduces the electronic donation from the support to Ru nanoparticles. In contrast, $LaH_{2.5}O_{0.25}$ has high resistance toward nitridation during ammonia synthesis, which results in high and stable catalytic activity. Thus, the H^- ion in the oxyhydride support serves as an electron source to promote ammonia synthesis on Ru catalyst, that is, electron injection from the oxyhydride with electrons trapped at the H^- vacancy sites to Ru greatly enhances N_2 cleavage at low reaction temperatures. Although the TM–LiH composite catalysts are reported to show high ammonia synthesis activity at low reaction temperatures [60], N_2 dissociation is facilitated by the formation of metal-imide intermediate species at the TM–LiH interface. Therefore, the nitridation of the hydride, the formation of NH_x species, is a key process to enhance the catalytic activity of TM catalysts. As for the Ru/$LaH_{3-2x}O_x$ catalyst, the nitride formation is not effective, but electron donation from the oxyhydride is the dominant factor to promote Ru catalyst.

7.3.2 Oxynitride-Hydride

A novel oxynitride-hydride, $BaCeO_{3-x}N_yH_z$, has been developed as an ammonia synthesis catalyst.[85] It is the first experimental demonstration of the synthesis and catalytic application of the oxynitride-hydride. The $BaCeO_{3-x}N_yH_z$ has a typical ABO_3-type perovskite structure (space group: *Pnma*), where a part of O^{2-} sites are substituted by N^{3-} and H^- ions (Fig. 7.10). It can be directly synthesized by the solid-state reaction of $Ba(NH_2)_2$ with CeO_2 at 300–600 °C under an NH_3 gas atmosphere. Generally, perovskite oxynitrides are prepared by nitridation of oxide precursor under NH_3 atmosphere at very high temperatures (800–1300 °C) [86]. In the case of the perovskite oxyhydrides (e.g., $BaTiO_{3-x}H_x$), the O^{2-} sites can be replaced with H^- at relatively low temperatures (500–600 °C) for a long time (about 1 week) by using CaH_2 as a reducing agent [69]. In addition, the resultant oxyhydrides should be washed to remove unreacted CaH_2 and byproduct CaO. Therefore, such harsh conditions and complicated processes are required for the replacement of O^{2-} sites by N^{3-} and/or H^- ions because of the high stability of the perovskite-type oxides. In addition, the perovskite oxide itself is also prepared at high temperatures (≥800 °C) because the decomposition of alkaline earth carbonates, which are used as A site cation sources, requires high temperatures. In contrast, the $BaCeO_{3-x}N_yH_z$ can be obtained at lower temperatures since the decomposition of $Ba(NH_2)_2$ starts from 200 °C. The target product can be formed without any complicated cleaning processes because only gas components such as N_2, H_2, and NH_3 are released as byproducts. In addition, it is difficult to introduce high-density heteroanion into the oxide lattice since the replacement of O^{2-} sites by N^{3-} or H^- with different valences leads to defect generation. On the other hand, in the $BaCeO_{3-x}N_yH_z$ synthesis process, two O^{2-} sites can be ideally replaced with one N^{3-} and one H^-, and therefore, high concentration heteroanions can be doped into the oxide framework without defect formation.

XRD patterns for $BaCeO_{3-x}N_yH_z$ prepared at various temperatures are summarized in Fig. 7.10. The reaction of CeO_2 with $Ba(NH_2)_2$ to form perovskite structure proceeds even at 300 °C, and an almost single-phase perovskite $BaCeO_{3-x}N_yH_z$ with an orthorhombic structure is obtained at 550 °C. The unit cell volume of $BaCeO_{3-x}N_yH_z$ is determined to be 349.8 Å^3, which is much larger than that of $BaCeO_3$ (340.5 Å^3). This lattice expansion is caused by the presence of N^{3-} (1.46 Å) and Ce^{3+} (1.14 Å) with larger ionic radius than O^{2-} (1.38 Å) and Ce^{4+} (0.97 Å). XAFS and XPS analyses revealed that 45% of Ce^{4+} sites are reduced into Ce^{3+} during the synthesis of $BaCeO_{3-x}N_yH_z$. The chemical composition was determined to be $BaCeO_{1.80}N_{0.57}H_{0.23}$ by various elemental analyses, indicating that 27% of anion sites are substituted by heteroanions such as N^{3-} or H^-. Such a heavy heteroanion doping cannot be achieved by conventional methods. For instance, N^{3-} or H^- ions are not introduced into the lattice of $BaCeO_3$ by simple heating with NH_3 even at 900 °C. $BaCeO_3$ is well known as a proton conducting material, and many researchers focus on cation substitution of $BaCeO_3$ to improve the conductivity [37]. However, the study on the anion substituted $BaCeO_3$ is extremely rare.

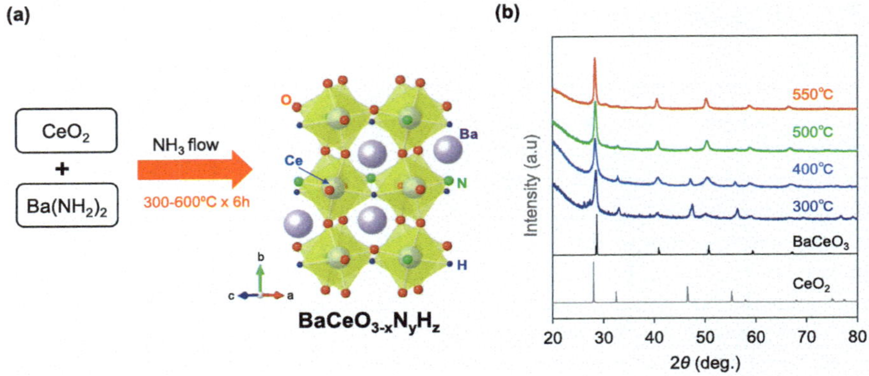

Fig. 7.10 **a** Synthesis of $BaCeO_{3-x}N_yH_z$ by the reaction of CeO_2 with $Ba(NH_2)_2$. **b** XRD patterns of $BaCeO_{3-x}N_yH_z$ prepared at various temperatures. Adapted with permission from Ref. [85]. Copyright 2019 American Chemical Society

The $BaCeO_{3-x}N_yH_z$ works as an efficient support of transition metal nanoparticle catalysts in ammonia synthesis. Although $Ru/BaCeO_3$ is reported to show high ammonia synthesis activity [88], the catalytic activity of $Ru/BaCeO_{3-x}N_yH_z$ is an order of magnitude higher than that of $Ru/BaCeO_3$ at low reaction temperatures (≤300 °C). The much higher catalytic activity of $Ru/BaCeO_{3-x}N_yH_z$ is attributed to the promotion effect by the oxynitride-hydride support because its Ru particle size (~5 nm) and the surface area (4.35 $m^2\ g^{-1}$) are almost the same as those of $Ru/BaCeO_3$. When the Co and Fe nanoparticles are supported on $BaCeO_{3-x}N_yH_z$, the ammonia synthesis rates are 40–200 times higher than those of Co and Fe-loaded $BaCeO_3$ (Fig. 7.11). In addition, the $BaCeO_{3-x}N_yH_z$ effectively promotes the activity of the supported Ni catalyst that is well known as an inefficient catalyst for ammonia synthesis because of its very low nitrogen adsorption energy. The promotion of Ni catalyst by nitride-based support has been also demonstrated by the Ni–LaN catalyst [89, 90], in which N_2 and H_2 molecules are activated separately at the nitride vacancy site on LaN and Ni metal surface, respectively. Such a unique reaction mechanism is considered to take place on the $Ni/BaCeO_{3-x}N_yH_z$ catalyst. It should be noted that $BaCeO_{3-x}N_yH_z$ without any transition metal nanoparticles has a stable catalytic activity for NH_3 synthesis, in contrast to no catalytic activity of $BaCeO_3$ (Fig. 7.11). This result means that lattice N^{3-} and H^- ions work as active sites in the NH_3 synthesis reaction, which is clarified by ammonia synthesis from isotopically labeled nitrogen ($^{15}N_2$) and hydrogen (D_2).

The difference in the reaction mechanism for ammonia synthesis between TM/ $BaCeO_{3-x}N_yH_z$ and conventional oxide-supported TM catalysts is summarized in Fig. 7.12. In the case of conventional catalyst, both N_2 and H_2 molecules are activated only on the TM surface to form NH_3 (Langmuir-Hinshelwood type mechanism). Therefore, NH_3 formation reaction efficiently proceeds on Ru with optimum nitrogen binding energy, but it hardly occurs on the Fe, Co, and Ni under mild reaction

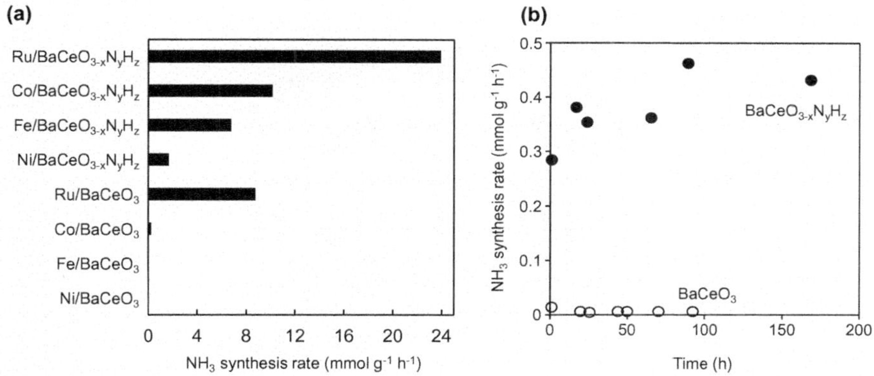

Fig. 7.11 **a** NH_3 synthesis rates for various transition metals supported on $BaCeO_{3-x}N_yH_z$ and $BaCeO_3$ at 400 °C and 0.9 MPa. **b** Stability test for NH_3 synthesis over $BaCeO_{3-x}N_yH_z$ at 400 °C and 0.9 MPa. Adapted with permission from Ref. [85]. Copyright 2019 American Chemical Society

conditions. As for TM/$BaCeO_{3-x}N_yH_z$ catalyst, the activation of N_2 and H_2 proceeds not only on the TM surface but also on the heteroanion sites on $BaCeO_{3-x}N_yH_z$ (Mars-van Krevelen-type mechanism). Such a unique reaction mechanism significantly promotes the activity of TM catalysts with much lower activation energy (46–62 kJ mol^{-1}) than conventional TM catalysts. As described in Fig. 7.12, the lattice nitrogen is hydrogenated by lattice H^- ion or spill-over hydrogen from the TM surface to form NH_3, and subsequently electrons are generated at anion vacancy sites (V_a). The electron at V_a facilitates the N_2 cleavage at the TM-support interface, resulting in the regeneration of lattice nitrogen in $BaCeO_{3-x}N_yH_z$. An associative mechanism for ammonia formation is also considered as another possible route, where N_2 molecules are activated at V_a sites followed by hydrogenation to form NNH species. Then, the dissociation of N–N bond is enhanced by stepwise hydrogenation, and finally, the V_a site is occupied by nitrogen. After the lattice nitrogen is hydrogenated into NH_3, electrons are captured at V_a sites, which further react with N_2 and H_2 to regenerate N^{3-} and H^- ions in the lattice. In both cases, H_2 poisoning over the TM surface can be prevented by the reversible exchange between electrons and H^- ions occurring at the metal-support interface as with the case of Ru-loaded electride catalysts.

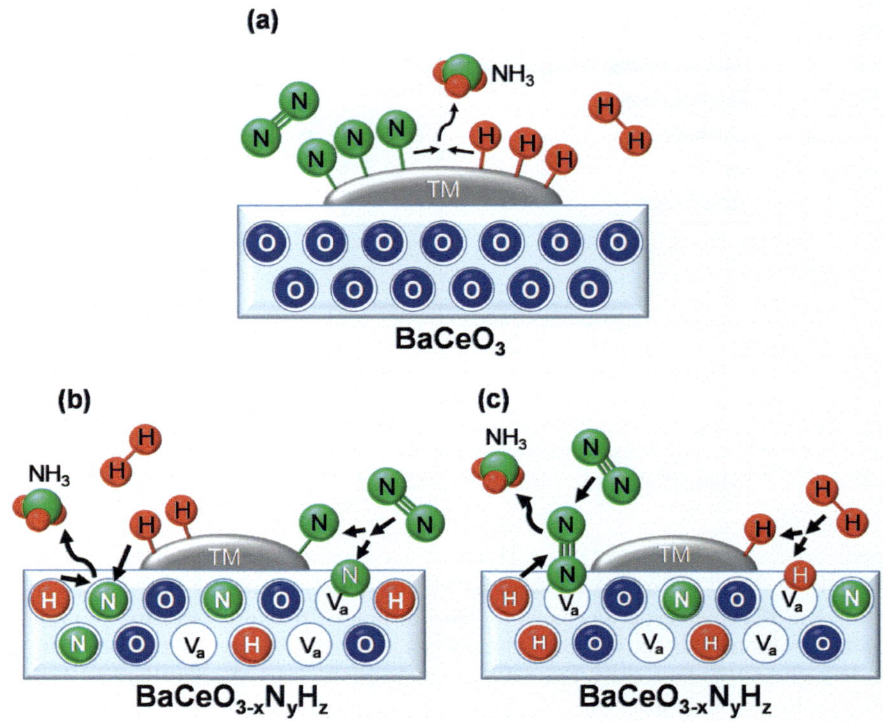

Fig. 7.12 Possible reaction mechanism for NH_3 synthesis over transition metal (TM) loaded **a** $BaCeO_3$ and (**b** and **c**) $BaCeO_{3-x}N_yH_z$. Adapted with permission from Ref. [85]. Copyright 2019 American Chemical Society

References

1. Beller M, Renken A, van Santen RA. Catalysis: from principles to applications. Wiley-VCH;2012.
2. Thomas, J. M.; Thomas, W. J. *Principles and practice of heterogeneous catalysis*; 2nd, revised ed ed.; Wiley-VCH;2015.
3. Munnik P, de Jongh PE, de Jong KP. Recent developments in the synthesis of supported catalysts. Chem Rev. 2015;115(14):6687.
4. Liu L, Corma A. Metal catalysts for heterogeneous catalysis: from single atoms to nanoclusters and nanoparticles. Chem Rev. 2018;118(10):4981.
5. Haruta M, Tsubota S, Kobayashi T, Kageyama H, Genet MJ, Delmon B. Low-temperature oxidation of Co over gold supported on TiO_2, a-Fe_2O_3, and Co_3O_4. J Catal. 1993;144(1):175.
6. Ishida T, Murayama T, Taketoshi A, Haruta M. Importance of size and contact structure of gold nanoparticles for the genesis of unique catalytic processes. Chem Rev. 2020;120(2):464.
7. Haruta M, Kobayashi T, Sano H, Yamada N. Novel gold catalysts for the oxidation of carbon-monoxide at a temperature far below 0°C. Chem Lett. 1987;16(2):405.
8. Gong XQ, Selloni A, Dulub O, Jacobson P, Diebold U. Small au and pt clusters at the anatase TiO_2(101) surface: behavior at terraces, steps, and surface oxygen vacancies. J Am Chem Soc. 2008;130(1):370.

9. Matthey D, Wang JG, Wendt S, Matthiesen J, Schaub R, Laegsgaard E, Hammer B, Besenbacher F. Enhanced bonding of gold nanoparticles on oxidized $TiO_2(110)$. Science. 2007;315(5819):1692.
10. Wan JW, Chen WX, Jia CY, Zheng LR, Dong JC, Zheng XS, Wang Y, Yan WS, Chen C, Peng Q, et al. Defect Effects on TiO_2 nanosheets: stabilizing Single atomic site Au and promoting catalytic properties. Adv Mater. 2018;30(11):1705369.
11. Qiao BT, Wang AQ, Yang XF, Allard LF, Jiang Z, Cui YT, Liu JY, Li J, Zhang T. Single-atom catalysis of CO oxidation using Pt_1/FeO_x. Nat Chem. 2011;3(8):634.
12. Wang AQ, Li J, Zhang T. Heterogeneous single-atom catalysis. Nat Rev Chem. 2018;2(6):65.
13. Lang R, Du X, Huang Y, Jiang X, Zhang Q, Guo Y, Liu K, Qiao B, Wang A, Zhang T. Single-atom catalysts based on the metal-oxide interaction. Chem Rev. 2020;120(21):11986.
14. Qin R, Liu K, Wu Q, Zheng N. Surface coordination chemistry of atomically dispersed metal catalysts. Chem Rev. 2020;120(21):11810.
15. Widmann D, Behm RJ. Activation of molecular oxygen and the nature of the active oxygen species for CO oxidation on oxide supported Au catalysts. Acc Chem Res. 2014;47(3):740.
16. Campbell CT. Catalyst-support interactions electronic perturbations. Nat Chem. 2012;4(8):597.
17. Pacchioni G. Electronic interactions and charge transfers of metal atoms and clusters on oxide surfaces. Phys Chem Chem Phys. 2013;15(6):1737.
18. Bruix A, Rodriguez JA, Ramirez PJ, Senanayake SD, Evans J, Park JB, Stacchiola D, Liu P, Hrbek J, Illas F. A new type of strong metal-support interaction and the production of H_2 through the transformation of water on $Pt/CeO_2(111)$ and $Pt/CeO_x/TiO_2(110)$ catalysts. J Am Chem Soc. 2012;134(21):8968.
19. Tauster SJ, Fung SC, Garten RL. Strong metal-support interactions. Group 8 noble metals supported on TiO_2. J Am Chem Soc. 1978;100:170.
20. Tauster SJ. Strong metal-support interactions. Acc Chem Res. 1987;20(11):389.
21. Lykhach Y, Kozlov SM, Skala T, Tovt A, Stetsovych V, Tsud N, Dvorak F, Johanek V, Neitzel A, Myslivecek J, et al. Counting electrons on supported nanoparticles. Nat Mater. 2016;15(3):284.
22. Yoon B, Hakkinen H, Landman U, Worz AS, Antonietti JM, Abbet S, Judai K, Heiz U. Charging effects on bonding and catalyzed oxidation of CO on Au_8 clusters on MgO. Science. 2005;307(5708):403.
23. Kim SW, Hosono H. Synthesis and properties of $12CaO{\cdot}7Al_2O_3$ electride: review of single crystal and thin film growth. Philos Mag. 2012;92(19–21):2596.
24. Hayashi K, Hirano M, Hosono H. Thermodynamics and kinetics of hydroxide ion formation in $12CaO{\cdot}7Al_2O_3$. J Phys Chem B. 2005;109(24):11900.
25. Hayashi K, Hirano M, Matsuishi S, Hosono H. Microporous crystal $12CaO{\cdot}7Al_2O_3$ encaging abundant O^- radicals. J Am Chem Soc. 2002;124(5):738.
26. Jeevaratnam J, Glasser FP, Glasser LSD. Anion substitution and structure of $12CaO{\cdot}7Al_2O_3$. J Am Ceram Soc. 1964;47(2):105.
27. Hayashi K, Matsuishi S, Kamiya T, Hirano M, Hosono H. Light-induced conversion of an insulating refractory oxide into a persistent electronic conductor. Nature. 2002;419(6906):462.
28. Hayashi K, Sushko PV, Hashimoto Y, Shluger AL, Hosono H. Hydride ions in oxide hosts hidden by hydroxide ions. Nat Commun. 2014;5:3515.
29. Miyakawa M, Kamioka H, Hirano M, Kamiya T, Sushko PV, Shluger AL, Matsunami N, Hosono H. Photoluminescence from Au ion-implanted nanoporous single-crystal $12CaO{\cdot}7Al_2O_3$. Phys Rev B. 2006;73(20): 205108.
30. Zhmoidin GI, Chatterjee AK. Conditions and mechanism of interconvertibility of compounds $12CaO{\cdot}7Al_2O_3$ and $5CaO{\cdot}3Al_2O_3$. Cem Concr Res. 1984;14(3):386.
31. Hayashi F, Tomota Y, Kitano M, Toda Y, Yokoyama T, Hosono H. NH^{2-} dianion entrapped in a nanoporous $12CaO{\cdot}7Al_2O_3$ crystal by ammonothermal treatment: reaction pathways, dynamics, and chemical stability. J Am Chem Soc. 2014;136(33):11698.
32. Nakao T, Ogasawara K, Kitano M, Matsuishi S, Sushko PV, Hosono H. Ship-in-a-bottle synthesis of high concentration of N_2 molecules in a cage-structured electride. J Phys Chem Lett. 2021;12(4):1295.

33. Matsuishi S, Toda Y, Miyakawa M, Hayashi K, Kamiya T, Hirano M, Tanaka I, Hosono H. High-density electron anions in a nanoporous single crystal: $[Ca_{24}Al_{28}O_{64}]^{4+}(4e^-)$. Science. 2003;301(5633):626.
34. Ellaboudy A, Dye JL, Smith PB. Cesium 18-crown-6 compounds. A crystalline ceside and a crystalline electride. J Am Chem Soc. 1983;105 (21):6490.
35. Hosono H, Kitano M. Advances in materials and applications of inorganic electrides. Chem Rev. 2021;121(5):3121.
36. Toda Y, Matsuishi S, Hayashi K, Ueda K, Kamiya T, Hirano M, Hosono H. Field emission of electron anions clathrated in subnanometer-sized cages in $[Ca_{24}Al_{28}O_{64}]^{4+}(4e^-)$. Adv Mater. 2004;16(8):685.
37. Hayashi K, Hirano M, Hosono H. Functionalities of a nanoporous crystal $12CaO{\cdot}7Al_2O_3$ riginating from the Incorporation of active anions. Bull Chem Soc Jpn. 2007;80(5):872.
38. Kitano M, Kanbara S, Inoue Y, Kuganathan N, Sushko PV, Yokoyama T, Hara M, Hosono H. Electride support boosts nitrogen dissociation over ruthenium catalyst and shifts the bottleneck in ammonia synthesis. Nat Commun. 2015;6:6731.
39. Ertl G, Weiss M, Lee SB. Role of potassium in the catalytic synthesis of ammonia. Chem Phys Lett. 1979;60(3):391.
40. Ertl G, Knözinger H, Schüth F, Weitkamp J. Handbook of heterogeneous catalysis; 2nd, completely rev. and enlarged ed ed.; Wiley-VCH;2008.
41. Ozaki A, Aika K, Hori H. New catalyst system for ammonia synthesis. Bull Chem Soc Jpn. 1971;44(11):3216.
42. Saadatjou N, Jafari A, Sahebdelfar S. Ruthenium nanocatalysts for ammonia synthesis: a review. Chem Eng Commun. 2015;202(4):420.
43. Aika K. Role of alkali promoter in ammonia synthesis over ruthenium catalysts-Effect on reaction mechanism. Catal Today. 2017;286:14.
44. Kitano M, Inoue Y, Yamazaki Y, Hayashi F, Kanbara S, Matsuishi S, Yokoyama T, Kim SW, Hara M, Hosono H. Ammonia synthesis using a stable electride as an electron donor and reversible hydrogen store. Nat Chem. 2012;4(11):934.
45. Kanbara S, Kitano M, Inoue Y, Yokoyama T, Hara M, Hosono H. Mechanism switching of ammonia synthesis over Ru-loaded electride catalyst at metal-insulator transition. J Am Chem Soc. 2015;137(45):14517.
46. Dahl S, Logadottir A, Egeberg RC, Larsen JH, Chorkendorff I, Tornqvist E, Norskov JK. Role of steps in N_2 activation on Ru(0001). Phys Rev Lett. 1999;83(9):1814.
47. Jacobsen CJH, Dahl S, Hansen PL, Tornqvist E, Jensen L, Topsoe H, Prip DV, Moenshaug PB, Chorkendorff I. Structure sensitivity of supported ruthenium catalysts for ammonia synthesis. J Mol Catal A. 2000;163(1–2):19.
48. Hansen TW, Hansen PL, Dahl S, Jacobsen CJH. Support effect and active sites on promoted ruthenium catalysts for ammonia synthesis. Catal Lett. 2002;84(1–2):7.
49. Sushko PV, Shluger AL, Hirano M, Hosono H. From insulator to electride: a theoretical model of nanoporous oxide $12CaO{\cdot}7Al_2O_3$. J Am Chem Soc. 2007;129(4):942.
50. Matsuishi S, Kim SW, Kamiya T, Hirano M, Hosono H. Localized and delocalized electrons in room-temperature stable electride $[Ca_{24}Al_{28}O_{64}]^{4+}(O^{2-})_{2-x}(e^-)_{2x}$: analysis of optical reflectance spectra. J Phys Chem C. 2008;112(12):4753.
51. Hosono H. Electron transfer from support/promotor to metal catalyst: requirements for effective support. Catal Lett. 2022;152(2):307.
52. Sushko PV, Shluger AL, Hayashi K, Hirano M, Hosono H. Electron localization and a confined electron gas in nanoporous inorganic electrides. Phys Rev Lett. 2003;91(12): 126401.
53. Siporin SE, Davis RJ. Use of kinetic models to explore the role of base promoters on Ru/MgO ammonia synthesis catalysts. J Catal. 2004;225(2):359.
54. Kammert J, Moon J, Cheng Y, Daemen L, Irle S, Fung V, Liu J, Page K, Ma X, Phaneuf V, et al. Nature of reactive hydrogen for ammonia synthesis over a Ru/C12A7 electride catalyst. J Am Chem Soc. 2020;142(16):7655.
55. Weber S, Schafer S, Saccoccio M, Ortner N, Bertmer M, Seidel K, Berendts S, Lerch M, Glaser R, Kohlmann H, et al. Mayenite-based electride $C12A7e^-$: a reactivity and stability study. Catalysts. 2021;11(3):334.

56. Inoue Y, Kitano M, Tokunari M, Taniguchi T, Ooya K, Abe H, Niwa Y, Sasase M, Hara M, Hosono H. Direct activation of cobalt catalyst by $12CaO{\cdot}7Al_2O_3$ electride for ammonia synthesis. ACS Catal. 2019;9(3):1670.
57. Jacobsen CJH, Dahl S, Clausen BS, Bahn S, Logadottir A, Nørskov JK. Catalyst design by interpolation in the periodic table: bimetallic ammonia synthesis catalysts. J Am Chem Soc. 2001;123(34):8404.
58. Nørskov JK, Bligaard T, Rossmeisl J, Christensen CH. Towards the computational design of solid catalysts. Nat Chem. 2009;1(1):37.
59. Kojima R, Aika K. Cobalt molybdenum bimetallic nitride catalysts for ammonia synthesis. Chem Lett. 2000;5:514.
60. Wang PK, Chang F, Gao WB, Guo JP, Wu GT, He T, Chen P. Breaking scaling relations to achieve low-temperature ammonia synthesis through LiH-mediated nitrogen transfer and hydrogenation. Nat Chem. 2017;9(1):64.
61. Zeinalipour-Yazdi CD, Hargreaves JSJ, Catlow CRA. Nitrogen activation in a mars-van krevelen mechanism for ammonia synthesis on Co_3Mo_3N. J Phys Chem C. 2015;119(51):28368.
62. Hayashi F, Toda Y, Kanie Y, Kitano M, Inoue Y, Yokoyama T, Hara M, Hosono H. Ammonia decomposition by ruthenium nanoparticles loaded on inorganic electride C12A7:e^-. Chem Sci. 2013;4(8):3124.
63. Sharif MJ, Kitano M, Inoue Y, Niwa Y, Abe H, Yokoyama T, Hara M, Hosono H. Electron donation enhanced CO oxidation over Ru-loaded $12CaO{\cdot}7Al_2O_3$ electride catalyst. J Phys Chem C. 2015;119(21):11725.
64. Ye TN, Li J, Kitano M, Sasase M, Hosono H. Electronic interactions between a stable electride and a nano-alloy control the chemoselective reduction reaction. Chem Sci. 2016;7(9):5969.
65. Ye TN, Xiao Z, Li J, Gong Y, Abe H, Niwa Y, Sasase M, Kitano M, Hosono H. Stable single platinum atoms trapped in sub-nanometer cavities in $12CaO{\cdot}7Al_2O_3$ for chemoselective hydrogenation of nitroarenes. Nat Commun. 2020;11 (1):1020.
66. Huang BQ, Corbett JD. Ba_3AlO_4H: synthesis and structure of a new hydrogen-stabilized phase. J Solid State Chem. 1998;141(2):570.
67. Hayward MA, Cussen EJ, Claridge JB, Bieringer M, Rosseinsky MJ, Kiely CJ, Blundell SJ, Marshall IM, Pratt FL. The hydride anion in an extended transition metal oxide array: $LaSrCoO_3H_{0.7}$. Science. 2002;*295* (5561):1882.
68. Iimura S, Matuishi S, Sato H, Hanna T, Muraba Y, Kim SW, Kim JE, Takata M, Hosono H. Two-dome structure in electron-doped iron arsenide superconductors. Nat Commun. 2012;3:943.
69. Kobayashi Y, Hernandez OJ, Sakaguchi T, Yajima T, Roisnel T, Tsujimoto Y, Morita M, Noda Y, Mogami Y, Kitada A, et al. An oxyhydride of $BaTiO_3$ exhibiting hydride exchange and electronic conductivity. Nat Mater. 2012;11(6):507.
70. Kobayashi G, Hinuma Y, Matsuoka S, Watanabe A, Iqbal M, Hirayama M, Yonemura M, Kamiyama T, Tanaka I, Kanno R. Pure H- conduction in oxyhydrides. Science. 2016;351(6279):1314.
71. Kageyama H, Hayashi K, Maeda K, Attfield JP, Hiroi Z, Rondinelli JM, Poeppelmeier KR. Expanding frontiers in materials chemistry and physics with multiple anions. Nat Commun. 2018;9:772.
72. Kitano M, Inoue Y, Ishikawa H, Yamagata K, Nakao T, Tada T, Matsuishi S, Yokoyama T, Hara M, Hosono H. Essential role of hydride ion in ruthenium-based ammonia synthesis catalysts. Chem Sci. 2016;7(7):4036.
73. Mizoguchi H, Okunaka M, Kitano M, Matsuishi S, Yokoyama T, Hosono H. Hydride-based electride material, LnH_2 (Ln = La, Ce, or Y). Inorg Chem. 2016;55(17):8833.
74. Ooya K, Li J, Fukui K, Iimura S, Nakao T, Ogasawara K, Sasase M, Abe H, Niwa Y, Kitano M, et al. Ruthenium catalysts promoted by lanthanide oxyhydrides with high hydride-ion mobility for low-temperature ammonia synthesis. Adv Energy Mater. 2021;11:2003723.
75. Kobayashi Y, Tang Y, Kageyama T, Yamashita H, Masuda N, Hosokawa S, Kageyama H. Titanium-based hydrides as heterogeneous catalysts for ammonia synthesis. J Am Chem Soc. 2017;139(50):18240.

76. Yamashita H, Broux T, Kobayashi Y, Takeiri F, Ubukata H, Zhu T, Hayward MA, Fujii K, Yashima M, Shitara K, et al. Chemical pressure-induced anion order-disorder transition in LnHO enabled by hydride size flexibility. J Am Chem Soc. 2018;140(36):11170.
77. Cao Y, Saito A, Kobayashi Y, Ubukata H, Tang Y, Kageyama H. Vanadium hydride as an ammonia synthesis catalyst. ChemCatChem. 2021;13(1):191.
78. Hattori M, Iijima S, Nakao T, Hosono H, Hara M. Solid solution for catalytic ammonia synthesis from nitrogen and hydrogen gases at 50 °C. Nat Commun. 2020;11(1):2001.
79. Azofra LM, Morlanes N, Poater A, Samantaray MK, Vidjayacoumar B, Albahily K, Cavallo L, Basset JM. Single-site molybdenum on solid support materials for catalytic hydrogenation of N_2-into-NH_3. Angew Chem Int Ed. 2018;57(48):15812.
80. Wang QR, Pan J, Guo JP, Hansen HA, Xie H, Jiang L, Hua L, Li HY, Guan YQ, Wang PK, et al. Ternary ruthenium complex hydrides for ammonia synthesis via the associative mechanism. Nat Catal. 2021;4(11):959.
81. Yajima T, Takeiri F, Aidzu K, Akamatsu H, Fujita K, Yoshimune W, Ohkura M, Lei SM, Gopalan V, Tanaka K, et al. A labile hydride strategy for the synthesis of heavily nitridized $BaTiO_3$. Nat Chem. 2015;7(12):1017.
82. Tang Y, Kobayashi Y, Masuda N, Uchida Y, Okamoto H, Kageyama T, Hosokawa S, Loyer F, Mitsuhara K, Yamanaka K, et al. Metal-dependent support effects of oxyhydride-supported Ru, Fe, Co catalysts for ammonia synthesis. Adv Energy Mater. 2018;1801772.
83. Fukui K, Iimura S, Tada T, Fujitsu S, Sasase M, Tamatsukuri H, Honda T, Ikeda K, Otomo T, Hosono H. Characteristic fast H- ion conduction in oxygen-substituted lanthanum hydride. Nat Commun. 2019;10:2578.
84. Fukui K, Iimura S, Iskandarov A, Tada T, Hosono H. Room-temperature fast H^- conduction in oxygen-substituted lanthanum hydride. J Am Chem Soc. 2022;144(4):1523.
85. Kitano M, Kujirai J, Ogasawara K, Matsuishi S, Tada T, Abe H, Niwa Y, Hosono H. Low-temperature synthesis of perovskite oxynitride-hydrides as ammonia synthesis catalysts. J Am Chem Soc. 2019;141(51):20344.
86. Ebbinghaus SG, Abicht HP, Dronskowski R, Muller T, Reller A, Weidenkaff A. Perovskite-related oxynitrides-recent developments in synthesis, characterisation and investigations of physical properties. Prog Solid State Chem. 2009;37(2–3):173.
87. Medvedev D, Murashkina A, Pikalova E, Demin A, Podias A, Tsiakaras P. $BaCeO_3$: materials development, properties and application. Prog Mater Sci. 2014;60:72.
88. Yang XL, Zhang WQ, Xia CG, Xiong XM, Mu XY, Hu B. Low temperature ruthenium catalyst for ammonia synthesis supported on $BaCeO_3$ nanocrystals. Catal Commun. 2010;11(10):867.
89. Ye TN, Park SW, Lu Y, Li J, Sasase M, Kitano M, Tada T, Hosono H. Vacancy-enabled N_2 activation for ammonia synthesis on an Ni-loaded catalyst. Nature. 2020;583(7816):391.
90. Ye TN, Park SW, Lu Y, Li J, Sasase M, Kitano M, Hosono H. Contribution of nitrogen vacancies to ammonia synthesis over metal nitride catalysts. J Am Chem Soc. 2020;142(33):14374.

Chapter 8
Crystalline Metal Oxide Catalysts for Organic Synthesis

Keigo Kamata and Takeshi Aihara

8.1 Catalyst Design of Crystalline Metal Oxides for Organic Synthesis

The chemical industry based on fossil resources such as oil and coal has contributed to global economic development over the past century through the synthesis of various useful chemicals; however, it can still cause many serious environmental problems. Therefore, the establishment of environmentally friendly chemical processes requires the development of innovative and cost-effective approaches to prevent environmental pollution. Green and sustainable chemistry is one of the most fascinating concepts for reducing the impact of chemicals on environmental pollution and human health. To this end, Anastas and Warner developed 12 principles in 1991 [1]. A simple definition of these principles is to reduce and/or eliminate the use or production of hazardous substances in the design, manufacture, and application of chemical products. Advances in green chemistry contribute not only to a reduction of the risks associated with global problems (e.g., climate change, energy production, availability of safe and adequate water supplies, food production, and the presence of toxic substances in the environment), but also to new technologies that provide reliable chemical products for society in an environmentally friendly manner [1–4].

In the production of chemical substances (in particular, high value-added chemicals), C–C bond formation reactions and functional group transformation reactions that involve reduction/oxidation are indispensable. However, conventional synthetic methods that use non-recoverable and non-recyclable stoichiometric reagents (e.g., acid–base reactions using mineral acids (H_2SO_4, etc.), Lewis acids ($AlCl_3$, etc.), and inorganic bases (NaOH, etc.), reduction reactions that use metals (Na, etc.) and metal hydrides ($LiAlH_4$, etc.), and oxidation reactions using permanganate) are still

K. Kamata (✉) · T. Aihara
Laboratory for Materials and Structures, Institute of Innovative Research, Tokyo Institute of Technology, Nagatsuta-cho 4259, Midori-ku, Yokohama 226-8503, Kanagawa, Japan
e-mail: kamata.k.ac@m.titech.ac.jp

W. Ueda (ed.), *Crystalline Metal Oxide Catalysts*,
https://doi.org/10.1007/978-981-19-5013-1_8

widely used. Therefore, the replacement of these stoichiometric methodologies with greener catalytic methods could reduce the use and production of toxic and dangerous substances [1–7]. To date, many soluble homogeneous catalysts with high catalytic activity and selectivity have been developed for the green synthesis of various chemicals because their structural and electronic properties can be controlled at the atomic and molecular levels. Despite these advantages, efficient and cost-effective catalyst/product separation and the reuse of expensive catalysts are difficult to achieve efficiently and cost-effectively; therefore, the share of homogeneous catalysts is only approximately 20% [8–12]. In contrast, the development of easily recoverable and recyclable solid catalysts has attracted interest in the environmentally friendly synthesis of high-value-added chemicals.

The uniform construction of structurally controlled active sites on solid materials is essential for achieving efficient and selective functional groups for transformation reactions. However, the fine-tuning of heterogeneous catalyst structures is typically difficult, which results in a problem where the catalytic performance is restrained, in sharp contrast to homogeneous catalysts with well-defined active sites. Therefore, the design and development of new high-performance all-inorganic heterogeneous catalysts that are applicable to various organic reactions remain a strongly desired goal and a challenging subject of research. In this context, crystalline metal oxides are good candidates as multifunctional heterogeneous catalysts because they have versatile crystal structures depending on the elemental composition, polymorphism, and oxidation states of the elements, in contrast to metallic compounds, and cover various types of reactions such as acid–base reactions, selective oxidation, complete oxidation, exhaust gas purification, biomass conversion, electrocatalysis, and photocatalysis [13–39]. In this chapter, recent studies on crystalline metal oxide catalysts for liquid-phase organic reactions are comprehensively summarized with a focus on the structure–activity relationship.

Transition metal oxides exist in many different crystallographic forms; however, there seems to be no simple generalization that associates a structure with stoichiometry or the type of elements [39]. The ionic radii of transition metals are smaller than that of O^{2-}; therefore, oxygen ions are typically close-packed with smaller metal ions in octahedral and tetrahedral holes. Metal oxides commonly studied as catalysts are corundum, rocksalt, wurtzite, spinel, perovskite, rutile, and layered structures, as shown in Fig. 8.1. The rocksalt structure consists of a three-dimensional array of alternating cations and anions, where each ion is in the center of an octahedron and the vertices are the opposite type of ions, which results in corner-sharing octahedra. Wurtzite consists of a three-dimensional net of corner-sharing tetrahedra. Oxygen ions assume a hexagonal close-packed arrangement in the corundum and rutile structures, whereas they have a cubic close-packed arrangement in the spinel structure. In an ideal rutile structure (general formula MO_2), half of the octahedral holes are filled with cations and the tetrahedral holes are empty. In the corundum structure (general formula M_2O_3), two-thirds of the octahedral holes are filled and the tetrahedral holes are empty. In an ideal spinel structure (general formula M_3O_4), half of the tetrahedral holes and half of the octahedral holes are filled. The most common oxidation states of metals are +2 and +3, and all M^{2+} and M^{3+} ions are in the tetrahedral and octahedral holes, respectively, in a normal spinel, whereas all M^{2+} ions are in the octahedral holes and M^{3+} ions are distributed equally between the octahedral and tetrahedral holes in an inverse spinel. The ideal perovskite structure (general formula ABO_3) consists of a cubic network with shared octahedra. The small, highly charged B cation is in the center of the octahedron, and the large A cation is in the center of the cavity defined by eight octahedra. Therefore, this latter cation coordinates to 12 oxygen ions. The B ion is typically a transition metal ion and the A ion is an alkaline, alkaline earth, or lanthanide ion.

Among crystalline complex metal oxides with a wide variety of structures, several examples of perovskite, spinel, metal phosphate, and other materials (hydrotalcite, montmorillonite, garnet, pyrochlore, murdochite) are highlighted with a focus on their structures, synthesis, and catalytic applications to liquid-phase environmentally-friendly functional group transformations. Typical examples for heterogeneously catalyzed liquid-phase organic reactions over crystalline metal oxides-based materials are listed in Fig. 8.2. In acid–base reactions, not only the replacement of a conventional homogeneous stoichiometric acid and/or base catalysts, but also specific functional group transformation reactions through the cooperative action of adjacent acid–base sites on solid materials have been reported. In the oxidation reaction, the development of heterogeneous catalysts that can selectively introduce oxygen-based functional groups under mild conditions has been achieved by control of their compositions and structures, the activation of oxidants, and the reactivity of the active oxygen species. In C–C bond formation reactions (mainly cross-coupling reactions with precious metals) and reduction reactions such as hydrogenation, metal nanoparticles formed on complex metal oxides under reductive conditions act as dominant active sites, and the specific effect of metal oxides as supports is observed in some cases. One-pot reactions typically require sequential multicomponent reactions over one solid catalyst; therefore, multifunctional complex metal oxides with both appropriate acid–base and redox abilities are effective, depending on the nature of the one-pot reactions. Figure 8.3 shows the catalyst design of crystalline complex metal oxides for liquid-phase organic reactions. The selection of chemical compositions and appropriate synthesis methods leads to the control of crystalline structure, oxygen vacancy formation, metal nanoparticle (NP) stabilization, and multifunction including acid–base and redox properties. The catalytic application of crystalline complex metal oxides to electrochemical, photocatalytic, and high-temperature gas-phase reactions has been examined in several excellent books and review articles [1–39]. In addition, catalysis by zeolites and polyoxometalates is comprehensively summarized in other chapters.

8.2 Synthesis Method of Crystalline Metal Oxides

Liquid-phase organic reactions are typically conducted at lower reaction temperatures (e.g., from room temperature to ca. 200 °C) than gas-phase heterogeneous catalytic systems; therefore, it is generally recognized that chemical reaction occurs

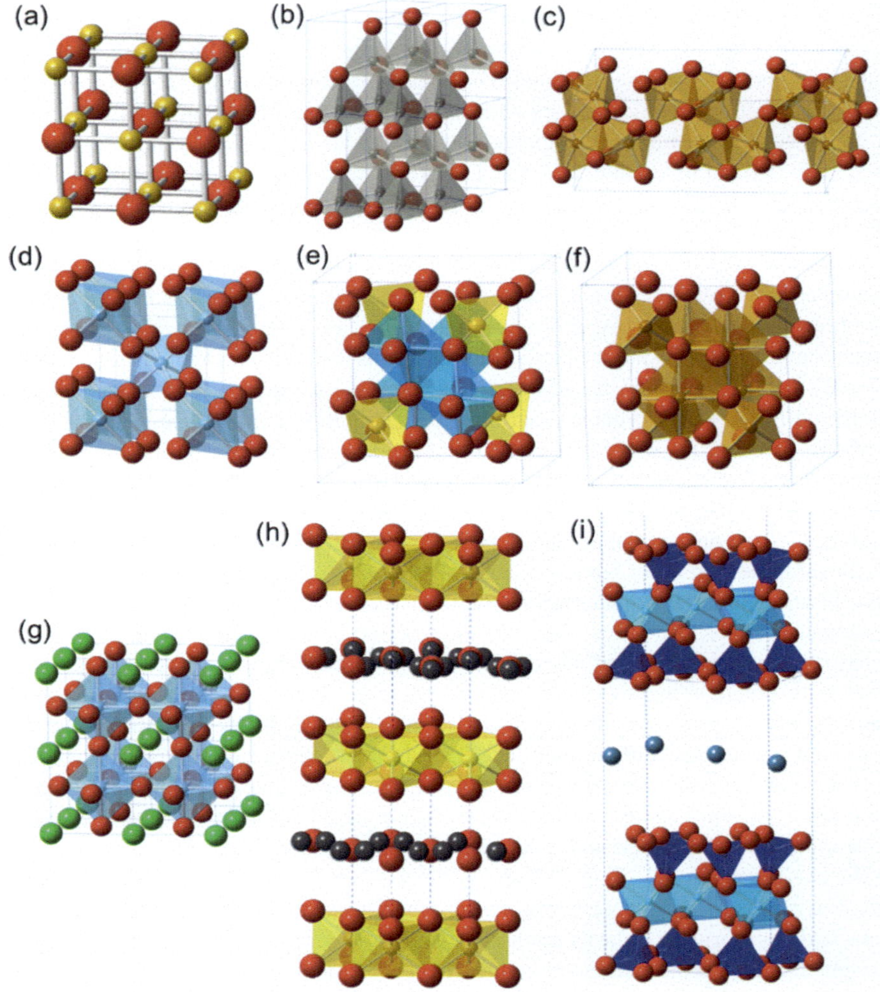

Fig. 8.1 Crystal structures of representative metal oxides. **a** rocksalt (MgO), **b** wurtzite (ZnO), **c** corundum (α-Fe_2O_3), **d** rutile (TiO_2), **e** normal spinel ($MgAl_2O_4$), **f** inverse spinel (Fe_3O_4), **g** perovskite ($SrTiO_3$), and layered compounds (**h** hydrotalcite and **i** montmorillonite))

on (or near) the catalyst surface, so that an increase in surface area is an important factor in the synthesis of complex metal oxide catalysts for liquid-phase organic reactions. To date, many research groups have reported that nanosized and/or porous crystalline complex oxide catalysts synthesized by co-precipitation, sol–gel, solution combustion, and soft/hard templating methods could be used as heterogeneous, recyclable, and durable catalysts for liquid-phase organic reactions (Fig. 8.3). Complex metal oxides are multicomponent and complex systems that consist of multiple

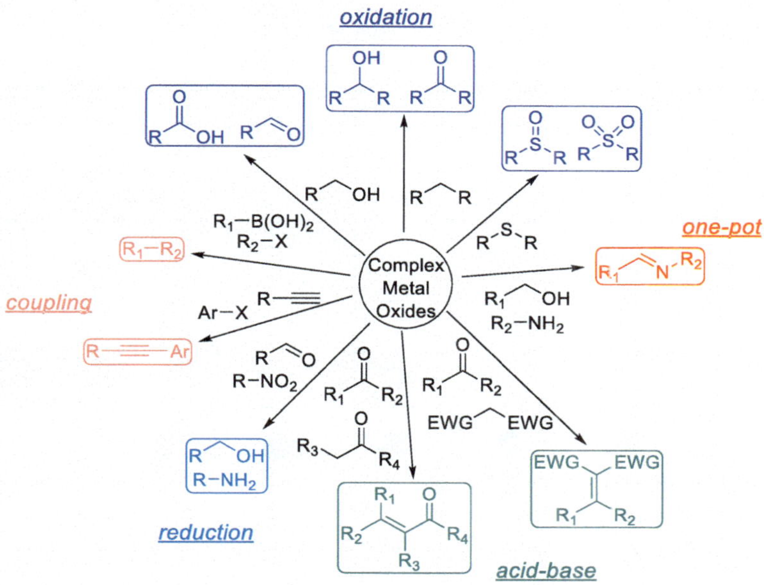

Fig. 8.2 Representative examples for heterogeneously catalyzed liquid-phase organic reactions over crystalline complex metal oxides-based materials

metals; therefore, the synthesis methods employed have a significant effect on the crystallinity, purity, size/shape of particles, surface area, and the size/amounts of pores. The solid-state synthesis method is typically employed for electronic and electrical applications because highly pure crystalline complex metal oxides can be synthesized from metal precursors such as oxides, hydroxides, and carbonates; however, the extremely low surface area of the product often limits the overall performance of the bulk catalyst. The co-precipitation method is one of the most useful techniques for the facile synthesis of crystalline complex metal oxides. The addition of suitable precipitants (NaOH, NH_3, amine, etc.) to an aqueous solution containing two or more metal species (metal nitrates are typically used as starting materials) gives precursors with high homogeneity. Another simple and useful technique to synthesize crystalline complex metal oxides with relatively high surface areas is the sol–gel method. The gradual formation of a solid-phase gel from the solution sol containing the metal sources and additives (in some cases) is observed by heating, pH control, and drying. The dried gel precursors are decomposed by calcination, which leads to the formation of the desired homogeneous complex metal oxides with accurately controlled chemical composition. Solution combustion synthesis using self-sustaining exothermic chemical reactions of metal nitrates with suitable organic fuels (urea, citric acid, glycine, alanine, glycerol, etc.) can also be applied to the synthesis of complex metal oxide nanoparticles. The reaction itself supplies the heat required for phase formation, so that the mixture is heated at lower temperatures than those used for conventional synthesis routes, i.e., there is no requirement for

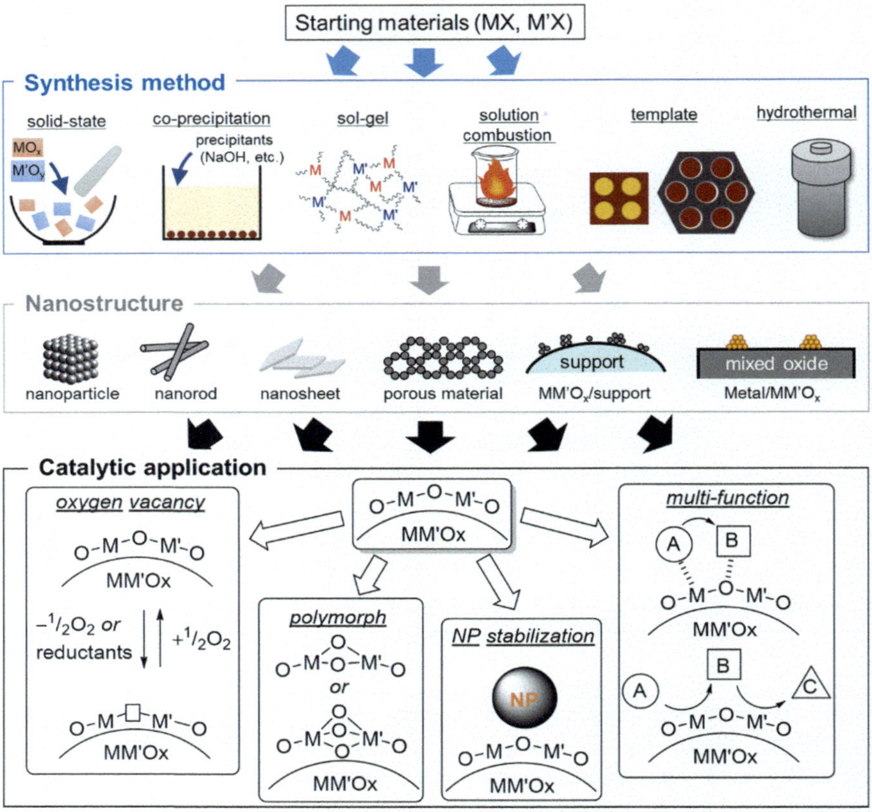

Fig. 8.3 Synthesis and catalyst design of various crystalline complex metal oxides for liquid-phase organic reactions

the decomposition of the intermediates and/or calcination steps. The reaction occurs only on the surface of non-porous metal oxide nanoparticles, which limits the accessibility of large substrates and thus the catalytic performance. The synthesis of porous complex oxide materials has been extensively investigated using polymeric materials and porous silicates as soft and hard templates, respectively. Hydrothermal synthesis is also an important inorganic synthesis technique. An aqueous mixture of precursors is heated above the boiling point of water in a sealed stainless-steel autoclave, which results in a significant increase in pressure in the reaction autoclave above atmospheric pressure. Under these conditions, a one-step process to produce highly crystalline materials is realized without the need for post calcination processes, and hydrothermal strategies have been utilized to synthesize a broad range of materials including single crystal materials, zeolites and related microporous materials, and inorganic–organic hybrid materials.

8.3 Catalytic Application to Liquid-Phase Organic Reactions

8.3.1 Simple Oxide

Simple oxides are widely used not only as catalysts for various types of reactions, but also as supports for metal-supported catalysts [37–40]. Their acid–base and redox properties are significantly dependent on the constituent elements and crystal structures (polymorphs), and in particular, the surface properties are also significantly affected by the catalyst preparation method. The preparation, properties, and catalyst activity of simple oxides have been reported in many excellent review articles and books; therefore, only typical examples and our recently reported polymorphic dependence of MnO_2 are introduced in this chapter.

Simple oxides are classified into two categories of main group oxides and transition metal oxides. Among the main group oxides, alkali or alkaline earth oxides (MgO, CaO, Rb_2O, etc.) and rare earth oxides (La_2O_3, etc.) are representative solid base catalysts that are used for many base-catalyzed reactions. For example, MgO could efficiently promote some liquid-phase base-catalyzed reactions such as the nitroaldol reaction, Michael dimerization of methyl crotonate, addition of methanol to 3-buten-2-one, and Michael addition of nitromethane [38]. The surface of alumina (Al_2O_3) is generally regarded as acidic rather than basic; however, the participation of both acidity and basicity in the catalysis is indispensable. Similarly, the existence of both acidity and basicity has been frequently observed experimentally for other simple oxides such as TiO_2, ZrO_2, ZnO, and CeO_2. Tomishige and coworkers have focused on the acid–base bifunctional properties of ZrO_2 for the selective synthesis of dimethyl carbonate from methanol and CO_2 (Eq. 8.1) [41, 42]. ZrO_2 prepared by the calcination of $Zr(OH)_4$ at 400 °C exhibited the highest catalyst activity, and the bulk and near-surface structures were mainly tetragonal and monoclinic phases, respectively, based on X-ray diffraction (XRD) and Raman measurements. According to the results of NH_3- and CO_2-temperature programmed desorption (TPD) measurements, the number of adjacent acid–base sites has been proposed to be related to the activity toward dimethyl carbonate formation (Fig. 8.4). Acidic hydroxyl groups are present on the surface of certain metal oxides such as Nb_2O_5 and WO_3, and the number of acidic hydroxyl groups is largely dependent on the calcination temperature for these oxide materials.

$$\underset{(250\ \text{mmol})}{CH_3OH} + \underset{(82\ \text{mmol})}{CO_2} \xrightarrow[170^\circ\text{C, 16 h}]{ZrO_2} \tag{8.1}$$

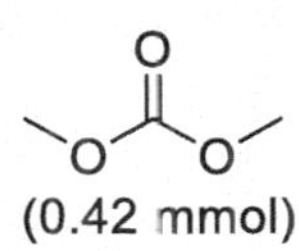

Redox-active metal oxides could catalyze oxidation reactions through the removal of lattice oxygen and its reinsertion, as well as one-pot reactions using a combination of the acid–base and redox properties of metal oxides. Yamaguchi and co-workers

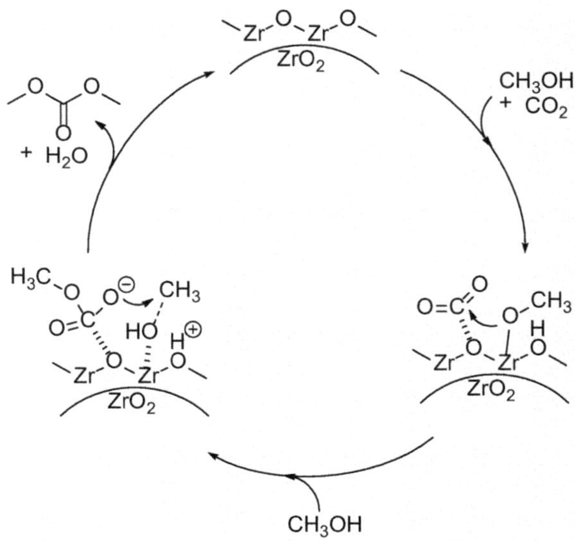

Fig. 8.4 Proposed reaction mechanism for the selective synthesis of dimethyl carbonate from methanol and CO_2 over ZrO_2. The catalytic cycle was created based on the description in reference [42]

reported that a cryptomelane-type manganese oxide-based octahedral molecular sieve (OMS-2) could act as a recyclable heterogeneous catalyst for the aerobic oxidative amidation of various primary alcohols or aldehydes, including aromatic, olefinic, heteroaromatic, and aliphatic species into the corresponding primary amides with ammonia (Eq. 8.2) [43]. It has been proposed that the redox and acid-base properties of OMS-2 promote the following four steps: (i) oxidation of alcohols, (ii) condensation of aldehydes with ammonia, (iii) oxidation of aldimines, and (iv) hydration of nitriles (Fig. 8.5). Tamura and co-workers developed heterogeneous catalyst systems based on CeO_2 for aerobic imine formation from various combinations of benzyl alcohols and anilines [44]. CeO_2 exhibited much higher catalytic activity than other metal oxides and could be reused three times without significant loss of activity for the reaction of benzyl alcohol and aniline (Eq. 8.3). Mechanistic studies including kinetic measurements, mass spectrometry, and IR analyses indicated that reactive oxygen species at the redox sites on CeO_2 results in the high activity, and that CeO_2 promoted both alcohol oxidation and imine formation, probably due to the redox and acid–base bifunctional properties of CeO_2 (Fig. 8.6).

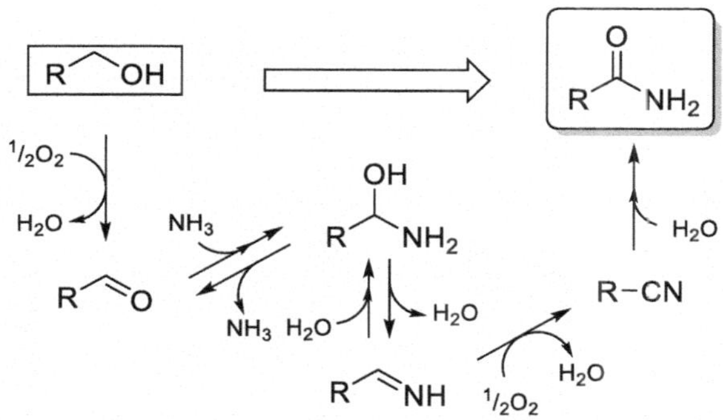

Fig. 8.5 Proposed reaction pathways for the synthesis of primary amides from primary alcohols over OMS-2 [43]

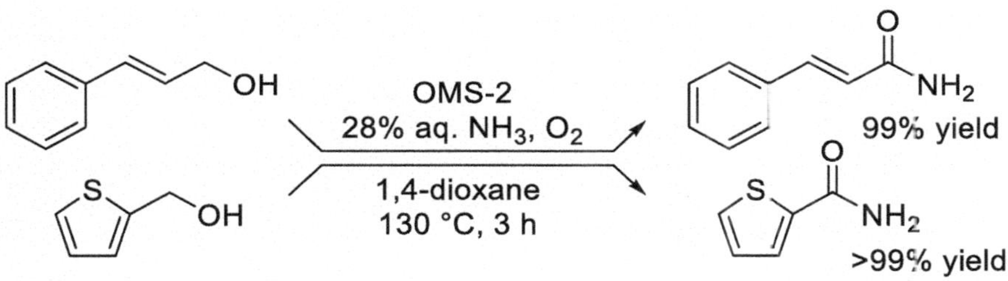

(8.2)

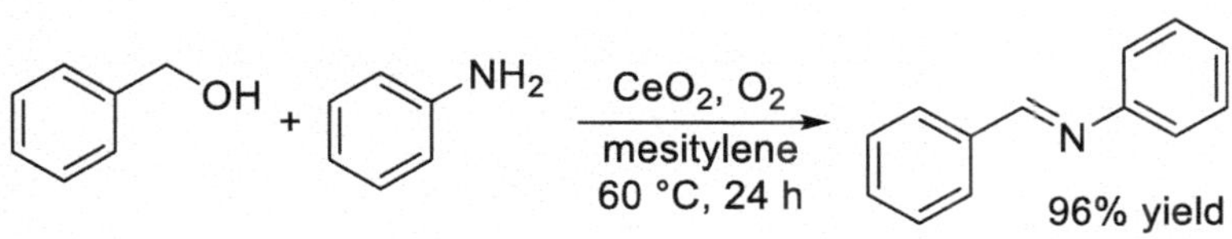

(8.3)

The polymorphism of metal oxides is an important factor in determining the catalytic performance of various reactions (e.g., N_2O decomposition [45], photocatalytic CO_2 hydrogenation [46], and electrocatalytic oxygen evolution reaction (OER), and oxygen reduction reaction (ORR) [47]). We have also investigated the polymorph dependence of MnO_2 for the aerobic oxidation of 5-hydroxymethylfurfural (HMF) to 2,5-furandicarboxylic acid (FDCA) as a bioplastic monomer by a simple system based on a non-precious-metal catalyst of MnO_2 and $NaHCO_3$ (Eq. 8.4) [48, 49]. We revealed that β-MnO_2 was the most active oxidation catalyst among the MnO_2 polymorphs (α-MnO_2, γ-MnO_2, δ-MnO_2, ε-MnO_2, and λ-MnO_2) through combined computational and experimental studies. Density functional theory (DFT) calculations indicated that β-MnO_2 consists of only planar oxygen sites with lower vacancy formation energies compared with α- and γ-MnO_2 (Fig. 8.7). In addition, the reaction rates per surface area for the slowest step (5-formyl-2-furancarboxylic acid (FFCA) oxidation to FDCA) decreased in the order of β-MnO_2 > λ-MnO_2 > γ-MnO_2 ≈ α-MnO_2 > δ-MnO_2 > ε-MnO_2, and the order was in good agreement not only with the

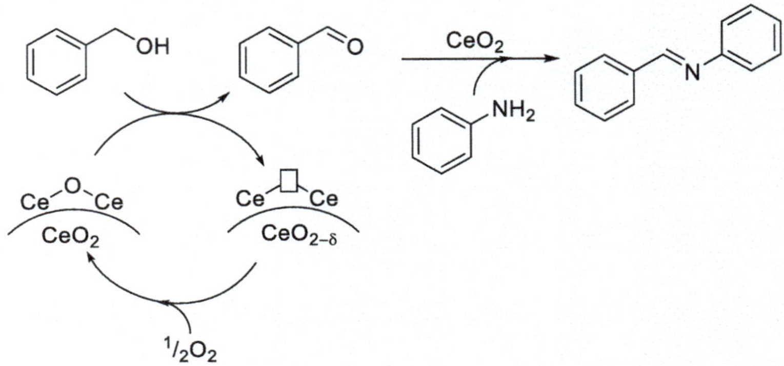

Fig. 8.6 Proposed reaction mechanism for the synthesis of imine from alcohol and aniline over CeO_2 [44]

DFT calculation results, but also with the reduction rates per surface area determined by H_2-temperature-programmed reduction (TPR) measurements for MnO_2 catalysts. Furthermore, we have developed a new synthesis method (low-temperature crystallization of precursors) to produce mesoporous β-MnO_2 nanoparticles with high surface areas, which results in significant improvement of the catalytic activity for the aerobic oxidation of HMF to FDCA [50].

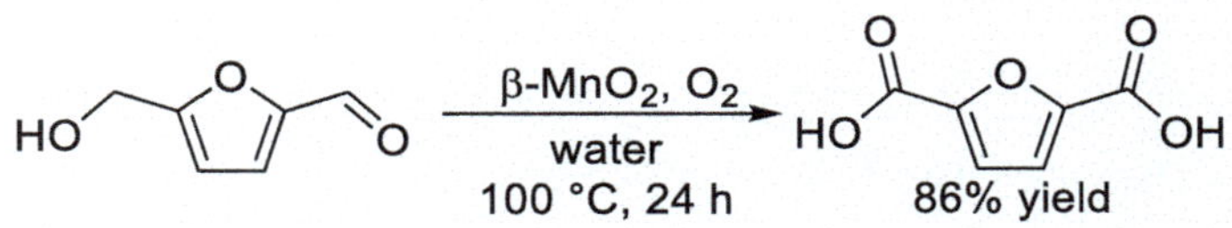

(8.4)

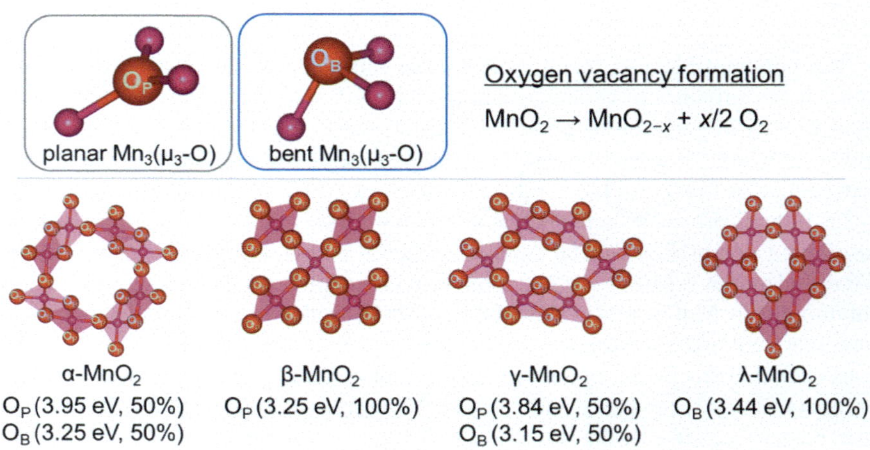

Fig. 8.7 Oxygen vacancy formation energy of MnO_2 [48]

These mesoporous β-MnO_2 nanoparticles were also active for the one-pot synthesis of sulfonamides, which are an important class of organic compounds widely used as medicines, plasticizers for fiber-reinforced composites materials, and intermediates for dyes, from aromatic thiols, O_2, and NH_3 (Eq. 8.5) [51]. The solvent effect on the one-pot oxidative sulfonamidation of benzenethiol with 28% aqueous NH_3 (5 equivalents with respect to the substrate) in various solvents catalyzed by β-MnO_2 at 90 °C was essential, and the mixed solvents of dimethyl formamide (DMF)/water (3/2, v/v) was the most effective because the presence of DMF and water may increase the solubility of both benzenethiol and NH_3, which likely improves the catalytic performance. The used β-MnO_2 catalyst could be recovered from the reaction mixture by simple filtration with negligible leaching of Mn species into the filtrate, and it could then be reused without significant loss of activity or selectivity. The present β-MnO_2-based catalytic system was applicable to the one-pot oxidative sulfonamidation of various aromatic and heteroaromatic thiols using O_2 and NH_3 and a larger scale one-pot oxidative sulfonamidation of *p*-toluenethiol (20 mmol scale) to give 3.05 g of the corresponding sulfonamide as an industrially important plasticizer for fiber-reinforced composite materials. Infrared (IR) spectroscopy measurements of β-MnO_2 with adsorbed pyridine and *S*-phenyl benzenethiosulfonate (intermediate) most likely indicate the interaction of this intermediate with Lewis acid sites on the β-MnO_2 surface, which would facilitate the nucleophilic attack of NH_3 on this intermediate (Fig. 8.8). Therefore, the strong oxidizing ability and Lewis acidity of β-MnO_2 are essential for the aerobic oxidation and the nucleophilic substitution with NH_3, respectively, and such dual functionality of β-MnO_2 results in highly efficient one-pot aerobic sulfonamidation from thiols, O_2, and NH_3.

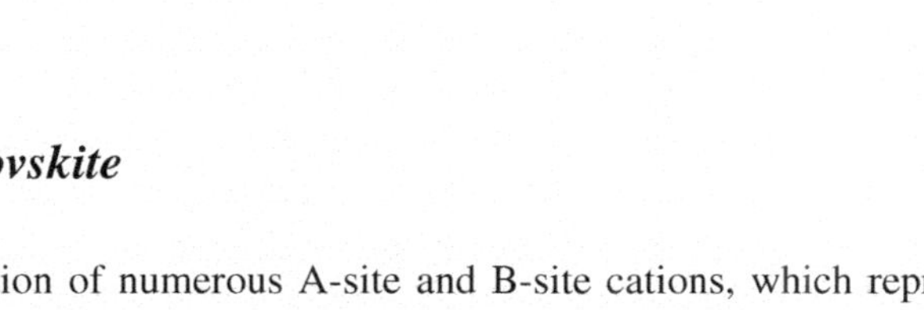

(8.5)

8.3.2 *Perovskite*

The combination of numerous A-site and B-site cations, which represents ≈90% of the metal elements in the periodic table, allows for the formation of the ABO_3 perovskite oxide structure. It is also possible to fine-tune the oxidation state of constituent B-site cations and oxygen vacancies of multi-component perovskite by partial substitution of A- or B-site cations with other metals. Such compositional and structural flexibility of perovskites makes it possible to control their physicochemical properties, which leads to unique functionalities. In Sect. 8.1, the ideal crystal structure of perovskite oxide was described as cubic, with large ionic radius A-site cations

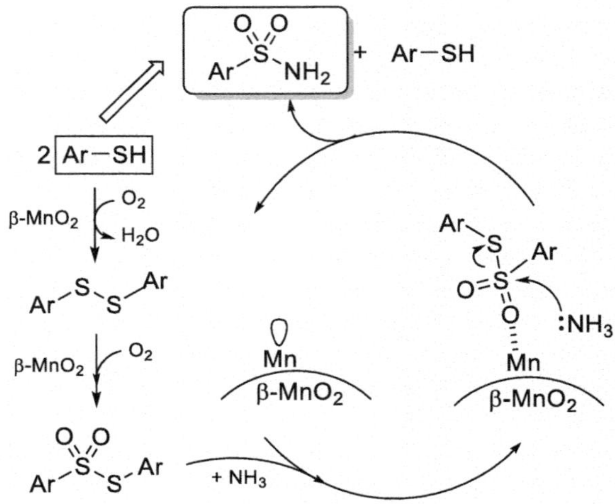

Fig. 8.8 Proposed reaction mechanism for the one-pot synthesis of sulfonamidies from thiol, NH_3, and O_2 over β-MnO_2 [51]

having 12-fold coordination and small ionic radius B-site cations having sixfold coordination. The tolerance factor (t) is typically defined by $t = (r_A + r_O)/\sqrt{2}\,(r_B + r_O)$ (where r_A, r_B, and r_O are the ionic radii for the A cations, B cations, and O^{2-} anions, respectively) as an indicator of the deviation from this ideal cubic structure [52–56]. The ideal cubic perovskite t value is 1, and the formation of perovskite structures with lower t values in the range of 0.75 to 1 is also observed. The cubic structure is distorted to tetragonal, rhombohedral, or other lower symmetry structures with a decrease in t. On the other hand, the use of large alkaline earth cations or small B-site cations results in a hexagonal perovskite structure that consists of face-sharing BO_6 octahedra with $t > 1$ (Fig. 8.9) [54]. A layered perovskite structure that consists of cation-interleaved two-dimensional perovskite slabs has also been reported [52].

Due to their unique structural diversity and controllable physicochemical properties, perovskite oxides and related materials have been extensively studied as piezoelectric, (multi)ferroelectric, magnetic, superconducting, and catalytic materials. In the field of catalytic chemistry with perovskite oxides, reports on liquid-phase organic reactions for application in green chemistry are still few compared to the electrochemical [57–59], photocatalytic [60, 61], and high-temperature gas-phase reactions (total oxidation of hydrocarbons, NO_x decomposition, etc.) because of the low surface areas of perovskite oxides. Heterogeneous and recyclable catalytic systems based on porous and nanosized perovskite oxides have been recently developed for liquid-phase organic reactions such as cross-coupling reactions [62–66], acid/base-catalyzed reactions [67–75], and selective oxidation reactions [76–92]. In this section, we comprehensively describe the unique catalytic properties of perovskite oxides for these three types of reactions, including the proposed reaction mechanism. Although some perovskite oxides are active in the hydrogenation of aldehydes, alkenes, and carboxylic acids, they will not be discussed in detail in this chapter because of

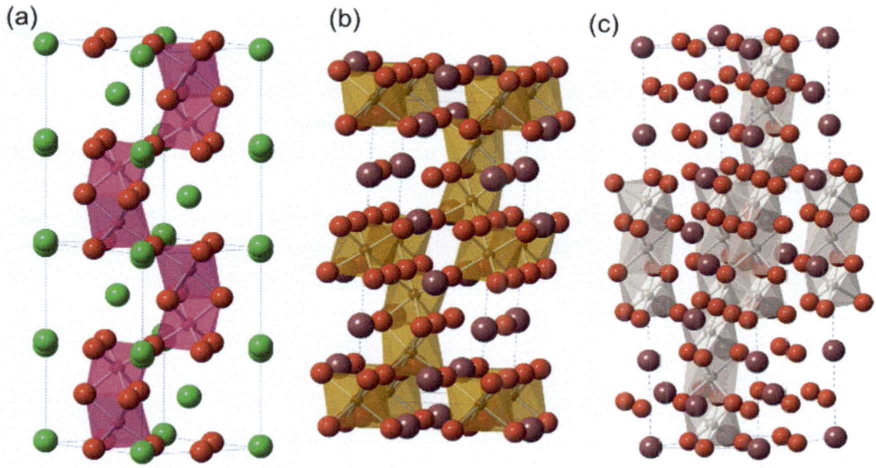

Fig. 8.9 Crystal structure of hexagonal perovskite oxides. **a** 4H-$SrMnO_3$, **b** 6H-$BaFeO_3$, and **c** 9R-$BaRuO_3$

the structural change to a metal-supported catalyst under reductive conditions (e.g., Ni/La_2O_3 from $LaNiO_3$) and the difficulty in the reuse of the catalyst [93, 94].

Transition metal-catalyzed cross-coupling reactions are among the most useful tools in organic synthesis for the production of many important products such as pharmaceuticals, materials, and optical devices [95–99]. Efficient C–C bond-forming cross-coupling reactions with homogeneous Pd catalysts that are workable under mild conditions are largely unaffected by the presence of water, and exhibit a wide range of substrate applicability. There is a strong demand to develop Pd-based heterogeneous catalysts that are easily recoverable and recyclable for industrial use; therefore, many approaches such as encapsulation and impregnation have been used to immobilize Pd species. However, leaching of Pd species from the support materials can occur, so the development of truly heterogeneous Pd catalysts remains an important research topic. On the other hand, Tanaka and co-workers have developed an innovative "intelligent catalyst" based on a Pd-containing perovskite oxide, which has significantly improved durability due to the self-regeneration of Pd nanoparticles, in a study on a three-way catalyst for automobiles that can simultaneously convert the emissions of three pollutants (CO, NO_x, and unburned hydrocarbons) in engine exhaust gas [100–102]. Whereas conventional noble metal-supported catalysts reduce catalytic activity due to aggregation and growth of the noble metal particles in use, self-regeneration is achieved through a cycle between the solid solution and the segregation of Pd in perovskite crystals. Attempts to apply the specific behavior of these Pd-containing perovskite catalysts to liquid-phase cross-coupling reactions have been reported by Ley, Smith, and co-workers [62–65]. The formation of active Pd species in the liquid phase and recapture of Pd^0 by the bulk inorganic phase are proposed to be key steps in a similar manner to that of intelligent catalyst.

Pd-containing multicomponent perovskite oxides (0.05 mol% Pd) such as $LaFe_{0.57}Co_{0.38}Pd_{0.05}O_3$, $La_{0.9}Ce_{0.1}Fe_{0.57}Co_{0.38}Pd_{0.05}O_3$, and $LaFe_{0.95}Pd_{0.05}O_3$ efficiently promote Suzuki reactions of aryl halides with boronic acids at 80 ºC in the presence of 3 equivalents of a base (K_2CO_3, K_3PO_4, or Cs_2CO_3) [62]. The corresponding biaryl derivatives are obtained in high yields from various combinations of aryl halides and boronic acids, and the addition of tetra-*n*-butylammonium bromide (TBABr) improves a few difficult cases such as 3,4,5-trimethoxyphenylboronic acid (Eq. 8.6). Substrates with substituents in the *ortho* position, heteroatom-containing compounds, vinyl boronic acids, and inactive aryl chlorides are tolerated in this catalytic system (Eq. 8.7).

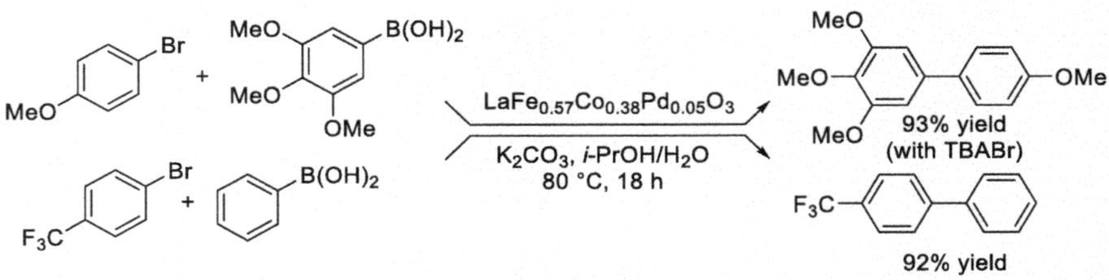

(8.6)

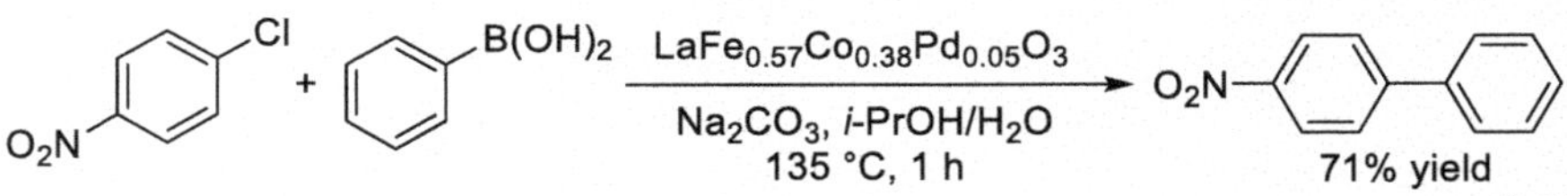

(8.7)

Similar multicomponent Cu- and Pd-containing perovskite oxides are applicable to the Ullmann and Sonogashira reactions [63]. In the presence of a Cu-containing perovskite, $La_{0.9}Ce_{0.1}Co_{0.6}Cu_{0.4}O_3$, the Ullmann condensation of aryl halides with phenols and thiols efficiently proceeds to give the corresponding biaryl ethers and sulfides, respectively, in high yields. In the case of ether synthesis, the presence of Cs_2CO_3 and ethyl acetate as additives results in the successful coupling of a variety of phenols with aryl halides (Eq. 8.8). The use of 2 equivalents of sodium *tert*-butoxide (NaO*t*Bu) and 10 mol% neocuproine (2,9-dimethyl-1,10-phenanthroline) is effective for the synthesis of sulfides from aryl halides and thiols (Eq. 8.9). The same research group reported that Pd-containing perovskites such as $LaFe_{0.57}Cu_{0.38}Pd_{0.05}O_3$, $LaFe_{0.57}Co_{0.38}Pd_{0.05}O_3$, and $Na_{2.04}Cu_{0.95}Pd_{0.05}O_4$ promoted Sonogashira coupling reactions between aryl halides and alkynes in a mixed solvent of 5% H_2O/DMF or H_2O/dimethylamine (DMA) at 120 °C. In the presence of 2.5 mol% catalyst and 4 equivalents of triethylamine (Et_3N), the disubstituted alkynes from various combinations (7 examples) were obtained in moderate to good yields (Eq. 8.10). Taking advantage of the remarkable features of a Pd-containing perovskite oxide as a solid catalyst, the same research group conducted a flow chemistry procedure for the Sonogashira and Heck cross-coupling reactions using Pd-containing perovskite $LaFe_{0.95}Pd_{0.05}O_3$ deposited on CeO_2 [64]. Under optimized conditions, a 10 mmol-scale Sonogashira cross-coupling reaction of 1-butyl-4-ethynylbenzene with iodobenzene was achieved and the catalyst could be used continuously for at least 24 h without significant loss of catalytic efficiency.

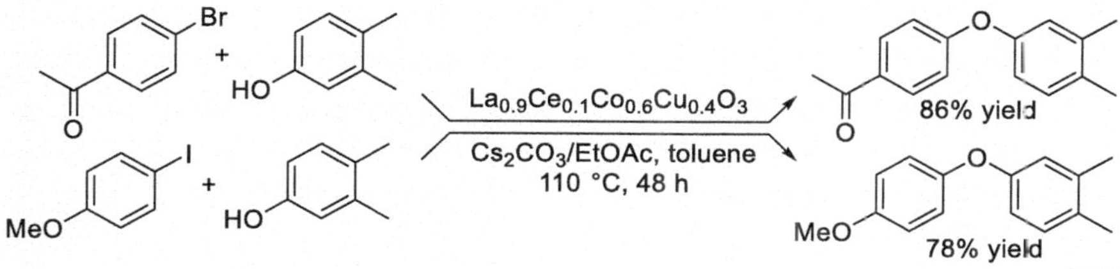

(8.8)

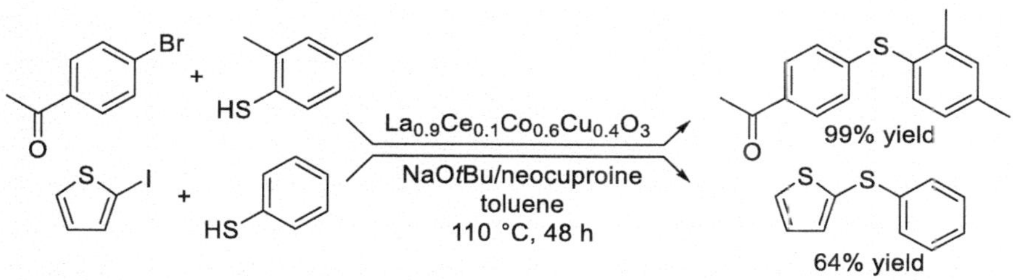

(8.9)

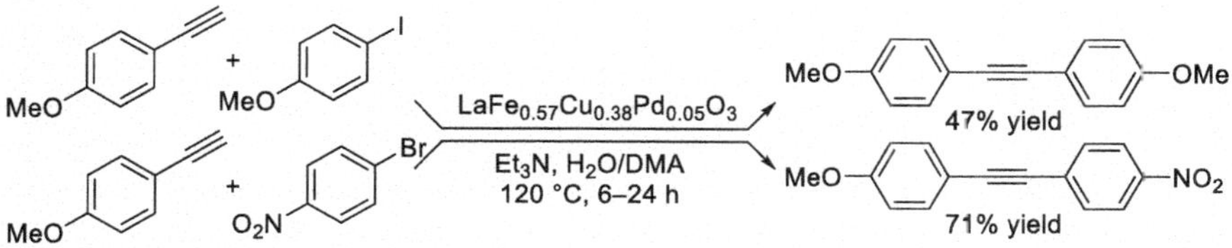

(8.10)

Ley and co-workers conducted detailed mechanistic studies on the Suzuki coupling reaction over Pd-containing perovskite oxides based on the reaction kinetics, catalyst poisoning experiments, electron microscopy, and three-phase tests [65]. The reaction of 4-bromoanisole with phenylboronic acid in the presence of $LaFe_{0.57}Co_{0.38}Pd_{0.05}O_3$ proceeded efficiently after an induction period that corresponded to the formation of the active Pd species in the liquid phase. After removal of the perovskite solid from the reaction mixture via a 0.45 μm syringe filter, the filtrate was determined to be catalytically active, which indicated the formation of a soluble Pd species. Strong catalyst poisoning was observed with soluble CS_2 and insoluble polymer-supported thiophenols. However, the addition of triphenylphosphine (Ph_3P) extended the induction period and the reaction proceeded efficiently after this induction period. This phenomenon is due to the competitive ligation of phosphine and bromoanisole to Pd^0 species. A three-phase study using solid-supported aryl iodide revealed that aryl halides are required to produce active Pd species in the liquid phase. Based on these results, a possible reaction mechanism was proposed, as shown in Fig. 8.10. First, Pd species in perovskite are reduced to Pd^0 bound to the solid surface, probably by

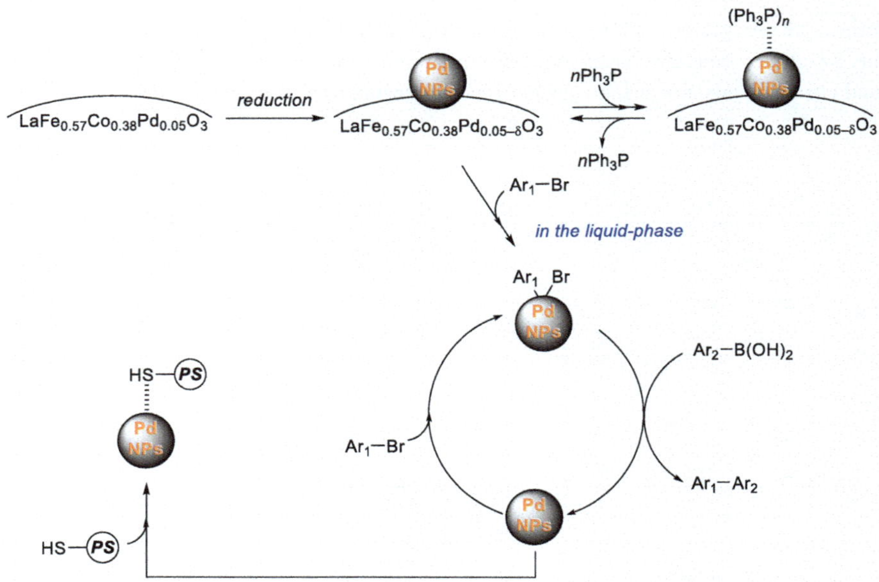

Fig. 8.10 Proposed reaction mechanism for Suzuki reaction catalyzed by $LaFe_{0.57}Co_{0.38}Pd_{0.05}O_3$ [65]

an aqueous alcohol solvent. Oxidative addition of aryl halides produces a soluble Pd species (probably halide-stabilized Pd^0 nanoparticles), which facilitates the Suzuki coupling reaction according to a standard mechanism. No formation of Pd black was observed in transmission electron microscopy (TEM) measurements and the leaching amounts of Pd species in the reaction mixture was negligible (ca. 2 ppm), which indicated that Pd^0 was recaptured by the bulk inorganic phase.

In contrast to the Suzuki reactions over Pd-containing perovskite oxides, the reaction mechanisms, including metal leaching, reusability, catalyst durability, and kinetics, have yet to be clarified [63]. Recently, Friedrich and co-workers reported that $LaCo_{0.9}Ni_{0.05}Pd_{0.05}O_3$ (Pd 1 mol%) could act as an efficient heterogeneous catalyst for the Sonogashira cross-coupling reaction between various substituted aryl iodides and phenylacetylenes (12 examples) in a mixed solvent of water and acetonitrile at 100 °C in the presence of 3 equivalents of Et_3N (Eq. 8.11); however, this system was inactive for bromobenzene and chlorobenzene [66]. The catalyst could be reused six times for the coupling reaction of phenylacetylene and iodobenzene without significant loss of catalytic activity. On the basis of catalyst poisoning experiments with Hg, hot-filtration, leaching of Pd species, and TEM measurements of the recovered catalyst, nanosized Pd^0 sites anchored by colloidal organic layers (e.g., iodobenzene) to the perovskite surfaces have been proposed to act as heterogeneous active sites for the present Sonogashira reaction (Fig. 8.11), whereas the catalytic role of Pd^0 species in the liquid phase cannot be completely ruled out.

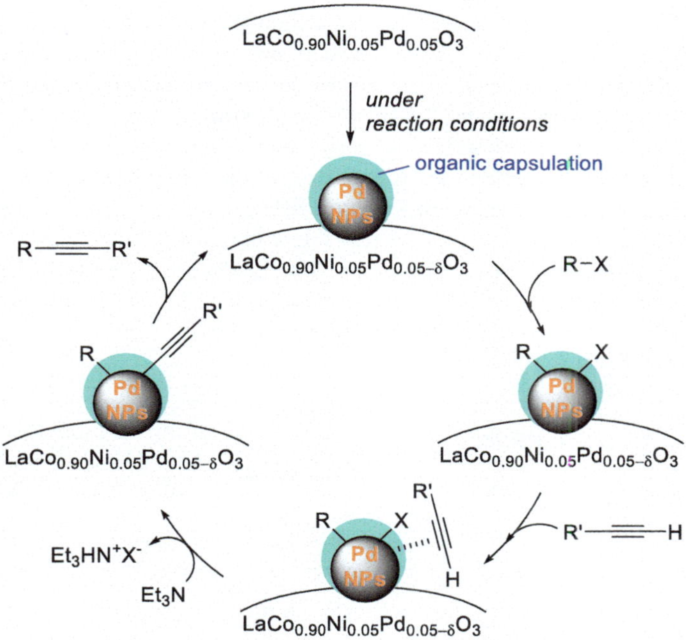

Fig. 8.11 Proposed reaction mechanism for Sonogashira reaction catalyzed by $LaCo_{0.90}Ni_{0.05}Pd_{0.05}O_3$ [66]

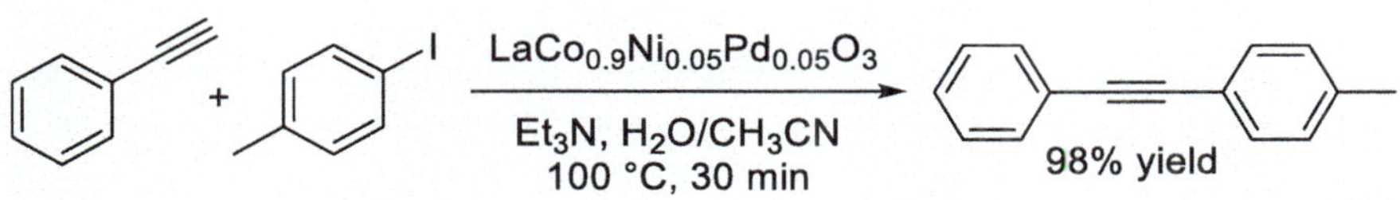

(8.11)

Solid acid–base catalysts have been used in many important industrial processes such as petroleum refining and the production of chemicals [37, 38]. In contrast to well-studied acid–base catalysts (e.g., zeolites, heteropolyacids, clays, resins, and complex metal oxides), the acid–base catalytic properties of perovskite oxides are still underexplored. Wu and co-workers recently gave a comprehensive summary of the acid–base properties of perovskites, including their catalytic performance [103]. The relationship between the acid–base properties (strength, density, and type of acid–base sites, surface reconstruction, and exposed facets) and the catalytic activity has been discussed mainly in terms of gas-phase probe reactions such as 2-propanol conversion. On the other hand, detailed studies on liquid-phase acid–base catalysis and its reaction mechanism are still limited.

Several iron-based perovskites, such as $SmFeO_3$ and $BiFeO_3$, promote liquid-phase acid–base reactions, although the acid–base sites on these catalysts are not well characterized. Yahiro and co-workers reported that $SmFeO_3$ exhibited higher catalytic activity based on weight for the cyanosilylation of benzaldehyde with trimethylsilyl cyanide (TMSCN) than typical solid acid catalysts such as SiO_2–Al_2O_3 and Amberlyst 15 DRY (Eq. 8.12) [67]. On the basis of catalyst poisoning experiments using pyridine and 2,6-di-*tert*-butylpyridine, the Brønsted acid sites on $SmFeO_3$ have been proposed as the active sites. Singh and co-workers reported the one-pot synthesis of various dihydro-2-oxypyrroles from dimethyl acetylenedicarboxylate, anilines, and aldehydes using a $BiFeO_3$ catalyst (Eq. 8.13) [68]. The magnetically recoverable catalyst was reusable without significant loss of catalytic performance or leaching of metal species. The Lewis acid sites on $BiFeO_3$ have been proposed to activate an imine intermediate, which facilitates a Mannich-type reaction with a hydroamination intermediate followed by subsequent cyclization reactions (Fig. 8.12).

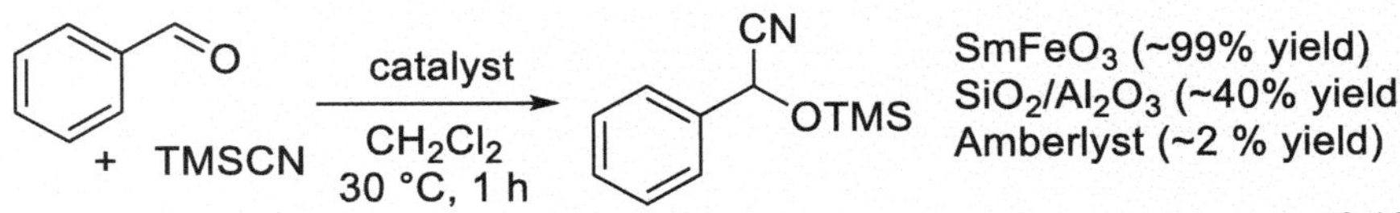

(8.12)

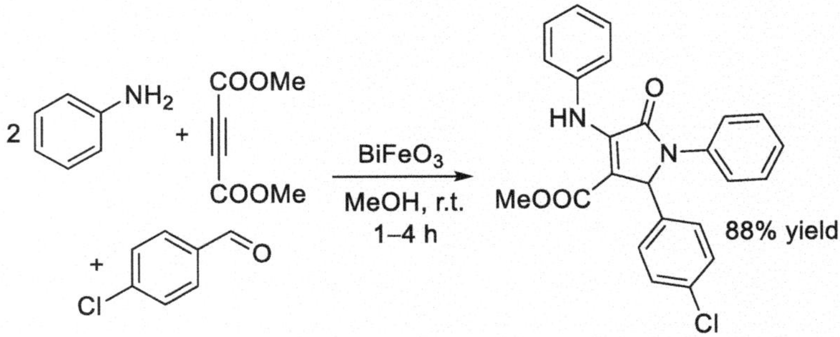

(8.13)

It has been well known that d_0-transition metals such as Ti^{4+} and Zr^{4+} act as acid catalysts; therefore, acid–base catalysis over d_0-transition metal-containing perovskite oxides have also been investigated. Bhaumik, Mukhopadhyay, and coworkers reported that mesoporous $ZnTiO_3$ nanoparticles could act as acid–base heterogeneous catalysts for Friedel–Crafts alkylation with benzyl chloride, esterification of C_{12}–C_{18} carboxylic acids with methanol, and the synthesis of 1,6-naphthyridine from various ylidines and thiols (Eqs. 8.14–8.16) [69, 70]. In each case, a possible reaction mechanism was through the activation of substrates by coordination to the acid–base sites (e.g., carboxylic acid with the Lewis acid site, cooperative action of Ti^{4+} (Lewis acid) and O^{2-} (base), etc.); however, detailed characterization of the acid–base sites, including their interaction with substances, has not yet been reported. Palkovits and co-workers reported the aldol condensation of isobutyraldehyde with formaldehyde to hydroxypivaldehyde catalyzed by Ti, Zr, and Ce-containing ABO_3 perovskites (A = Ca, Sr, Ba; B = Ti, Zr, and Ce) (Eq. 8.17) [71]. Although mixed phases of simple BO_2 metal oxides and ABO_3 perovskite oxides

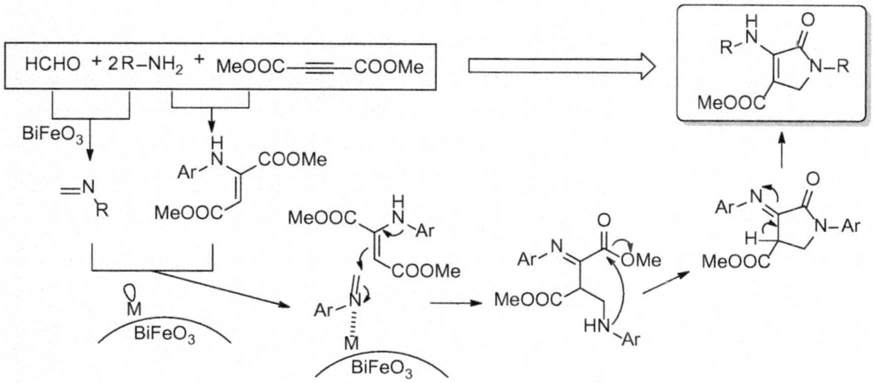

Fig. 8.12 Proposed reaction mechanism for the one-pot synthesis of various dihydro-2-oxypyrroles from dimethyl acetylenedicarboxylate, anilines, and formaldehyde over $BiFeO_3$ [68]

were formed due to the calcination of a simple mixture prepared from aqueous solutions of the metal sources, Zr-based $SrZrO_3$ and $BaZrO_3$ showed the highest catalytic activity for aldol condensation. The degree of distortion and the presence of other phases are important, but the structure and properties of base sites have not yet been clarified in a similar way to those for $ZnTiO_3$.

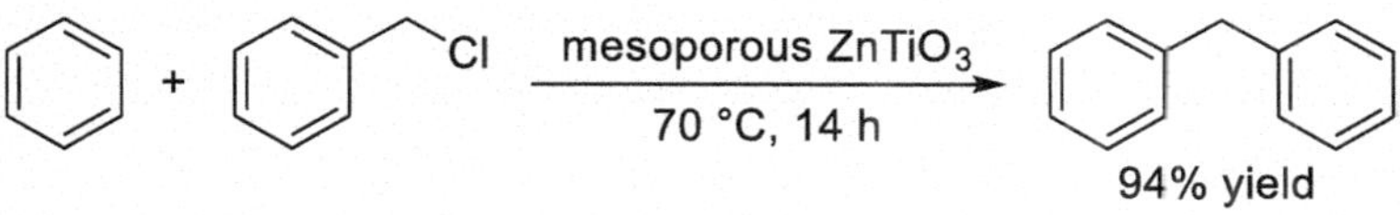

(8.14)

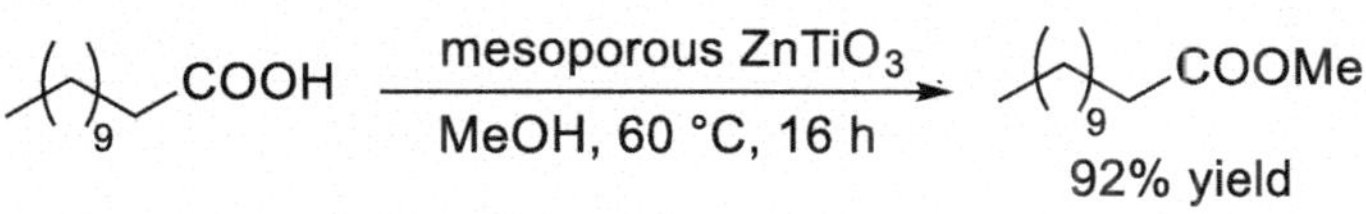

(8.15)

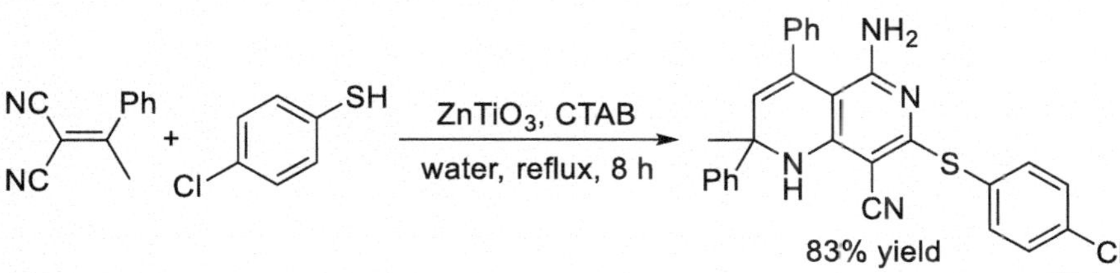

(8.16)

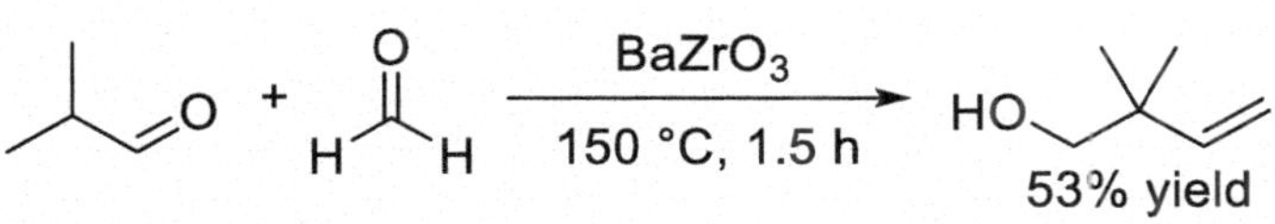

(8.17)

On the other hand, the presence of strong Brønsted acid sites on $HTiNbO_5$ nanosheets was directly determined by NH_3-TPD and ^{1}H magic angle spinning nuclear magnetic resonance spectroscopy (MAS NMR) measurements by Domen and co-workers [72]. Exfoliated $HTiNbO_5$ and $HSr_2Nb_3O_{10}$ nanosheets, which were prepared by exfoliation of cation-exchangeable layered metal oxides such as $HTiNbO_5$ and the layered perovskite oxide $HSr_2Nb_3O_{10}$ in aqueous solutions followed by precipitation under acidic conditions, could act as heterogeneous acid catalysts for the esterification of acetic acid, cracking of cumene, and dehydration of isopropanol (Eq. 8.18). However, layered perovskite $HSr_2Nb_3O_{10}$ nanosheets are almost inactive for these reactions, although approaches based on the exfoliation of layered perovskite catalysts may lead to high catalytic performance. Acid–base catalysis of perovskite oxides–organic composites was also reported. $HCa_2Nb_{3-x}Ta_xO_{10}$ ($x = 0$–3), known as Dion–Jacobson-type layered perovskite oxides, were prepared by the solid-state method following ion-exchange [73]. The $HCa_2Nb_3O_{10}$ layered perovskite combined with organic compounds such as dodecyltributylphosphonium bromide (C_{12}TBPBr), dodecyltriphenylphosphonium bromide (C_{12}TPPBr), and dodecyltrimethylammonium chloride (C_{12}TMACl), and showed activity for the Knoevenagel condensation of ethyl cyanoacetate with benzaldehyde. In contrast, the yield of the product was negligible over only the layered perovskite and only the organic compounds, which indicated the importance of the combination of the perovskite–organic compound for basic catalysis. Furthermore, the catalytic activity toward Knoevenagel condensation over the $HCa_2Nb_{3-x}Ta_xO_{10}$-C_{12}TMACl composites decreased with increasing Ta content. Although the expansion of the interlayer space of the composites with high Ta content was not observed, the low Ta content composites showed expansion of the layer space. These results indicated that the hydrophobicity of the perovskites affects the intercalation of organic compounds and the basicity. The authors also investigated deacetalization–Knoevenagel condensation over the $HCa_2Nb_3O_{10}$-organic compounds and the sequential reaction proceeded to form the desired product, which suggests that these composite catalysts could be used as acid–base catalysts (Eq. 8.19).

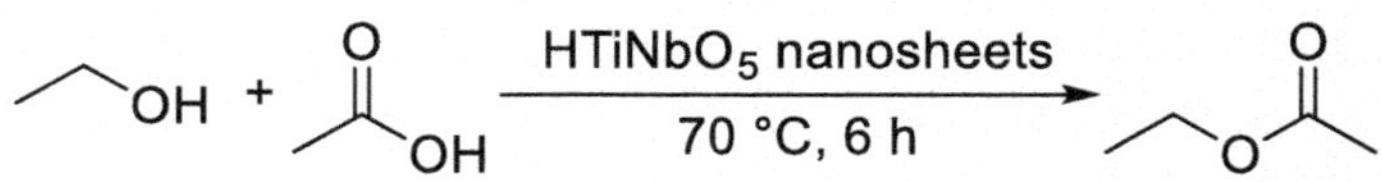

(8.18)

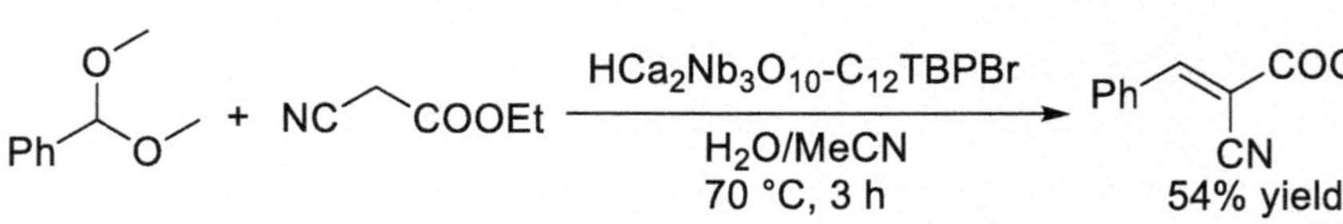

(8.19)

Qiao and co-workers recently synthesized $LaMO_3$ (M = Fe, Co, Ni, Mn) mesoporous perovskite oxides via a resol-assisted cationic coordinative co-assembly approach [74]. Among them, mesoporous $LaMnO_3$ rich in oxygen vacancies introduced by post-treatment with $NaBH_4$ exhibited higher catalytic activity than nonporous bulk $LaMnO_3$ and mesoporous $LaMnO_3$ without post-treatment for the Meerwein–Ponndorf–Verley (MPV) reduction of furfural to furfuryl alcohol in isopropanol solvent (Eq. 8.20). On the basis of DFT calculations combined with experimental results (NH_3-, CO_2-, and O_2-TPD and X-ray photoelectron spectroscopy (XPS)), porous structures based on nanosized $LaMnO_3$ crystals related to the oxygen deficiency and oxygen vacant sites have been proposed to facilitate the adsorption and activation of furfural on oxygen vacancy-rich $LaMnO_3$ to result in a lower activation barrier for the hydrogenation process. The $LaFeO_3$-catalyzed hydrogenation of various ortho- and para-substituted nitrobenzenes to the corresponding anilines in 2-propanol and in the presence of KOH as a promoter was similarly reported by Jayaram and co-workers (Eq. 8.21) [75, 76]. From the effect of the strontium content in $La_{1-x}Sr_xFeO_3$ on the hydrogenation of nitrobenzene, the same research group found that $La_{0.8}Sr_{0.2}FeO_3$ was the most active and recyclable catalyst without any loss in activity, in contrast to other metal oxides such as a Mg–Fe hydrotalcite precursor.

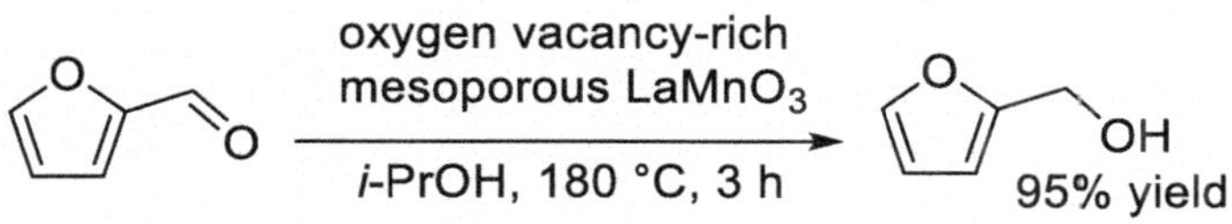

(8.20)

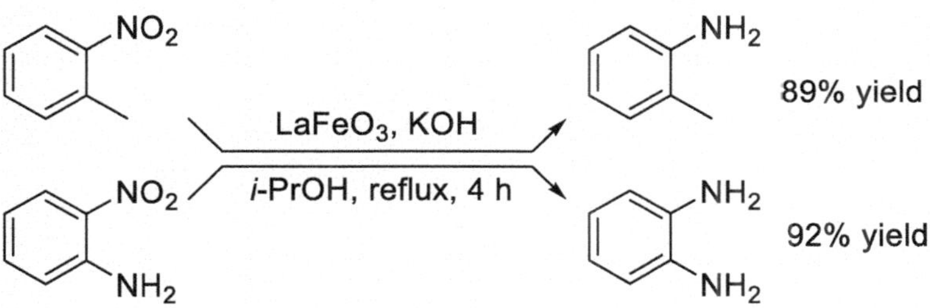

(8.21)

Useful oxygenated products such as alcohols, carbonyl compounds, epoxides, and carboxylic acids are widely used as platform commodities and specialty chemicals; therefore, the selective oxidation of petroleum-based feedstocks into these products is an important reaction in the chemical industry [104–112]. Although efficient perovskite oxide-catalyzed liquid-phase selective oxidation reactions with hydroperoxides (tert-butyl hydroperoxide (TBHP) and hydrogen peroxide (H_2O_2)) have been reported (e.g., oxidation of benzyl alcohol with TBHP by Cu/$LaFeO_3$ [77], oxidation of alkylarenes with TBHP by $LaCrO_3$ [78], oxidation of 2-(methylthio)benzothiazole with H_2O_2 by titanium-substituted potassium niobates [79], and oxidation of phenol with H_2O_2 by $Y_2BaCuO_{5\pm x}$ [80, 81]), catalytic oxidation with O_2 (or air) as an ideal and environmentally-friendly oxidant is more desirable due to the high content of active oxygen species with often only water as a byproduct. In this section, we focus on unique aerobic oxidation catalysis over perovskite oxides and the corresponding reaction mechanism. The catalytic performance of noble metal-supported catalysts is mainly derived from the specific properties of noble metals [82, 83]; therefore, reports on the simple use of perovskite oxides as supports for metal-supported catalysts are omitted in this section.

Selective oxidation of alcohols to the corresponding aldehydes, ketones, and carboxylic acids is one of the most important organic reactions because these products are useful building blocks for the synthesis of polymers and fine chemicals. In particular, the replacement of nonrenewable fossil resources with renewable biomass feedstocks, including lignocellulose and triglyceride, is a key issue for building a sustainable society, and the development of catalysts for the oxidation of biomass-derived substrates to useful chemicals has been extensively studied [113–117]. Hamid and co-workers found that a mixture of $CuZrO_3$ and CuO prepared with a Cu/Zr molar ratio of 1/2 could act as a heterogeneous catalyst for the aerobic oxidation of vanillyl alcohol, which is a model compound for lignin [84]. With 21 bar of air as the oxidant, vanillyl alcohol was converted to vanillin at 91% conversion with 76% selectivity, whereas a CuO catalyst was almost inactive under the same reaction conditions (Eq. 8.22). Despite the slight decrease in the catalytic activity during catalyst recycling experiments, the mixed catalyst was essentially reusable with no change in selectivity toward vanillin and no leaching of metal species into the reaction mixture. High concentrations of Cu–O–Zr bonds as the active phase have been proposed to improve the redox properties of the catalyst. Saffari and co-workers reported that La_2CuO_4 nanoparticles, which were synthesized via an ultrasonic-assisted co-precipitation route, efficiently catalyzed the conversion of various benzaldehydes with electron-donating/-withdrawing substituents to the corresponding carboxylic acids (Eq. 8.23) [85]. In this case, the catalytic activity also gradually decreased with subsequent recycling; therefore, the mechanistic aspects, including the purity/stability of these Cu-based catalysts and the structure–activity relationship, should be studied further.

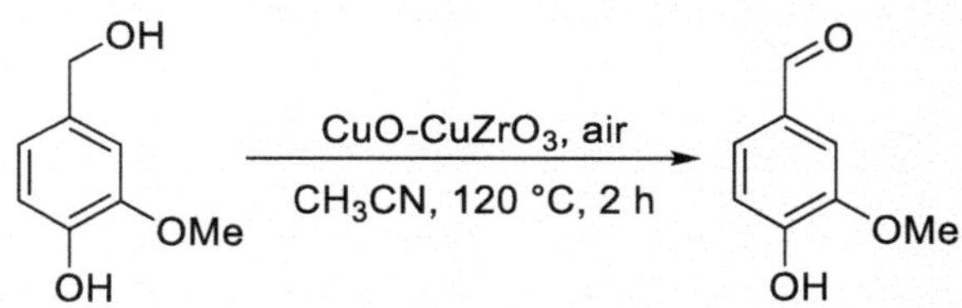

(8.22)

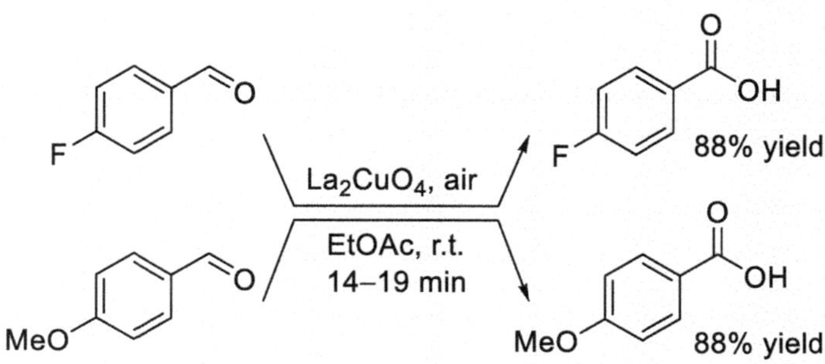

(8.23)

In the fields of homogeneous and bio-inspired catalysis based on well-defined metal complexes and enzymes, active oxygen species bound on mono- and polynuclear high-valency metal species formed by the reductive activation of O_2 can oxidize various substrates under mild reaction conditions [109, 118]. Against such a background, we have recently focused on hexagonal and/or rhombohedral perovskite oxides such as $SrMnO_3$, $BaRuO_3$, and $BaFeO_{3-\delta}$ because these materials have unique face-sharing octahedral units that consist of high valency metal species [86–90]. In contrast to lanthanum-based perovskites with corner-sharing BO_6 octahedra, these hexagonal perovskite catalysts efficiently promote the selective oxidation of alcohols, alkylarenes, alkanes, alkenes, and sulfides with O_2 as the sole oxidant under mild reaction conditions. Herein, we briefly introduce our recent examples of aerobic oxidation with hexagonal perovskite catalysts.

First, hexagonal $SrMnO_3$ ($SrMnO_3$-PC) perovskite nanoparticles were successfully synthesized by the polymerized complex method [86]. $SrMnO_3$-PC could act as an efficient and reusable heterogeneous catalyst for the liquid-phase selective oxidation of various aromatic, allylic, and heteroatom-containing primary and secondary alcohols to the corresponding carbonyl compounds with O_2 (Eq. 8.24). For the aerobic oxidation of 1-phenylethanol to acetophenone, $SrMnO_3$-PC exhibited higher catalytic activity than typical manganese oxide-based catalysts. Mechanistic studies, including the atmosphere effect (O_2 vs. Ar), and kinetics studies, indicated that O_2 activation is most likely involved in $SrMnO_3$-PC-catalyzed oxidation. In addition, the reversible formation of an Mn-superoxo species on the surface Mn species by the reaction of $SrMnO_3$-PC with O_2 was directly confirmed by IR spectroscopy measurements of a sample with adsorbed O_2. This O_2-activation $SrMnO_3$-PC system could also be applied to oxidative homocoupling reactions of amines and phenols, the selective oxidation of alkylarenes, and isobutyraldehyde-assisted electrophilic oxidation of thioanisole and cyclooctene. Although computational studies on the reductive activation of O_2 reduction over $La_{1-x}Sr_xMnO_3$ have been reported [119, 120], this experimental study provides evidence for the detection of a superoxide over $SrMnO_3$ and its catalytic application. After our report, Türkmen and co-workers reported the formation of a superoxide species over $LaMnO_3$ with the reduction pretreatment in H_2 at 400–800 °C, and such oxygen species promoted the oxidation of alkylarenes and oxidative dimerization of 2-napthol with O_2 as the sole oxidant (Eq. 8.25) [91]. The activity of the recovered catalyst decreased considerably, even with the additional reductive treatment, and such catalyst deactivation has been proposed to be due to the adsorption of reactants/products/intermediates and/or surface chemical alterations on the catalyst. Alonso and co-workers similarly reported that the oxygen-deficient orthorhombic $(La,Sr)_{0.5}(Mn,Co)_{0.5}O_{3-\delta}$ perovskite oxide obtained by the heat-treatment of $(La,Sr)_{0.5}(Mn,Co)_{0.5}O_3$ in N_2/H_2 (95/5) exhibited higher selectivity than $(La,Sr)_{0.5}(Mn,Co)_{0.5}O_3$ toward terephthalic acid for the oxidation of *p*-xylene under 20 bar of O_2 in the presence of *N*-hydroxyphthalimide (NHPI) (Eq. 8.26) [92]. The presence of a high number of oxygen vacancies has been proposed to favor the activation of NHPI to form the phthalimide *N*-oxyl radical (PINO), which results in the promotion of hydrogen abstraction from the substrate, although a detailed mechanistic investigation that includes evaluation of metal species leaching is required.

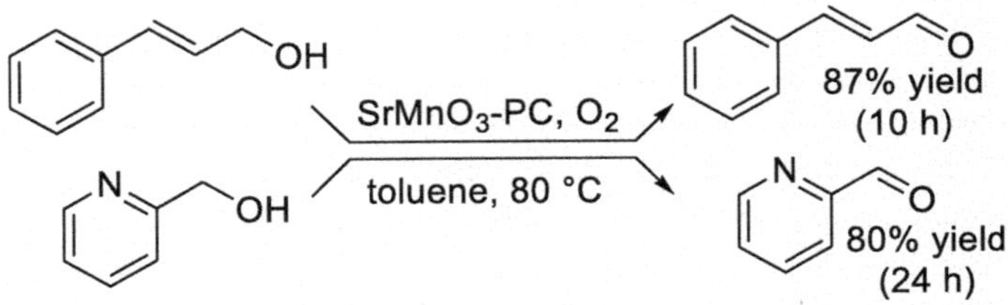

(8.24)

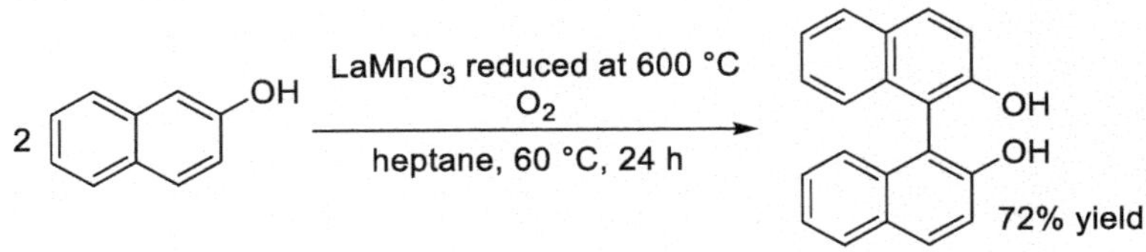

(8.25)

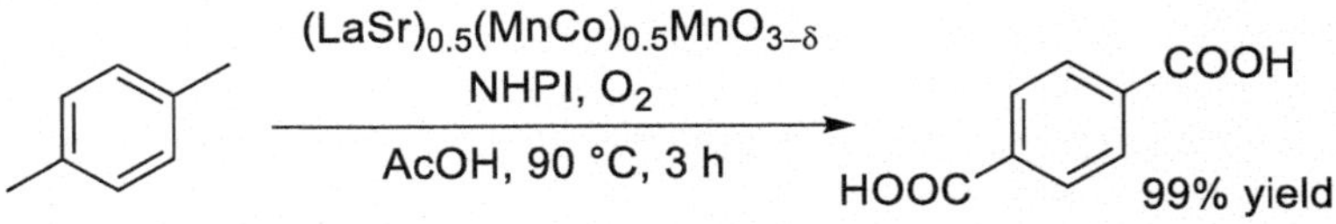

(8.26)

Further high functionalization of various hexagonal perovskite oxides by an increase in the surface area can improve the catalytic oxidation performance as well as application to other oxidative functional transformation reactions; however, the synthesis of versatile hexagonal perovskite oxide nanoparticles is not possible by typical sol–gel methods such as the polymerized complex method or hydroxyl acid-assisted method using nitrates as metal sources [121–123]. In particular, amorphous precursors are carefully prepared using toxic ethylene glycol, and significant amounts of organic reagents via complicated procedures including metal complex and polymer gel formation followed by pyrolysis, which requires high-temperature calcination to remove alkaline-earth carbonates easily formed from the carbonaceous precursors and sometimes leads to low surface areas. In contrast, an alternative, facile, and effective synthetic route to high-surface-area hexagonal perovskite oxides was achieved by simple calcination of an amorphous precursor with low carbon content using dicarboxylic acids (aspartic acid and malic acid) and metal acetates without the need for pH adjustment [87]. These amorphous precursors are easily decomposed at lower temperatures, which results in the suppression of nanoparticle sintering. The respective specific surface areas of $SrMnO_3$ prepared using aspartic acid ($SrMnO_3$-AA) and malic acid ($SrMnO_3$-MA) were 47 and 42 $m^2\ g^{-1}$, which are much larger than those obtained for previously reported $SrMnO_3$ including $SrMnO_3$-PC [86]. The catalytic activity of $SrMnO_3$-AA was the highest among the catalysts tested for the aerobic oxidation of alkylarenes (Eq. 8.27). The mechanism for O_2 activation on $SrMnO_3$ has also been proposed for the oxidation of fluorene on the basis of the catalyst effect, and kinetic and mechanistic studies including ^{18}O-labeling experiments.

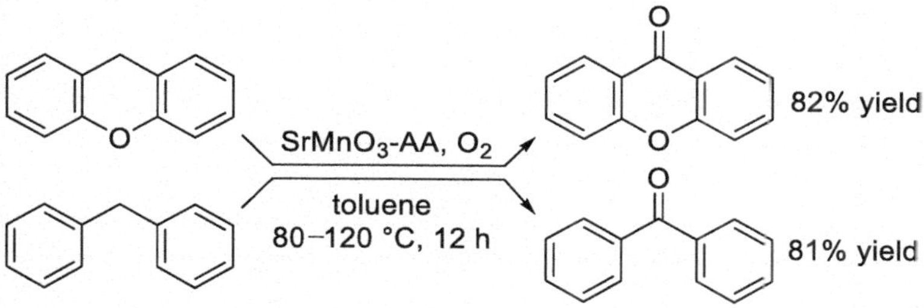

(8.27)

According to this method using aspartic or malic acids [87], various types of highly pure hexagonal perovskite oxides with relatively high surface areas could be synthesized and applied to more difficult oxidation reactions such as the aerobic oxidation of sulfides and alkanes [88]. Reports on the aerobic oxidation of inactive aryl sulfides without the need for additives are still limited, in contrast to the H_2O_2-based selective oxidation of sulfides using organic hydroperoxides and H_2O_2. Rhombohedral $BaRuO_3$ nanoparticles could function as an efficient, recyclable, and heterogeneous catalyst for the aerobic oxidation of various aromatic and aliphatic sulfides to the corresponding sulfoxides and sulfones, including application to the large-scale oxidation of 4-methoxy thioanisole (Eq. 8.28). The catalytic activity of $BaRuO_3$ was larger than those of other catalysts such as $ARuO_3$ (A = Ca and Sr), RuO_2, $Ru(OAc)_x$, $Ru(acac)_3$, $SrMnO_3$, activated MnO_2, and Mo-doped α-MnO_2 under mild reaction conditions. Mechanistic studies, such as examination of Hammett plots, the atmosphere effect, ^{18}O-labeling experiments, and kinetics have suggested that substrate oxidation likely proceeds with an oxygen species originating from the solid via the Mars–van Krevelen mechanism. The intrinsic reactivity per surface area of ruthenium-based oxide was dependent on the crystal structure and decreased in the order of $BaRuO_3 > SrRuO_3 > CaRuO_3 > RuO_2$. To investigate the structure–activity relationship, the energy for oxygen vacancy formation was determined by DFT calculations because the values can be a good descriptor of the oxidizing power of an oxide. Not only the vacancy formation energy of the face-sharing octahedra in $BaRuO_3$, but also the calculated energy change for the reduction of $BaRuO_3$ into metallic Ru with H_2 ($BaRuO_3 + 2H_2 \rightarrow BaO + Ru + 2H_2O$) were the lowest compared to $BaRuO_3$, $SrRuO_3$, $CaRuO_3$, and RuO with corner-sharing octahedra (Fig. 8.13), which suggests that oxygen transfer from $BaRuO_3$ to a sulfide and reoxidation of the partially reduced $BaRuO_{3-x}$ species with O_2 proceeds much more smoothly than with the other Ru-based oxides. In the present oxidation reaction, μ_2-face-sharing oxygen atoms and/or related adsorbed oxygen species would be possible active sites (Fig. 8.14).

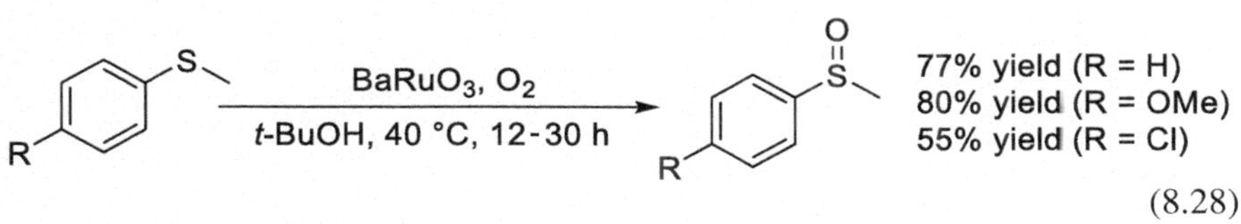

(8.28)

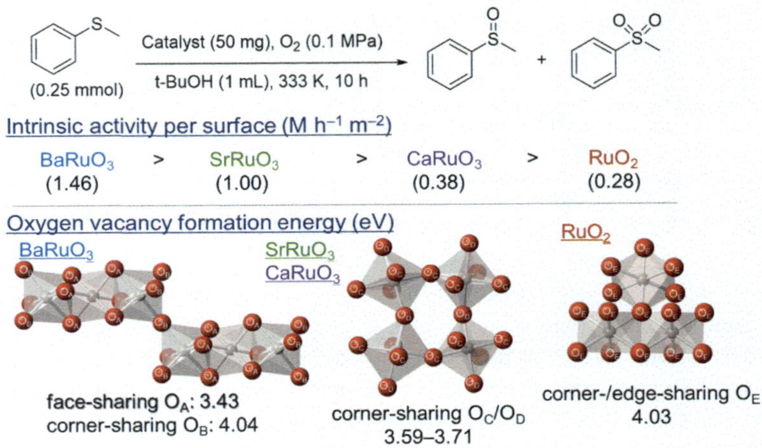

Fig. 8.13 Correlation between intrinsic activity per surface and calculated energy for the formation of an oxygen vacancy for $BaRuO_3$, $SrRuO_3/CaRuO_3$, and RuO_2 [88]

High valency iron-oxo species have been reported to be efficient active oxidants for various hydrocarbon oxidation reactions under mild conditions, especially in the fields of bioinorganic catalysis; therefore, we focused on perovskite oxides containing high valency iron species as potentially efficient heterogeneous oxidation catalysts. Hexagonal 6H-$BaFeO_{3-\delta}$ (δ = ca. 0.1) could act as an efficient and reusable solid catalyst for the oxidation of various hydrocarbon substrates, including alkanes and alkylarenes, using O_2 at atmospheric pressure as the sole oxidant (Eq. 8.29) [89]. For the aerobic oxidation of adamantane as a model reaction, $BaFeO_{3-\delta}$ exhibited much higher activity than typical Fe^{3+}/Fe^{2+}-containing iron-oxide-based catalysts ($CaFeO_{2.5}$, $LaFeO_3$, $BaFe_2O_4$, FeO, Fe_3O_4, and Fe_2O_3), Ba- and Mn-based perovskites ($BaBO_3$ (B = Mn, Co, and Ru), and $AMnO_3$ (A

Fig. 8.14 Proposed reaction mechanism for aerobic oxidation of sulfides catalyzed by $BaRuO_3$ [85]

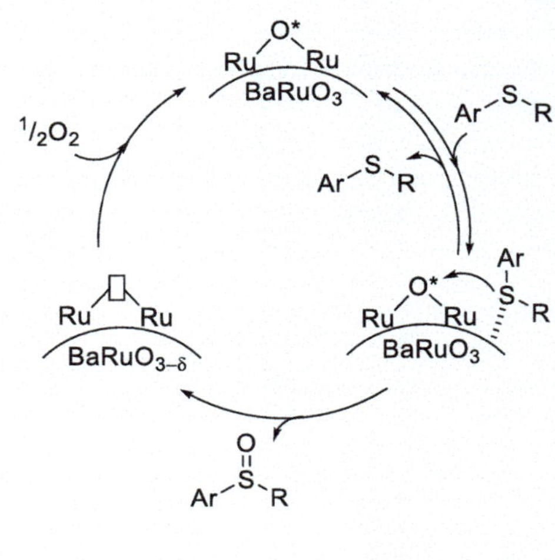

= Ca and Sr)). Although some efficient catalytic systems for the aerobic oxidation of adamantane have been reported, most of these systems are homogeneous and typically require additives or photo- or microwave irradiation. From the perspective of recoverable and reusable heterogeneous catalysts with O_2 as the sole oxidant, reported examples have been limited to Ru- and V-based catalysts [124, 125]; therefore, this $BaFeO_{3-\delta}$-based system was the first example of a naturally abundant iron oxide-based heterogeneous catalyst for the aerobic oxidation of adamantane without the need for additives. $BaFeO_{3-\delta}$-based oxidation has been proposed to proceed through a radical-mediated oxidation mechanism, and H-abstraction by $BaFeO_{3-\delta}$ is the rate-determining step according to the following results (Fig. 8.15); (i) an induction period was observed, (ii) oxidation was completely suppressed by the addition of a radical scavenger (2,6-di-*tert*-butyl-4-methylphenol), (iii) a kinetic isotope effect ($k_H/k_D = 5.0$) between fluorene and fluorene-d_{10} was observed, (iv) the selectivity ratio (29) of tertiary/secondary (3°/2°) C–H activation of adamantane C–H bonds was comparable to those of metal-catalyzed radical-mediated oxidation catalysts with O_2, (v) the formation and the substantial inversion of a 9-decalyl radical intermediate was proposed from a stereoisomeric mixture of 9-decalol from *cis*-decalin, (vi) high ^{18}O contents (96–97%) in products were observed for the oxidation of adamantane with $^{18}O_2$. A good correlation between the reaction rate and C–H bond dissociation energy (BDE) supported this reaction mechanism.

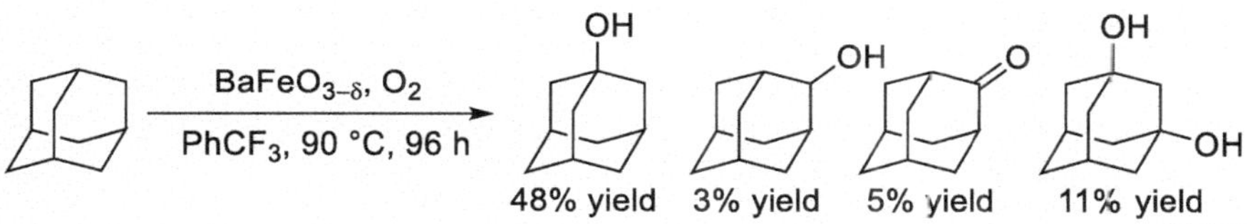

(8.29)

This high valency iron containing perovskite system could also be applied to the aerobic oxidative cleavage of various aromatic alkenes, such as styrenes with electron-donating and -withdrawing *p*-substituents, and o-disubstituted styrenes to the corresponding carbonyl compounds (Eq. 8.30) [90]. The catalyst effect for the oxidative C = C bond cleavage of styrene was similar to that for the aerobic oxidation

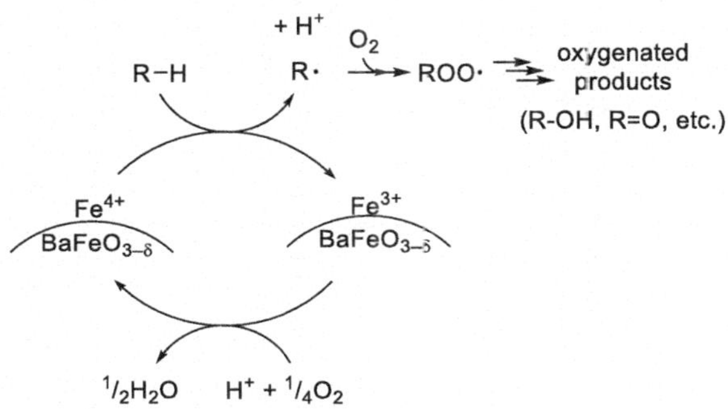

Fig. 8.15 Proposed reaction mechanism for aerobic oxidation of alkanes catalyzed by $BaFeO_{3-\delta}$ [89]

of adamantane, and $BaFeO_{3-\delta}$ was the most active catalyst. The selectivity toward benzaldehyde increased with the O_2 pressure, while the selectivity toward styrene oxide as a byproduct decreased. Mechanistic studies, including the effect of radical scavenger/initiator, the atmosphere, and hot filtration experiments, indicated that the oxidative cleavage reaction of alkenes likely involved a radical mechanism in a similar manner to alkane oxidation (Fig. 8.15), and radical species have been proposed to react with O_2 to form peroxy intermediates followed by rearrangement to benzaldehyde at a high O_2 pressure. In this system, the gram-scale oxidation of 1,1-diphenylethylene was achieved to isolate 2.71 g (75% yield) of the analytically pure ketone.

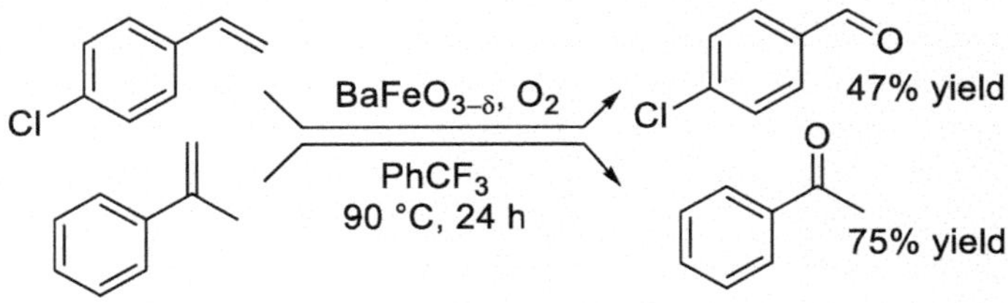

(8.30)

8.3.3 *Spinel*

Spinel has a general formula of AB_2O_4, and crystallizes in a cubic (isometric) crystal system. Almost all main group metals and transition metals have been observed in spinels, and A (at the center of the tetrahedrally coordinated position) and B (at the center of the octahedrally coordinated position) are typically divalent and trivalent metal cations, respectively, including Mn, Zn or Fe, Mg, Al, Cr, Ti, and Si. Such versatile chemical compositions of spinels can control their structures, electron configurations, and valence states, which lead to magnetic, optical, electrical, and catalytic applications [126]. In particular, various catalytic reactions, such as gas-phase reactions (NO_x and CO_2 reduction, CO oxidation, methane combustion) and electrocatalytic reactions (hydrogen evolution reaction, oxygen reduction reaction, and oxygen evolution reaction), have been developed using the advantages of spinels and related compounds [126–128]. In this chapter, liquid-phase organic reactions using spinel catalysts are examined.

The most extensively studied spinels are the magnetically recyclable catalysts because inverse spinel Fe_3O_4 and spinel ferrites MFe_2O_4 can be simply and efficiently separated from reaction mixtures with an external magnetic field, in contrast to catalyst recovery through filtration or centrifugation methods [129]. These magnetically recoverable catalysts of combined catalyst of copper isophthalate-based MOF and $CoFe_2O_4$ [130], $FeAl_2O_4$ [131], $MnFe_2O_4$ [132], and $CoFe_2O_4$ [133] efficiently catalyze 2-substituted benzimidazole synthesis through oxidative cross-coupling among substituted anilines, amines, and sodium azide with TBHP, benzo[*a*]pyrano[2,3-c]phenazine and polyhydroquinoline derivative through a multi-component reaction, nitrorarene reduction with $NaBH_4$, and Knoevenagel condensation reaction between aldehydes and ethyl cyanoacetate, respectively (Eq. 8.31). For example, a general mechanism involving the activation of carbonyl compounds by Lewis acid sites on $FeAl_2O_4$ has been proposed; however, detailed mechanistic studies on the active sites and activation mode of the substrate have not been conducted, because the main purpose of those studies was to achieve efficient organic reactions.

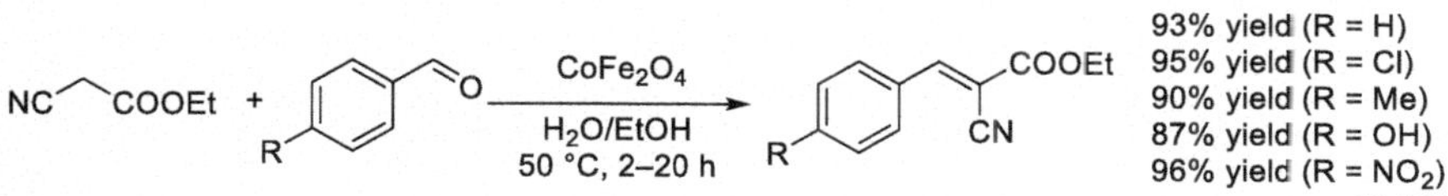

(8.31)

Another focus of attention in spinel catalyst research is the role of oxygen-vacancy (defect) sites. A Cu–Mn oxide is one of the well-studied spinels, and active sites (e.g., surface-adsorbed oxygen, oxygen vacancies, and radical species) can be formed over the Cu–Mn surface because a redox cycle between Cu and Mn ions can easily occur. Based on this concept, Cu–Mn spinel catalytic systems such as the aerobic oxidation of HMF [134, 135], amide formation via aminyl radical cations [136], ligand-free Huisgen [3 + 2] cycloaddition [137], and the regioselective halogenation of phenols and *N*-heteroarenes [138], have been strategically developed. $Cu_{1.4}Mn_{1.6}O_4$ spinel nanofiber was synthesized by electrospinning and tailored thermal treatment [136]. XPS measurements indicated the presence of Cu^+ sites and oxygen species on the surface of the spinel nanofiber, and these species could be tuned by thermal treatment. Among the synthesized catalysts, the spinel treated at 500 °C with the highest content of surface oxygen species showed the highest catalytic activity for the oxidation of HMF to form FDCA (96% yield) at 120 °C for 24 h in alkali-bearing aqueous solution under 0.3 MPa of O_2. Note that, the FDCA yield over the spinel catalyst was higher than that over monometallic oxides (e.g. CuO and Mn_2O_3) and their physical mixtures. These results indicated that the oxygen vacancies in the spinels had an important role, and synergistic effects of the Cn–Mn couple may enhance the oxidation of HMF to produce FDCA (Fig. 8.16).

Shah and co-workers developed a Cu–Mn spinel catalyst (Cu/Mn = 3:0.25) for the aminolysis of ethyl benzoate by pyrrolidine to form the desired product (93% yield) at 80 °C for 6 h in tetrahydrofuran (THF) under an air atmosphere (Eq. 8.32) [136]. The catalyst could be used for a wide range of esters and amines to produce the corresponding products. To reveal the reaction mechanism, the authors investigated the reaction of ethyl benzoate with pyrrolidine under Ar and with the addition of 2,2,6,6-Tetramethylpiperidine 1-oxyl (TEMPO) used as a radical inhibiter. The reactions did not completely proceed under either condition, which suggested that the Cu–Mn spinel catalyzes the aminolysis of esters through a radical pathway. This spinel catalyst also showed activity for the dehydrogenative homo- and cross-coupling of anilines (Eq. 8.33). The authors concluded that the high activity of the

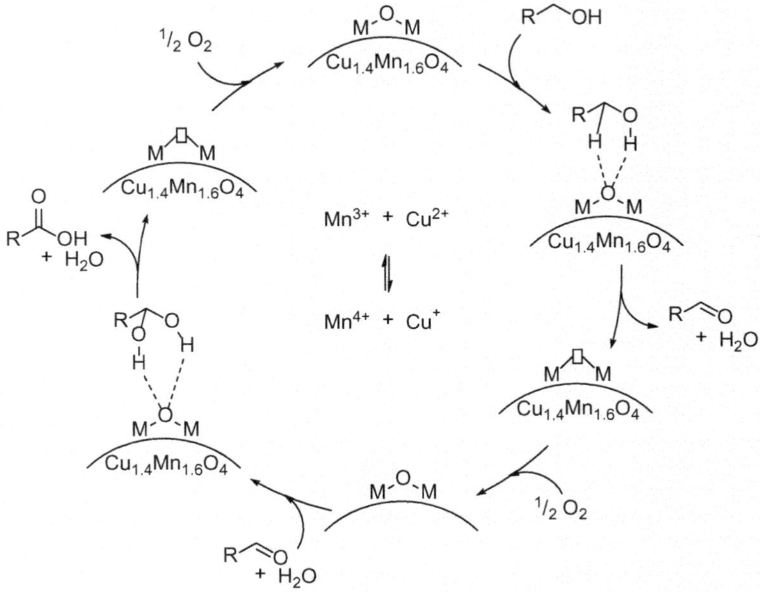

Fig. 8.16 Proposed reaction mechanism for the oxidation of alcohols to carboxylic acid over $Cu_{1.4}Mn_{1.6}O_4$ spinel oxide [136]

Cu–Mn spinel for these radical reactions was caused by the switchability of the oxidation states between $Cu^{+/2+}$ and $Mn^{3+/4+}$ (Fig. 8.17). However, the relationship between the catalyst structure and catalytic activity is unclear.

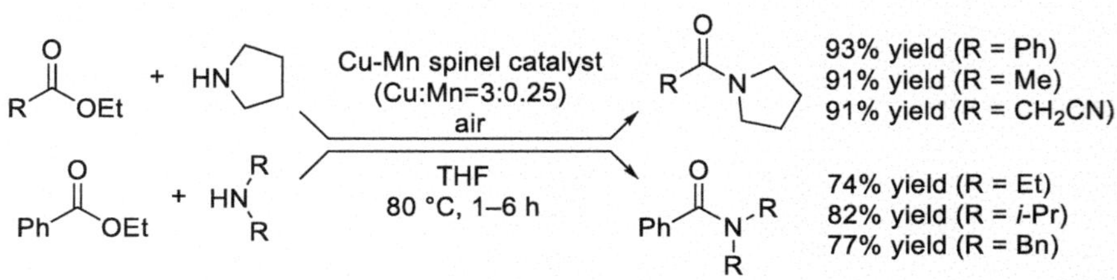

(8.32)

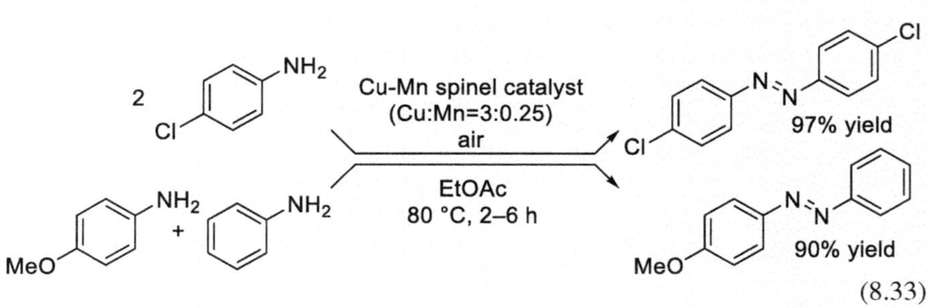

(8.33)

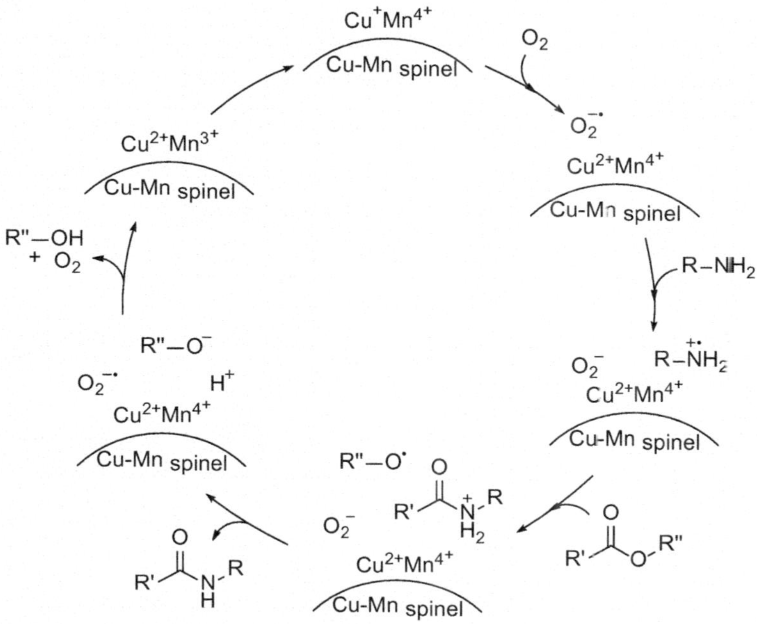

Fig. 8.17 Proposed reaction mechanism synthesis of amides via aminyl radical cations [136]

Defect sites in $NiCo_2O_4$ play an important role in controlling the selectivity for the hydrogenation of α,β-unsaturated carbonyl compounds. Pd nanoparticles supported on defect-born $NiCo_2O_4$ ($NiCo_2O_{4-x}$) prepared by the reduction of $NiCo_2O_4$ with $NaBH_4$ could act as a recyclable heterogeneous catalyst for the selective hydrogenation of cinnamaldehyde to hydrocinnamaldehyde under mild reaction conditions (30 °C, H_2 1 atm). The selectivity toward hydrocinnamaldehyde was maintained even with an increase in the reaction time and temperature. On the basis of DFT calculations, the adsorption of the C=C double bond of cinnamaldehyde on electron deficient Pd is more preferable to that of the C=O bond. In addition, the strong adsorption of the C=O bond of hydrocinnamaldehyde on the oxygen defects of $NiCo_2O_{4-x}$ is considered to suppress further hydrogenation of hydrocinnamaldehyde.

Acharyya and co-workers reported the catalytic activity of $CuCr_2O_4$ spinel nanoparticles for the oxidation of toluene to form benzaldehyde (Eq. 8.34) [139]. $CuCr_2O_4$ nanoparticles were synthesized by the hydrothermal method using cetyltrimethylammonium bromide (CTAB) as a surfactant. The specific surface area of $CuCr_2O_4$ was estimated to be 92 $m^2\ g^{-1}$ with particle sizes of 20–40 nm. Although commercially obtained CuO, Cu_2O, Cr_2O_3, and $CuCr_2O_4$ showed almost no activity for toluene oxidation, the synthesized $CuCr_2O_4$ nanoparticles promoted benzaldehyde formation with a high yield (49%) at 75 °C for 10 h in acetonitrile with H_2O_2 used as an oxidant. Cu clusters supported on a Cr_2O_3 catalyst also showed some product yield (15%), which indicated the importance of the small size of Cu^{2+} and the spinel phase for efficient catalytic oxidation. The same group also reported catalysis over the $CuCr_2O_4$ spinel for liquid phase oxidation of benzenes with electron donor groups to the corresponding phenols and ketones [140], anilines to the corresponding azoxybenzenes [141], the oxyamination of benzene to aniline [142], and ethylbenzene to acetophenone (Eqs. 8.35–8.38) [143].

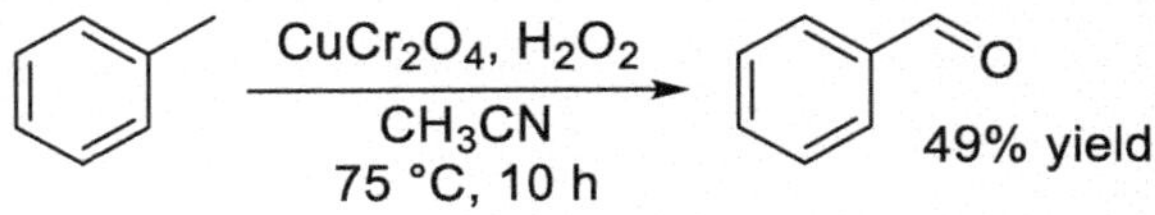

(8.34)

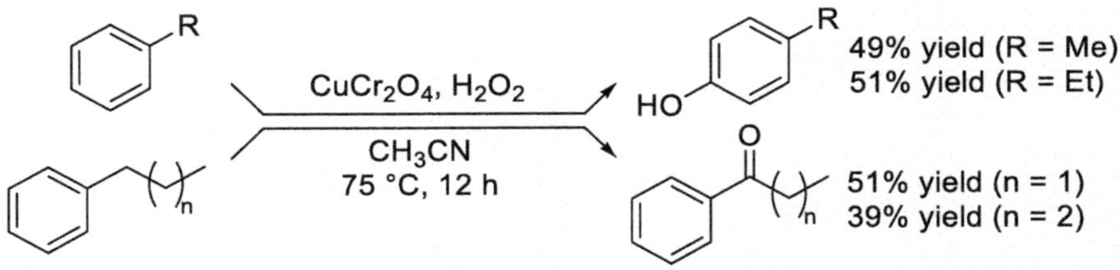

(8.35)

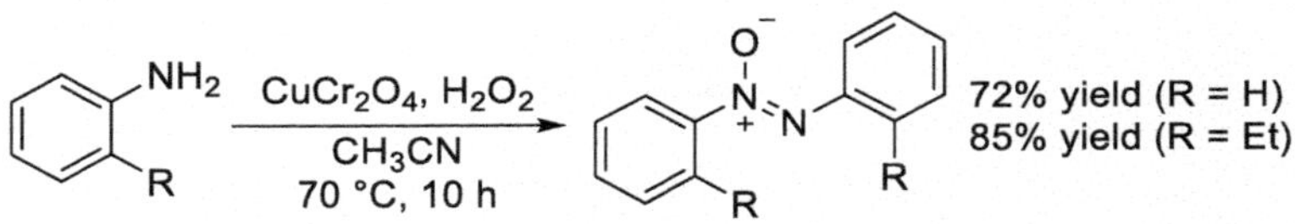

(8.36)

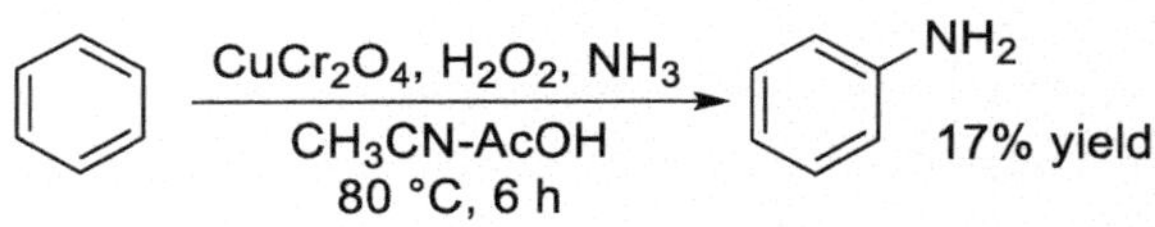

(8.37)

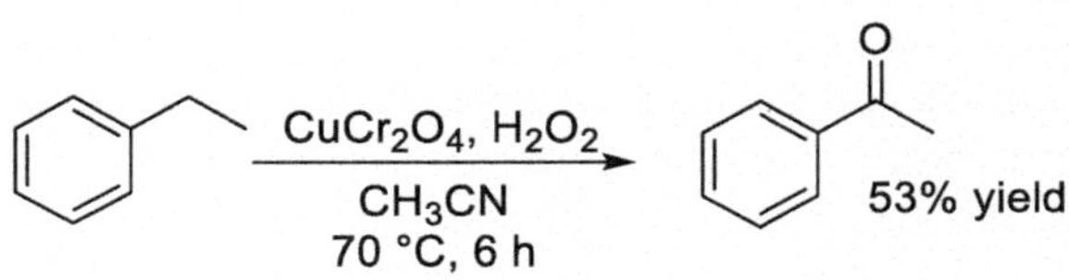

(8.38)

A spinel oxide also showed catalytic activity for the hydrogenation of aldehydes. In this case, metal nanoparticles of the constituent elements of the spinel oxide are typically formed as catalytically active sites under reductive conditions, in a similar manner to that of some perovskite catalysts. Köhler and co-workers reported catalytic activity for the dehydration of butyraldehyde to form butanol over a $CuAl_2O_4$ spinel oxide that was prepared by a co-precipitation method [144]. Detailed characterization, including in situ X-ray absorption spectroscopy (XAS) revealed that the Cu species in the spinel structure was reduced in the presence of H_2 to migrate and form Cu metal nanoparticles on the catalyst surface. This reduced spinel lattice was stabilized by protons. The authors proposed that H_2 molecules were activated over Cu nanoparticles to form protons and electrons, followed by the hydrogenation of aldehydes to produce alcohols (Fig. 8.18).

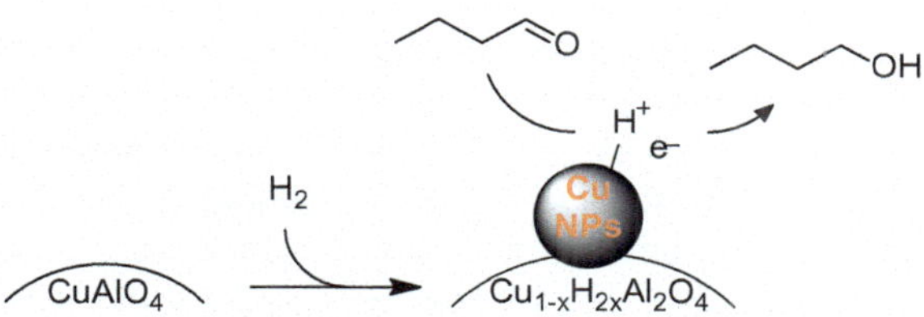

Fig. 8.18 Proposed reaction mechanism for the activation of spinel and hydrogenation of aldehydes over $CuAl_2O_4$ [144]

Apesteguía and co-workers also applied the formation of surface metal species by reduction of the spinel oxide to the hydrogenation of α,β-unsaturated carbonyl compounds [145]. The rate of cinnamaldehyde hydrogenation over Cu metal particles that were formed on the Cu–M–Al spinels (M = Co, Ni, or Zn) during the reaction was faster than that over Cu/SiO_2. Moreover, the introduction of metal cations (e.g., Co, Ni, or Zn) in a spinel structure promoted the selective hydrogenation of carbonyl groups. The authors concluded that α,β-unsaturated carbonyls prefer to adsorb onto M^{2+} cations via carbonyl groups than via carbon–carbon double bonds, and are selectively hydrogenated to yield unsaturated alcohols by activated hydrogen species on the highly dispersed neighboring Cu nanoparticles (Fig. 8.19).

Some reports have focused on acid–base catalysis over spinel oxides. Fourier transform infrared (FT-IR) spectroscopy measurements of pyridine and CO_2 as probe molecules revealed the surface acidity and basicity of $Cu_{1-x}Co_xFe_2O_4$ spinel oxides (x = 0–1), respectively [146]. The amount of acid sites on the $Cu_{1-x}Co_xFe_2O_4$ spinels increased with the Co to Cu ratio. In contrast, the number of base sites on $Cu_{1-x}Co_xFe_2O_4$ decreased with an increase in the Co to Cu ratio. The author concluded that the surface state of Fe^{3+} can be tuned by the ratio of Cu^{2+} to Co^{2+}, whereby the acid–base properties of the spinel are changed. Among the synthesized $Cu_{1-x}Co_xFe_2O_4$ spinels, the $Cu_{0.5}Co_{0.5}Fe_2O_4$ catalyst exhibited the best catalytic activity for phenol methylation, which suggests that moderate the acid–base properties tuned by the Cu–Co synergistic effect enhance the catalytic activity.

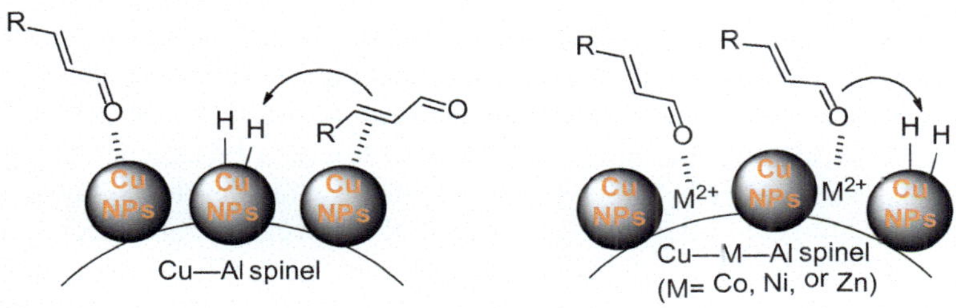

Fig. 8.19 Proposed adsorption models of α,β-unsaturated carbonyl compound over Cu–Al and Cu–M–Al spinel catalysts [145]

Lu and co-workers successfully developed a mesoporous Co–Al spinel oxide [147] by the thermal decomposition of a Co–Al hydrotalcite-like compound. This spinel had a pore volume of 0.25 cm^3 g^{-1} and its surface area was estimated to be 256 m^2 g^{-1}. The authors suggested that the Co ions in the precursor coordinated with adjacent oxygen atoms and acted as pillars between interlayers during the thermal decomposition process to produce the mesoporous spinel with a high surface area. The mesoporous spinel catalyzed the self-aldol condensation of acetone to form diacetone alcohol with subsequent dehydration to obtain mesityl oxide, which suggested the ability of the spinel to act as a solid-base catalyst.

8.3.4 Metal Phosphate

Metal phosphates are complex oxide materials that consist of various types of metals and phosphate units (PO_4^{3-}, $P_2O_7^{4-}$, etc.), and many crystal structures have been reported by the selection of versatile combinations of metals and metal/P molar ratios [148–150]. For example, Al- and B-containing phosphates are extensively studied as solid catalysts for various acid-catalyzed reactions such as the dehydration of alcohols and alkylation [149], and vanadium-containing phosphates are used as industrial oxidation catalysts for the oxidation of *n*-butane to maleic anhydride [150]. In addition, microporous and mesoporous metal phosphates (e.g., aluminophosphate molecular sieves (AlPO) and silicoaluminoposphate molecular sieves (SAPO)) can be modified by the introduction of various elements (Li, Co, Fe, Mg, Mn, Ze, Ni, Ga, Cr, Ti, V, etc.), which leads to their wide application to catalytic reactions such as acid–base, redox, and electrocatalytic reactions [151–154]. For example, microporous ZrAPO-5 and CoAPO-11 can promote the isomerization of *m*-xylene oxidation of cyclohexane, and the active sites are typically proposed to be an isolated monomeric species in the metal phosphate frameworks [151]. In this chapter, we will discuss and introduce liquid-phase organic synthesis reactions using some crystalline metal phosphates from the viewpoint of the structure–activity relationship.

Among the crystalline metal phosphate catalysts, apatites ($Ca_{10}(PO_4)_6X_2$, X = OH^- (hydroxyapatite (HAP)), F^- (fluoroapatite (FAP)), Cl^- (chloroapatite) are the most extensively studied [155–157]. As long as the charge balance of the apatite crystal structure is maintained, the calcium ion and the corresponding counter anion are exchangeable. In particular, HAP, the main component of bones and teeth, has received much attention in view of its potential use as a biomaterial, adsorbent, and ion-exchanger. Kaneda and co-workers systematically developed highly functionalized apatite-based heterogeneous catalysts through cation-, anion-, and ligand-exchange for various types of organic reactions (Fig. 8.20) [15]. A monomeric Ru phosphate complex (RuHAP) with Ru–O and Ru–Cl bonds was prepared on the HAP surface by the exchange of Ca^{2+} for Ru^{3+}, and was characterized by XRD, XPS, energy dispersive X-ray spectroscopy (EDX), and Ru K-edge XAS measurements [158]. RuHAP could act as a recyclable solid catalyst for the aerobic oxidation of alcohols to carbonyl compounds and amines to nitriles (Eq. 8.39) [158, 159]. In particular, RuHAP exhibited a wide substrate scope for the oxidation of alcohols, and various types of benzylic, allylic, and nitrogen- and sulfur-containing heterocyclic alcohols were converted into the corresponding aldehydes or ketones in excellent yields. According to the reaction mechanism for the RuHAP-catalyzed oxidation of alcohols, β-hydride elimination has been proposed to be the rate-limiting step. The oxidation of silanes to silanols was also promoted by RuHAP using water and atmospheric oxygen [160], and the presence of both water and molecular oxygen was essential to obtaining high yields of silanols (Eq. 8.40). The origin of oxygen atoms in silanols was not O_2, but water based, according to ^{18}O-labeling experiments using $H_2{}^{18}O$. The replacement of the Cl ligand on RuHAP with water by treatment with AgX (X = SbF_6^- and TfO^-) led to the formation of a cationic ruthenium species on the HAP surface [161]. This cationic RuHAP could be applied to Diels–Alder and aqueous aldol-type reactions of carbonyl compounds with nitriles to α,β-unsaturated nitriles (Eq. 8.41) through the activation of substrates on the Lewis acidic Ru species (and the basic phosphate ligand, in the case of aldol-type reactions).

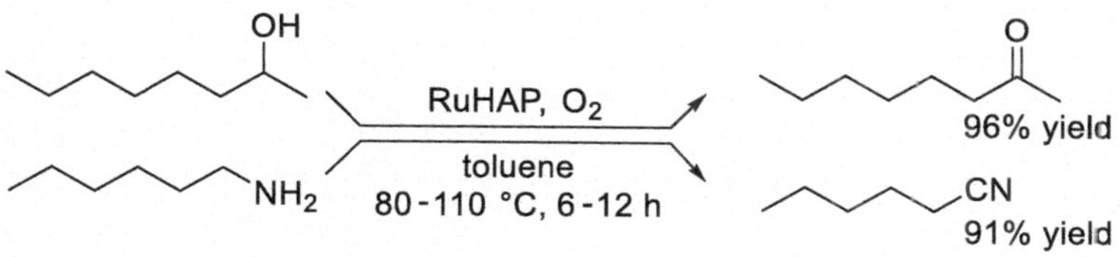

(8.39)

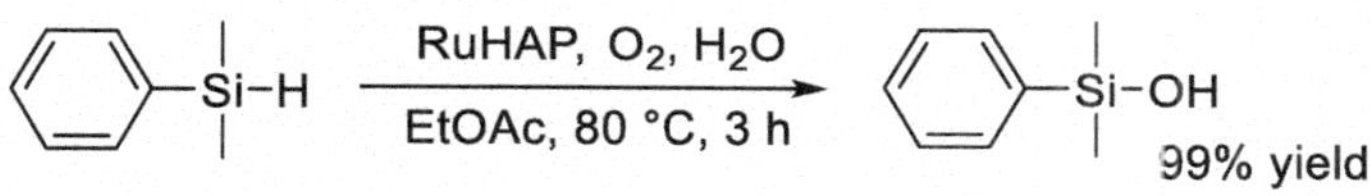

(8.40)

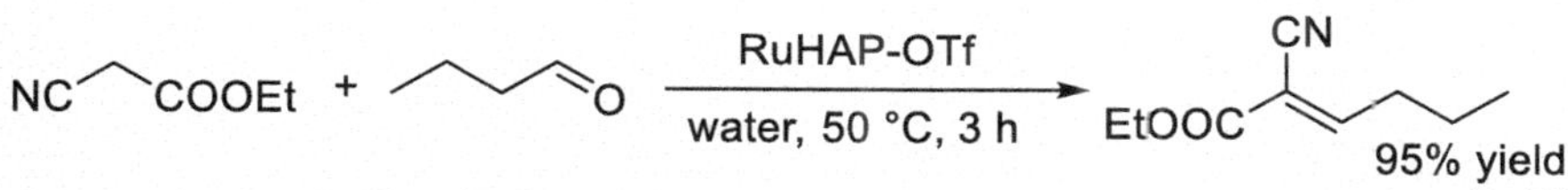

(8.41)

Zn- and La-exchanged HAP (ZnHAP and LaHAP, respectively) were also prepared by Kaneda and co-workes [162, 163]. ZnHAP and LaHAP could function as acid–base catalysts without further ligand exchange 3by post-treatment in the case of RuHAP. The active site structure on ZnHAP was confirmed to be a monomeric Zn^{2+} phosphate complex with tetrahedral coordination based on XPS, Zn K-edge X-ray absorption near edge structure (XANES), and Zn K-edge extended X-ray absorption fine structure (EXAFS) analyses. ZnHAP efficiently catalyzed the cycloaddition of CO_2 with various aliphatic and aromatic terminal epoxides in the presence of a Lewis basic cocatalyst such as 4-(dimethylamino)pyridine (DMAP) (Eq. 8.42). Based on the results for retention of the epoxide configuration in the corresponding carbonates obtained from enantiomerically pure epoxides, the cooperative action of a monomeric Lewis acidic Zn site and a Lewis base has been proposed to facilitate epoxide activation and epoxide ring-opening by attacking the less sterically hindered carbon atom. LaHAP with a monomeric La^{3+} species surrounded by hydroxyl, phosphate,

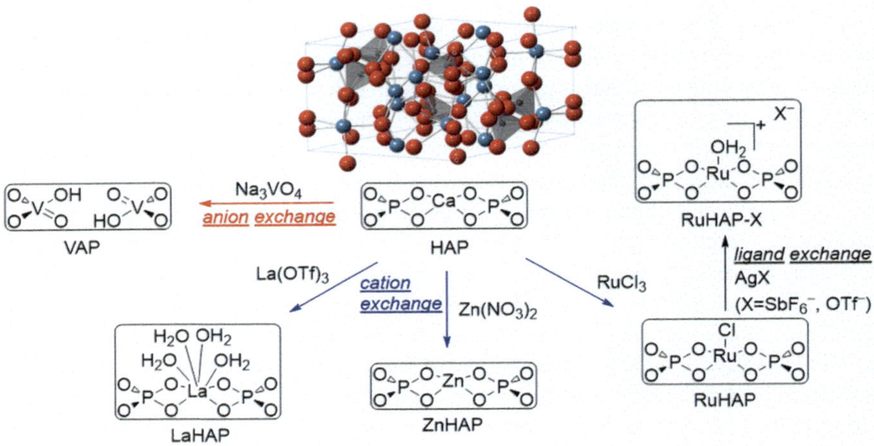

Fig. 8.20 Catalyst design of HAP-based functionalized heterogeneous catalysts [157]

and weakly coordinated aqua ligands could act as a reusable heterogeneous catalyst for the Michael reaction between 1,3-dicarbonyls and enones (Eq. 8.43). Additional coordination of chiral ligands to La species with vacant coordination sites led to the formation of an enantioselective sphere on the solid surface. Tartaric acid (TA) modified FAP exchanged with La^{3+}, TA-LaFAP showed high catalytic activity for the reaction of methyl 1-oxoindan-2-carboxylate with methyl vinyl ketone in toluene to give the Michael adduct with an *S* configuration in 97% yield and 60% ee.

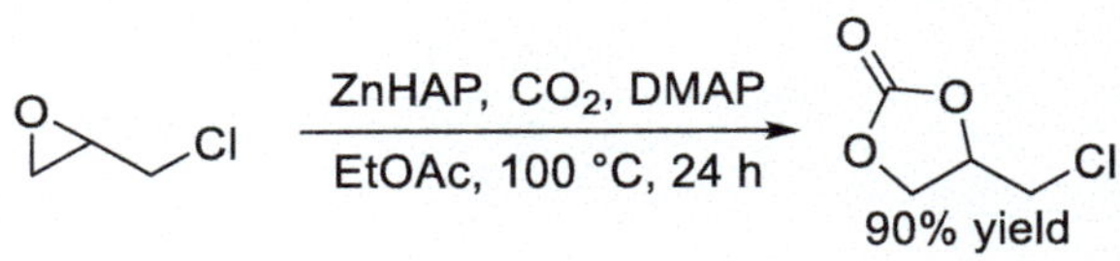

(8.42)

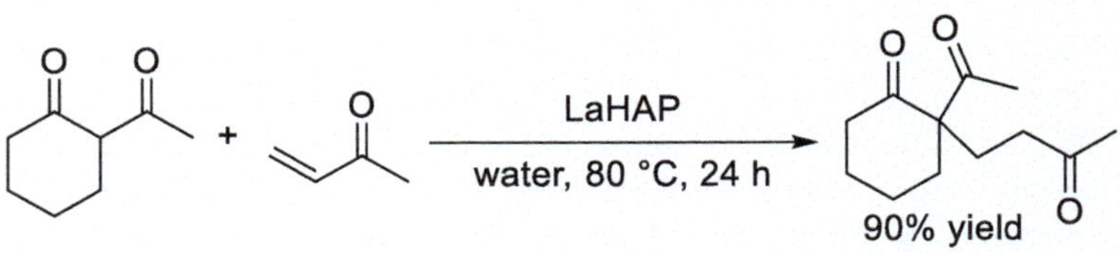

(8.43)

High functionalization of HAP was achieved by anion exchange of PO_4^{3-} with VO_4^{3-}. Calcium vanadate apatite (VAP; $Ca_{10}(VO_4)_6(OH)_2$) exhibited high catalytic activity for C–C bond formation reactions such as the Michael and Knoevenagel reactions, and the one-pot syntheses of 1,3-dinitro compounds or pyrazolinones in water [164]. For example, a 200-mmol-scale Michael reaction of 2-oxo-cyclopentane carboxylic acid ethyl ester with methyl vinyl ketone at 40 °C in water rapidly proceeded to give the corresponding Michael adduct (41.6 g) in 92% isolated yield under triphasic conditions (Eq. 8.44). In this case, the turnover number reached up to 260,400 with an excellent turnover frequency of 48 s^{-1} based on the surface vanadium content. The three-phase reaction mixture allowed for a straightforward workup procedure, and the product was easily separated by extraction after the removal of the solid catalyst by filtration.

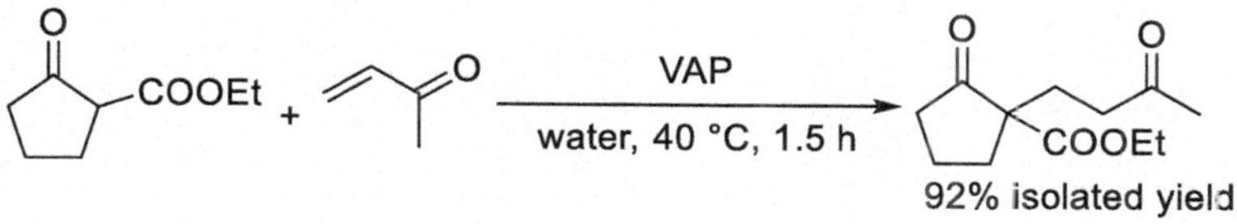

(8.44)

We found that a monomeric phosphate $[PO_4]^{3-}$ with a high pK_a value could act as an efficient homogeneous catalyst for the selective N-alkylation of indoles with α,β-unsaturated compounds (Eq. 8.45) [165, 166]. The catalytic activity of $[(n\text{-}C_4H_9)_4N]_3[PO_4]$ was much higher than those of other polyoxometalates ($[MO_4]^{2-}$ (M = W and Mo) [167, 168], $[\gamma\text{-}H_2GeW_{10}O_{36}]^{6-}$ [169–171], $[H_2\{Y(H_2O)_2\}_2\{\gamma\text{-}SiW_{10}O_{36}\}_2]^{8-}$) [172], and strong inorganic and organic bases. The formation of the indolyl anion through the activation of indole by $[PO_4]^{3-}$, which plays an important role in the N-alkylation, has been proposed based on NMR and IR spectroscopy results. Against such a background, we reported the synthesis of a new acid–base bifunctional solid catalyst by the combination of rare earth elements and $[PO_4]^{3-}$ as Lewis acid and base sites, respectively. A monoclinic $CePO_4$ catalyst synthesized by the hydrothermal method exhibited high catalytic performance for the chemoselective acetalization of HMF, a versatile carbonyl compound with sensitive functional groups derived from biomass resources, with alcohols, in sharp contrast to other homogeneous and heterogeneous acid and/or base catalysts [173]. In the presence of $CePO_4$, various combinations of carbonyl compounds and alcohols are efficiently converted into the corresponding acetal derivatives in good to excellent yields (Eq. 8.46). IR spectroscopy measurements for a sample with adsorbed probe molecules (pyridine, chloroform, acetone, and methanol) indicated the presence of uniform Lewis acid and weak base sites on $CePO_4$. The interaction of the uniform Lewis acid and weak base sites on $CePO_4$ with carbonyl compounds and alcohols, respectively, has been proposed to promote acetalization (Fig. 8.21). Aberration-corrected high-angle annular dark-field scanning transmission electron microscopy (HAADF-STEM) images and fast Fourier transform (FFT) patterns clarified the detailed surface structure of $CePO_4$ [174]. From the HAADF-STEM results, the dominant facet exposed on the $CePO_4$ nanorods was (110), and the amount of surface Ce Lewis acid sites at the (110) facet was estimated to be 1.6 nm^{-2}, which is in good agreement with the experimental value (1.6 nm^{-2}) calculated from the Brunauer–Emmett–Teller (BET) surface area of $CePO_4$ (37 $m^2\,g^{-1}$) and the surface Ce cations with Lewis acid sites measured using pyridine-IR (96 μmol g^{-1}).

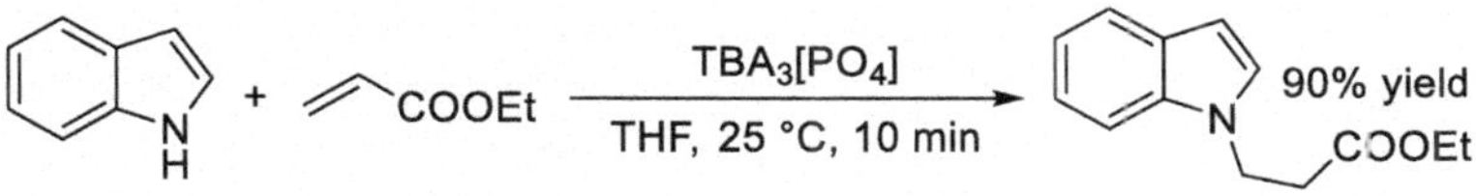

(8.45)

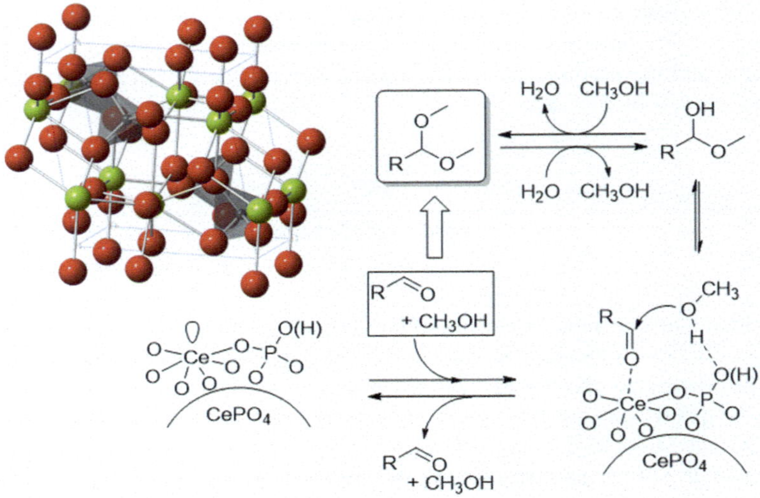

Fig. 8.21 Proposed reaction mechanism for the acetalization of carbonyl compounds with methanol over $CePO_4$ [173]

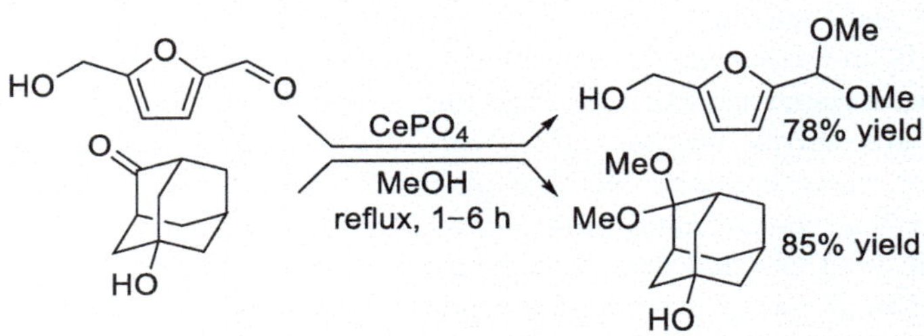

(8.46)

These specific acid–base properties of metal phosphates (i.e., uniform Lewis acid sites and weak base sites) made them effective not only as catalysts for oxidative methane conversion [174, 175], but also as supports for gold-supported catalysts [176]. The $CePO_4$ nanorod catalyst could function as a stable solid catalyst for the oxidative coupling of methane in an electric field without the need for external heating, to give the highest C_2 yield of 18% [174]. In addition, $FePO_4$ nanoparticles with redox-active Lewis acidic Fe^{3+} sites and weak base sites were also effective for the direct oxidation of methane into formaldehyde with O_2 as the sole oxidant [175]. Deposition–precipitation with aqueous ammonia enabled small gold nanoparticles to be deposited onto a series of metal phosphates with high dispersity and density, and $Au/GaPO_4$ exhibited higher catalytic activity for the hydroamination of alkynes with amines than other catalysts, including Au/ZrP_2O_7 and Au/TiO_2 [176].

8.3.5 *Others*

In the previous sections, we have comprehensively summarized some examples of the catalytic reactions and reaction mechanisms for crystalline complex oxides such as perovskites, spinels, and metal phosphates, with a focus on the correlation between structure and reactivity. There are many crystal structures and catalytic reactions of complex oxide materials apart from perovskites, spinels, and metal phosphates; therefore, it is not possible to cover all their catalytic applications. In this chapter, we will introduce several examples of organic synthesis reactions using layered compounds and crystalline complex oxides with garnet, pyrochlore, and murdochite-type structures.

Single- or multi-layer transition metal oxides have been extensively studied as functional materials in the fields of electronics, superconductivity, energy storage, catalysis, and sensors because of the different physical and chemical properties of two-dimensional transition metal oxides and intercalation ability of molecules and/or ions between layers [177]. In catalysis for liquid-phase organic synthesis, two representative layered oxides of hydrotalcite and montmorillonite are frequently investigated. Hydrotalcite (HT; typical example $Mg_6Al_2(OH)_{16}CO_3{\cdot}4H_2O$) is an anionic layered double hydroxide clay comprised of positively charged brucite-like layers with various inorganic and organic anions located between layers along with water to compensate for the positive charges [178, 179]. On the other hand, montmorillonites (mont; $(Na,Ca)_{0.33}(Al,Mg)_2(Si_4O_{10})(OH)_2{\cdot}nH_2O$) are layered clay minerals composed of alumina octahedral sheets sandwiched by two silica tetrahedral layers [17, 180, 181]. The partial replacement of aluminum by magnesium or iron atoms results in cation deficiency, so that cations and protons are present within the interlayer spaces. In this context, HT and mont are classified as anionic and cationic clays, respectively.

Based on features such as the cation-exchange of brucite-like layers, anion-exchange of interlayers, and a memory effect, high-performance HT-based solid catalysts have been developed for liquid phase organic reactions such as oxidation (epoxidation, alcohol oxidation, Bayer–Villiger oxidation, etc.), C–C bond formation (cyanosilylation, alkylation, aldol-type reaction, Michael reaction, etc.), deoxygenation, and hydrogenation [178, 179]. One of the most interesting features of HT is the memory effect. The thermal decomposition of HT into a mixed oxide proceeds at around 450–500 °C with subsequent rehydration in an inert atmosphere to regenerate the layered HT structure [182]. The reconstructed HT (re-HT) containing OH^- anions can function as an alternative to conventional, homogeneous strong bases. Kaneda and co-workers reported that the reconstructed HT with an Mg/Al ratio of 3 promoted the cross-aldol reaction with aliphatic aldehydes using excess ketones to avoid self-aldolization of the aldehyde (Eq. 8.47).

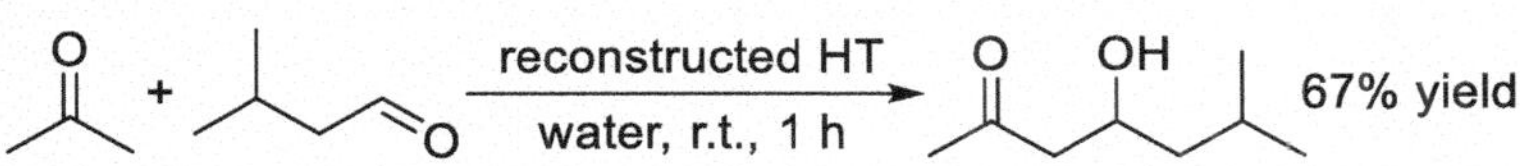

(8.47)

Kaneda and co-workers developed various types of acid-catalyzed organic reactions using cation-exchanged mont [180]. In the case of Ti^{4+} introduction within the interlayer, a two-dimensional titanium oxide species has been proposed as a Brønsted acid site [183]. On the other hand, monomeric metal aqua complexes are formed between the layers in the case of Cu^{2+} and Sc^{3+} [184, 185]. The Brønsted acid properties of Ti^{4+}-mont could be applied to the synthesis of 9,9-bis[4-(2-hydroxyethoxy)phenyl]fluorene, which is a valuable starting material for highly functionalized polymers used in optical products, by the aromatic alkylation of phenoxyethanol with fluoren-9-one (Eq. 8.48). In addition, Cu^{2+}- and Sc^{3+}-monts could be efficient solid Lewis acid catalysts for a variety of carbon–carbon bond-forming reactions, such as the Michael reaction, Sakurai–Hosomi allylation, and the Diels–Alder reaction, even under solvent-free and aqueous conditions (Eqs. 8.49 and 8.50).

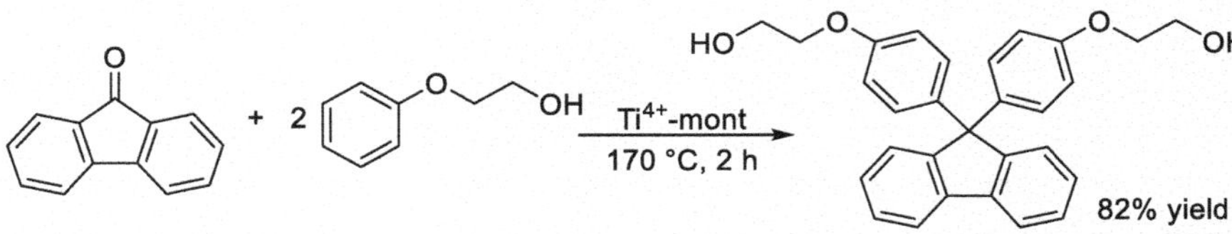

(8.48)

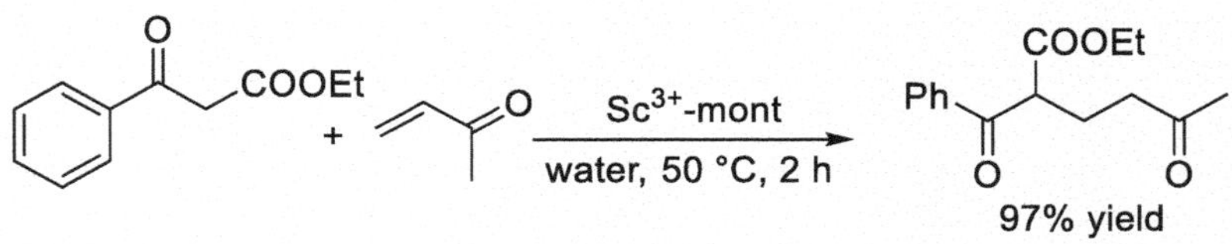

(8.49)

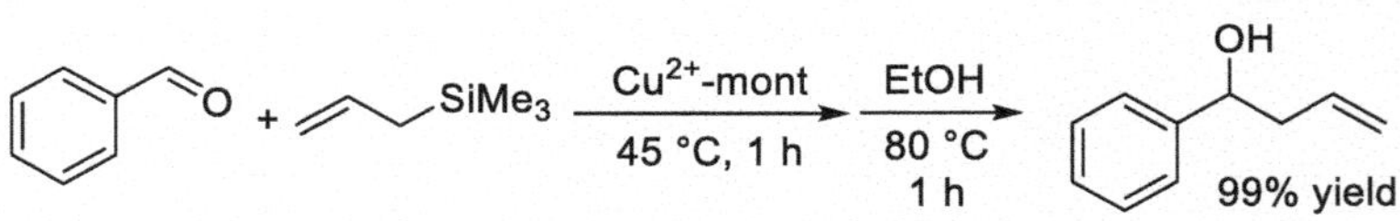

(8.50)

Rodríguez-Aguado and co-workers synthesized two series of yttrium iron garnets (e.g., $Y_3(Fe_{1-x}M_x)_5O_{12}$, M = Zn or Ni) by the conventional sol–gel method, and applied these catalysts to the oxidation of ethylbenzene to form acetophenone at 323 K for 5 h in acetone and in the presence of H_2O_2 as a promoter (Eq. 8.51) [186]. The $Y_3Fe_{4.97}Ni_{0.03}O_{12-\gamma}$ catalyst showed the highest activity for the oxidation of ethylbenzene among the tested catalysts. Electron paramagnetic resonance (EPR) and XPS measurements revealed that the electron transfer in the Fe^{2+}/Fe^{3+} redox couple enhanced the catalytic activity by surface reaction and the ease of access of reactants to the active sites. Yttrium iron garnet ($Y_3Fe_5O_{12}$) nanoparticles prepared by the sol–gel method catalyzed the intermolecular dehydration of alcohols to afford the corresponding ethers under solvent-free conditions [187]. This $Y_3Fe_5O_{12}$ catalyst also showed high catalytic performance for the one-pot synthesis of pyrano[2,3-*c*]pyrazole derivatives, which are used as anticancer, anti-HIV, anti-inflammatory, analgesic, and antiviral drugs, and could be reused without significant loss of product yield (Eq. 8.52).

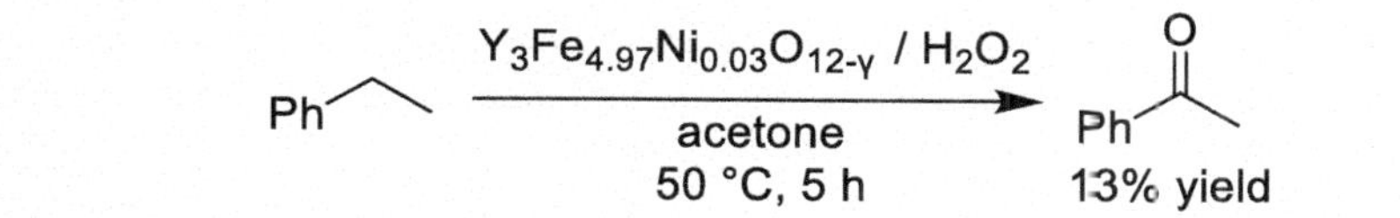

(8.51)

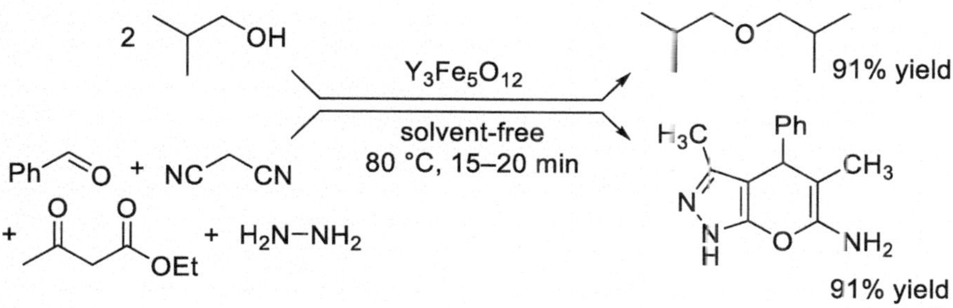

(8.52)

Proton exchanged pyrochlore oxides, $(H_3O^+)_xSb_xTe_{2-x}O_3$ (x = 1, 1.1, and 1.25) showed catalytic activity for fructose dehydration to form HMF in H_2O/methyl isobutyl ketone [188]. NH_3-TPD measurements indicated that $(H_3O^+)_{1.25}Sb_{1.25}Te_{0.75}O_3$ had the strongest and largest amount of acid sites of the prepared pyrochlores. The authors concluded that the strongly covalent subnetwork constructed by octahedral $Sb^{5+}O_6$ and $Te^{6+}O_6$ units resulted in a strong acid character for this catalyst and the formation of HMF with a high yield (Eq. 8.53). Felthouse and co-workers reported the catalytic activity of ruthenium pyrochlore oxides for the oxidative cleavage of vicinal diols with molecular oxygen at low temperature (Eq. 8.54) [189, 190]. A cubic-$Pb_{2.62}Ru_{1.38}O_{6.5}$ pyrochlore was synthesized by the precipitation method with sufficient O_2 bubbling to stabilize Pb^{4+}, which partially occupied the octahedral sites. The surface area of this pyrochlore was estimated to be 60 $m^2\ g^{-1}$, although there was no investigation of the correlation between catalytic activity and the structure.

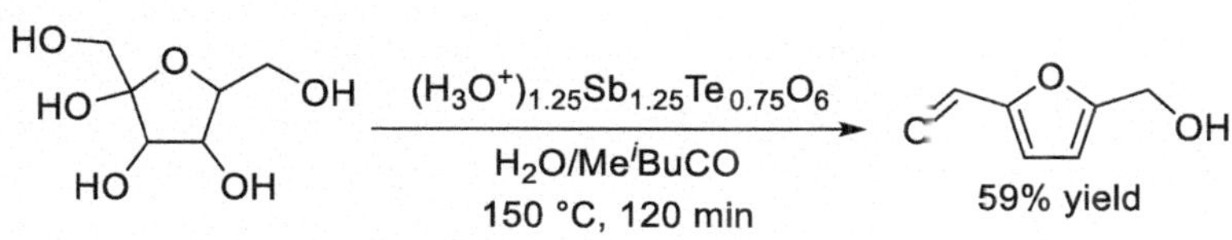

(8.53)

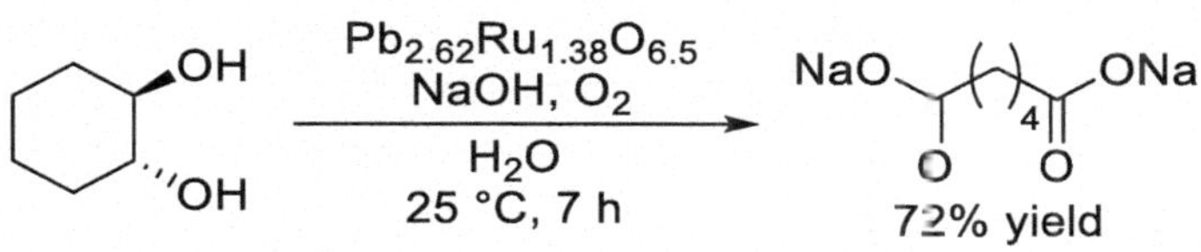

(8.54)

Structural and electronic control of the active sites is an important factor related to the intrinsic activity of catalysts. In addition, the catalytic performance in both homogeneous and heterogeneous oxidation systems is also affected by the acid–base properties, and the basicity of transition metal oxo complexes and supported metal catalysts sometimes promotes oxidative C–H activation under mild reaction conditions. We have recently focused on isolated high-valency Mn species located in a crystalline basic matrix as a recyclable heterogeneous catalyst for aerobic oxidative C–H functionalization of various types of alkylarenes with O_2 as the sole oxidant (Eq. 8.55) [191]. This approach to catalyst design is distinct from the incorporation of isolated Mn species into acidic matrices (aluminophosphates, zeolites) and/or crystal structure control of manganese oxides with Mn–O–Mn structures such as β-MnO_2 and $SrMnO_3$. Nanosized murdochite-type oxide Mg_6MnO_8, which can be considered as the rock-salt structure of MgO with the replacement of one-eighth of the Mg^{2+} ions with Mn^{4+} ions and one-eighth with vacancies, was successfully synthesized by the sol–gel method using malic acid. The specific surface area (104 $m^2\ g^{-1}$) was the highest among Mg_6MnO_8 synthesized by previously reported methods (2–15 $m^2\ g^{-1}$). In addition, Mg_6MnO_8 nanoparticles exhibited higher catalytic performance than other Mn- and Mg-based oxides, including manganese oxides with Mn–O–Mn active sites such as β-MnO_2 and $SrMnO_3$ for the oxidation of fluorene with O_2, and the oxidation reaction proceeded even under mild conditions (40 °C). The order of oxidation activity (Mg_6MnO_8-MA > β-MnO_2 ~ $SrMnO_3$) was not so correlated with that of the reduction rate determined by TPR (β-MnO_2 > $SrMnO_3$ > Mg_6MnO_8-MA), but with that of basicity determined by $CHCl_3$-adsorbed IR and CO_2 TPD results. Furthermore, the correlation between the reactivity and pK_a of the substrates and kinetic isotope effects suggests a basicity-controlled mechanism of hydrogen atom transfer (Fig. 8.22). On the basis of ^{18}O-labeling experiments, kinetics, and mechanistic studies, the H abstraction of alkylarenes proceeds via a mechanism involving O_2 activation. The structure of Mg_6MnO_8 consists of isolated Mn^{4+} species located in a basic MgO matrix, which likely plays an important role in this oxidation reaction.

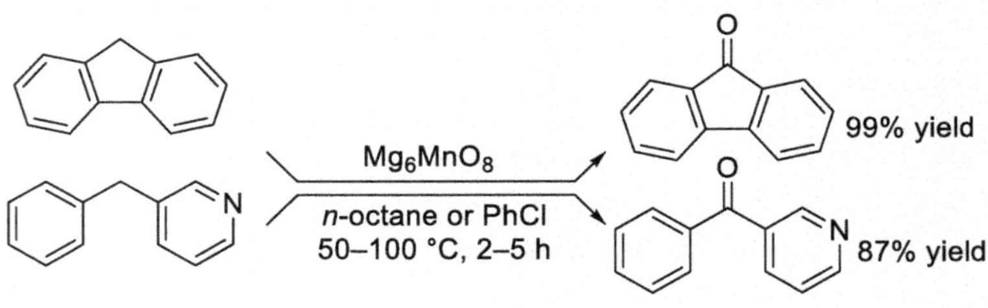

(8.55)

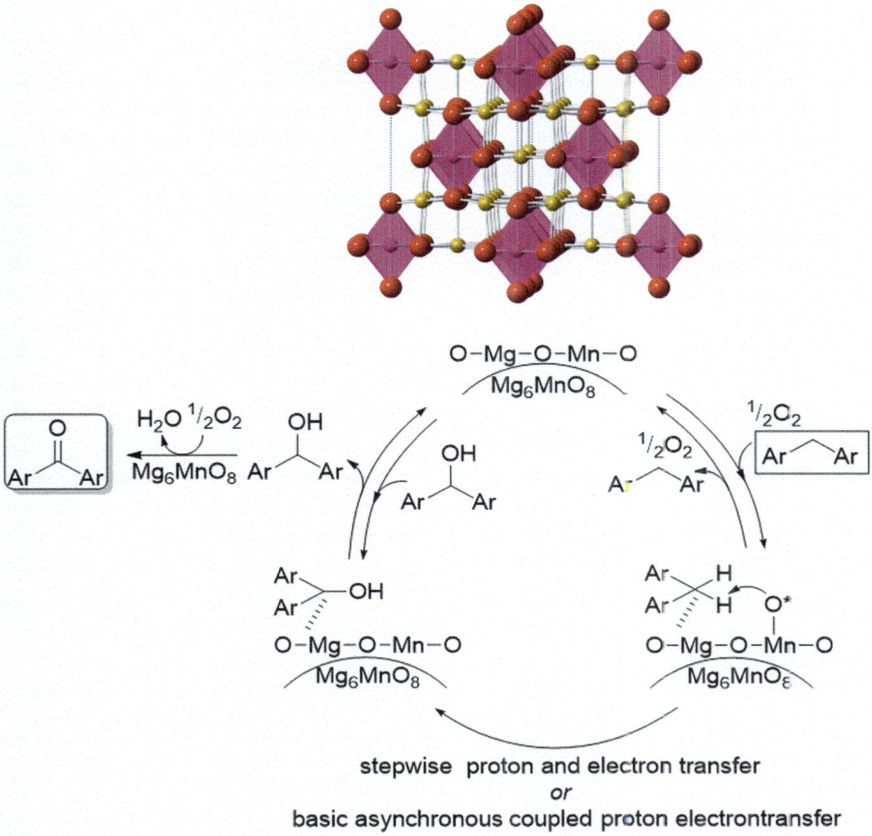

Fig. 8.22 Proposed reaction mechanism for the aerobic oxidation of alkylarenes over Mg_6MnO_8 [191]

8.4 Concluding Remarks

Taking advantage of metal oxides than can cover various types of reactions, structurally controlled complex metal oxides exhibit unique catalytic properties for liquid phase organic reactions. Various types of nanosized and porous crystalline metal oxides can be synthesized by several techniques, such as the co-precipitation, sol–gel, solution combustion, and soft/hard templating methods, which results in catalysts that achieve the atom-efficient synthesis of value-added chemicals. Durable and multifunctional simple oxides with both acid–base and redox properties are effective for one-pot sequential reactions and selective transformation of biomass-derived substrates. Multicomponent Pd- and Cu-containing perovskite oxides can efficiently catalyze C–C bond-forming coupling reactions (e.g., Suzuki, Sonogashira, and Ullmann-type cross-coupling reactions). Despite detailed mechanistic studies, the inclusion of metal nanoparticles in solution or metal nanoparticles formed on perovskite oxides as active species have been proposed, and further discussion will be continued. Some acid–base perovskite catalysts can efficiently promote aldol condensation, cyanosilylation, alkylation, esterification, and one-pot synthesis; however, there are still unclear points with regard to the relationship between the catalytic performance and acid–base sites. Selective oxidation reactions of various substrates (e.g., alkanes, arenes, alcohols, sulfides, etc.) using O_2 as an ideal oxidant are achieved by the development of hexagonal and oxygen-deficient perovskite oxides, and their application to more difficult reactions and biomass conversion is expected. Spinels also function as effective solid catalysts for oxidation, reduction, cycloaddition, and halogenation, and defect sites play an important role in the promotion of the catalytic activity and selectivity. Apatite- and hydrotalcite-based heterogeneous catalytic systems (in particular, functionalized with various transition metals) are powerful tools for the organic synthesis of various chemical products through their unique acid–base and redox properties. The dual functions of $CePO_4$ with both acid and base sites, and Mg_6MnO_8 with both base and redox-active sites are respectively effective for the selective acetalization of biomass-derived HMF and oxidative C–H activation under mild conditions.

The present synthesis methods have fundamental disadvantages such as the need for complicated procedures and inapplicability to versatile chemical compositions. In this context, more simple and efficient synthesis methods applicable to a variety of multicomponent complex oxides with high surface areas and without the need for specific conditions and reagents are required. In the development of efficient liquid-phase organic synthesis reactions, the main target is to improve the catalytic performance (i.e., yield, selectivity, stability, etc.); therefore, the purity of the crystalline metal oxide catalysts and/or characterization of the active site structures are still insufficient in some cases. In addition, in the case of multicomponent complex oxide catalysts, there are only a limited number of reports on systematic and detailed mechanistic studies to understand the role of the elements in catalysis. Therefore, further elucidation of the mechanistic aspects (especially, the correlation between the structure and catalytic performance) in combination with advanced inorganic synthesis, operando spectroscopy, and computational approaches will be important, along with the development of organic synthesis reactions. These approaches can lead to the precise design and development of highly active crystalline complex oxides that efficiently catalyze various types of liquid-phase organic reactions under mild reaction conditions.

References

1. Anastas PT, Warner JC. Green. Chemistry: theory and practice. Oxford University Press, Oxford, 1998.
2. Sheldon RA. Fundamentals of green chemistry: efficiency in reaction design. Chem Soc Rev. 2012;41(4):1437–51.
3. Grigoropoulou G, Clark JH, Elings JA. Recent developments on the epoxidation of alkenes using hydrogen peroxide as an oxidant. Green Chem. 2003;5(1):1–7.
4. Thomas JM, Raja R. Designing catalysts for clean technology; green chemistry; and sustainable development. Annu Rev Mater Res. 2005;35(1):315–50.
5. Tundo P, Anastas P, Black DS, Breen J, Collins TJ, Memoli S, Miyamoto J, Polyakoff M, Tumas W. Synthetic pathways and processes in green chemistry. Introductory overview Pure Appl Chem. 2000;72(7):1207–28.
6. Horvath IT. Green chemistry. Acc Chem Res. 2002;35(9):685.
7. Sheldon RA. Atom efficiency and catalysis in organic synthesis. Pure Appl Chem. 2000;72(7):1233–46.
8. Cornils B, Herrmann WA, Beller M, Paciello R. Applied homogeneous catalysis with organometallic compounds: a comprehensive handbook in Four Volumes. 3rd ed. Weinheim: Wily-VCH; 2017.
9. Hagen J. Industrial catalysis: a practical approach. Weinheim: Wily-VCH; 1999.
10. Ertl G, Knözinger H, Weitkamp J. Handbook of heterogeneous catalysis. 2nd ed. Weinheim: Wily-VCH; 2008.
11. An K, Somorjai GA. Size and shape control of metal nanoparticles for reaction selectivity in catalysis. ChemCatChem. 2012;4(10):1512–24.
12. Mizuno N. Modern heterogeneous oxidation catalysis. Weinheim: Wily-VCH; 2009.
13. Martínez C, Corma A. Inorganic molecular sieves: Preparation; modification and industrial application in catalytic processes. Coord Chem Rev. 2011;255(13–14):1558–80.
14. Liang J, Liang Z, Zou R, Zhao Y. Heterogeneous catalysis in zeolites; mesoporous silica; and metal-organic frameworks. Adv Mater. 2017;29(30):1701139.
15. Yamaguchi K, Mizuno N. Green functional group transformations by supported ruthenium hydroxide catalysts. Synlett. 2010;16:2365–82.
16. Takagaki A, Tagusagawa C, Hayashi S, Hara M, Domen K. Nanosheets as highly active solid acid catalysts for green chemical syntheses. Energy Environ Sci. 2010;3(1):82–93.
17. Hechelski M, Ghinet A, Louvel B, Dufrenoy P, Rigo B, Daich A, Waterlot C. From conventional lewis acids to heterogeneous montmorillonite k10: eco-friendly plant-based catalysts used as green lewis acids. ChemSusChem. 2018;11(8):1249–1277.
18. Ishikawa S, Zhang Z, Ueda W. Unit synthesis approach for creating high dimensionally structured complex metal oxides as catalysts for selective oxidations. ACS Catal. 2018;8(4):2935–43.
19. Mizuno N, Yamaguchi K, Kamata K. Molecular design of Polyoxometalate-based compounds for environmentally-friendly functional group transformations, from molecular catalysts to heterogeneous catalysts. Catal Surv Asia. 2011;15(2):68–79
20. Enferadi-Kerenkan A, Do T-O, Kaliaguine S. Heterogeneous catalysis by tungsten-based heteropoly compounds. Catal Sci Technol. 2018;8(9):2257–84.
21. Védrine JC. Heterogeneous catalysis on metal oxides Catalysts. 2017;7(11):341.
22. Védrine JC. Recent developments and prospectives of acid-base and redox catalytic processes by metal oxides. Appl Catal A. 2019;575:170–9.
23. Védrine JC. Metal oxides in heterogeneous oxidation catalysis, state of the art and challenges for a more sustainable world. Chemsuschem. 2019;12(3):577–88.
24. Wang Y, Arandiyan H, Scott J, Bagheri A, Dai H, Amal R. Recent advances in ordered meso/macroporous metal oxides for heterogeneous catalysis, a review. J Mater Chem A. 2017;5(19):8825–46.

25. Hong WT, Risch M, Stoerzinger KA, Grimaud A, Suntivich J, Shao-Horn Y. Toward the rational design of non-precious transition metal oxides for oxygen electrocatalysis. Energy Environ Sci. 2015;8(5):1404–27.
26. Maeda K. Photocatalytic water splitting using semiconductor particles: history and recent developments. J Photochem Photobiol C. 2011;12(4):237–68.
27. Kamata K. Perovskite oxide catalysts for liquid-phase organic reactions. Bull Chem Soc Jpn. 2019;92(1):133–51.
28. Arandiyan H, Wang Y, Sun H, Rezaei M, Dai H. Ordered meso- and macroporous perovskite oxide catalysts for emerging applications. Chem Commun. 2018;54(50):6484–502.
29. Zhu J, Li H, Zhong L, Xiao P, Xu X, Yang X, Zhao Z, Li J. Perovskite oxides, preparation; characterizations; and applications in heterogeneous catalysis. ACS Catal. 2017;4(9):2917–40.
30. Wang W, Tade MO, Shao Z. Research progress of perovskite materials in photocatalysis- and photovoltaics-related energy conversion and environmental treatment. Chem Soc Rev. 2015;44(15):5371–408.
31. Domen K, Kondo JN, Hara M, Takata T. Photo- and Mechano-catalytic overall water splitting reactions to form hydrogen and oxygen on heterogeneous catalysts. Bull Chem Soc Jpn. 2000;73(6):1307–31.
32. Zhu H, Zhang P, Dai S. Recent advances of lanthanum-based perovskite oxides for catalysis. ACS Catal. 2015;5(11):6370–85.
33. Royer S, Duprez D, Can F, Courtois X, Batiot-Dupeyrat C, Laassiri S, Alamdari H. Perovskites as substitutes of noble metals for heterogeneous catalysis: dream or reality. Chem Rev. 2014;114(20):10292–368.
34. Konsolakis M. Recent advances on nitrous oxide (N_2O) decomposition over non-noble-metal oxide catalysts: catalytic performance; mechanistic considerations; and surface chemistry aspects. ACS Catal. 2015;5(11):6397–421.
35. Chen D, Chen C, Baiyee ZM, Shao Z, Ciucci F. Nonstoichiometric oxides as low-cost and highly-efficient oxygen reduction/evolution catalysts for low-temperature electrochemical devices. Chem Rev. 2015;115(18):9869–921.
36. Lee DU, Xu P, Cano ZP, Kashkooli AG, Park MG, Chen Z. Recent progress and perspectives on bi-functional oxygen electrocatalysts for advanced rechargeable metal-air batteries. J Mater Chem A. 2016;4(19):7107–7034.
37. Hattori H, Ono Y. Solid acid catalysis: from fundamentals to applications. New York: Jenny Stanford Publishing; 2015.
38. Ono Y, Hattori H. Solid base catalysis. Heidelberg: Springer; 2011.
39. Kung HH. Transition metal oxides: surface chemistry and catalysis. Amsterdam: Elsevier; 1989.
40. Dimitrov V, Komatsu T. Classification of simple oxides: a polarizability approach. J Solid State Chem. 2002;163(1):100–12.
41. Tomishige K, Ikeda Y, Sakaihori T, Fujimoto K. Catalytic properties and structure of zirconia catalysts for direct synthesis of dimethyl carbonate from methanol and carbon dioxide. J Catal. 2000;192(2):355–62.
42. Tomishige K, Gu Y, Chang T, Tamura M, Nakagawa Y. Catalytic function of CeO_2 in non-reductive conversion of CO_2 with alcohols. Mater Today Sustain. 2020;9: 100035.
43. Yamaguchi K, Kobayashi H, Wang Y, Oishi T, Ogasawara Y, Mizuno N. Green oxidative synthesis of primary amides from primary alcohols or aldehydes catalyzed by a cryptomelane-type manganese oxide-based octahedral molecular sieve, OMS-2. Catal Sci Technol. 2013;3(2):318–27.
44. Tamura M, Tomishige K. Redox properties of CeO_2 at low temperature: the direct synthesis of imines from alcohol and amine. Angew Chem Int Ed. 2015;54(3):864–7.
45. Grzybek G, Ciura K, Wójcik S, Gryboś J, Indyka P, Inger M, Antoniak-Jurak K, Kowalik P, Kotarba A, Sojka Z. On the selection of the best polymorph of Al_2O_3 carriers for supported cobalt nano-spinel catalysts for N_2O abatement: an interplay between preferable surface spreading and damaging active phase-support interaction. Catal Sci Technol. 2017;7(23):5723–32.

46. Yan T, Wang L, Liang Y, Makaremi M, Wood TE, Dai Y, Huang B, Jelle AA, Dong Y, Ozin GA. Polymorph selection towards photocatalytic gaseous CO_2 hydrogenation. Nat Commun. 2019;10(1):2521.
47. Gond R, Vanam SP, Barpanda P. $Na_2MnP_2O_7$ polymorphs as efficient bifunctional catalysts for oxygen reduction and oxygen evolution reactions. Chem Commun. 2019;55(77):11595–8.
48. Hayashi E, Yamaguchi Y, Kamata K, Tsunoda N, Kumagai Y, Oba F, Hara M. Effect of MnO_2 crystal structure on aerobic oxidation of 5-hydroxymethylfurfural to 2,5-furandicarboxylic acid. J Am Chem Soc. 2019;141:890–900.
49. Hayashi E, Komanoya T, Kamata K, Hara M. Heterogeneously-catalyzed aerobic oxidation of 5-hydroxymethylfurfural to 2;5-furandicarboxylic acid with MnO_2. Chemsuschem. 2017;10(4):654–8.
50. Yamaguchi Y, Aono R, Hayashi E, Kamata K, Hara M. Template-free synthesis of mesoporous β-MnO_2 nanoparticles: structure, formation mechanism, and catalytic properties. ACS Appl Mater Interfaces. 2020;12(32):36004–13.
51. Hayashi E, Yamaguchi Y, Kita Y, Kamata K, Hara M. One-pot aerobic oxidative sulfonamidation of aromatic thiols with ammonia by a dual-functional beta-MnO_2 nanocatalyst. Chem Commun. 2020;56(14):2095–8.
52. Granger P, Parvulescu VI, Kaliaguine S, Prellier W. Perovskites and related mixed oxides: concepts and application. Weinheim: Wiley-VCH; 2016.
53. Athayde DD, Souza DF, Silva AMA, Vasconcelos D, Nunes EHM, Diniz da Costa JC, Vasconcelos WL. Review of perovskite ceramic synthesis and membrane preparation methods. Ceram Int. 2016;42(6):6555–6671.
54. Darriet J, Subramanian MA. Structural relationships between compounds based on the stacking of mixed layers related to hexagonal perovskite-type structures. J Mater Chem. 1995;5(4):543–52.
55. Centi G, Perathoner S. Catalysis by layered materials: a review. Microporous Mesoporous Mater. 2007;107(1–2):3–15.
56. Peña MA, Fierro JLG. Chemical structures and performance of perovskite oxides. Chem Rev. 2001;101(7):1981–2018.
57. Zhu Y, Zhou W, Shao Z. Perovskite/carbon composites: applications in oxygen Electrocatalysis. Small. 2017;13(12):1603793.
58. Ge X, Sumboja A, Wuu D, An T, Li B, Goh FWT, Hor TSA, Zong Y, Liu Z. Oxygen reduction in alkaline media: from mechanisms to recent advances of catalysts. ACS Catal. 2015;5(8):4643–67.
59. Tan P, Liu M, Shao Z, Ni M. Recent advances in perovskite oxides as electrode materials for Nonaqueous Lithium-Oxygen batteries. Adv Energy Mater. 2017;7(13):1602674.
60. Grabowska E. Selected perovskite oxides: characterization, preparation and photocatalytic properties-a review. Appl Catal B. 2016;186:97–126.
61. Kubicek M, Bork AH, Rupp JLM. Perovskite oxides—a review on a versatile material class for solar-to-fuel conversion processes. J Mater Chem A. 2016;5(24):11983–2C00.
62. Smith MD, Stepan AF, Ramarao C, Brennan PE, Ley SV. Palladium-containing perovskites: recoverable and reusable catalysts for Suzuki couplings. Chem Commun. 2003;21:2652–3.
63. Lohmann S, Andrews SP, Burke BJ, Smith MD, Attfield JE, Tanaka H, Kaneko K, Ley SV. Copper- and palladium-containing perovskites: catalysts for the ullmann and Sonogashira reactions. Synlett. 2005;8:1291–5.
64. Battilocchio C, Bhawal BN, Chorghade R, Deadman BJ, Hawkins JM, Ley SV. Flow-based, cerium oxide enhanced, low-level palladium Sonogashira and Heck coupling reactions by perovskite catalysts. Isr J Chem. 2014;54(4):371–80.
65. Andrews SP, Stepan AF, Tanaka H, Ley SV, Smith MD. Heterogeneous or homogeneous? a case study involving palladium-containing perovskites in the Suzuki reaction. Adv Synth Catal. 2005;347(5):647–54.
66. Alapour S, Farahani MD, Ramjugernath D, Koorbanally NA, Friedrich HB. Ligand free heterogeneous Sonogashira cross-coupling reaction over an in situ Organoiodine capsulized palladium anchored to a perovskite catalyst. ACS Sustain Chem Eng. 2019;7(15):12697–706.

67. Yamaguchi S, Okuwa T, Wada H, Yamaura H, Yahiro H. Cyanosilylation of benzaldehyde with TMSCN over perovskite-type oxide catalyst prepared by thermal decomposition of heteronuclear cyano complex precursors. Res Chem Intermed. 2015;41(12):9551–60.
68. Singh H, Rajput JK. Chelation and calcination promoted preparation of perovskite-structured $BiFeO_3$ nanoparticles: a novel magnetic catalyst for the synthesis of dihydro-2-oxypyrroles. J Mater Sci. 2018;53(5):3163–88.
69. Pal N, Paul M, Bhaumik A. New mesoporous perovskite $ZnTiO_3$ and its excellent catalytic activity in liquid phase organic transformations. Appl Catal A. 2011;393(1–2):153–60.
70. Ray S, Das P, Banerjee B, Bhaumik A, Mukhopadhyay C. Cubic Perovskite $ZnTiO_3$ Nanopowder as a recyclable heterogeneous catalyst for the Synthesis of 1,6-Naphthyridines in water. ChemPlusChem. 2015;80(4):731–9.
71. Kleineberg H, Eisenacher M, Lange H, Strutz H, Palkovits R. Perovskites and metal nitrides as catalysts in the base-catalysed aldol addition of isobutyraldehyde to formaldehyde. Catal Sci Technol. 2016;6(15):6057–65.
72. Takagaki A, Sugisawa M, Lu D, Kondo JN, Hara M, Domen K, Hayashi S. Exfoliated nanosheets as a new strong solid acid catalyst. J Am Chem Soc. 2003;125(18):5479–85.
73. Ogasawara M, Ban T, Saito K, Kato S. Preparation of inorganic organic composites as acid base catalysts using $HCa_2Nb_{3-x}Ta_xO_{10}$ and quaternary onium salts. J Ceram Soc Jpn. 2020;128(1):51–5.
74. Zheng Y, Zhang R, Zhang L, Gu Q, Qiao ZA. A Resol-Assisted cationic coordinative co-assembly approach to mesoporous ABO_3 perovskite oxides with rich oxygen vacancy for enhanced hydrogenation of furfural to Furfuryl Alcohol. Angew Chem Int Ed. 2021;60(9):4774–81.
75. Kulkarni AS, Jayaram RV. Liquid phase catalytic transfer hydrogenation of aromatic nitro compounds on perovskites prepared by microwave irradiation. Appl Catal A. 2003;252(2):225–30.
76. Kulkarni AS, Jayaram RV. Liquid phase catalytic transfer hydrogenation of aromatic nitro compounds on $La_{1-x}Sr_xFeO_3$ perovskites prepared by microwave irradiation. J Mol Catal A. 2004;223(1–2):107–10.
77. Mistri R, Das D, Llorca J, Dominguez M, Mandal TK, Mohanty P, Ray BC, Gayen A. Selective liquid phase benzyl alcohol oxidation over Cu-loaded $LaFeO_3$ perovskite. RSC Adv. 2016;6(6):4469–77.
78. Singh SJ, Jayaram RV. Oxidation of alkylaromatics to benzylic ketones using TBHP as an oxidant over $LaMO_3$ (M=Cr; Co; Fe; Mn; Ni) perovskites. Catal Commun. 2009;10(15):2004–7.
79. Saux C, Marchena CL, Dinamarca R, Pecchi G, Pierella L. Active potassium niobates and titanoniobates as catalysts for organic sulfide remediation. Catal Commun. 2016;76:58–61.
80. Liu C-B, Zhao Z, Yang X-G, Ye X-K, Wu Y. Y-Ba-Cu-O mixed oxides for production of diphenols in liquid-solid phase. Chin J Chem. 1996;4(6):516–24.
81. Liu C, Zhao Z, Yang X, Ye X, Wu Y. Superconductor mixed oxides $La_{2-x}Sr_xCuO_{4\pm\lambda}$ for catalytic hydroxylation of phenol in the liquid-solid phase. Chem Commun. 1996;9:1019–20.
82. Evans CD, Kondrat SA, Smith PJ, Manning TD, Miedziak PJ, Brett GL, Armstrong RD, Bartley JK, Taylor SH, Rosseinsky MJ, Hutchings GJ. The preparation of large surface area lanthanum based perovskite supports for AuPt nanoparticles: tuning the glycerol oxidation reaction pathway by switching the perovskite B site. Faraday Discuss. 2016;188:427–50.
83. Adilina IB, Hara T, Ichikuni N, Kumada N, Shimazu S. Recyclable Pd-incorporated perovskite-titanate catalysts synthesized in molten salts for the liquid-phase oxidation of alcohols with molecular oxygen. Bull Chem Soc Jpn. 2013;86(1):146–52.
84. Saha S, Hamid SBA. $CuZrO_3$ nanoparticles catalyst in aerobic oxidation of vanillyl alcohol. RSC Adv. 2017;7(16):9914–25.
85. Rahmani A, Saffari J. Preparation, structure and selected catalytic properties of La_2CuO_4 nano mixed metal oxides. J Nanostruct. 2016;6(4):301–6.
86. Kawasaki S, Kamata K, Hara M. Dioxygen activation by a hexagonal $SrMnO_3$ perovskite catalyst for aerobic liquid-phase oxidation. ChemCatChem. 2016;8(20):3247–53.

87. Sugahara K, Kamata K, Muratsugu S, Hara M. Amino acid-aided synthesis of a hexagonal $SrMnO_3$ Nanoperovskite catalyst for aerobic oxidation. ACS Omega. 2017;2(4):1608–16.
88. Kamata K, Sugahara K, Kato Y, Muratsugu S, Kumagai Y, Oba F, Hara M. Heterogeneously catalyzed aerobic oxidation of sulfides with a $BaRuO_3$ Nanoperovskite. ACS Appl Mater Interfaces. 2018;10(28):23792–801.
89. Shibata S, Sugahara K, Kamata K, Hara M. Liquid-phase oxidation of alkanes with molecular oxygen catalyzed by high valent iron-based perovskite. Chem Commun. 2018;54(50):6772–5.
90. Shibata S, Kamata K, Hara M. Aerobic oxidative C=C bond cleavage of aromatic alkenes by a high valency iron-containing perovskite catalyst. Catal Sci Technol. 2021;11(7):2369–73.
91. Sahin Y, Sika-Nartey AT, Ercan KE, Kocak Y, Senol S, Ozensoy E, Türkmen YE. Precious metal-free $LaMnO_3$ perovskite catalyst with an optimized nanostructure for aerobic C–H bond activation reactions: Alkylarene oxidation and Naphthol dimerization. ACS Appl Mater Interfaces. 2021;13(4):5099–110.
92. Aguadero A, Falcon H, Campos-Martin JM, Al-Zahrani SM, Fierro JL, Alonso JA. An oxygen-deficient perovskite as selective catalyst in the oxidation of alkyl benzenes. Angew Chem Int Ed. 2011;50(29):6557–61.
93. Bewana S, Ndolomingo MJ, Carleschi E, Doyle BP, Meijboom R, Bingwa N. Inorganic Perovskite-Induced Synergy on Highly Selective Pd-Catalyzed Hydrogenation of Cinnamaldehyde. ACS Appl Mater Interfaces. 2019;11(36):32994–3005.
94. Chen C, Fan R, Gong W, Zhang H, Wang G, Zhao H. The catalytic behavior in aqueous-phase hydrogenation over a renewable Ni catalyst derived from a perovskite-type oxide. Dalton Trans. 2018;47(48):17276–84.
95. Corbet JP, Mignani G. Selected patented cross-coupling reaction technologies. Chem Rev. 2006;106(7):2651–710.
96. Chemler SR, Trauner D, Danishefsky SJ. The B-Alkyl Suzuki-Miyaura cross-coupling reaction: development, mechanistic study, and applications in natural product synthesis. Angew Chem Int Ed. 2001;40(24):4544–68.
97. Miyaura N, Suzuki A. Palladium-catalyzed cross-coupling reactions of Organoboron compounds. Chem Rev. 1995;95(7):2457–83.
98. Rossi R, Bellina F, Carpita A. Palladium catalysts for the Suzuki cross-coupling reaction: an overview of recent advances. Synthesis. 2004;15:2419–40.
99. Miyaura N, Yanagi T, Suzuki A. The palladium-catalyzed cross-coupling reaction of Phenylboronic acid with Haloarenes in the presence of bases. Synth Commun. 2006;11(7):513–9.
100. Nishihata Y, Mizuki J, Akao T, Tanaka H, Uenishi M, Kimura M, Okamoto T, Hamada N. Self-regeneration of a Pd-perovskite catalyst for automotive emissions control. Nature. 2002;418:164–7.
101. Tanaka H. An intelligent catalyst: the self-regenerative palladium-perovskite catalyst for automotive emissions control. Catal Surv Asia. 2005;9(2):63–74.
102. Jarrige I, Ishii K, Matsumura D, Nishihata Y, Yoshida M, Kishi H, Taniguchi M, Uenishi M, Tanaka H, Kasai H, Mizuki J. Toward optimizing the performance of self-regenerating Pt-based perovskite catalysts. ACS Catal. 2015;5(2), 1112–1118.
103. Polo-Garzon F, Wu Z. Acid-base catalysis over perovskites: a review. J Mater Chem A. 2018;6(7):2877–94.
104. Punniyamurthy T, Velusamy S, Iqbal J. Recent advances in transition metal catalyzed oxidation of organic substrates with molecular oxygen. Chem Rev. 2005;105(6):2329–63.
105. Cavani F, Teles JH. Sustainability in catalytic oxidation: an alternative approach or a structural evolution? Chemsuschem. 2009;2(6):508–34.
106. Kamata K. Design of highly functionalized Polyoxometalate-based catalysts. Bull Chem Soc Jpn. 2015;88(8):1017–128.
107. Mizuno N, Kamata K. Catalytic oxidation of hydrocarbons with hydrogen peroxide by vanadium-based polyoxometalates. Coord Chem Rev. 2011;255(19–20):2358–70.
108. Mizuno N, Yamaguchi K, Kamata K. Oxidation of olefins with hydrogen peroxide catalyzed by polyoxometalates. Coord Chem Rev. 2005;249(17–18):1944–56.

109. Engelmann X, Monte-Perez I, Ray K. Oxidation Reactions with Bioinspired Mononuclear Non-Heme Metal-Oxo Complexes. Angew Chem Int Ed. 2016;55(27):7632–49.
110. Stahl SS. Palladium oxidase catalysis. Selective oxidation of organic chemicals by direct dioxygen-coupled turnover. Angew Chem Int Ed. 2004;43(26):3400–3420.
111. Weinstock IA, Schreiber RE, Neumann R. Dioxygen in Polyoxometalate mediated reactions. Chem Rev. 2018;118(5):2680–717.
112. Ishikawa S, Ueda W. Microporous crystalline Mo-V mixed oxides for selective oxidations. Catal Sci Technol. 2016;6(3):617–29.
113. Komanoya T, Kinemura T, Kita Y, Kamata K, Hara M. Electronic effect of ruthenium nanoparticles on efficient reductive amination of carbonyl compounds. J Am Chem Soc. 2017;139(33):11493–9.
114. Climent MJ, Corma A, Iborra S. Conversion of biomass platform molecules into fuel additives and liquid hydrocarbon fuels. Green Chem. 2014;16(2):516–47.
115. Hara M, Nakajima K, Kamata K. Recent progress in the development of solid catalysts for biomass conversion into high value-added chemicals. Sci Technol Adv Mater. 2015;16(3): 034903.
116. Alonso DM, Wettstein SG, Dumesic JA. Bimetallic catalysts for upgrading of biomass to fuels and chemicals. Chem Soc Rev. 2012;41(24):8075–98.
117. Besson M, Gallezot P, Pinel C. Conversion of biomass into chemicals over metal catalysts. Chem Rev. 2014;114(3):1827–70.
118. Que JL, Tolman WB. Bis(μ-oxo)dimetal "Diamond" cores in copper and iron complexes relevant to Biocatalysis. Angew Chem Int Ed. 2002;41(7):1114–37.
119. Choi Y, Lin MC, Liu M. Computational study on the catalytic mechanism of oxygen reduction on $La_{0.5}Sr_{0.5}MnO_3$ in solid oxide fuel cells. Angew Chem Int Ed. 2007;46(38):7214–7219.
120. Choi Y, Mebane DS, Lin MC, Liu M. Oxygen reduction on $LaMnO_3$-based cathode materials in solid oxide fuel cells. Chem Mater. 2007;19(7):1690–9.
121. Kakihana M. Invited review "sol-gel" preparation of high temperature superconducting oxides. J Sol-Gel Sci Technol. 1996;6(1):7–55.
122. Zhang H-M, Teraoka Y, Yamazoe N. Preparation of perovskite-type oxides with large surface area by citrate process. Chem Lett. 1987;16(4):665–8.
123. Teraoka Y, Kakebayashi H, Moriguchi I, Kagawa S. Hydroxy acid-aided synthesis of perovskite-type oxides of cobalt and manganese. Chem Lett. 1991;20(4):673–6.
124. Yamaguchi K, Mizuno N. Heterogeneously catalyzed liquid-phase oxidation of alkanes and alcohols with molecular oxygen. New J Chem. 2002;26(8):972–4.
125. Mitsudome T, Nosaka N, Mori K, Mizugaki T, Ebitani K, Kaneda K. Liquid-phase epoxidation of alkenes using molecular oxygen catalyzed by vanadium cation-exchanged montmorillonite. Chem Lett. 2005;34(12):1626–7.
126. Zhao Q, Yan Z, Chen C, Chen J. Spinels: controlled preparation; oxygen reduction/evolution reaction application, and beyond. Chem Rev. 2017;117(15):10121–211.
127. Dey S, Dhal GC. Catalytic conversion of carbon monoxide into carbon dioxide over spinel catalysts: an overview. Mater Sci Energy Technol. 2019;2(3):575–88.
128. Zhou Y, Sun S, Wei C, Sun Y, Xi P, Feng Z, Xu ZJ. Significance of engineering the octahedral units to promote the oxygen evolution reaction of spinel oxides. Adv Mater. 2019;31(41): e1902509.
129. Wang D, Astruc D. Fast-growing field of magnetically recyclable nanocatalysts. Chem Rev. 2014;114(14):6949–85.
130. Sharma RK, Yadav S, Sharma S, Dutta S, Sharma A. Expanding the horizon of multi-component oxidative coupling Reaction via the design of a unique, 3D copper Isophthalate MOF-based catalyst decorated with mixed spinel $CoFe_2O_4$ nanoparticles. ACS Omega. 2018;3(11):15100–11.
131. Ghorbani-Choghamarani A, Mohammadi M, Shiri L, Taherinia Z. Synthesis and characterization of spinel $FeAl_2O_4$ (hercynite) magnetic nanoparticles and their application in multicomponent reactions. Res Chem Intermed. 2019;45(11):5705–23.

132. Ibrahim I, Ali IO, Salama TM, Bahgat AA, Mohamed MM. Synthesis of magnetically recyclable spinel ferrite (MFe_2O_4; M = Zn; Co; Mn) nanocrystals engineered by sol gel-hydrothermal technology: High catalytic performances for nitroarenes reduction. Appl Catal B. 2016;181:389–402.
133. Senapati KK, Borgohain C, Phukan P. Synthesis of highly stable $CoFe_2O_4$ nanoparticles and their use as magnetically separable catalyst for Knoevenagel reaction in aqueous medium. J Mol Catal A. 2011;339(1–2):24–31.
134. Dong X, Wang W, Song H, Zhang Y, Yuan A, Guo Z, Wang Q Yang F. Enabling efficient aerobic 5-hydroxymethylfurfural oxidation to 2,5-furandicarboxylic acid in water by interfacial engineering reinforced Cu–Mn oxides hollow nanofiber. Chemsuschem. 2022. https://doi.org/10.1002/cssc.202200076.
135. Wan A, Tang N, Xie Q, Zhao S, Zhou C, Dai Y, Yang Y. A $CuMn_2O_4$ spinel oxide as a superior catalyst for the aerobic oxidation of 5-hydroxymethylfurfural toward 2,5-furandicarboxylic acid in aqueous solvent. Catal Sci Technol. 2021;11(4):1497–509.
136. Sultan S, Kumar M, Devari S, Mukherjee D, Shah BA. Copper-manganese spinel oxide catalyzed synthesis of amides and Azobenzenes via Aminyl Radical Cations ChemCatChem. 2016;8(4):703–7.
137. Yousuf SK, Mukherjee D, Singh B, Maityc S, Taneja SC. Cu-Mn bimetallic catalyst for Huisgen [3+2]-cycloaddition. Green Chem. 2010;12(9):1568–72.
138. Singh PP, Thatikonda T, Kumar KAA, Sawant SD, Singh B, Sharma AK, Sharma PR, Singh D, Vishwakarma RA. Cu−Mn Spinel Oxide catalyzed Regioselective halogenation of phenols and N-Heteroarenes. J Org Chem. 2021;77(13):5823–8.
139. Acharyya SS, Ghosh S, Tiwari R, Sarkar B, Singha RK, Pendem C, Sasaki T, Bal R. Preparation of the $CuCr_2O_4$ spinel nanoparticles catalyst for selective oxidation of toluene to benzaldehyde. Green Chem. 2014;16(5):2500–8.
140. Acharyya SS, Ghosh S, Siddiqui N, Konathala LNS, Bal R. Cetyl alcohol mediated synthesis of $CuCr_2O_4$ spinel nanoparticles: a green catalyst for selective oxidation of aromatic C–H bonds with hydrogen peroxide. RSC Adv. 2015;5(7):4838–43.
141. Acharyya SS, Ghosh S, Bal R. Catalytic oxidation of aniline to Azoxybenzene Over $CuCr_2O_4$ spinel nanoparticle catalyst. ACS Sustain Chem Eng. 2014;2(4):584–9.
142. Acharyya SS, Ghosh S, Bal R. Direct catalytic oxyamination of benzene to aniline over Cu(II) nanoclusters supported on $CuCr_2O_4$ spinel nanoparticles via simultaneous activation of C–H and N–H bonds. Chem Commun. 2014;50(87):13311–4.
143. Acharyya SS, Ghosh S, Bal R. Surfactant promoted synthesis of $CuCr_2O_4$ spinel nanoparticles: a recyclable catalyst for One-Pot synthesis of Acetophenone from ethylbenzene. Ind Eng Chem Res. 2014;53(51):20056–63.
144. Dörfelt C, Hammerton M, Martin D, Wellmann A, Aletsee CC, Tromp M, Köhler K. Manganese containing copper aluminate catalysts: genesis of structures and active sites for hydrogenation of aldehydes. J Catal. 2021;395:80–90.
145. Marchi AJ, Gordo DA, Trasarti AF, Apesteguía CR. Liquid phase hydrogenation of cinnamaldehyde on Cu-based catalysts. Appl Catal A. 2003;249(1):53–67.
146. Mathew T, Tope BB, Shiju NR, Hegde SG, Rao BS, Gopinath CS. Acid-base properties of $Cu_{1-x}Co_xFe_2O_4$ ferrospinels. FTIR investigations Phys Chem Phys. 2002;4(17):4260–7.
147. Xu W, Liu X, Ren J, Liu H, Ma Y, Wang Y, Lu G. Synthesis of nanosized mesoporous Co–Al spinel and its application as solid base catalyst. Microporous Mesoporous Meter. 2011;141(1):251–7.
148. Natarajan S, Mandal S. Open-framework structures of transition metal compounds. Angew Chem Int Ed. 2008;47(26):4798–828.
149. Moffat JB. Phosphates as catalysts. Catal Rev. 1978;18(2):199–258.
150. Hutchings GJ. Vanadium phosphate: a new look at the active components of catalysts for the oxidation of butane to maleic anhydride. J Mater Chem. 2004;14(23):3385–95.
151. Bhanja P, Na J, Jing T, Lin J, Wakihara T, Bhaumik A, Yamauchi Y. Nanoarchitectured metal phosphates and phosphonates: a new material horizon toward emerging applications. Chem Mater. 2019;31(15):5343–62.

152. Shivhare A, Kumar A, Srivastava R. Metal phosphate catalysts to upgrade lignocellulose biomass into value-added chemicals and biofuels. Green Chem. 2021;23(11):3818–41.
153. Zhao H, Yuan ZY. Insights into transition metal phosphate materials for efficient Electrocatalysis. ChemCatChem. 2020;12(15):3797–37810.
154. Kanan MW, Surendranath Y, Nocera DG. Cobalt-phosphate oxygen-evolving compound. Chem Soc Rev. 2009;38(1):109–14.
155. Izumi Y, Onaka M. Liquid-phase organic reactions catalyzed by inorganic solid acids and bases. J Mol Catal. 1992;74(1–3):35–42.
156. Gruselle M. Apatites: a new family of catalysts in organic synthesis. J Org Chem. 2015;793:93–101.
157. Kaneda K, Mizugaki T. Design of high-performance heterogeneous catalysts using apatite compounds for liquid-phase organic syntheses. ACS Catal. 2017;7(2):920–35.
158. Yamaguchi K, Mori K, Mizugaki T, Ebitani K, Kaneda K. Creation of a monomeric Ru species on the surface of hydroxyapatite as an efficient heterogeneous catalyst for aerobic alcohol oxidation. J Am Chem Soc. 2000;122(29):7144–5.
159. Mori K, Yamaguchi K, Mizugaki T, Ebitani K, Kaneda K. Catalysis of a hydroxyapatite-bound Ru complex: efficient heterogeneous oxidation of primary amines to nitriles in the presence of molecular oxygen. Chem Commun. 2001;5:461–2.
160. Mori K, Tano M, Mizugaki T, Ebitani K, Kaneda K. Efficient heterogeneous oxidation of organosilanes to silanols catalyzed by a hydroxyapatite-bound Ru complex in the presence of water and molecular oxygen. New J Chem. 2002;26(11):1536–8.
161. Mori K, Hara T, Mizugaki T, Ebitani K, Kaneda K. Hydroxyapatite-bound cationic ruthenium complexes as novel heterogeneous Lewis acid catalysts for Diels-alder and aldol reactions. J Am Chem Soc. 2003;125(38):11460–1.
162. Mori K, Mitani Y, Hara T, Mizugaki T, Ebitani K, Kaneda K. A single-site hydroxyapatite-bound zinc catalyst for highly efficient chemical fixation of carbon dioxide with epoxides. Chem Commun. 2005;26:3331–3.
163. Mori K, Oshiba M, Hara T, Mizugaki T, Ebitani K, Kaneda K. Creation of monomeric La complexes on apatite surfaces and their application as heterogeneous catalysts for Michael reactions. New J Chem. 2006;30(1):44–52.
164. Hara T, Kanai S, Mori K, Mizugaki T, Ebitani K, Jitsukawa K, Kaneda K. Highly efficient C–C bond-forming reactions in aqueous media catalyzed by monomeric vanadate species in an apatite framework. J Org Chem. 2006;71(19):7455–62.
165. Sunaba H, Kamata K, Mizuno N. Selective N-Alkylation of indoles with α, β-unsaturated compounds catalyzed by a monomeric phosphate. ChemCatChem. 2014;6(8):2333–2238.
166. Kamata K, Sugahara K. Base catalysis by mono- and Polyoxometalates. Catalysts. 2017;7(11):345.
167. Kimura T, Kamata K, Mizuno N. A bifunctional tungstate catalyst for chemical fixation of CO_2 at atmospheric pressure. Angew Chem Int Ed. 2012;51(27):6700–3.
168. Kimura T, Sunaba H, Kamata K, Mizuno N. Efficient [WO4]2–catalyzed chemical fixation of carbon dioxide with 2-aminobenzonitriles to quinazoline-2,4(1H,3H)-diones. Inorg Chem. 2012;51(23):13001–8.
169. Sugahara K, Satake N, Kamata K, Nakajima T, Mizuno N. A basic germanodecatungstate with a −7 charge: efficient chemoselective acylation of primary alcohols. Angew Chem Int Ed. 2014;53(48):13248–52.
170. Sugahara K, Kuzuya S, Hirano T, Kamata K, Mizuno N. Reversible deprotonation and protonation behaviors of a tetra-protonated gamma-Keggin silicodecatungstate. Inorg Chem. 2012;51(14):7932–9.
171. Sugahara K, Kimura T, Kamata K, Yamaguchi K, Mizuno N. A highly negatively charged gamma-Keggin germanodecatungstate efficient for Knoevenagel condensation. Chemical Commun. 2012;48(67):8422–4.
172. Kikukawa Y, Suzuki K, Sugawa M, Hirano T, Kamata K, Yamaguchi K, Mizuno N. Cyanosilylation of carbonyl compounds with trimethylsilyl cyanide catalyzed by an yttrium-pillared silicotungstate dimer. Angew Chem Int Ed. 2012;51(15):3686–90.

173. Kanai S, Nagahara I, Kita Y, Kamata K, Hara M. A bifunctional cerium phosphate catalyst for chemoselective acetalization. Chem Sci. 2017;8(4):3146–53.
174. Sato A, Ogo S, Kamata K, Takeno Y, Yabe T, Yamamoto T, Matsumura; Hara M, Sekine Y. Ambient-temperature oxidative coupling of methane in an electric field by a cerium phosphate nanorod catalyst. Chem Commun. 2019;55(28):4019–4022.
175. Matsuda A, Tateno H, Kamata K, Hara M. Iron phosphate nanoparticle catalyst for direct oxidation of methane into formaldehyde: effect of surface redox and acid-base properties. Catal Sci Technol. 2021;11(21):6987–98.
176. Nishio H, Miura H, Kamata K, Shishido T. Deposition of highly dispersed gold nanoparticles onto metal phosphates by deposition-precipitation with aqueous ammonia. Catal Sci Technol. 2021;11(21):7141–50.
177. Kalantar-zadeh K, Ou JZ, Daeneke T, Mitchell A, Sasaki T, Fuhrer MS. Two dimensional and layered transition metal oxides. Appl Mater Today. 2016;5:73–89.
178. Kaneda K, Mizugaki T. Development of concerto metal catalysts using apatite compounds for green organic syntheses. Energy Environ Sci. 2009;2(6):655–73.
179. Sels BF, De Vos DE, Jacobs PA. Hydrotalcite-like anionic clays in catalytic organic reactions. Catal Rev. 2001;43(4):443–88.
180. Kaneda K. Cation-exchanged montmorillonites as solid acid catalysts for organic synthesis. Synlett. 2007;7:0999–1015.
181. Kumar BS, Dhakshinamoorthy A, Pitchumani K. K10 montmorillonite clays as environmentally benign catalysts for organic reactions. Catal Sci Technol. 2014;4(8):2378–96.
182. Ebitani K, Motokura K, Mori K, Mizugaki T, Kaneda K. Reconstructed Hydrotalcite as a highly active heterogeneous base catalyst for carbon-carbon bond formations in the presence of water. J Org Chem. 2006;71(15):5440–7.
183. Ebitani K, Kawabata T, Nagashima K, Mizugaki T, Kaneda K. Simple and clean synthesis of 9,9-bis[4-(2-hydroxyethoxy)phenyl]fluorene from the aromatic alkylation of phenoxyethanol with fluoren-9-one catalysed by titanium cation-exchanged montmorillonite. Green Chem. 2000;2(4):157–60.
184. Kawabata T, Mizugaki T, Ebitani K, Kaneda K. A novel montmorillonite-enwrapped scandium as a heterogeneous catalyst for Michael reaction. J Am Chem Soc. 2003;125(35):10486–7.
185. Kawabata T, Kato M, Mizugaki T, Ebitani K, Kaneda K. Monomeric metal aqua complexes in the interlayer space of montmorillonites as strong Lewis acid catalysts for heterogeneous carbon-carbon bond-forming reactions. Chem Eur J. 2005;11(1):288–97.
186. do Carmo JVC, Pinheiro ALG, Oliveira AC, de Castro MO, Soares JM, Padron-Hernandez E, Peña-Garcia R, Saraiva GD, Rodríguez-Castellón E, Rodríguez-Aguado E. Comparison of the catalytic performance of YIG garnets and Fe-containing oxides catalysts for oxidation of ethylbenzene. Ceram Int. 2021;47(5):6279–6289.
187. Sedighinia E, Badri R, Kiasat AR. Application of Yttrium Iron Garnet as a Powerful and Recyclable Nanocatalyst for One-Pot Synthesis of Pyrano[2,3-*c*]pyrazole Derivatives under Solvent-Free Conditions. Russ J Org Chem. 2019;55(11):1755–63.
188. Mayer SF, Falcón H, Dipaola R, Ribota P, Moyano L, Morales-delaRosa S, Mariscal R, Campos-Martín JM, Alonso JA, Fierro JLG. Dehydration of fructose to HMF in presence of $(H_3O)_xSb_xTe_{(2-x)}O_6$ (x=1, 1.1, 1.25) in H_2O-MIBK. Mol Catal. 2020;481:110276.
189. Felthouse TR. Catalytic oxidative cleavage of vicinal diols and related oxidations by Ruthenium Pyrochlore oxides: new catalysts for low-temperature oxidations with molecular oxygen. J Am Chem Soc. 1987;109:7566–8.
190. Felthouse TR, Fraundorf PB, Friedman RM, Schosser CL. Expanded lattice ruthenium pyrochlore oxide catalysts I. Liquid-phase oxidations of vicinal diols, primary alcohols, and related substrates with molecular oxygen. J Catal. 1991;127(1):393–420.
191. Hayashi E, Tamura T, Aihara T, Kamata K, Hara M. Base-assisted aerobic C–H Oxidation of Alkylarenes with a Murdochite-type oxide Mg_6MnO_8 nanoparticle catalyst. ACS Appl Mater Interfaces. 2022;14(5):6528–37.

Chapter 9
Crystal and Band Engineering in Photocatalytic Materials

Ryu Abe and Hajime Suzuki

9.1 History of Photocatalytic Water Splitting Using Metal Oxide-Based Semiconductors

Photocatalytic water splitting using semiconductor particles is regarded as a promising technology for clean and low-cost hydrogen production from water by utilizing abundant solar light energy and thus has been studied extensively [1–3]. Such technologies for solar-to-hydrogen production were initially highlighted by a report on photoelectrochemical water splitting under UV light using a photoanode of single-crystalline rutile-TiO_2 (band gap: 3.0 eV) coupled with a Pt cathode, demonstrated by Fujishima and Honda in 1972 (Fig. 9.1a) [4]. After this pioneering work, many researchers started to develop photocatalytic water splitting systems (Fig. 9.1b), in which small particles of metal (e.g., Pt) are loaded directly onto the surface of a particulate semiconductor (e.g., TiO_2) as the cocatalyst for promoting water reduction. However, the demonstration of simultaneous and stoichiometric evolution of H_2 and O_2 on such simple photocatalysts (e.g., Pt/TiO_2) has been hampered for many years mainly by the occurrence of the backward reaction ($2H_2 + O_2 \rightarrow 2H_2O$) on a metal cocatalyst such as Pt. In 1980, Sato et al. demonstrated that the coating of Pt-loaded anatase-TiO_2 (bandgap: 3.2 eV) by a deliquesced NaOH layer enabled it to split water vapor into H_2 and O_2 in a stoichiometric ratio under UV-light irradiation [5]. As for the splitting of liquid water, Domen et al. first succeeded at it in 1982 by employing $SrTiO_3$ (bandgap: 3.2 eV) loaded with a NiO_x cocatalyst as schematically illustrated in Fig. 9.2. [6] (All crystal structures in this chapter were drawn using the VESTA program [7]. Herein, a part of the Ni metal surface was

R. Abe (✉)
Graduate School of Engineering, Kyoto University, Nishikyo-ku, Katsura 615-8510, Kyoto, Japan
e-mail: ryu-abe@scl.kyoto-u.ac.jp

H. Suzuki
Graduate School of Engineering, Kyoto University, Nishikyo-ku, Katsura 615-8510, Kyoto, Japan
e-mail: suzuki.hajime.7x@kyoto-u.ac.jp

W. Ueda (ed.), *Crystalline Metal Oxide Catalysts*,
https://doi.org/10.1007/978-981-19-5013-1_9

oxidized under a mild condition to effectively suppress the backward reaction while maintaining substantially high catalytic activity for H_2 production, thus endowing the $SrTiO_3$ photocatalyst with the ability to generate H_2 and O_2 simultaneously from liquid water at steady rates. This successful demonstration of liquid water splitting further encouraged researchers to develop various photocatalysis materials for achieving efficient H_2 production.

Then, as will be described later in this chapter, various metal oxides with specific crystal structures such as layers or tunnels have been proven to function as efficient photocatalysts for water splitting under UV light. In 2003, Kudo et al. developed a NiO-loaded $NaTaO_3$:La (bandgap: 4.0 eV) photocatalyst that exhibited the astonishingly high apparent quantum efficiency (AQE) of 56% under monochromatic light irradiation at 270 nm [8]. Note that, the AQE described hereafter in this chapter will represent the values calculated using the following equation:

$$AQE(\%) = 100 \times A \times R/I$$

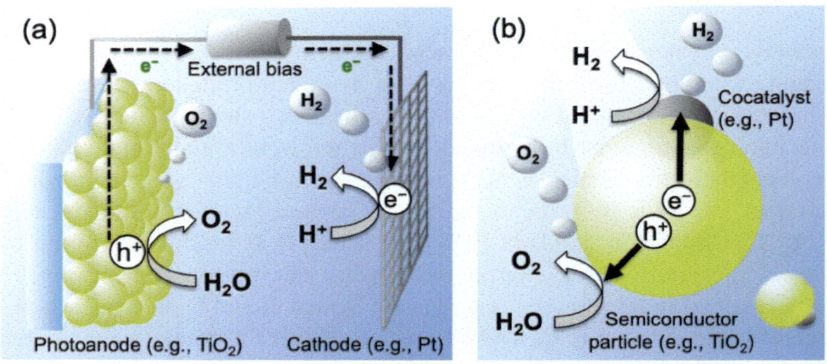

Fig. 9.1 Illustration of **a** photoelectrochemical and **b** photocatalytic water splitting

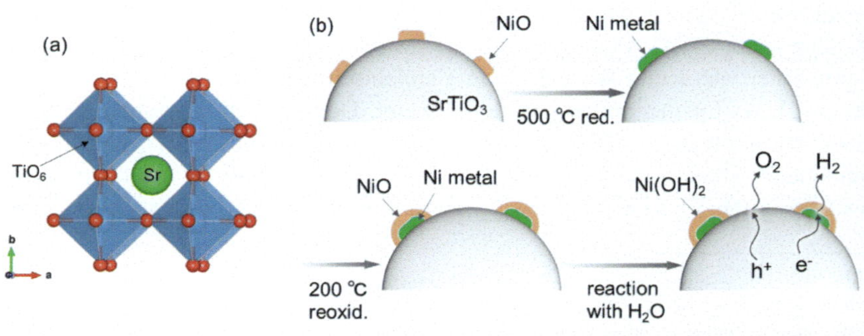

Fig. 9.2 **a** Crystal structure of $SrTiO_3$ and **b** Schematic view of the structures of NiO_x-$SrTiO_3$ photocatalyst after various treatments

where A is the number of photons required to generate one molecule of the product (i.e., two for H_2 and four for O_2). R is the amount of evolved gas (H_2 or O_2) molecules, and I is the number of incident photons.

Although various metal oxide photocatalysts have been developed as briefly introduced above, they all can only function under UV light irradiation with a wavelength shorter than 400 nm due to having bandgaps wider than 3.0 eV. Since almost half of all incident solar energy at the Earth's surface falls in the visible region, the efficient utilization of visible light remains indispensable for realizing practical H_2 production based on photocatalytic water splitting. Thus, the search for and development of visible-light-responsive photocatalysts for water splitting has become one of the most active research areas since the late 1990s [9]. However, the demonstration of visible-light-induced water spitting using a single photocatalyst material (i.e., via one-step photoexcitation) has been quite difficult primarily due to the strict conditions (e.g., appropriate band levels) required for the semiconductors employed. Then, photocatalytic water splitting under visible light was first demonstrated in 2001 using a Z-scheme photocatalytic system (i.e., via two-step photoexcitation) that consists of $SrTiO_3$ doped with Cr and Ta for H_2 evolution, WO_3 for O_2 evolution, and an iodate/iodide (IO_3^-/I^-) redox couple as an electron mediator between them [10]. Since then, various photocatalyst materials—not only metal oxides but also various mixed-anion materials such as oxynitrides—have been developed for both one- and two-step water splitting under visible light. For example, a scalable and highly efficient photocatalyst sheet was recently demonstrated based on the Z-scheme mechanism between La- and Rh-co-doped $SrTiO_3$ and Mo-doped $BiVO_4$ photocatalysts; exhibiting a solar-to-hydrogen (STH) energy conversion efficiency of 1.1% and an AQE of > 30% at 419 nm [11]. To establish higher STH beyond 5%, which is necessary for achieving cost-competitive H_2 production [12], the search for and development of photocatalyst materials continues. This chapter introduces some representative examples of crystal and band engineering in oxide-based materials to develop efficient photocatalysts.

9.2 Crystal Engineering of Metal Oxides for Efficient Water Splitting Under UV Light

9.2.1 Layered Metal Oxides with Cation-Exchangeable Interlayer Spaces

As described above, the first successful example of the photocatalytic splitting of liquid water was demonstrated by using a particulate $SrTiO_3$ semiconductor, although the AQE was probably about 0.01% (not measured), estimated by the gas evolution rates. Since $SrTiO_3$ consists of Ti^{4+} cations with d_0 electron configuration, metal oxides containing Ti^{4+}, Zr^{4+}, Nb^{5+} or Ta^{5+} with d_0 configuration were extensively investigated in the early stage of research, along with focusing specifically on crystal engineering for promoting charge separation and following the redox reaction with H_2O molecules.

The first example of such efficient photocatalysts developed based on crystal engineering is Ni-intercalated $K_4Nb_6O_{17}$ (bandgap: 3.3 eV), also developed by Domen et al., that can split water efficiently based on its unique nanostructure [13]. As shown in Fig. 9.3a, a potassium niobate $K_4Nb_6O_{17}$ consists of octahedral units of NbO_6 that form a two-dimensional layers via bridging oxygen anions. The two-dimensional layers are negatively charged (i.e., $[Nb_6O_{17}]^{4-}$), and K^+ cations exist between them to compensate for the negative charges, thus forming an ion-exchangeable layered structure. One characteristic of $K_4Nb_6O_{17}$ is the spontaneous hydration of its interlayer spaces even under atmospheric conditions, which is further enhanced when suspended in an aqueous solution. This facile hydration property was expected to provide extended reaction spaces inside layers not only on the outer surface, in stark contrast to conventional bulk semiconductors such as $SrTiO_3$. Two types of interlayer exist (interlayer I and II) and, interestingly, they differ from each other in their properties of hydration and cation exchange with K^+. For instance, divalent Ni^{2+} cations can only replace the K^+ in interlayer I, while no intercalation of Ni^{2+} occurs in interlayer II. By taking advantage of this unique cation-exchange property, the ultrafine particles of Ni metal (smaller than 1 nm in diameter) were selectively loaded inside interlayer I through the partial exchange of K^+ by Ni^{2+} and the following reduction with H_2 stream at high temperature. It has been suggested that the Ni cocatalyst in interlayer I can effectively capture the photoexcited electrons from the niobate sheets through the significantly shortened migration length, while the photogenerated holes can efficiently react with the H_2O molecules intercalated into interlayer II (Fig. 9.3b). The utilization of interlayer spaces as reaction sites undoubtedly increases the usage efficiency of the photogenerated carriers by suppressing the possibility of their recombination during their migration from bulk to the surface. In addition, the alternative arrangement of reduction (interlayer I) and oxidation (interlayer II) sites is advantageous for suppressing the backward reaction ($2H_2 + O_2 \rightarrow 2H_2O$) on the Ni metal cocatalysts, providing high-efficiency water splitting with an AQE of 3.5% at 330 nm [14].

The second example of a highly efficient photocatalyst with cation-exchangeable interlayer spaces is NiO_x-loaded $A_2La_2Ti_3O_{10}$ (A = K, Rb, band gap: 3.4–3.5 eV) with a layered perovskite structure that was also developed by Domen et al. [16] Fig. 9.4a shows the crystal structure of $A_2La_2Ti_3O_{10}$; the interlayer spaces of $A_2La_2Ti_3O_{10}$ and $Rb_2La_2Ti_3O_{10}$ are spontaneously hydrated similar to $K_4Nb_6O_{17}$. However, most of the loaded NiO_x was found to exist as large particles (10–20 nm in diameter) at the external surface of the photocatalyst as illustrated in Fig. 9.4b. Thus, it was suggested that the photoexcited electrons travel efficiently in the perovskite layer to the NiO_x cocatalyst particles outside and then reduce H_2O, whereas photogenerated holes oxidize the H_2O molecules that are intercalated. Importantly, the authors synthesized various analogous compounds (e.g., $RbLa_2Ti_2NbO_{10}$) to control the hydration property (i.e., hydration number), and thereby testified that the high hydration ability of the interlayer space is indispensable for achieving efficient water

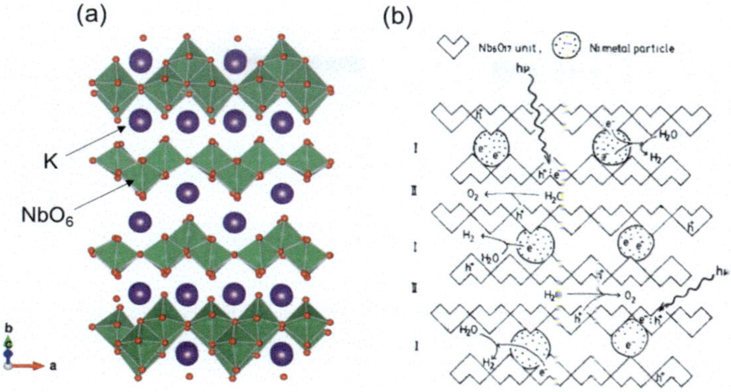

Fig. 9.3 **a** Crystal structure of $K_4Nb_6O_{17}$ and **b** the reaction mechanism of H_2O decomposition into H_2 and O_2. Reprinted with permission from Ref. [15]. Copyright 1989, Elsevier

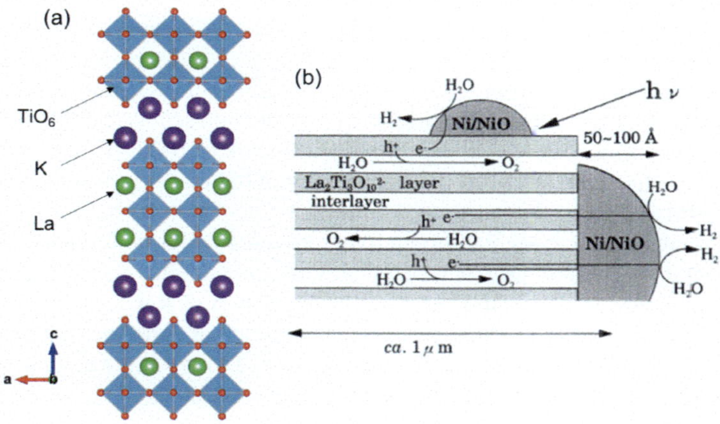

Fig. 9.4 **a** Crystal structure of $K_2La_2Ti_3O_{10}$ and **b** the reaction mechanism of H_2O decomposition into H_2 and O_2. Reprinted with permission from Ref. [16] Copyright 1997, American Chemical Society

splitting. The optimized NiO_x-loaded $Rb_2La_2Ti_3O_{10}$ exhibited an AQE of 5% at around 330 nm.

9.2.2 Metal Oxides with Tunnel Structures

Inoue et al. developed various metal oxide photocatalysts with tunnel structures [17]; a representative example is the RuO_2-loaded $BaTi_4O_9$ (bandgap: 3.0 eV) with pentagonal prism tunnel structure as shown in Fig. 9.5a. One specific feature of

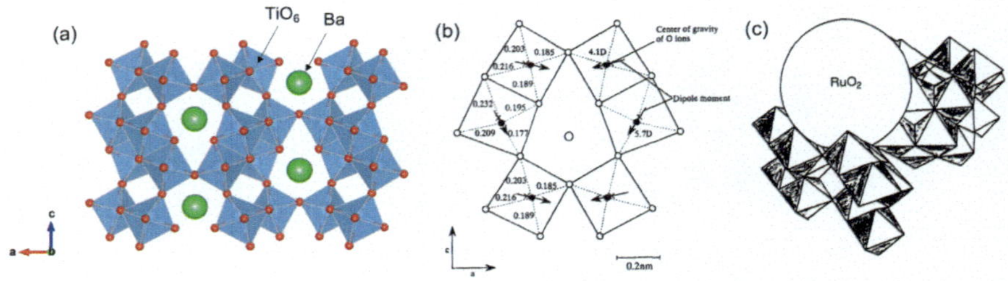

Fig. 9.5 **a** Crystal structure of $BaTi_4O_9$, **b** the dipole moments present in the distorted TiO_6 octahedra, and **c** a "nest" effect of a pentagonal-prism tunnel in the RuO_2-loaded $BaTi_4O_9$ photocatalyst. Reprinted with permission from Refs. [17, 18]. Copyright 1997, Royal Society of Chemistry. Copyright 1997, Elsevier

$BaTi_4O_9$ for achieving efficient photocatalysis is the dipole moments generated in the distorted TiO_6 octahedra. The X-ray diffraction study on a single crystalline $BaTi_4O_9$ sample indicated the presence of two types of largely distorted TiO_6 octahedra as seen in Fig. 9.5b. This structure produces two kinds of dipole moments in the pentagonal-prism tunnel, one at 5.7 D and the other at 4.1 D. Based on the experimental results of EPR measurement, it was strongly suggested that these dipole moments elongated the charge separation between photoexcited electrons and holes, thus resulting in highly efficient water splitting [18]. Another important feature of $BaTi_4O_9$ material is that the tunnel structure certainly helped the generation of highly active RuO_2 cocatalysts with uniformly small particle sizes (1.4–3.0 nm in diameter) and high dispersion. It was suggested that the pentagonal-prism space provides the form of a "nest" site, i.e., a concave site with a ridge (Fig. 9.5c). The presence of nest sites probably prevented the RuO_2 particles, which were loaded via impregnation of $RuCl_3$aq. and followed calcination in air, from aggregating and growing into larger particles. The optimized RuO_2-loaded $BaTi_4O_9$ exhibited AQE of $10 \pm 3\%$ at 330 nm. Note that the AQE of $BaTi_4O_9$ photocatalyst has been improved very recently via the combination of the optimized synthesis conditions of $BaTi_4O_9$ particles and the appropriate choice of cocatalysts, achieving 41% at 313 nm [19].

A similar effect of dipole moment for improving photocatalytic activity has been also suggested in studies on other metal oxides such as $Na_2Ti_6O_{13}$ with a rectangular tunnel structure (Fig. 9.6) [20].

9.2.3 Highly Efficient Water Splitting on Perovskite-Type Metal Oxides

In the 1990s, significant attention was paid to developing metal oxides with specific structures such as cation-exchangeable layered ones, as described above. This was

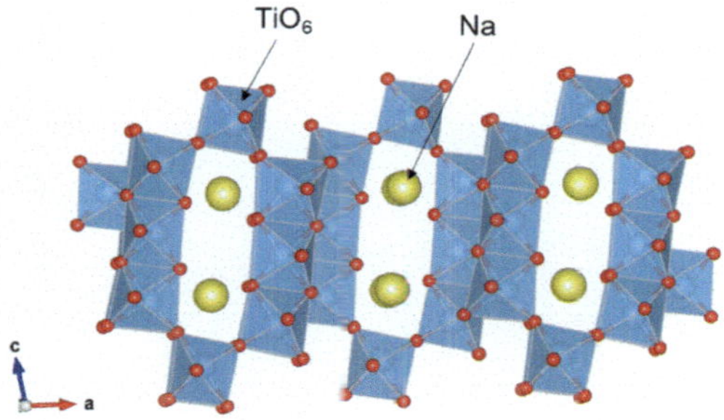

Fig. 9.6 Crystal structure of $Na_2Ti_6O_{13}$

probably based on the hypothesis that such nano-porous reaction sites are indispensable for achieving highly efficient water splitting. In 2003, Kudo et al. demonstrated an astonishingly high AQE (56% at 270 nm) by using the non-porous metal oxide $NaTaO_3$ (bandgap: 4.0 eV) with a conventional perovskite structure shown in Fig. 9.7b [8]. Note that the authors first found that the ionic radius of A in $ATaO_3$ (A = Li, Na, K) significantly affects TaO_6 octahedral tilting, and consequently does the photocatalytic activities for water splitting. Among them, $NaTaO_3$ with moderate TaO_6 octahedral tilting showed the highest activity due to the suitable conduction band level consisting of Ta-5d and energy delocalization caused by the small distortion of TaO_6 [21]. Then, it was revealed that the doping of an appropriate amount of lanthanum (1–2 mol%) into $NaTaO_3$ drastically improved the water-splitting activity. The addition of La_2O_3 into other precursors (Na_2CO_3 and Ta_2O_5) during SSR synthesis was found to provide the ordered surface nanostructure with many characteristic steps (Fig. 9.8) along with the decreased particle size of $NaTaO_3$ (0.1–0.7 μm) compared to the non-doped one (2–3 μm). The combination of TEM and EXAFS revealed that the NiO cocatalysts were loaded on the edge of the nanostep structure as illustrated in Fig. 9.8, thus promoting the separation of photoexcited electrons and holes. Note that the NiO cocatalyst in this system, which was loaded via simple impregnation of $Ni(NO_3)_2$aq. and following calcination in air, differs from the previously mentioned NiO_x cocatalyst with core–shell structure (Fig. 9.2). Although this $NaTaO_3$:La photocatalyst can harvest only a fraction of UV light shorter than 310 nm due to the wide band gap, such astonishingly high AQE has inspired researchers to search and develop non-porous compounds including perovskites.

Very recently, another surprising result was reported by Domen et al., in which an aluminum-doped $SrTiO_3$ ($SrTiO_3$:Al) photocatalyst demonstrated overall water splitting into H_2 and O_2 with AQE of up to 96% at 360 nm wavelength, which is equivalent to an internal quantum efficiency of almost 100% [22]. Although the initial AQE of the $NiO_x/SrTiO_3$ photocatalyst in 1980 was quite low (ca. 0.01%) as described previously, various refinements have drastically improved the AQE up to 69% [23]. Notably, the activity of $SrTiO_3$ was dramatically improved by $SrCl_2$ flux treatment at 1100 °C in an alumina crucible mainly due to the doping of Al species derived from

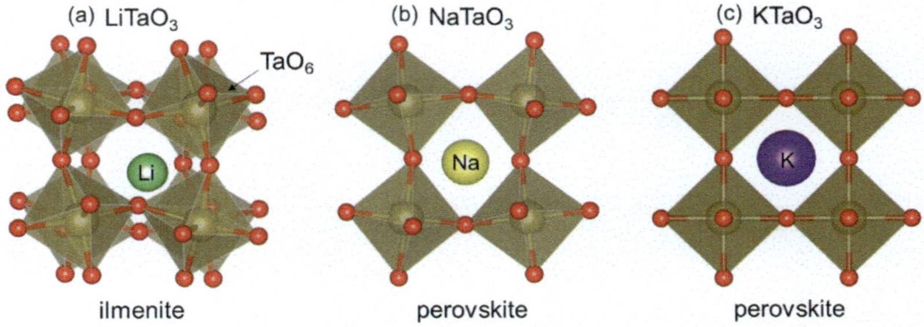

Fig. 9.7 Crystal structure of $LiTaO_3$, $NaTaO_3$, and $KTaO_3$

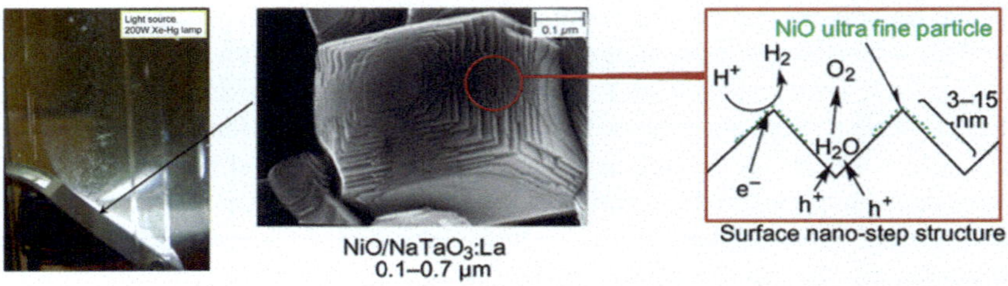

Fig. 9.8 Proposed mechanism of highly efficient photocatalytic water splitting over $NiO/NaTaO_3$:La. Reprinted with permission from Refs. [1, 8]. Copyright 2009, Royal Society of Chemistry. Copyright 2003, American Chemical Society

the Al_2O_3 crucibles, while external doping was also effective [24]. Further improvement was achieved by employing a modified $SrTiO_3$:Al photocatalyst on which two different cocatalysts, i.e., an Rh/Cr_2O_3 core–shell one for H_2 production and CoOOH for O_2 evolution, respectively, were loaded selectively on {100} and {110} facets via photodeposition (Fig. 9.9). Simulation of photocarrier distribution on $SrTiO_3$:Al particles indicated that the anisotropic charge separation (i.e., e^- to {100} while h^+ to {110}) is derived from the difference in work functions of the exposed facets. The combination of such anisotropic charge transfer in the $SrTiO_3$:Al bulk and the presence of favorable cocatalysts at each facet enabled multiple consecutive charge transfers without suffering from recombination, thus achieving the upper limit of AQE for water splitting. Although $SrTiO_3$:Al still requires UV light irradiation, the present result paves the way toward scalable and practical photocatalysis systems with STH > 5% by introducing some newly developed visible-light-responsive photocatalyst materials based on the guiding principles obtained herein.

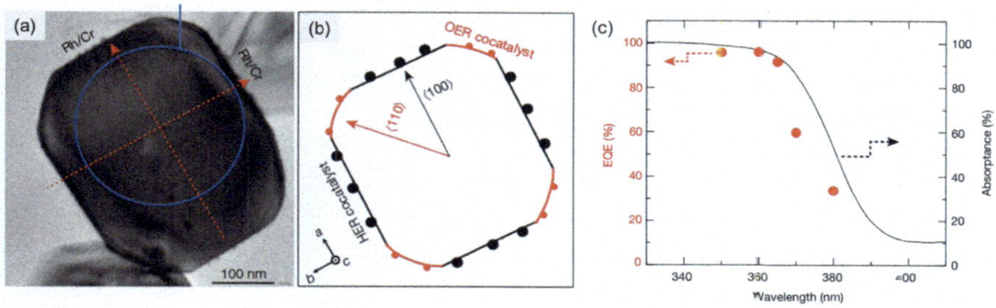

Fig. 9.9 **a** Transmission electron microscopy image of $SrTiO_3$:Al loaded with Rh (0.1 wt%)/Cr_2O_3 (0.05 wt%)/CoOOH (0.05 wt%). **b** Particle morphology and crystal orientation. **c** Diffuse reflectance spectrum of bare $SrTiO_3$:Al and wavelength dependence of external quantum efficiency (EQE) during water splitting. Reprinted with permission from Ref. [22]. Copyright 2020, Springer Nature

9.3 Band Engineering of Metal Oxides for Achieving Visible-Light-Induced Water Splitting

9.3.1 The Difficulty of Achieving Visible-Light-Induced Water Splitting Using Pristine Metal Oxides

As described above, various metal oxides are reportedly active photocatalysts for water splitting; however, most of them only function under UV light due to the large bandgaps. Since nearly half of the solar energy incident on the Earth's surface lies in the visible region ($400 < \lambda < 800$ nm), it is essential to use visible light efficiently to realize large-scale practical H_2 production via photocatalytic water splitting. The maximum solar-to-hydrogen (STH) energy conversion efficiency assuming 100% quantum efficiency can be calculated using the standard solar spectrum (Fig. 9.10). Even if all of the UV light up to 400 nm were utilized, the solar conversion efficiency would be only 2%, which is insufficient for the commercialization of the produced H_2. However, utilizing the visible light up to 600 nm would drastically improve the efficiency to 16%; a further extension up to 800 nm would result in 32% conversion efficiency. Therefore, achieving efficient water splitting under visible light has been a challenging goal since the discovery of the photoelectrochemical water splitting using a TiO_2-photoanode in 1972. Despite many years of intensive efforts by researchers around the world, no reproducible demonstration was achieved before 2000.

When a particulate semiconductor is used as a photocatalyst for water splitting, the conduction band minimum (CBM) must be more negative than the water reduction potential (0 V vs. SHE) to produce H_2, while the valence band maximum (VBM) must be more positive than the water oxidation potential (1.23 V vs. SHE) to produce O_2 (Fig. 9.11a). Furthermore, the photocatalyst material itself must be stable both chemically and photochemically. Metal oxide semiconductors are generally highly stable against photocorrosion and were thus extensively studied in the early stages of research as previously described. However, Scaife noted in 1980 that it is intrinsically

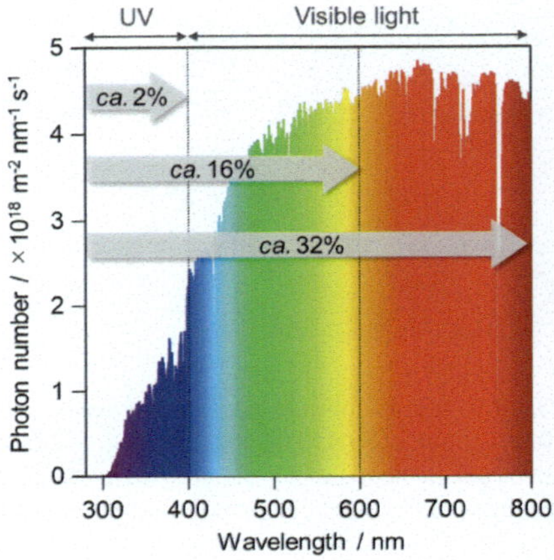

Fig. 9.10 Solar spectrum and maximum solar light conversion efficiencies for water splitting reaction with 100% of quantum efficiency

difficult to develop an oxide semiconductor photocatalyst that has both a sufficiently negative CBM for H_2 production and a sufficiently narrow band gap (i.e., <3.0 eV) for visible light absorption [25]. As seen in Fig. 9.12, one can find an almost straight-line relationship between the flat band potentials and the bandgaps of various oxides consisting of different cations. This relationship is derived from the fixed levels of VBMs that are formed exclusively by the O-2p orbitals at the deeply positive value of +3.0 V (vs. SHE). Consequently, visible-light-responsive oxides such as WO_3 cannot produce H_2 from water due to their CBMs being more positive than the water reduction potential (Fig. 9.11b). Although certain non-oxide semiconductors such as sulfides and nitrides possess appropriate band levels for water splitting under visible light (Fig. 9.11c), they are generally unstable and readily become deactivated through either photocorrosion or self-oxidation rather than evolving O_2.

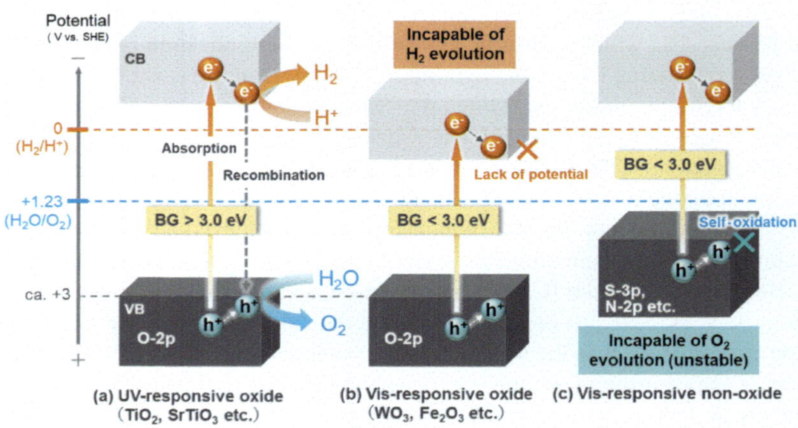

Fig. 9.11 Band energy levels of various semiconductors

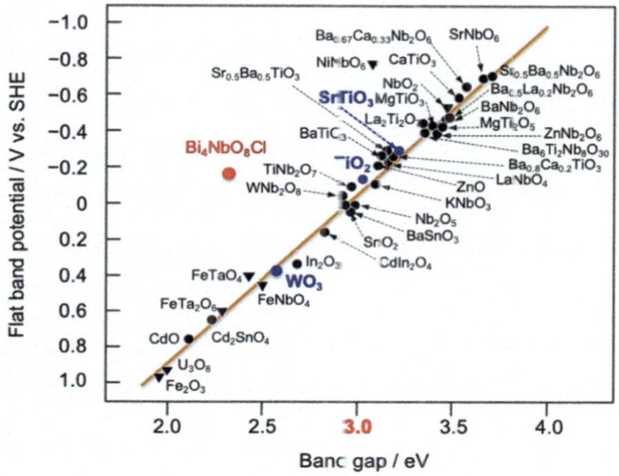

Fig. 9.12 Scaife's plot reproduced from his original paper [25]

As described above, the dominant contribution of O-2p orbitals to the density of states (DOS) around the VBMs of conventional metal oxides has hampered the demonstration of visible-light-induced water splitting by using pristine metal oxides for many years. Thus, band engineering has emerged and been extensively studied since the late 1990s for developing new oxide-based photocatalysis materials for visible-light-induced water splitting.

9.3.2 Cation-Doping into Perovskite-Type Metal Oxides

Doping of metal cations (e.g., Cr^{3+}, Mn^{4+}, and Fe^{3+}) into the lattice of UV-responsive metal oxides such as TiO_2 has been extensively investigated to introduce donor or acceptor levels within the band gap and thereby endow them with visible light absorption. Although such doping actually provided color to originally white TiO_2 samples, almost all of them showed negligibly low activity under visible light irradiation. Note that such cation doping reportedly decreases the activity of host material (e.g., TiO_2) drastically, even under bandgap excitation by UV light, which strongly suggests that the doped species function as recombination site between photoexcited electrons and holes via the redox cycle in transition metal cations (e.g., Fe^{3+}/Fe^{2+}).

Kudo et al. revealed that the appropriate doping of Cr^{3+} or Rh^{3+} into the Ti^{4+} sites of $SrTiO_3$ crystal provides intense absorption bands in the visible light region that are indeed available for photocatalytic reactions under visible light [26, 27]. For example, the active response of Rh-doped $SrTiO_3$ ($SrTiO_3$:Rh) is due to a new donor level induced by Rh^{3+} (Fig. 9.13a). An as-prepared $SrTiO_3$:Rh sample exhibits two onsets of the absorption band, namely a shoulder one at 520 nm and a broad band at 580–1,000 nm derived from Rh^{3+} and Rh^{4+} species, respectively. However, the Rh^{4+} could be easily reduced to Rh^{3+} by photoexcited electrons in the initial stage of

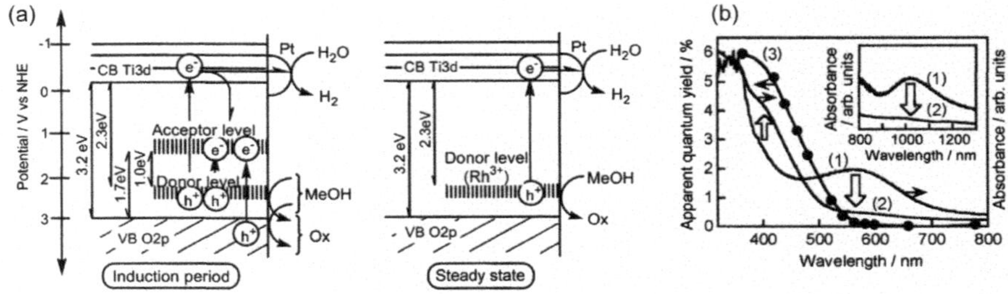

Fig. 9.13 **a** Proposed band structure and visible light response of Rh-doped $SrTiO_3$ photocatalyst. **b** Diffuse reflectance spectra of (1) before and (2) after photocatalytic reaction, and (3) action spectrum for H_2 evolution from an aqueous methanol solution over a $SrTiO_3$:Rh(1%) photocatalyst. Reprinted with permission from Ref. [26]. Copyright 2004, American Chemical Society

the photocatalytic reaction, thus maintaining the steady state at which electrons are excited from the Rh^{3+} donor level to the CB of $SrTiO_3$ while leaving the oxidized Rh^{4+} species. Then, the photoexcited electrons in CB reduce water to H_2 and the holes (positive charge) left on Rh^{4+} oxidize an electron donor such as methanol or iodide anion (I^-). Action spectra analysis on the $SrTiO_3$:Rh photocatalyst confirmed that visible light up to 550 nm can be utilized for H_2 evolution (Fig. 9.13b).

Photocatalytic water splitting under visible light was first demonstrated in 2001 by Abe et al. using a Z-scheme photocatalytic system (Fig. 9.14) that consisted of $SrTiO_3$ doped with Cr and Ta ($SrTiO_3$:Cr/Ta) for H_2 evolution, WO_3 for O_2 evolution, and an iodate/iodide (IO_3^-/I^-) redox couple as an electron mediator [10, 28]. Although the rate of H_2 evolution over only the Pt/$SrTiO_3$:Cr/Ta photocatalyst decreased remarkably with irradiation time due to the backward reaction (re-reduction of IO_3^- to I^- by photoexcited electrons), using both Pt/$SrTiO_3$:Cr/Ta and Pt/WO_3 photocatalysts resulted in the simultaneous and steady evolution of H_2 and O_2 from an aqueous NaI solution (5 mM, pH 6.5 without adjustment) under visible light irradiation ($\lambda >$ 410 nm). This is because both the reduction of IO_3^- to I^- and the oxidation of water to O_2 occurred efficiently and selectively over the Pt/WO_3 photocatalyst due to the preferential adsorption of IO_3^- to I^- on the WO_3 surface. The AQE for overall water splitting was determined to be ca. 1% at 420 in this system.

The $SrTiO_3$:Rh has been employed in various Z-scheme systems as a highly active H_2-evolving photocatalyst, not only in redox-mediated systems (e.g., with Fe^{3+}/Fe^{2+}) but also in interparticle electron transfer systems (i.e., without an electron mediator in solution) [29, 30], which will be introduced in the next section.

9.3.3 *Valence Band Engineering via s–p Interaction*

A bismuth vanadate $BiVO_4$ is a known polymorphic compound with three crystal structures: tetragonal zircon, monoclinic scheelite, and tetragonal scheelite phases. Kudo et al. first revealed that monoclinic scheelite $BiVO_4$ showed much higher

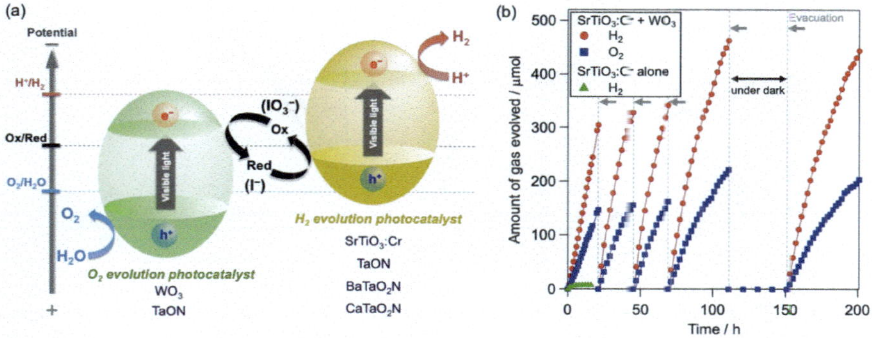

Fig. 9.14 **a** Overview of water splitting on Z-scheme photocatalysis system with a redox couple. **b** Time course of gas evolution of H_2 and O_2 using a mixture of Pt(0.3 wt%)/$SrTiO_3$ (Cr, Ta 4 mol% doped) and Pt(0.5 wt%)/WO_3 photocatalysts suspended in 5 mM of NaI aqueous solution under visible light irradiation

activity for water oxidation than conventional WO_3 under visible light in the presence of an electron acceptor such as Ag^+ [31]. They also pointed out a significant difference in bandgaps and photocatalytic activity depending on the crystal structures (Fig. 9.15). Monoclinic $BiVO_4$ has a smaller bandgap (2.4 eV) than tetragonal zircon (2.9 eV) and exhibited much higher O_2 evolution activity. The authors suggested that the monoclinic one possesses more negative VBM compared to the tetragonal zircon because the valence band of the monoclinic one is formed by Bi-6s or a hybrid orbital of Bi-6s and O-2p, whereas the valence band of tetragonal zircon is mainly formed by O-2p, similar to other conventional metal oxides. The more negative VBM and the contribution of Bi-6s orbitals were later confirmed by the combination of XPS, DFT calculation, and Mott–Schottky analysis in 2006 by Sayama et al. [32]. It was proposed that the monoclinic one possesses CBM located at −0.46 V (vs. SHE at pH 0) and VBM at +2.86 V (vs. SHE at pH 0).

Although the more negative VBM of monoclinic $BiVO_4$ compared to conventional oxides (basically fixed at +3.0 V vs. SHE) should be attributed to the hybridization between the O-2p and Bi-6s orbitals, the reason for such unusual hybridization (i.e., interaction) between the electron-filled O-2p and Bi-6s was elusive. Although the Bi-$6s^2$ electrons have been regarded as an inert lone pair, Walsh et al. recently proposed that there is strong s–p interaction between $6s^2$ electrons of Bi^{3+} and O-2p electrons in specific oxides such as $BiVO_4$ [33]. According to this theory, called the Revised Lone Pair (RLP) model, bonding and anti-bonding orbitals are formed from Bi-6s and O-2p orbitals, an unstable situation. However, the resultant anti-bonding orbital of (Bi-6s + O-2p)* is concurrently stabilized through the interaction with the unoccupied Bi-6p orbital, leading to an elevated energy level relative to the original O-2p orbital. Although even the monoclinic $BiVO_4$ with elevated VBM cannot generate H_2 from water due to the more positive CBM than the water reduction potential, it has been employed as a highly active O_2-evolving photocatalyst in various Z-scheme water splitting systems [34].

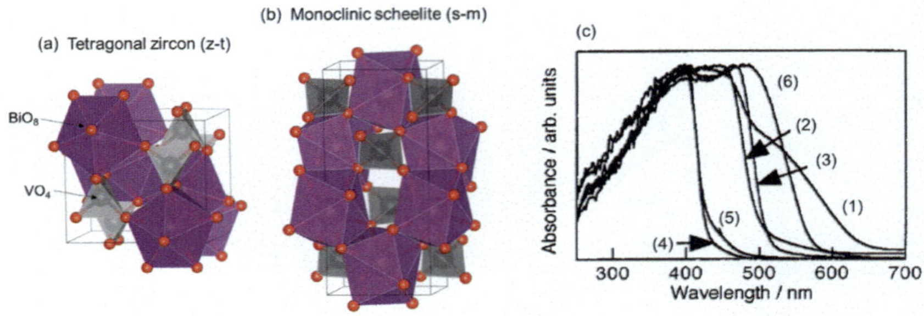

Fig. 9.15 Crystal structure of $BiVO_4$ with **a** tetragonal zircon and **b** monoclinic scheelite structure. **c** Diffuse reflectance spectra of $BiVO_4$ synthesized from KV_3O_8 at different ratios of vanadium to bismuth in the starting materials: (1) monoclinic $BiVO_4$ + KV_3O_8, (2, 3) monoclinic $BiVO_4$, (4, 5) tetragonal $BiVO_4$, and (6) KV_3O_8. Reprinted with permission from Ref. [31]. Copyright 1999, American Chemical Society

As described above, various oxide-based photocatalysts that can function under visible light have been developed based on band engineering. Their combination in Z-scheme reaction mechanisms afforded efficient water splitting by harvesting a considerable fraction of visible light. For example, both La- and Rh-co-doped $SrTiO_3$ and Mo-doped $BiVO_4$ photocatalysts were physically embedded into the gold layer (Fig. 9.16) to form an effective photocatalyst sheet that exhibited 1.1% solar-to-hydrogen (STH) energy conversion efficiency and > 30% AQE at 419 nm [11].

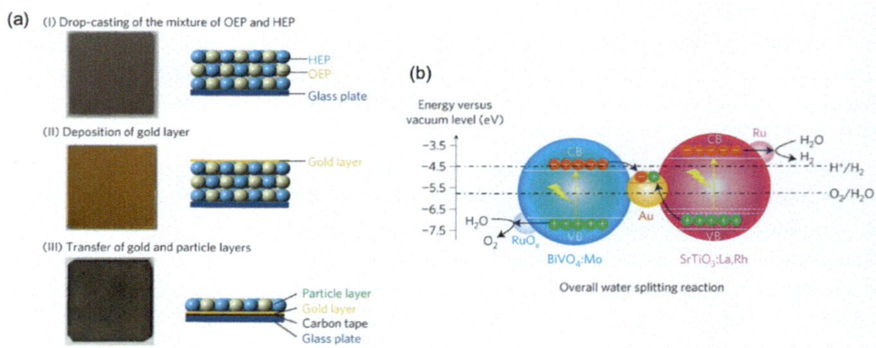

Fig. 9.16 **a** Illustration of the preparation of the $SrTiO_3$:La,Rh/Au/$BiVO_4$:Mo sheet by the particle transfer method, and **b** schematic of overall water splitting on the Ru-modified $SrTiO_3$:La,Rh/Au/$BiVO_4$:Mo sheet. Reproduced with permission from Ref. [11]. Copyright 2016, Springer Nature

9.3.4 Mixed-Anion Oxides for Z-scheme Water Splitting Under Visible Light

Then, various mixed-anion oxides such as oxynitrides and oxysulfides have been eagerly studied as promising photocatalysts because many of them have suitable band levels for both H_2 and O_2 evolution and narrow band gaps for visible-light response. In these mixed-anion oxides, the mixed VB formed via the hybridization of non-oxide p-orbitals and O-2p orbitals endows them with higher durability against oxidation than non-oxides (e.g., nitrides and sulfides), as well as elevated VBM leads to a narrower bandgap than corresponding oxides. Domen et al. reported that some oxynitride powder photocatalysts exhibit activity for H_2 production from water in the presence of a sacrificial electron donor such as methanol under visible light irradiation. Since N-2p orbitals have much more negative energy levels than O-2p orbitals, the VBMs of oxynitrides and nitrides generally locate a more negative potential than those of the corresponding metal oxides. For example, the VBM potentials of Ta-based semiconductors increase in the order Ta_2O_5 (3.6 V) < TaON (2.0 V) < Ta_3N_5 (1.6 V vs. SHE at pH 0) [35]. However, their DOS around CBMs consist of predominantly empty 5-d orbitals of Ta^{5+}, providing a similar CBM potential to Ta_2O_5 (in the range –0.3 to –0.5 V vs. SHE at pH 0), which are suitable for H_2 production.

In 2005, Abe and Domen et al. demonstrated water splitting under visible light using tantalum oxynitride (TaON) as an H_2 photocatalyst in the presence of an IO_3^-/I^- shuttle redox mediator [36]. The combination of Pt/TaON and PtO_x/WO_3 photocatalysts allows simultaneous H_2 and O_2 evolution with a stoichiometric ratio from NaI aqueous solution under visible light as seen in Fig 9.17, while H_2 evolution over Pt/TaON alone terminated rapidly due to the significant backward reaction (the re-reduction of IO_3^- to I^-). Although the (Pt/TaON)–(PtO_x/WO_3) system showed relatively low AQE (0.4% at 420 nm) in the initial stage, the surface modification of TaON with ZrO_2 significantly increased the apparent quantum yield to 6.3% at 420 nm [37]. In 2008, the same authors employed a series of mixed oxynitrides $ATaO_2N$ (A = Ca, Sr, Ba) as the H_2-evolving photocatalyst in the IO_3^-/I^--mediated Z-scheme system to harvest a wider range of visible light [38]. As shown in Fig. 9.18a, the absorption edges of $ATaO_2N$ shift to longer wavelengths as the ionic radius of A^{2+} (Ca^{2+}: 1.34 Å; Sr^{2+}: 1.44 Å; Ba^{2+}: 1.61 Å) increases, providing smaller band gaps (Ca^{2+}: 2.4 eV; Sr^{2+}: 2.1 eV; Ba^{2+}: 1.8 eV). The combination of experimental results and DFT calculations indicated that the CBMs of $ATaO_2N$ are significantly changed by the difference in ionic radius of A^{2+} while VBMs are sensitive to the anion ratio (N^{3-} to O^{2-}) [39]. As the radius of A^{2+} decreases, the tilting of the octahedra increases (Fig. 9.18b) thereby narrowing the conduction bands. Consequently, the decreased dispersion of the conduction bands raises CBMs. The (Pt/$BaTaO_2N$)–(PtO_x/WO_3) was found to be photoactive at wavelengths up to 660 nm, representing the first example of a photocatalytic and overall water-splitting that effectively utilizes visible light at wavelengths longer than 600 nm [40].

Although the VBs hybridized from non-oxide p-orbitals (e.g., N-2p) and O-2p orbitals endow them with greater durability against oxidation than simple non-oxides,

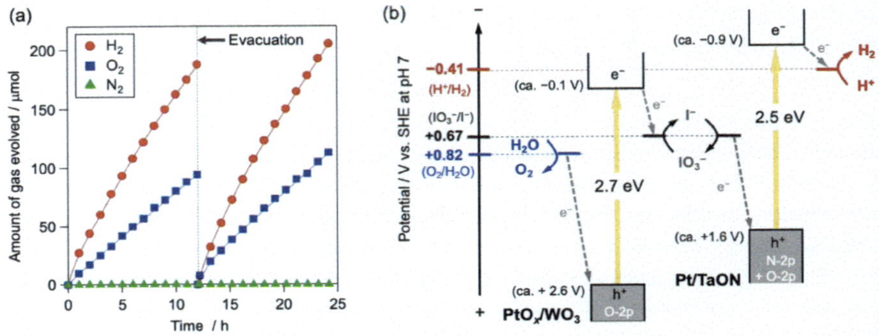

Fig. 9.17 **a** Time courses of gas evolution over a mixture of Pt/TaON and Pt/WO_3 under visible light irradiation from an aqueous NaI solution (5 mM, pH 6.5). **b** Speculated reaction mechanism

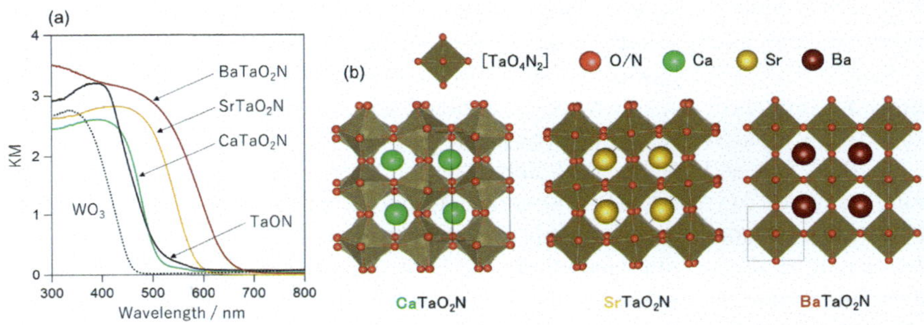

Fig. 9.18 **a** Diffused reflectance spectra and **b** crystal structures of $ATaO_2N$ (A = Ca, Sr, Ba)

it was pointed out that most oxynitrides still undergo self-oxidative decomposition whereby N^{3-} are oxidized to N_2 by photogenerated holes ($2N^{3-} + 6\ h^+ \rightarrow N_2$). Fortunately, no significant N_2 generation was evolved when such oxynitrides (e.g., TaON and $BaTaO_2N$) were employed as an H_2-evolving photocatalyst in the Z-scheme system with IO_3^-/I^-, as seen in Fig. 9.17a. It was revealed that I^- ions readily adsorb onto the surface of the TaON photocatalyst powder, thus enabling the efficient oxidation of I^- by holes and the effective suppression of self-oxidative deactivation of TaON [41]. Meanwhile, appreciable deactivation accompanied by N_2 generation was observed when these oxynitrides were used as an O_2-evolving photocatalyst in the presence of sacrificial (e.g., Ag^+) or reversible electron acceptors (e.g., IO_3^-, Fe^{3+}). Abe et al. have demonstrated that this unfavorable deactivation of oxynitride materials can be solved by loading cocatalyst nanoparticles (e.g., IrO_x, CoO_x) onto the surface of oxynitrides that promote water oxidation by holes, as illustrated in Fig. 9.19 [37, 42, 43]. The loading of effective cocatalysts enables the construction of Z-scheme systems employing oxynitrides as an O_2-evolving photocatalyst [44, 45].

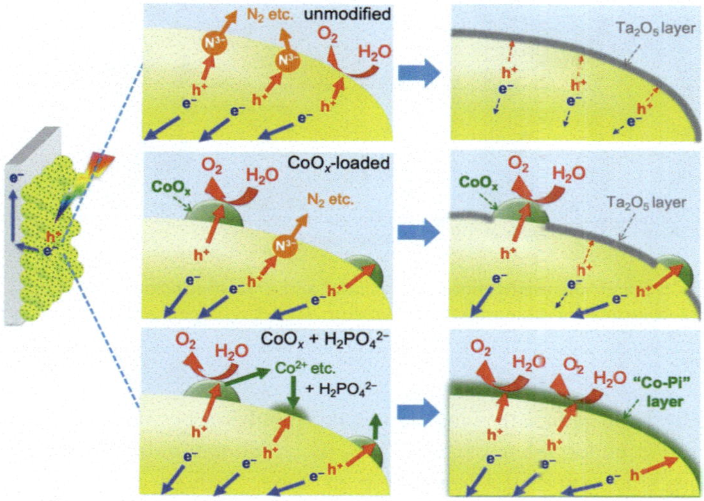

Fig. 9.19 Proposed mechanism for suppressed self-oxidative deactivation on TaON surface by cocatalyst loading

9.3.5 *Mixed-Anion Oxides for One-Step Water Splitting Under Visible Light*

As mentioned previously, the demonstration of one-step water splitting (i.e., using a single photocatalyst material) under visible light was difficult because of its strict requirements that the photocatalyst material must have a narrow bandgap, be stable under photoirradiation, and have suitable conduction and valence band levels for H_2 and O_2 production. In 2005, Domen et al. demonstrated water splitting into H_2 and O_2 under visible light using a GaN-rich solid solution of GaN and ZnO as a single photocatalyst material, namely $(Ga_{1-x}Zn_x)(N_{1-x}O_x)$ [46]. The most interesting and important aspect of this solid solution material is that it can absorb visible light, whereas pure GaN (bandgap: 3.4 eV) and ZnO (bandgap: 3.2 eV)—both of which have wurtzite structures with similar lattice parameters—cannot. As seen in Fig. 9.20, the product obtained by heating a mixture of Ga_2O_3 and ZrO (1:2) under NH_3 flow (100–500 mL/min) at 1,123 K for 5–20 h exhibited a yellow color.

Although the $(Ga_{1-x}Z_x)(N_{1-x}O_x)$ samples with > 75% ZnO concentration can absorb long-wavelength visible light, only the GaN-rich one with a ZnO concentration < 22% exhibited activity for overall water splitting. For example, the $(Ga_{1-x}Z_x)(N_{1-x}O_x)$ (x = 0.13) sample with 2.6 eV band gap exhibited an average AQE of 0.14% in the range of 300–480 nm after the loading the RuO_2 cocatalyst. The application of a newly developed co-catalyst, Rh-Cr mixed oxide $(Rh_{2-y}Cr_yO_3)$ nanoparticles, drastically improved the efficiency of water splitting on the $(Ga_{1-x}Z_x)(N_{1-x}O_x)$ photocatalyst up to about 2.5% under irradiation in the range 420–440 nm [48]. The origin of visible light absorption emerged by solid solution formation was at first

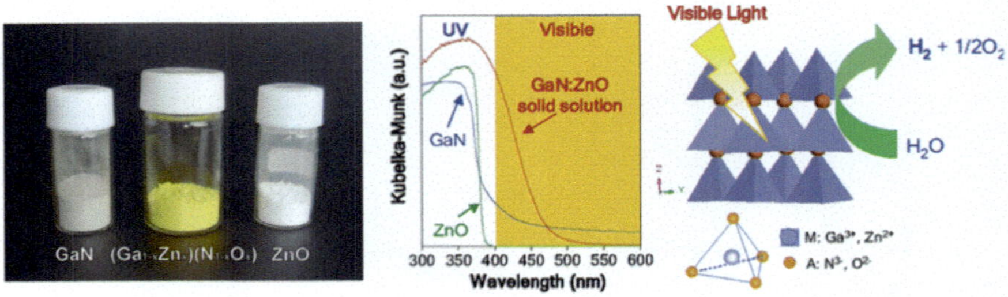

Fig. 9.20 Photograph, diffused reflectance spectra, and crystal structure of $(Ga_{1-x}Z_x)(N_{1-x}O_x)$ photocatalyst. Reprinted with permission from Ref. [46]. Copyright 2005, American Chemical Society

explained by band gap narrowing due to the repulsion between Zn-3d and N-2p electrons in the upper valence band. Then, the combination of photoluminescence spectroscopy studies and DFT calculation on GaN-rich $(Ga_{1-x}Z_x)(N_{1-x}O_x)$ materials suggest that anomalous visible light absorption is derived from the electron transitions from the Zn acceptor level to the conduction band, as illustrated in Fig. 9.21 [47]. Since the degree of Zn doping has a higher atomic density (10%) than conventional doped photocatalysts (a few percent or less), the doping level likely functions as an impurity band with a high density of states, enabling the Zn acceptor level to be filled with electrons derived from O donor levels or thermal excitation. Consequently, unusually visible light absorption emerges via electron transition from the Zn acceptor level to the CB.

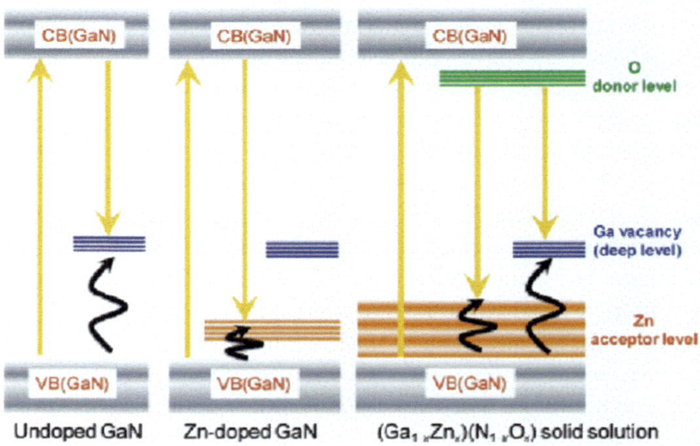

Fig. 9.21 Expected energy-level diagram for impurity levels in undoped GaN, Zn-doped GaN, and GaN-rich $(Ga_{1-x}Zn_x)(N_{1-x}O_x)$ solid solutions. Arrows denote photoabsorption, photoluminescence, and thermal relaxation. Reprinted with permission from Ref. [47]. Copyright 2010, American Chemical Society

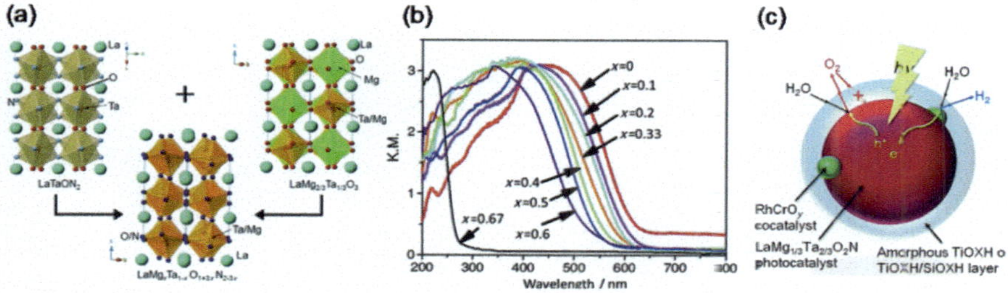

Fig. 9.22 **a** Crystal structure and **b** diffused reflectance spectra of $LaMg_xTa_{1-x}O_{1+3x}N_{2-3x}$. **c** Reaction mechanism for water splitting on a surface-coated photocatalyst. Reprinted with permission from Ref. [49]. Copyright 2015, John Wiley and Sons

Another successful example of a photocatalyst for one-step water splitting is $LaMg_xTa_{1-x}O_{1+3x}N_{2-3x}$, which is the solid solution between two perovskite materials with comparable lattice constants (Fig. 9.22a), i.e., a narrow band gap oxynitride $LaTaON_2$ (band gap: 1.9 eV) and a wide band gap oxide $LaMg_{2/3}Ta_{1/3}O_3$ (band gap: 4.6 eV) [49]. Neither of them can split water under visible light; however, fine-tuning of band levels based on solid solution formation enabled them to be active photocatalysts. The $LaMg_xTa_{1-x}O_{1+3x}N_{2-3x}$ solid solution sample with x = 1/3 (i.e., $LaMg_{1/3}Ta_{2/3}O_2N$) has a moderate band gap of 2.1 eV (Fig. 9.22b) and exhibited the highest activity among them for overall water splitting (AQE: 0.03% at 440 nm). On these particles, amorphous oxyhydroxide layers, not only a cocatalyst for H_2 evolution, works to control the surface redox reactions. Coating the water oxidation site with an amorphous oxyhydroxide layer clearly suppressed N_2 evolution, implying that the prompt h^+ injection into oxygen species in the amorphous layer (i.e., OH groups or hydrated water) can prevent the excessive accumulation of h^+ at the N^{3-} anions around the surface and thereby suppress their oxidation to N_2, as illustrated in Fig. 9.22c.

Very recently, an oxysulfide $Y_2Ti_2O_5S_2$ (Fig. 9.23a) has also been demonstrated to be an active photocatalyst for overall water splitting under visible light [50]. Similar to oxynitrides, many oxysulfides have smaller bandgaps compared to their corresponding oxides due to the significant contribution of high-energy S-3p orbitals to VB formation hybridized with O-2p. For example, Domen et al. reported in 2002 that an oxysulfide $Sm_2Ti_2O_5S_2$ possessed appropriate band levels both for H_2 and O_2 evolution along with a narrow bandgap (1.9 eV), and indeed exhibited activity for H_2 or O_2 evolution in the presence of a sacrificial electron donor or acceptor, respectively [51]. However, the demonstration of overall water splitting has probably been hampered by the instability of S^{2-} anions during water oxidation. The same authors demonstrated that the loading of IrO_2 and Rh/Cr_2O_3 core–shell particles as O_2 and H_2 evolution co-catalysts, respectively, along with fine-tuning the reaction conditions (e.g., pH), enabled the simultaneous and stoichiometric generation of O_2 and H_2 evolution by harvesting a relatively wide range of visible light up to 600 nm (Fig. 9.23b).

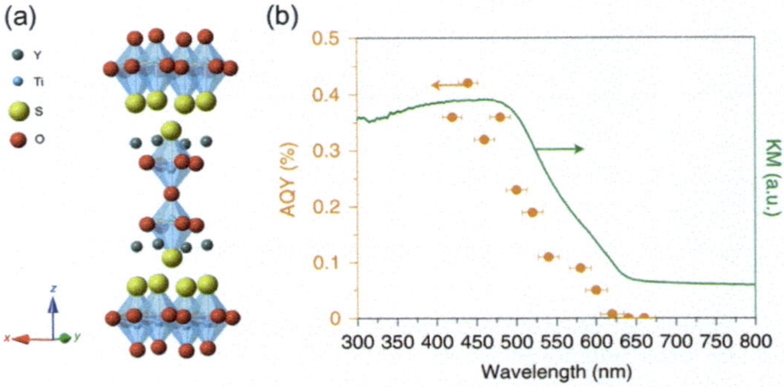

Fig. 9.23 **a** Crystal structure of $Y_2Ti_2O_5S_2$ and **b** dependence of AQE on incident light wavelength for overall water splitting along with diffused reflectance spectra of $Y_2Ti_2O_5S_2$. Reprinted with permission from Ref. [50]. Copyright 2019, Springer Nature

9.3.6 Layered Oxyhalides as Promising Photocatalysts for Visible-Light-Induced Water Splitting

Recently, the present authors have revealed that a series of layered perovskite oxyhalides such as Bi_4MO_8X (M = Nb, Ta, X = Cl, Br) fulfill the necessary conditions for visible-light-induced water splitting (sufficiently high stability and appropriate band levels both for H_2 and O_2 evolution) due to their unique VB structures [52]. For example, Bi_4NbO_8Cl belongs to the $n = 1$ members of the Sillén–Aurivillius perovskite family $[(Bi_2O_2)_2X]^{3+}[A_{n\text{-}1}BnO_{3n+1}]^3$, which is composed of fluorite $[Bi_2O_2]$, perovskite $[NbO_4]$, and halogen [Cl] layers as depicted in Fig. 9.24. DFT calculations (Fig. 9.25a) show that the VBM mainly consists of O-2p orbitals instead of Cl-3p, but its position determined by electrochemical measurement is much more negative (ca. 2.2 V vs. SHE) than those of conventional metal oxides and even the abovementioned monoclinic $BiVO_4$. The significantly elevated VBMs of Bi_4NbO_8Cl can be also understood by the RLP model in which strong hybridization between electron-filled Bi-6 s and O-2p occurs with the help of empty Bi-6p orbitals (Fig. 9.25b) [53]. In addition, Madelung site potential analysis for Bi_4NbO_8Cl indicated that the oxygen sites in the fluorite layer were electrostatically destabilized, which also certainly contributed to the elevation of the VBM that mainly consists of O-2p [54]. Consequently, Bi_4NbO_8Cl possesses both a narrow bandgap (2.4 eV) for visible light absorption up to ca. 500 nm and a more negative CBM (ca. –0.2 V vs. SHE) than the water reduction potential. This appropriate band level was testified by the substantial generation of H_2 or O_2 in the presence of an electron donor or acceptor, respectively. The substantially high durability against self-oxidative deactivation was also confirmed by the long-term water splitting in a Z-scheme system in which Bi_4NbO_8Cl was employed as the O_2-evolving photocatalyst. We attributed the observed high stability to the occupation of the O-2p orbitals around the VBM. Since O^- anions are known to be relatively stable, photogenerated holes populated at the O-2p orbitals will not lead to self-oxidative decomposition but rather to the oxidation of water.

More oxyhalides with various perovskite layers (n) have been explored. However, reports on the synthesis of double-layered ($n = 2$) perovskites were limited compared to single-layered ones, probably due to the severe conditions required for their synthesis in the conventional solid-state reaction. To solve this problem, a new two-step synthesis method including polymerized complex synthesis was developed [55]. Briefly, fine particles of a precursor oxide (e.g., $SrBi_2Ta_2O_9$) were first prepared via the polymerized complex method, followed by heating with an oxyhalide (e.g., $SrBiO_2Cl$) to form $Sr_2Bi_3Ta_2O_{11}Cl$. This two-step method enabled us to synthesize 14 compounds of $A_4A'M_2O_{11}Cl$ (A and A' = Sr^{2+}, Ba^{2+}, Pb^{2+} and Bi^{3+}; M = Ti^{4+}, Nb^{5+} and Ta^{5+}), including 10 unreported ones. The results of band-level analysis and DFT calculation suggested a great possibility of extending cation substitution for band engineering in these layered oxyhalides (Fig. 9.26).

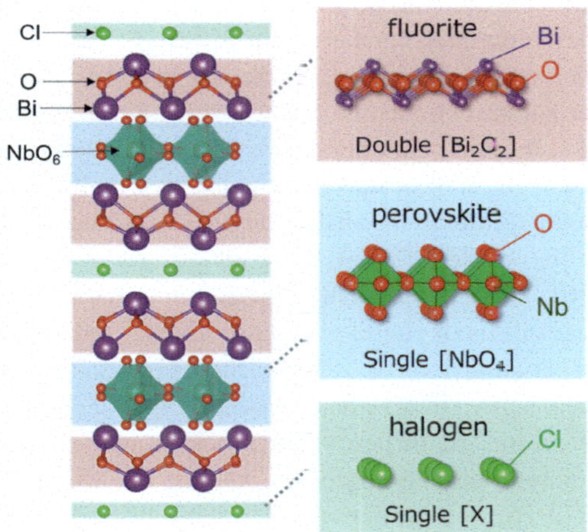

Fig. 9.24 Crystal structure of a Sillén–Aurivillius perovskite Bi_4NbO_8Cl

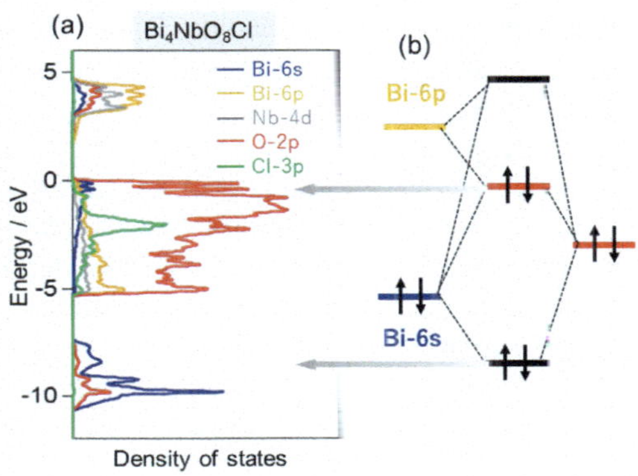

Fig. 9.25 **a** Density of states (DOS) of Bi_4NbO_8Cl. **b** Interactions among Bi-6 s, Bi-6p, and O-2p orbitals based on the revised lone pair (RLP) model

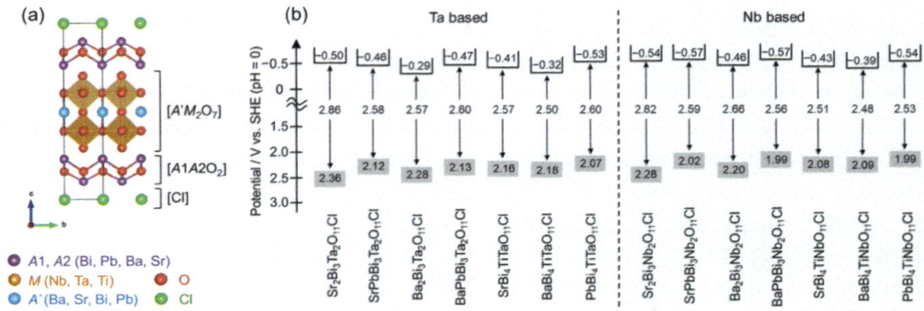

Fig. 9.26 **a** Crystal structure of double layered ($n = 2$) perovskites $A_4A'M_2O_{11}Cl$ (A and A' = Sr^{2+}, Ba^{2+}, Pb^{2+} and Bi^{3+}; M = Ti^{4+}, Nb^{5+} and Ta^{5+}). **b** Band edge positions of various Sillén Aurivillius-type double layered perovskites

Compounds containing iodide (I^-) anions, which have high polarizability, have attracted considerable attention not only due to their narrow band gaps but also due to their elongated carrier lifetime, as typified by halide perovskite solar cells. However, such compounds (e.g., layered oxyiodides) have thus far been regarded as unsuitable for harsh photocatalytic water splitting due to the facile occurrence of self-oxidation of I^- by photogenerated holes. Our previous study on Sillén oxyhalides $PbBiO_2X$ (X = Cl, Br, I) indicated that the iodide $PbBiO_2X$ was unstable for water oxidation due to the dominant contribution of I-5p orbitals to DOS around VBM, in stark contrast to the stable Cl^- and Br^- counterparts of which O-2p mainly formed their VBMs [56]. Then, the present authors found that $Ba_2Bi_3Nb_2O_{11}I$, one of the $n = 2$ Sillén – Aurivillius oxyhalides, not only possessed a smaller band gap (2.3 eV) than Cl^- and Br^- counterparts (2.6 eV) but also functions as a stable photocatalyst without suffering from self-oxidative deactivation (Fig. 9.27) [57]. DFT calculations reveal that the O-2p orbitals in the perovskite block—rather than the fluorite Bi_2O_2 block as previously pointed out—anomalously push up the VBM. This unusually high VBM consisting of O-2p could be explained by a modified Madelung potential analysis that takes into account the high polarizability of iodide. In addition, the highly polarizable iodide contributes to a longer carrier lifetime of $Ba_2Bi_3Nb_2O_{11}I$, allowing for a significantly higher quantum efficiency than its Cl^- and Br^- counterparts.

We also attempted to synthesize Sillén–Aurivillius perovskite compounds with higher numbers of perovskite layers (n = 3–5) while their synthesis was much more difficult than $n = 2$. In general, the synthesis of layered perovskites requires high-temperature calcination with increasing n. However, oxyhalides suffer from volatilization of halogen species under such high-temperature treatment. Then, to solve the trade-off, we employed a "bricklaying" synthesis (Fig. 9.28) [58]. The first step is a reaction between an Aurivillius phase $(Bi_2O_2)(A_{n-1}Ti_nO_{3n+1})$ with $n = k$ and a plain perovskite $BaTiO_3$ to form an Aurivillius phase with $n = k + 1$. The second step is a reaction between a Sillén phase $BaBiO_2Cl$ and the pre-synthesized Aurivillius phase with $n = k + 1$ to form a targeted Sillén–Aurivillius

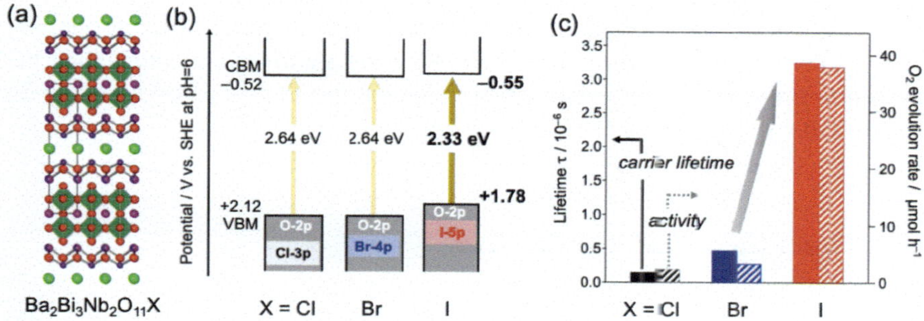

Fig. 9.27 **a** Crystal structure and **b** band structures of $Ba_2Bi_3Nb_2O_{11}X$ (X = Cl, Br, I). **c** Correlation between carrier lifetime and photocatalytic activity for O_2 evolution from $AgNO_3$aq. under visible light

phase with $n = k + 1$. By employing this "bricklaying" method, three new oxychlorides, $BaBi_5Ti_3O_{14}Cl$, $Ba_2Bi_5Ti4O_{17}Cl$, and $Ba_3Bi_5Ti_5O_{20}Cl$ with triple-, quadruple- and quintuple-perovskite layers, respectively, were successfully synthesized in pure phases. Their photocatalytic activity was enhanced by increasing the number of perovskite layers. DFT calculation for $BaBi_5Ti_3O_{14}Cl$ (n = 3) and $Ba_3Bi_5Ti_5O_{20}Cl$ (n = 5) indicates that their valence and conduction bands are separated spatially on the perovskite and fluorite layers, respectively, which probably promotes carrier separation and thereby enhances the photocatalytic activity.

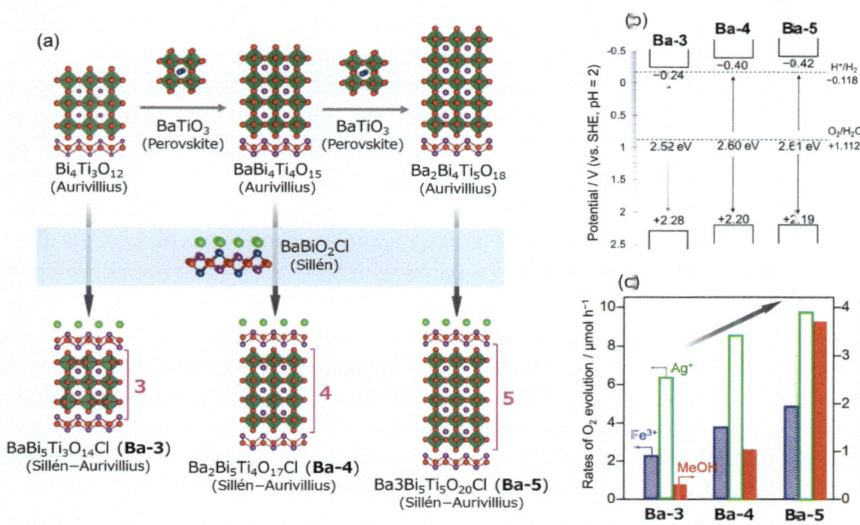

Fig. 9.28 **a** Illustration of the bricklaying synthesis for Sillén–Aurivillius phases $BaBi_5Ti_3O_{14}Cl$ (Ba-3), $Ba_2Bi_5Ti_4O_{17}Cl$ (Ba-4), and $Ba_3Bi_5Ti_5O_{20}Cl$ (Ba-5) as the representatives. **b** Their band levels and **c** photocatalytic activities

These studies on new oxyhalide materials provide new strategies for developing stable photocatalysts for visible-light-induced water splitting by manipulating the interaction between the s orbitals of post-transition metals (e.g., Bi-6 s) and O-2p orbitals in the specific crystal structure.

9.4 Conclusions and Perspective

In this chapter, we briefly looked back at the relatively long history (about 40 years) of the development of photocatalyst materials and systems toward highly efficient water splitting, particularly focusing on the pioneering works in crystal and band engineering for improved efficiency. In the early stage of research in 1980–2000, various metal oxides, starting from the well-known perovskite $SrTiO_3$, have been extensively explored primarily due to their high stability against self-oxidative deactivation, which is often observed in non-oxide semiconductors such as simple metal sulfides. Various metal oxides with specific crystal structures such as layered or tunnel ones have been developed as efficient photocatalysts for water splitting under UV light, as well as effective cocatalysts for boosting H_2 production. These early studies on oxide photocatalysts have provided some useful guiding principles for achieving efficient water splitting, not only using oxides but also other oxide-based ones. Meanwhile, it was recognized from an early time that the effective utilization of visible light, which accounts for almost half of the solar light spectrum on the surface of the Earth, is indispensable for achieving the high solar-to-hydrogen conversion efficiency required for practical applications. However, the dominant contribution of O-2p orbitals to the DOS around the VBMs of metal oxides has hampered the demonstration of visible-light-induced water splitting. Thus, band engineering has emerged and extensively studied since the late 1990s, including strategies such as cation-doping, valence band engineering via s–p interactions, or introducing high-energy p orbitals of non-oxide anions, as well as constructing a Z-scheme type water-splitting system. Then, water splitting under visible light has indeed been demonstrated using various heterogeneous photocatalytic systems after 2001, and highly efficient photocatalyst panes with STH $> 1\%$ have been developed based on Z-scheme-type electron transfer. However, the STH values must be further improved by at least 5% by introducing new photocatalyst materials with narrower band gaps and/or showing much higher AQE values. At present, the AQE values of visible-light-responsive photocatalysts are generally low ($<10\%$) due to the significant occurrence of the recombination of photoexcited electrons and holes inside the semiconductor particles, as well as at the surface in part. Therefore, introducing new and effective strategies for achieving the efficient separation of electrons and holes in powdered semiconductor particles is highly desirable. The construction of p–n heterojunctions between two different materials and/or designing a built-in separation mechanism—as typified by recent results for aluminum-doped $SrTiO_3$ showing almost 100% internal quantum efficiency—will be future important challenges to realize the dream of clean H_2 production.

References

1. Kudo A, Miseki Y. Heterogeneous photocatalyst materials for water splitting. Chem Soc Rev. 2009;38:253–78.
2. Osterloh FE. Inorganic nanostructures for photoelectrochemical and photocatalytic water splitting. Chem Soc Rev. 2013;42:2294–320.
3. Chen SS, Takata T, Domen K. Particulate photocatalysts for overall water splitting. Nat Rev Mater. 2017;2:17050.
4. Fujishima A, Honda K. Electrochemical photolysis of water at a semiconductor electrode. Nature. 1972;238:37–8.
5. Sato S, White JM. Photodecomposition of water over Pt/TiO_2 catalysts. Chem Phys Lett. 1980;72:83–6.
6. Domen K, Naito S, Onishi T, Tamaru K. Study of the photocatalytic decomposition of water vapor over a nickel(II) oxide-strontium titanate ($SrTiO_3$) catalyst. J Phys Chem. 1982;86:3657–61.
7. Momma K, Izumi F. VESTA 3 for three-dimensional visualization of crystal, volumetric and morphology data. J Appl Crystallogr. 2011;44:1272–6.
8. Kato H, Asakura K, Kudo A. Highly efficient water splitting into H_2 and O_2 over lanthanum-doped $NaTaO_3$ Photocatalysts with high crystallinity and surface nanostructure. J Am Chem Soc. 2003;125:3082–9.
9. Abe R. Recent progress on photocatalytic and photoelectrochemical water splitting under visible light irradiation. J Photoch Photobio C. 2010;11:179–209.
10. Sayama K, Mukasa K, Abe R, Abe Y, Arakawa H. Stoichiometric water splitting into H_2 and O_2 using a mixture of two different photocatalysts and an IO_3^-/I^- shuttle redox mediator under visible light irradiation. Chem Commun. 2001:2416–2417.
11. Wang Q, Hisatomi T, Jia QX, Tokudome H, Zhong M, Wang CZ, et al. Scalable water splitting on particulate photocatalyst sheets with a solar-to-hydrogen energy conversion efficiency exceeding 1%. Nat Mater. 2016;15:611–5.
12. Pinaud BA, Benck JD, Seitz LC, Forman AJ, Chen Z, Deutsch TG. Technical and economic feasibility of centralized facilities for solar hydrogen production via photocatalysis and photoelectrochemistry. Energy Environ Sci. 2013;6:1983–2002.
13. Domen K, Kudo A, Shibata M, Tanaka A, Maruya K, Onishi T. Novel photocatalysts, ion-exchanged $K_4Nb_6O_{17}$, with a layer structure. J Chem Soc Chem Commun. 1986:1706–1707.
14. Kudo A, Tanaka A, Domen K, Maruya K, Aika K, Onishi T. Photocatalytic decomposition of water over NiO-$K_4Nb_6O_{17}$ catalyst. J Catal. 1988;111:67–76.
15. Kudo A, Sayama K, Tanaka A, Asakura K, Domen K, Maruya K, et al. Nickel-loaded $K_4Nb_6O_{17}$ photocatalyst in the decomposition of H_2O into H_2 and O_2: structure and reaction mechanism. J Catal. 1989;120:337–52.
16. Takata T, Furumi Y, Shinohara K, Tanaka A, Hara M, Kondo JN, et al. Photocatalytic decomposition of water on spontaneously hydrated layered perovskites. Chem Mater. 1997;9:1063–4.
17. Inoue Y, Asai Y, Sato K. Photocatalysts with tunnel structures for decomposition of water. Part 1.—$BaTi_4O_9$, a pentagonal prism tunnel structure, and its combination with various promoters. J Chem Soc Faraday Trans. 1994;90:797–802.
18. Kohno M, Ogura S, Sato K, Inoue Y. Properties of photocatalysts with tunnel structures: formation of a surface lattice O− radical by the UV irradiation of $BaTi_4O_9$ with a pentagonal-prism tunnel structure. Chem Phys Lett. 1997;267:72–6.
19. Hiramachi Y, Fujimori H, Yamakata A, Sakata Y. Achievement of high photocatalytic performance to $BaTi_4O_9$ toward overall H_2O splitting. ChemCatChem. 2019;11:6213–7.
20. Ogura S, Kohno M, Sato K, Inoue Y. Photocatalytic properties of $M_2Ti_6O_{13}$ (M=Na, K, Rb, Cs) with rectangular tunnel and layer structures: behavior of a surface radical produced by UV irradiation and photocatalytic activity for water decomposition. Phys Chem Chem Phys. 1999;1:179–83.
21. Kato H, Kudo A. Water splitting into H_2 and O_2 on Alkali Tantalate Photocatalysts $ATaO_3$ (A = Li, Na, and K). J Phys Chem B. 2001;105:4285–92.

22. Takata T, Jiang J, Sakata Y, Nakabayashi M, Shibata N, Nandal V, et al. Photocatalytic water splitting with a quantum efficiency of almost unity. Nature. 2020;581:411–4.
23. Chiang TH, Lyu H, Hisatomi T, Goto Y, Takata T, Katayama M, et al. Efficient photocatalytic water splitting using al-doped $SrTiO_3$ Coloaded with molybdenum oxide and rhodium-chromium oxide. ACS Catal. 2018;8:2782–8.
24. Ham Y, Hisatomi T, Goto Y, Moriya Y, Sakata Y, Yamakata A, et al. Flux-mediated doping of $SrTiO_3$ photocatalysts for efficient overall water splitting. J Mater Chem A. 2016;4:3027–33.
25. Scaife DE. Oxide semiconductors in photoelectrochemical conversion of solar energy. Sol Energy. 1980;25:41–54.
26. Konta R, Ishii T, Kato H, Kudo A. Photocatalytic activities of noble metal ion doped $SrTiO_3$ under visible light irradiation. J Phys Chem B. 2004;108:8992–5.
27. Kato H, Kudo A. Visible-Light-response and photocatalytic activities of TiO_2 and $SrTiO_3$ Photocatalysts Codoped with antimony and chromium. J Phys Chem B. 2002;106:5029–34.
28. Abe R, Sayama K, Sugihara H. Development of New photocatalytic water splitting into H_2 and O_2 using two different semiconductor photocatalysts and a shuttle redox mediator IO_3^-/I^-. J Phys Chem B. 2005;109:16052–61.
29. Kato H, Hori M, Konta R, Shimodaira Y, Kudo A. Construction of Z-scheme type heterogeneous photocatalysis systems for water splitting into H_2 and O_2 under visible light irradiation. Chem Lett. 2004;33:1348–9.
30. Sasaki Y, Iwase A, Kato H, Kudo A. The effect of co-catalyst for Z-scheme photocatalysis systems with an Fe^{3+}/Fe^{2+} electron mediator on overall water splitting under visible light irradiation. J Catal. 2008;259:133–7.
31. Kudo A, Omori K, Kato H. A novel aqueous process for preparation of crystal form-controlled and highly crystalline $BiVO_4$ powder from layered Vanadates at room temperature and its photocatalytic and Photophysical properties. J Am Chem Soc. 1999;121:11459–67.
32. Sayama K, Nomura A, Arai T, Sugita T, Abe R, Yanagida M, et al. Photoelectrochemical decomposition of water into H_2 and O_2 on porous $BiVO_4$ thin-film electrodes under visible light and significant effect of ag ion treatment. J Phys Chem B. 2006;110:11352–60.
33. Walsh A, Watson GW, Payne DJ, Edgell RG, Guo J, Glans P-A, et al. Electronic structure of the α and δ phases of Bi_2O_3: Acombined ab initio and x-ray spectroscopy study. Phys Rev B. 2006;73: 235104.
34. Wang YO, Suzuki H, Xie JJ, Tomita O, Martin DJ, Higashi M, et al. Mimicking natural photosynthesis: solar to renewable h_2 fuel synthesis by Z-scheme water splitting systems. Chem Re. 2018;118:5201–41.
35. Chun WJ, Ishikawa A, Fujisawa H, Takata T, Kondo JN, Hara M, et al. Conduction and valence band positions of Ta_2O_5, TaON, and Ta_3N_5 by UPS and electrochemical methods. J Phys Chem B. 2003;107:1798–803.
36. Abe R, Takata T, Sugihara H, Domen K. Photocatalytic overall water splitting under visible light by TaON and WO_3 with an IO_3^-/I^- shuttle redox mediator. Chem Commun. 2005:3829–3831.
37. Maeda K, Higashi M, Lu D, Abe R, Domen K. Efficient Nonsacrificial water splitting through two-step photoexcitation by visible light using a modified Oxynitride as a hydrogen evolution Photocatalyst. J Am Chem Soc. 2010;132:5858–68.
38. Higashi M, Abe R, Teramura K, Takata T, Ohtani B, Domen K. Two step water splitting into H_2 and O_2 under visible light by $ATaO_2N$ (A=Ca, Sr, Ba) and WO_3 with IO_3^-/I^- shuttle redox mediator. Chem Phys Lett. 2008;452:120–3.
39. Balaz S, Porter SH, Woodward PM, Brillson LJ. Electronic structure of tantalum Oxynitride perovskite Photocatalysts. Chem Mater. 2013;25:3337–43.
40. Higashi M, Abe R, Takata T, Domen K. Photocatalytic overall water splitting under visible light using $ATaO_2N$ (A = Ca, Sr, Ba) and WO_3 in a IO_3^-/I^- shuttle redox mediated system. Chem Mater. 2009;21:1543–9.
41. Abe R, Higashi M, Domen K. Overall water splitting under visible light through a two-step photoexcitation between TaON and WO_3 in the presence of an iodate-iodide shuttle redox mediator. Chemsuschem. 2011;4:228–37.

42. Higashi M, Domen K, Abe R. Highly stable water splitting on Oxynitride TaON Photoanode system under visible light irradiation. J Am Chem Soc. 2012;134:6968–71.
43. Higashi M, Tomita O, Abe R. Porous TaON Photoanodes loaded with cobalt-based Cocatalysts for efficient and stable water oxidation under visible light. Top Catal. 2016;59:740–9.
44. Higashi M, Abe R, Ishikawa A, Takata T, Ohtani B, Domen K. Z-scheme overall water splitting on modified-TaON Photocatalysts under visible Light (λ<500 nm). Chem Lett. 2008;37:138–9.
45. Iwase Y, Tomita O, Higashi M, Nakada A, Abe R. Effective strategy for enhancing Z-scheme water splitting with the IO_3^-/I^- redox mediator by using a visible light responsive TaON photocatalyst co-loaded with independently optimized two different cocatalysts. Sustain Energy Fuels. 2019;3:1501–8.
46. Maeda K, Takata T, Hara M, Saito N, Inoue Y, Kobayashi H, et al. GaN:ZnO solid solution as a Photocatalyst for visible-light-driven overall water splitting. J Am Chem Soc. 2005;127:8286–7.
47. Yoshida M, Hirai T, Maeda K, Saito N, Kubota J, Kobayashi H, et al. Photoluminescence spectroscopic and computational investigation of the origin of the visible light response of $(Ga_{1-x}Zn_x)(N_{1-x}O_x)$ Photocatalyst for overall water splitting. J Phys Chem C. 2010;114:15510–5.
48. Maeda K, Teramura K, Lu DL, Takata T, Saito N, Inoue Y, et al. Photocatalyst releasing hydrogen from water. Nature. 2006;440:295.
49. Pan CS, Takata T, Nakabayashi M, Matsumoto T, Shibata N, Ikuhara Y, et al. A complex Perovskite-type oxynitride: the first Photocatalyst for water splitting operable at up to 600 nm. Angew Chem Int Edit. 2015;54:2955–9.
50. Wang Q, Nakabayashi M, Hisatomi T, Sun S, Akiyama S, Wang Z, et al. Oxysulfide photocatalyst for visible-light-driven overall water splitting. Nature. 2019;18:827–32.
51. Ishikawa A, Takata T, Kondo JN, Hara M, Kobayashi H, Domen K. Oxysulfide $Sm_2Ti_2S_2O_5$ as a stable Photocatalyst for water Oxidation and reduction under visible light irradiation ($\lambda \leq$ 650 nm). J Am Chem Soc. 2002;124:13547–53.
52. Fujito H, Kunioku H, Kato D, Suzuki H, Higashi M, Kageyama H, et al. Layered perovskite oxychloride Bi_4NbO_8Cl: a stable visible light responsive Photocatalyst for water splitting. J Am Chem Soc. 2016;138:2082–5.
53. Kunioku H, Higashi M, Tomita O, Yabuuchi M, Kato D, Fujito H, et al. Strong hybridization between Bi-6s and O-2p orbitals in Sillén-Aurivillius perovskite Bi_4MO_8X (M = Nb, Ta; X = Cl, Br), visible light photocatalysts enabling stable water oxidation. J Mater Chem A. 2018;6:3100–7.
54. Kato D, Hongo K, Maezono R, Higashi M, Kunioku H, Yabuuchi M, et al. Valence band engineering of layered bismuth Oxyhalides toward stable visible-light water splitting: madelung site potential analysis. J Am Chem Soc. 2017;139:18725–31.
55. Nakada A, Higashi M, Kimura T, Suzuki H, Kato D, Okajima H, et al. Band engineering of double-layered Sillén-Aurivillius Perovskite oxychlorides for visible-light-driven water splitting. Chem Mater. 2019;31:3419–29.
56. Suzuki H, Kunioku H, Higashi M, Tomita O, Kato D, Kageyama H, et al. Lead Bismuth Oxyhalides $PbBiO_2X$ (X = Cl, Br) as visible-light-responsive Photocatalysts for water oxidation: role of lone-pair electrons in valence band engineering. Chem Mater. 2018;30:5862–9.
57. Ogawa K, Suzuki H, Zhong CC, Sakamoto R, Tomita O, Saeki A, et al. Layered Perovskite Oxyiodide with narrow band gap and long lifetime carriers for water splitting Photocatalysis. J Am Chem Soc. 2021;143:8446–53.
58. Ozaki D, Suzuki H, Ogawa K, Sakamoto R, Inaguma Y, Nakashima K, et al. Synthesis, band structure and photocatalytic properties of Sillén-Aurivillius oxychlorides $BaBi_5Ti_3O_{14}Cl$, $Ba_2Bi_5Ti_4O_{17}Cl$ and $Ba_3Bi_5Ti_5O_{20}Cl$ with triple-, quadruple- and quintuple-perovskite layers. J Mater Chem A. 2021;9:8332–40.

Chapter 10
Metal Oxide Catalysts in Relation to Environmental Protection and Energy Conversion

Saburo Hosokawa and Teruki Motohashi

10.1 General Background for Environmental Protection and Energy Conversion

Global warming is becoming a serious problem, leading to an urgent need for efficient processes for the removal of harmful gases, as well as the development of renewable energy resources. The development of highly functional catalyst materials is therefore highly needed [1, 2]. Furthermore, from the viewpoint of resources and economy of platinum group metals (PGMs) such as Rh and Pd, it is also desired to develop catalyst materials that do not contain PGMs or reduce the use of PGMs minimized as much as possible [3–6]. Nowadays, base metal catalysts have drawn attention with researchers focusing on the development of metal oxides that exhibit excellent redox properties. This chapter introduces metal oxide catalysts that contribute to environmental protection and the effective utilization of methane. In particular, the chemical characteristics of CeO_2-based oxides with a fluorite-type structure and perovskite-type oxides are outlined from the perspective of the packing structure of O^{2-} ions. In addition, oxygen storage materials, which are applicable to various fields involving catalytic reactions, are introduced from the viewpoint of the structural changes associated with the insertion and extraction of O^{2-} ions. Finally, oxidative coupling on metal oxides, which contributes to the effective utilization of methane, is introduced.

S. Hosokawa (✉)
Faculty of Materials Science and Engineering, Kyoto Institute of Technology, Matsugasaki, Sakyo-ku, Kyoto 606-8585, Japan
e-mail: hosokawa@kit.ac.jp

T. Motohashi (✉)
Department of Material and Life Chemistry, Kanagawa University, 3-27-1, Rokkakubashi, Kanagawa-ku, Yokohama 221-8686, Japan
e-mail: t-mot@kanagawa-u.ac.jp

W. Ueda (ed.), *Crystalline Metal Oxide Catalysts*,
https://doi.org/10.1007/978-981-19-5013-1_10

10.2 CO and VOC Oxidations

10.2.1 *Overview*

Catalytic combustion is effective for removing CO and volatile organic compounds (VOCs). This process uses a catalyst material that purifies harmful gases to CO_2 and H_2O at temperatures typically below 300 °C. Such a procedure does not induce flame combustion and is much safer than simple combustion processes that ignite with a flame. The process is also advantageous for environmental protection because of the suppressed evolution of incomplete combustion products and thermal NO_x, which are formed by exposure to N_2 and O_2 in air at high temperatures. The use of PGMs is indispensable for improving catalytic activity in the low-temperature range [7], whereas many studies have been conducted on metal oxide catalysts without PGM species for catalytic combustion. Highly active metal oxide catalysts and their crystal structures have been the subject of many publications [8], including CeO_2-based oxides, perovskite-type oxides, and spinel-type oxides. In general, all catalyst materials utilize metal oxides containing elements with excellent redox properties.

This section begins by presenting the relationship between the redox ability of metal oxides and the reaction mechanisms, followed by the structural characteristics of each metal oxide for typical catalyst materials.

10.2.2 *Reaction Mechanism (Mars-Van Krevelen Mechanism)*

In the catalytic combustion on a PGM catalyst, adsorption of reaction substrates and activation of O_2 molecules are known to occur on the surface of PGM. That is, the catalytic reaction proceeds according to a reaction based on the Langmuir–Hinshelwood mechanism (or Eley–Rideal mechanism) (Fig. 10.1) [9]. Therefore, when designing supported PGM catalysts, the high dispersion of PGM particles and the control of the electronic state by forming PGM alloy nanoparticles should be studied [10, 11]. In contrast, regarding metal oxide catalysts, the combustion mechanism involving lattice oxygen based on the Mars-van Krevelen (MvK) mechanism has become common [12–14]. Here, CO oxidation (Eqs. 10.1 and 10.2) is presented as an example to explain this reaction mechanism. First, CO reacts with lattice oxygen (O_L) at the oxide surface to generate CO_2 followed by an O_2 molecule filling the generated oxygen vacancies (V_O).

$$CO + O_L \rightarrow CO_2 + V_O \tag{10.1}$$

$$O_2 + V_O \rightarrow 2O_L \tag{10.2}$$

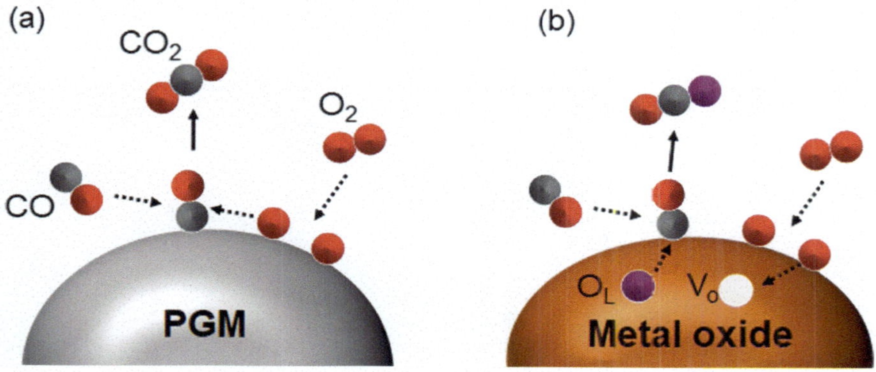

Fig. 10.1 CO oxidation based on **a** Langmuir–Hinshelwood and **b** Mars-van Krevelen mechanisms

The transfer of lattice oxygen in the metal oxide is a key step in this reaction mechanism. It is therefore preferable to optimize the redox properties of the constituent metal species when designing metal oxide catalysts.

10.2.3 CeO_2-Based Materials

Ce^{4+} ions contained in CeO_2 are easily reduced to Ce^{3+} and this redox plays a crucial role in the use of CeO_2 as a catalyst [15]. CeO_2 crystallizes in a fluorite-type structure in which Ce^{4+} ions form a face-centered cubic lattice, with O^{2-} ions occupying the tetrahedral voids (Fig. 10.2). In other words, the O^{2-} ions compose a primitive cubic structure, with Ce^{4+} ions located at the body-centered sites. In the case of transition-metal oxides such as NaCl-type MnO and spinel-type Co_3O_4, the ionic radius of transition metals such as Fe^{3+} (0.55 Å) and Co^{3+} (0.545 Å) is much smaller than that of the O^{2-} ion (1.40 Å). This difference in radius results in a close-packed structure composed of the O^{2-} ions (Fig. 10.3) where the transition metal ion occupies the tetrahedral or octahedral voids formed by the O^{2-} ions. For CeO_2, on the other hand, the size of the Ce^{4+} ions (0.97 Å) is larger than that of the transition metal ions, indicating that the O^{2-} ions do not form a close-packed structure and prefer a primitive cubic structure instead. This characteristic packing structure of the O^{2-} ions facilitates their fast diffusion in the CeO_2 crystal lattice, even though their diffusion depends on the defect chemistry as well [16]. Thus, the redox property of cerium and the transfer ability of O^{2-} ions in CeO_2 are promising features for the above-mentioned MvK mechanism.

Many studies on combustion catalysts using CeO_2 exist, analyzing the correlation between the oxygen release property and combustion activity and reporting that the reactivity of lattice oxygen in CeO_2 was drastically improved by PGM loading [17–19]. CeO_2 is often combined with CuO or Co_3O_4 as a PGM-free catalyst [20–23]; the

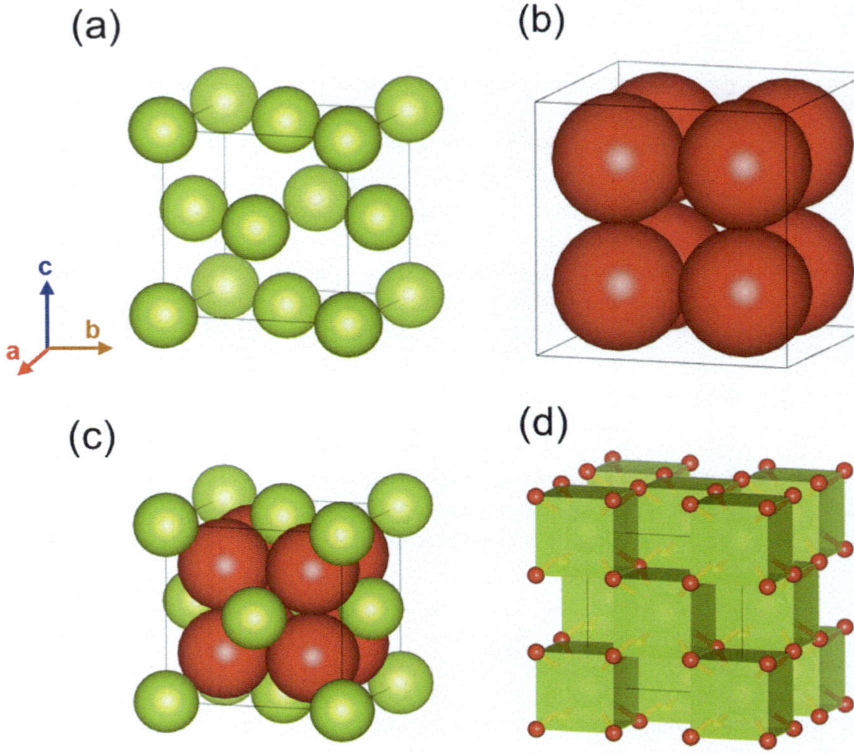

Fig. 10.2 Crystal structure of CeO_2 with a fluorite structure. **a** Positions of Ce ions and **b** O^{2-} ions in the unit cell. **c** Structure constructed by Ce and O^{2-} ions. **d** Crystal structure providing a clear view of the coordination environment of the Ce ions. Panels (**a–c**) are shown based on actual ionic radii: yellow balls, Ce^{4+} ions; red balls, O^{2-} ions. The schematic crystal structures in this chapter are drawn by VESTA program (K. Momma and F. Izumi. J. Appl. crystallogr., 2011, 44, 1272)

formed composites exhibit high mobility of lattice oxygen, resulting in high catalytic activity for VOCs combustion. Most of these reaction behaviors are explained by the MvK mechanism. Meanwhile, the reactivity of lattice oxygen in CeO_2 is closely related to the exposed crystal plane. Theoretical studies indicate that the (110) and (100) planes have high catalytic activity, emphasizing the importance of morphology control of the CeO_2 nanoparticles [24, 25]. In fact, rod-shaped CeO_2 particles, in which (110) planes are preferentially exposed, exhibit high VOC combustion activity. Furthermore, it has been demonstrated that the catalytic oxidation of CO with rod-shaped CeO_2 proceeds at lower temperatures than with cubic or octahedrally shaped CeO_2 grains.

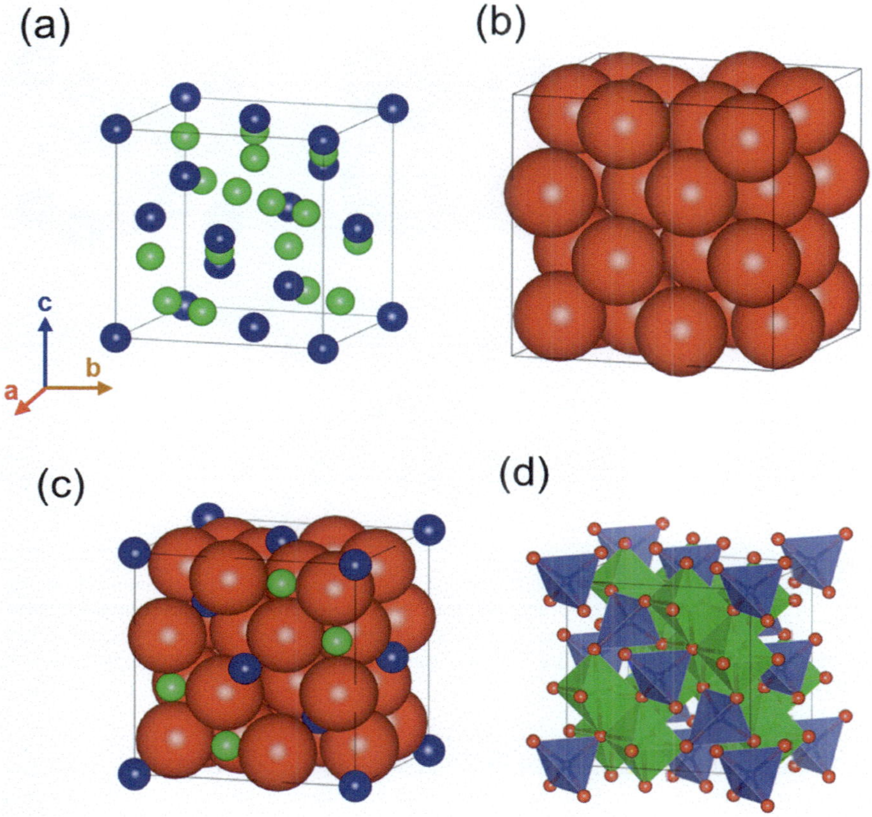

Fig. 10.3 Crystal structure of Co_3O_4 with a spinel structure. **a** Positions of Co ions and **b** O^{2-} ions in the unit cell. **c** Structure constructed by Co and O^{2-} ions. **d** Crystal structure providing a clear view of the coordination environment of the Co ions. Panels (**a–c**) are shown based on actual ionic radii: blue balls, Co^{2+} ions; green balls, Co^{3+} ions; red balls, O^{2-} ions

10.2.4 Perovskite Oxides

Since 3d transition metals can adopt a wide range of valence states, transition metal-based oxide catalysts have been extensively studied for combustion reactions from the perspective of the redox ability of their metal species. Many reports exist on perovskite oxides, which are generally represented by the chemical formula ABO_3 [6, 26–28]. The A site of the perovskite structure is often composed of alkali metals, alkaline-earth metals, or rare-earth metals, whereas the B site contains transition metals. Unlike the close-packed structure of the O^{2-} ions in NaCl-type and spinel-type transition metal oxides described in the previous section, the A-site cation and O^{2-} ion form a close-packed structure. This is due to the ionic radius of the A-site cation in the perovskite structure being similar to that of the O^{2-} ion (Fig. 10.4).

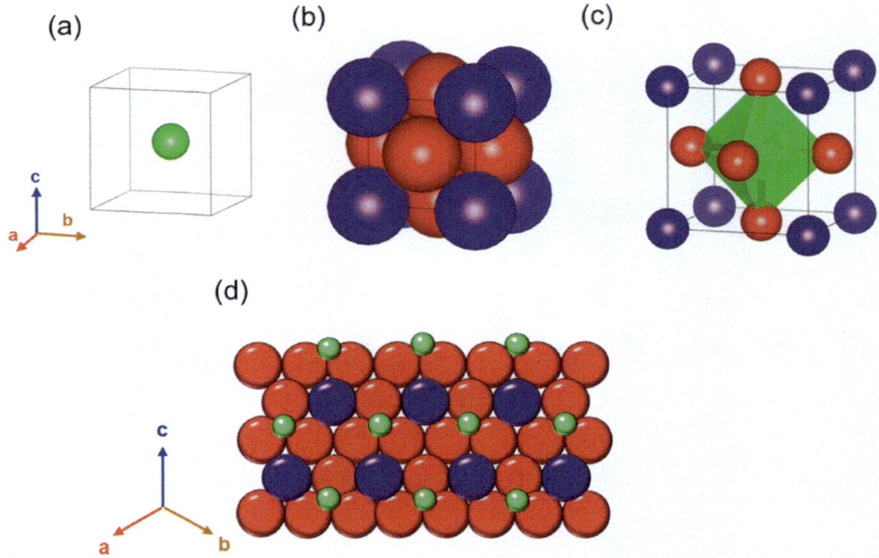

Fig. 10.4 Crystal structure of ABO_3 with a perovskite structure. **a** Positions of B ions and **b** A ions, and O^{2-} ions in the unit cell. **c** Crystal structure with a clear view of the coordination environment of the B ion. **d** Relationship between the A and B ions. Panels (**a**, **b**, and **d**) are shown based on actual ionic radii: Purple balls, A ions (alkali or alkali-earth elements); brown balls, B ions (transition metal elements); red balls, O^{2-} ions

The B-site metal occupies the octahedral interstice consisting of O^{2-} ions. The ideal perovskite structure belongs to the cubic system, whereas many perovskite oxides are known to exhibit tetragonal or orthorhombic distortion according to the ionic radii of the metals occupying the A and B sites.

An important advantage of perovskite oxides from the viewpoint of catalyst material design is that the constituent elements of the A and B sites can be chemically substituted. For example, partial substitutions of the A site can adjust the valence state of the B-site element or promote the formation of oxygen vacancies according to the charge balance. The substitution of Sr^{2+} for La^{3+} in $La^{3+}Co^{3+}O_3$ is expected to induce oxygen defects and control the Co valence state. In addition, partial substitutions of the B site can modulate the redox capacity of the B-site cation and the mobility of the O^{2-} ions. The possibility of adjusting these properties through the choice of dopants is a critical factor for efficient catalytic reactions based on the MvK mechanism.

Among the La-based transition metal oxides, $LaM^{3+}O_3$, $LaMnO_3$, and $LaCoO_3$ are known to be active in CO oxidation [27]. Furthermore, the Sr-for-La-substituted material, $La_{1-x}Sr_xMnO_3$, exhibits excellent catalytic performance. There are many other reports on the development of combustion catalysts based on perovskite structures, and the reader is referred to various review articles for further detail [6, 28, 29].

10.2.5 Other Metal Oxides

One of the most representative materials as a metal oxide catalyst for CO oxidation is Co_3O_4 with a spinel-type structure [30]. The Co_3O_4 catalyst having a rod-like morphology shows excellent catalytic ability even at room temperature. In general, the spinel structure presents a composition of AB_2O_4, where the A cation occupies the tetrahedral interstice formed in the close-packed structure of the O^{2-} ions, and the B cation occupies the octahedral interstice. The A and B cations are typically divalent and trivalent, respectively. For example, when clarifying the valence of Co species of Co_3O_4, it can be expressed as $Co^{2+}[Co^{3+}]_2O_4$. Since the 3d transition metal can be distributed at the A and B sites in the spinel structure, binary and ternary mixed oxides can be easily obtained by substituting various elements. Furthermore, it is possible to adjust the redox ability by selecting the metal species and multicomponent spinel oxides exhibit excellent catalytic ability. For example, spinel-type $CoMn_2O_4$ shows a higher ability to release lattice oxygen than a mixture of Co_3O_4 and MnO_x, a feature that is useful for toluene oxidation [31–33].

Other types of metal oxide catalysts are hexagonal rare-earth-iron mixed oxides ($REFeO_3$). $REFeO_3$ crystallizes in an orthorhombic perovskite structure with *Pbnm* being a stable form for all the rare-earth elements, while hexagonal $REFeO_3$ (*h*-$REFeO_3$) with $P6_3cm$ exists as a metastable phase [34]. In the unit cell of *h*-$REFeO_3$, layers of iron ions and RE ions are stacked alternately, with the iron ions having a unique structure with trigonal bipyramidal 5-coordination (Fig. 10.5). When *h*-$YbFeO_3$ is synthesized by a solvothermal reaction, *h*-$YbFeO_3$ nanocrystals with thin hexagonal plate morphology are obtained, in which the crystal growth in the *c*-axis direction is suppressed. *h*-$YbFeO_3$ can act as a high-performance catalyst for combustion reactions of hydrocarbons [35, 36], and when modified with Mn species, it exhibits higher CO oxidation activity than PGM catalysts [37, 38].

10.3 NO Oxidation and Reduction

10.3.1 Overview

NO_x abatement is becoming increasingly important due to the growing awareness of environmental issues, such as acid rain caused by NO_x emission from thermal power plants or automobiles [39–41]. Although the direct decomposition of NO is thermodynamically favorable, it is kinetically very difficult and often requires temperatures above 600 °C. Therefore, NO reduction to N_2 using hydrocarbons and NH_3 as reductants is widely applied. In contrast, since NO reduction is difficult in an oxygen-rich atmosphere, NO_x-trap materials are sometimes used. In such materials, as NO_2 is more easily adsorbed on the materials' surface than NO, a material with NO oxidation ability in addition to NO_x adsorption ability would be ideal. A typical NO_x-trap catalyst, $Pt/Ba/Al_2O_3$, oxidizes NO to NO_2 at the Pt site and traps NO_2 at the

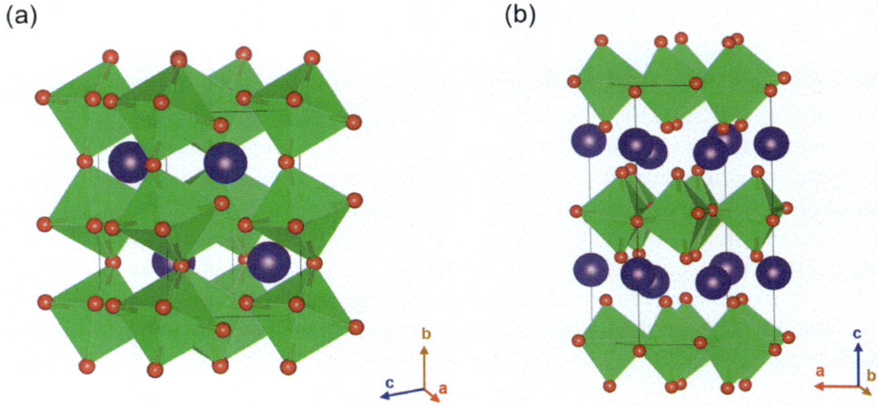

Fig. 10.5 Crystal structure of **a** orthorhombic perovskite $REFeO_3$ with *Pbnm* and **b** hexagonal $REFeO_3$ with $P6_3cm$. Crystal structure is shown as a clear view of the coordination environment of the Fe ion. Purple balls, RE^{3+} ions; green balls, Fe^{3+} ions; red balls; O^{2-} ions

BaO site [39, 42–44]. The NO_x species adsorbed on the catalyst are instantaneously reduced to N_2 by injecting a reductant such as hydrocarbons. Recently, PGM-free metal oxide catalysts that have both NO oxidation and NO_x storage capacity have been reported [45, 46].

In this section, NO_x storage materials involving NO oxidation mechanisms are firstly described, followed by catalyst materials for effective NO reduction based on the MvK-type reaction mechanism. Since most metal oxide catalysts following these reaction mechanisms are related to combustion catalysts [47], the characteristics of their crystal structures are included in Sect. 10.2.

10.3.2 Catalyst Materials

The reaction mechanism for NO oxidation on metal oxides is similar to the previously described MvK mechanism [47]. Many perovskite materials have been reported as catalyst materials for NO oxidation or NO_x storage, such as $La_{1-x}Sr_xCoO_3$ which are known to exhibit a NO oxidation activity comparable to Pt/Al_2O_3 [48, 49]. Ruddlesden-Popper type perovskite $Sr_3Fe_2O_{7-\delta}$, which has a layered structure consisting of a perovskite layer ($SrFeO_{3-\delta}$ layer) and a rock salt layer (SrO layer), shows excellent NO_x storage properties [45]. For these materials, the NO oxidation process involves MvK-type NO oxidation with lattice oxygen, as shown in Eq. (10.3) [50].

$$NO + O_L \rightarrow NO_2 \tag{10.3}$$

The chemical formula of $Sr_3Fe_2O_{7-\delta}$ can be rewritten as $SrO \cdot 2SrFeO_{3-\delta}$, where the SrO layer is incorporated in the structure. In fact, monodentate nitrate (NO_3^-) and ionic nitrate species were observed on the surface of $Sr_3Fe_2O_{7-\delta}$ after NO_x adsorption, suggesting that the SrO layer contributes to the adsorption of NO_x as shown in Eq. (10.4).

$$SrO + 2NO_2 + O^* \rightarrow Sr(NO_3)_2 \qquad (10.4)$$

The oxygen source (O*) in Eq. (10.4) is either lattice oxygen species or oxygen species produced by the disproportionation of NO_2. Apart from the NO oxidation properties derived from the $SrFeO_{3-\delta}$ layer, $Sr_3Fe_2O_{7-\delta}$ also presents NO_x adsorption properties derived from the SrO layer, resulting in a high NO_x storage capacity (Fig. 10.6).

These materials require reaction temperatures above 250 °C since NO oxidation triggers NO_x adsorption. In contrast, passive NO_x adsorber (PNA) materials, in which Pd species are highly dispersed in zeolite materials such as ZSM-5 and SSZ, are known to effectively trap NO_x at lower temperatures [51–53]. The NO molecule is adsorbed as it is on the Pd species without any NO oxidation process. However, since the NO-to-Pd ratio is 1, a large amount of Pd loading is essential for high NO_x storage capacity. It has been reported recently that perovskite-type $SrTi_{1-x}Mn_xO_3$ catalysts can efficiently adsorb NO at temperatures below 200 °C even without PGM loading [46].

Regarding NO reduction on metal oxides, the reduction mechanism has been recently proposed according to Eq. (10.5) (Fig. 10.7) [54, 55]:

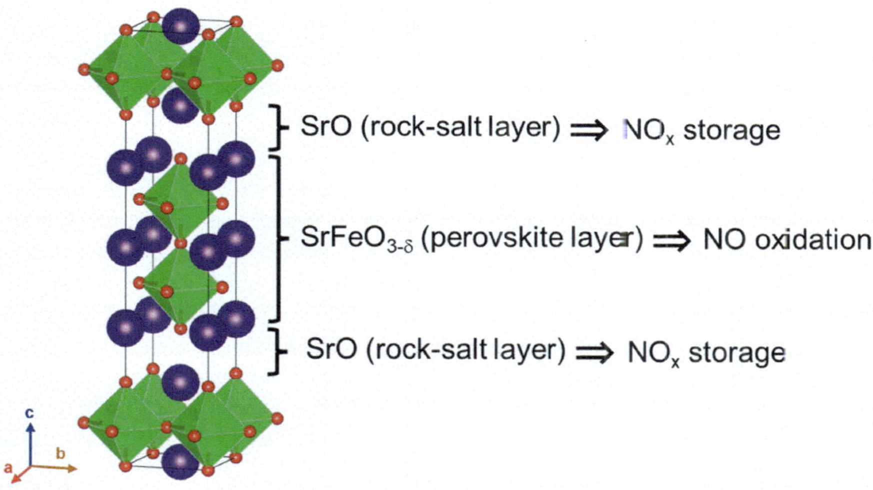

Fig. 10.6 Crystal structure of $Sr_3Fe_2O_{7-\delta}$. Crystal structure is shown as a clear view of the coordination environment of the Fe ions. Purple balls, Sr^{2+} ions; green balls Fe^{3+} or Fe^{4+} ions; red balls, O^{2-} ions

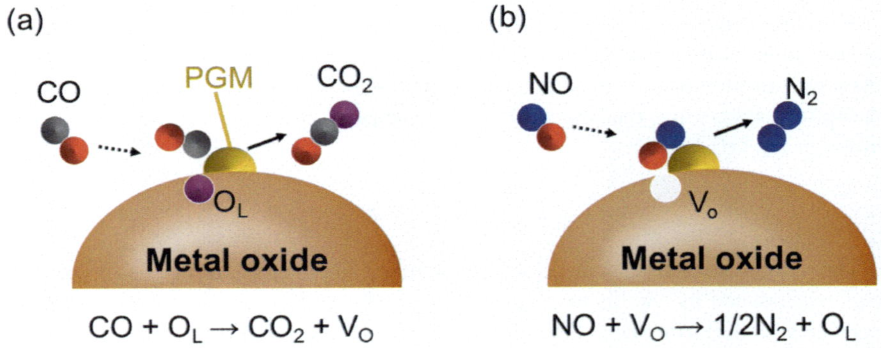

Fig. 10.7 Reaction mechanism of NO–CO reaction based on the MvK mechanism. **a** CO oxidation using lattice oxygen (O_L) and **b** NO reduction using oxygen vacancy (V_o)

$$NO + V_O \rightarrow 1/2N_2 + O_L \tag{10.5}$$

In this reaction, oxygen vacancies are regarded as active sites instead of lattice oxygen that was the case for the MvK mechanism described in Eq. (10.1). During this reduction mechanism, the continuous generation of oxygen vacancy sites is difficult. Therefore, a high temperature or a reducing agent is required for reducing the NO to N_2. For NO reduction processes using CO as a reducing agent, the overall reaction can be written as follows:

$$CO + O_L \rightarrow CO_2 + V_O \tag{10.6}$$

$$NO + V_O \rightarrow N + O_L \tag{10.7}$$

$$N + N \rightarrow N_2 \tag{10.8}$$

A supported PGM catalyst is often used for NO reduction. The reasons are that oxygen vacancy sites can be induced by the PGM loading, and that PGM sites can be used for NO/CO adsorption [38, 55]. For example, in procedures involving automotive exhaust purification, when a very small amount of Pd is supported on the Mn-modified hexagonal $YbFeO_3$, the NO reduction and CO oxidation are drastically enhanced at low temperatures. In this material, the arrangement of Mn cations on the MnO (111) plane is very close to that of Yb cations on the hexagonal $YbFeO_3$ (001); that is, the MnO (111) and $YbFeO_3$ (001) planes are bonded epitaxially. The role of Pd species was theoretically studied by the density functional theory (DFT) using a model in which Pd is supported by the MnO (111). It has been proven that the Pd loading promotes adsorption and dissociation of NO molecules. PGM-free materials have also been developed with Cu and Cr-modified CeO_2 catalysts exhibiting high catalytic activity following the MvK-type NO reduction mechanism [56]. It has also been found that spinel-type $NiFe_2O_4$ [57, 58] or multicomponent spinel-type catalysts containing four metallic elements (Cu, Ni, Al, Cr) are also effective for NO reduction [59].

10.4 Metal Oxides with Oxygen Storage Performance

10.4.1 Overview

Oxygen storage materials (OSMs) are oxides that can reversibly store and release lattice oxygens in response to variations in temperature and/or oxygen partial pressure. Therefore, this class of materials can be used in various oxygen-related technologies such as exhaust purification in automobiles, chemical looping reactions, and oxygen separation membranes/oxygen sensors [60–64]. OSMs show an interesting crystallographic feature which could be advantageous to rapid and reversible reactions; upon oxygen storage and release, slight structural changes take place. For many transition metal oxides such as spinel-type structures, the crystal structure is composed of close packing O^{2-} ions such that oxygen desorption may significantly deform the lattice framework. However, as described in Sect. 10.2, the packed structures of O^{2-} ions of fluorite-type CeO_2 and perovskite-type oxides are different from those of typical metal oxides. Materials with such characteristic O^{2-} ion-filled structures often store and release oxygen reversibly while maintaining the cationic arrangement. Chemical reactions in which the lattice framework is retained is called "topotactic" [65, 66]. The term "topotactic" is often used for Li intercalation/deintercalation of positive electrode materials in Li-ion secondary batteries. This section introduces the insertion/extraction processes of lattice oxygens based on topotactic reactions.

10.4.2 CeO_2–ZrO_2 Solid Solution

CeO_2–ZrO_2 solid solution (CZ) with fluorite structure is the most well-known oxygen storage material for automotive catalysts [67]. This material reversibly stores and releases lattice oxygens depending on the oxygen concentration in the exhaust gas, contributing to oxygen-pressure control around PGM species that act as active sites. It is known that CZ crystallizes in a fluorite-type structure when the material absorbs oxygen, and the oxygen storage capacity can be adjusted by the mixing ratio of Ce and Zr. For this material, oxygen storage/release proceeds via redox between Ce^{4+} and Ce^{3+}. It can therefore be expected that the oxygen storage capacity depends on the Ce content. However, the oxygen storage capacity of CZ with about 30 mol% Zr is reported to be much higher than that of Zr-free CeO_2. During redox processes, the ionic radii of Ce and Zr are responsible for this phenomenon; when oxygen is released from the crystal lattice involving a reduction of Ce^{4+} (0.97 Å) to Ce^{3+} (1.14 Å), the lattice volume expands, resulting in local lattice distortions. The presence of smaller Zr^{4+} ions (0.84 Å) at the cation site adjacent to Ce^{3+} mitigates the lattice distortions to form a stable crystal structure containing Ce^{3+} and oxygen deficiencies [68].

During the preparation of CZ exhibiting improved oxygen storage performance, the formation of pyrochlore-type $Ce_2Zr_2O_7$ consisting of an ordered arrangement of Ce and Zr should be formed under a reducing atmosphere at temperatures above 1000 °C [67]. Upon oxygen storage, a phase transition from pyrochlore to fluorite-type structure occurs. Although these two crystal structures are distinct from each other, structural analyses proved that the cationic arrangement is maintained during the phase transition [69]. In other words, topotactic reactions proceed upon oxygen storage and release.

10.4.3 Perovskite and Layered Perovskite Oxides

The topotactic oxygen storage/release also occurs in mixed oxides with perovskite-related structures [70–73]. For this class of materials, since a close-packed structure is formed by A-site cations and O^{2-} ions, the release of oxygen proceeds with minor structural changes. For example, when perovskite-type $SrFeO_{3-\delta}$ is deoxygenated under a hydrogen-containing atmosphere, $Sr_2Fe_2O_5$ with an oxygen-deficient perovskite structure named "brownmillerite-type" is formed without rearrangements at the cation sites. Details of the brownmillerite-type structure are discussed below. However, severe reductive treatments of $SrFeO_{3-\delta}$ at high temperatures lead to structural collapse and the formation of metallic iron. To suppress this structural collapse, a partial substitution of Fe with Ti was reported to be effective by enhancing the structural stability [72].

On the other hand, $Sr_3Fe_2O_{7-\delta}$ with a layered perovskite structure cannot decompose into metallic iron even by severe reductive treatments [66]. The improved phase stability of $Sr_3Fe_2O_{7-\delta}$ is suggested to originate from SrO double layers within the crystal lattice, which would stabilize the oxygen-deficient structure. During oxygen storage/release, the crystal structure remains essentially unchanged with the *I*4*mmm* tetragonal space group (Fig. 10.8). The Fe species in $Sr_3Fe_2O_{7-\delta}$ are Fe^{3+}/Fe^{4+} mixed valent, and the redox between Fe^{3+} and Fe^{4+} is responsible for the oxygen storage and release. The oxygen storage performances are higher than those of Pt/CZ (Fig. 10.9) and the partial replacement of Fe with Ni can improve the oxygen storage capacity of $Sr_3Fe_2O_{7-\delta}$. Structural analysis of deoxygenated $Sr_3(Fe_{0.8}Ni_{0.2})_2O_{7-\delta}$ showed that the reduction of Ni^{3+} to Ni^{2+} proceeds in addition to the reduction of Fe^{4+} to Fe^{3+} [74]. This phenomenon indicates that the Ni^{3+} substitution in $Sr_3Fe_2O_{7-\delta}$ induces the redox activity of the Fe^{3+} site. Indeed, the Pd/$Sr_3(Fe_{0.8}Ni_{0.2})_2O_{7-\delta}$ catalyst shows superior catalytic activity to Pd/$Sr_3Fe_2O_{7-\delta}$ and Pd/CZ for the purification of automotive exhaust gases under fluctuating oxygen concentration, most likely owing to the enhanced oxygen storage capacity of $Sr_3(Fe_{0.8}Ni_{0.2})_2O_{7-\delta}$.

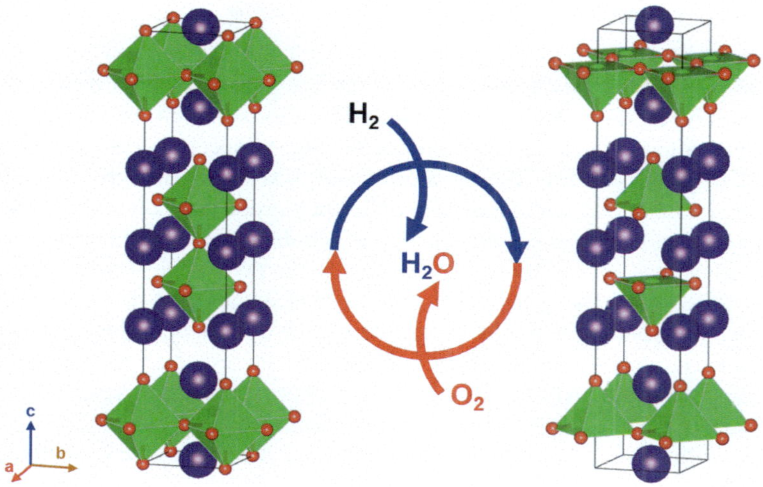

Fig. 10.8 Topotactic oxygen release and storage between $Sr_3Fe_2O_{7-\delta}$ and $Sr_3Fe_2O_6$. Crystal structure is shown as a clear view of the coordination environment of the Fe ions. Purple balls, Sr^{2+} ions; green balls, Fe^{3+} or Fe^{4+} ions; red balls, O^{2-} ions

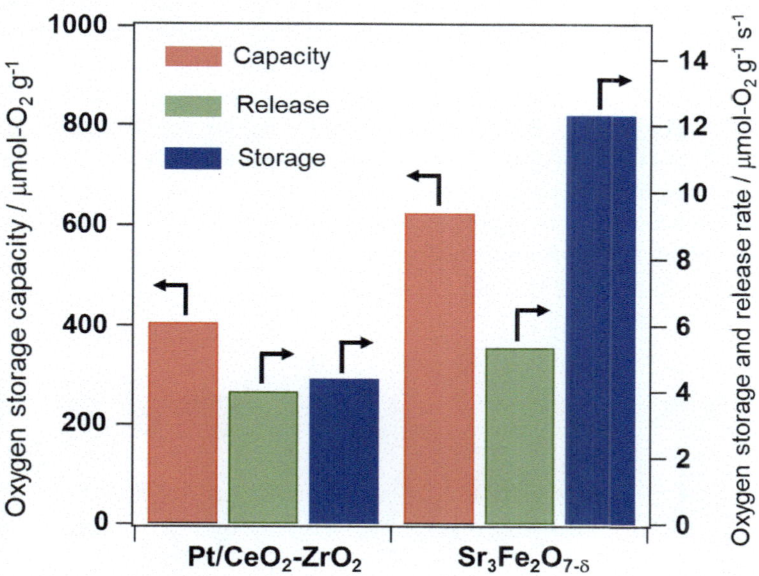

Fig. 10.9 Oxygen storage performances of $Pt/CeO_2\text{-}ZrO_2$ and $Sr_3Fe_2O_{7-\delta}$. These values were obtained using thermogravimetric analysis under H_2–O_2 cycles at 500 °C

10.4.4 Brownmillerite Oxides

As described in Sect. 10.4.3, perovskite oxides are potential OSMs owing to their topotactic oxygen storage/release capability. In the perovskite-type lattice, oxygen deficiencies often form long-range orders, leading to specific crystal phases that enhance structural stability. The brownmillerite (BM)-type structure is a member of the perovskite family with an oxygen deficiency-ordered structure [75]. The BM structure consists of a layered arrangement of tetrahedral BO_4 and octahedral BO_6 units, as illustrated in Fig. 10.10a, in contrast to the perovskite structure that only contains BO_6 units. The general formula for BM oxides is $A_2B_2O_5$ or $ABO_{2.5}$. This structural type is noteworthy owing to the high density of oxygen deficiencies among stable perovskite-derived lattices, that is, 1/6 with respect to the stoichiometric perovskite composition ABO_3. The high density of oxygen deficiencies in BM oxides is beneficial for enhancing their oxygen storage capacity.

Several BM-type OSMs have been reported including $(Sr,Ca)FeO_{2.5}$ and $Sr(Fe,Co)O_{2.5}$ [76–80]. These oxides have attracted attention as potential materials for oxygen absorption that can selectively separate oxygen gas from the atmosphere. This technology is promising as a novel method of oxygen gas production, being more efficient than cryogenic O_2/N_2 distillation and conventional pressure-swing-adsorption using zeolites as N_2 sorbents. Ca_2AlMnO_5 is another BM-type OSM presenting alternating stacking of tetrahedral AlO_4 and octahedral MnO_6 [81]. This oxide stores excess oxygen topotactically and is then transformed into an oxygen-rich form, $Ca_2AlMnO_{5.5}$. The excess oxygen can then be released in a highly reversible manner [82]. The resultant oxygen storage capacity exceeds 3.0 wt%, which is even larger than that of CZ. $Ca_2AlMnO_{5+\delta}$, with excess oxygen amounts δ ranging from 0 to 0.5, is promising in terms of elemental abundance, but its high operating temperatures (~550 °C) are an obstacle to future practical applications. Efforts have been made to lower the operating temperatures, such as Ga-for-Al or Sr-for-Ca substitutions and atomic-defect engineering [83–86].

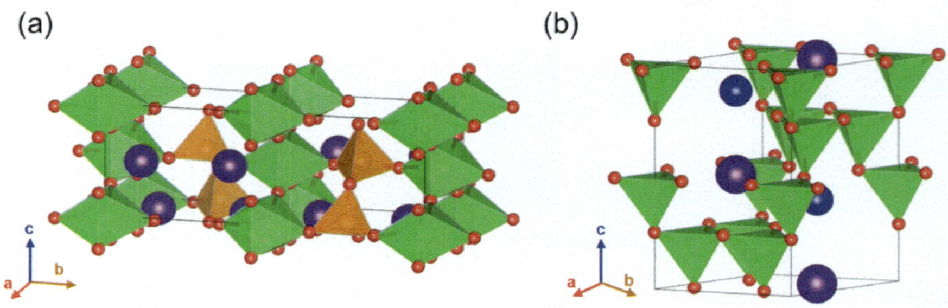

Fig. 10.10 Crystal structure of **a** Ca_2AlMnO_5 with a brownmillerite structure and **b** $YBaCo_4O_7$. Crystal structure is shown as a clear view of the coordination environments of the Mn, Al, and Co ions. Purple balls, Ca^{2+} or Ba^{2+} ions; blue balls, Y^{3+} ions; green balls, Mn^{3+} or Co^{2+} ions; orange balls, Al^{3+} ions; red balls, O^{2-} ions

10.4.5 Other Nonstoichiometric Oxides

Besides the aforementioned CeO_2-based materials and perovskite-related oxides, novel OSMs of other structural types have been extensively studied. Such materials are noteworthy because their unique atomic arrangements could induce characteristic redox behaviors and hence open up the possibility of new areas of applications. In this section, two OSMs, $YBaCo_4O_{7+\delta}$ and $YMnO_{3+\delta}$, are briefly introduced. Noticeably, these oxides have the following common features, which could be crucial factors for high-performance OSMs. (1) Both the oxides contain cationic sites with smaller coordination numbers (CN) than six. Such "coordinatively unsaturated ($CN < 6$)" sites act as reaction centers for oxygen storage and release. (2) These oxides incorporate excess oxygen topotactically to transform into distinct oxygen-rich forms. Remarkable oxygen storage/release triggered by structural transformations arises even with slight variations in temperature and/or oxygen partial pressure.

$YBaCo_4O_{7+\delta}$ (denoted as "YBCO") has a characteristic structure consisting of a 3D network of corner-sharing CoO_4 tetrahedra, as illustrated in Fig. 10.10b [87]. The corner-sharing framework favors excess-oxygen incorporation. Upon oxygen intake, a part of the CoO_4 tetrahedra transforms into CoO_6 octahedra, leading to large oxygen nonstoichiometry ranging $0 \le \delta \le 1.5$ [88, 89]. The oxygen storage/release processes are highly reversible, being controlled by both temperature and oxygen partial pressure. The maximum oxygen storage capacity reaches 4.2 wt%, corresponding to $\Delta\delta = 1.5$ [90]. This value is much larger than that for CZ, making this oxide a promising candidate for high-performance OSM. YBCO works at relatively low temperatures of 300–400 °C in air, and such low operating temperatures are crucial for this oxide to be applied to oxygen-gas production [91]. It should also be pointed out that YBCO shows compositional flexibility, and its oxygen storage characteristics can be widely controlled by chemical substitutions [92].

Oxygen-stoichiometric $YMnO_3$ crystallizes in a hexagonal structure consisting of an alternate stacking of corner-shared MnO_5 trigonal-bipyramids and layers of edge-shared YO_7 polyhedra. This structure is essentially identical to that of h-$YbFeO_3$ shown in Fig. 10.5b. This oxide can incorporate excess oxygen into interstitial positions in the MnO_5 layer, resulting in the oxygen-rich form [93]. Upon oxygen intake, various oxygen-rich phases appear in the $YMnO_{3+\delta}$ system (denoted as "YMO") depending on the amount of excess oxygen (δ). These phases are labeled as Hex1 (space group: $R3c$) with $\delta \approx 0.28$, Hex2 ($Pca2_1$) with $\delta \approx 0.41$, and Hex3 ($P6_3mc$) with $\delta \approx 0.45$ [94, 95]. YMO exhibits remarkable oxygen storage/release even at 250 °C or lower, with the maximum oxygen storage capacity reaching 2.5 wt%. The large oxygen storage capacity and low operating temperature ensure that YMO can be a promising OSM. Although YMO is effective as a material for oxygen-gas production, its sluggish oxygen storage/release kinetics are an obstacle to practical applications. To address this issue, various material tailoring protocols were implemented, and partial substitutions of lanthanoids at the Y site have proven effective. Materials with larger lanthanoids, such as $Sm_{0.25}Y_{0.75}MnO_{3+\delta}$ and $Ce_{0.15}Tb_{0.15}Y_{0.70}MnO_{3+\delta}$, were found to show enhanced oxygen storage kinetics at low temperatures [96, 97].

10.5 CH_4 Conversion: Oxidative Coupling of Methane

10.5.1 Overview

Concerns regarding the depletion of natural carbon resources have increased interest in developing ways to effectively utilize methane. Methane is the primary component of natural gas and is the most promising alternative to petroleum. However, methane is a chemically stable molecule with no functional groups, limiting its use to simple combustion for mobility and energy generation. The oxidative coupling of methane (OCM) to C_2 hydrocarbons, such as C_2H_6 and C_2H_4, is a possible route for facile conversion into commodity chemicals [98, 99]:

$$CH_4 + 1/4O_2 \rightarrow 1/2C_2H_6 + 1/2H_2O \quad \Delta H = -89\ \text{kJ mol}^{-1} \tag{10.9}$$

$$CH_4 + 1/2O_2 \rightarrow 1/2C_2H_4 + H_2O \quad \Delta H = -141\ \text{kJ mol}^{-1} \tag{10.10}$$

OCM is a fascinating reaction that is direct, exothermic, and unaffected by thermodynamic constraints. However, to date, no economically feasible process has been established. The reason for this is the absence of suitable catalysts with high C_2 yields (>30%) and long-term durability to meet the practical requirements. The deep oxidation of methane to CO and CO_2 hinders OCM, causing the deterioration of the C_2 selectivity.

10.5.2 Existing Potential Catalysts and OCM Reaction Mechanism

Since the pioneering work [100] of Keller and Bhasin in the 1980s, approximately 2300 reports on OCM have been published. Statistical analysis of the literature data on OCM catalysts suggests several key factors that positively affect the OCM catalytic activity [101]. Catalysts containing alkali, alkali-earth, and rare-earth metals tend to show high catalytic activity. In fact, Li-doped MgO (Li/MgO) [102] and a composite consisting of manganese oxides and sodium tungstates on silica support (Mn-Na_2WO_4/SiO_2) [103, 104] are known to be OCM-active. Unfortunately, both catalysts face technical drawbacks: in long-term operations, they suffer from deactivation due to the vaporization of alkali metals. Regarding the practical application of OCM technology, alternative concepts for catalyst development may help improve the catalytic performance, as reported recently.

The catalytic mechanism of OCM has been extensively investigated. Previous studies have indicated that the formation of C_2 hydrocarbons predominantly involves the gas-phase association of methyl radicals ($\cdot CH_3$) generated at oxygen sites on the catalyst surface [105, 106]. This finding implies that methane activation, that is, the hydrogen abstraction reaction by the catalyst, is one of the dominating factors affecting OCM activity. Several research groups have reported the importance of the electronic state of oxygen species on the catalyst surface. However, there is no concrete view of the role of such oxygen species, and the activation factors of highly active OCM catalysts are not completely understood.

10.5.3 Recent Progress in the Development of Complex Metal Oxide Catalysts

The aforementioned Mn-Na_2WO_4/SiO_2 catalyst is promising because of its relatively high catalytic activity; typically, C_2 yields are larger than 25% [103, 104]. Nevertheless, the primary activation factor for OCM is still debated owing to the chemical/structural complexity, causing difficulties in further improving its catalytic performance. To overcome this obstacle, single-phase oxides that contain multiple elements seem noteworthy. Single-phase catalysts provide well-defined active sites that are advantageous for mechanistic studies. On the lookout for novel OCM catalysts, few studies have been conducted based on such a strategy, and most of the previous works have focused on perovskite-type oxides [99]. Wang et al. characterized three model $La_2B_2O_7$ compounds with Ti^{4+}, Zr^{4+}, and Ce^{4+} at the B site to prove the phase structure-OCM activity relationship [107]. $La_2Ti_2O_7$, $La_2Zr_2O_7$, and $La_2Ce_2O_7$ crystallize in perovskite-related, pyrochlore, and defective fluorite structures, respectively. The OCM activity followed the order: $La_2Ce_2O_7 > La_2Zr_2O_7 > La_2Ti_2O_7$. The authors suggested that superoxide O_2^- is the active oxygen species, and that the structural type influences the concentration of O_2^- anions, emphasizing the importance of the crystallographic viewpoint.

Matsumoto et al. recently discovered the high OCM activity of crystalline Li_2CaSiO_4 (Fig. 10.11) [108]. The C_2 selectivity and CH_4 conversion reached 77.5% and 28.3%, respectively, at 750 °C with a feed gas ratio of $CH_4/O_2 = 4/1$. In addition, Li_2CaSiO_4 showed stable OCM performance in a durability test over 50 h without phase decomposition and compositional changes. The significant OCM activity of this catalyst is likely to originate from a combination of multiple cations in the crystal lattice; the local coordination around oxygen would give a dual effect of strong basicity and lattice stability, leading to simultaneous enhancements of the CH_4 conversion and C_2 selectivity. Notably, Li_4SiO_4 and Li_4GeO_4 were also found to be OCM-active, with performances comparable to Li_2CaSiO_4 [109]. The high OCM performance of these catalysts is believed to originate from the crystallographic similarities to Li_2CaSiO_4, where every oxygen site is surrounded by one Si/Ge atom and multiple Li atoms, which fulfill the OCM activation factors.

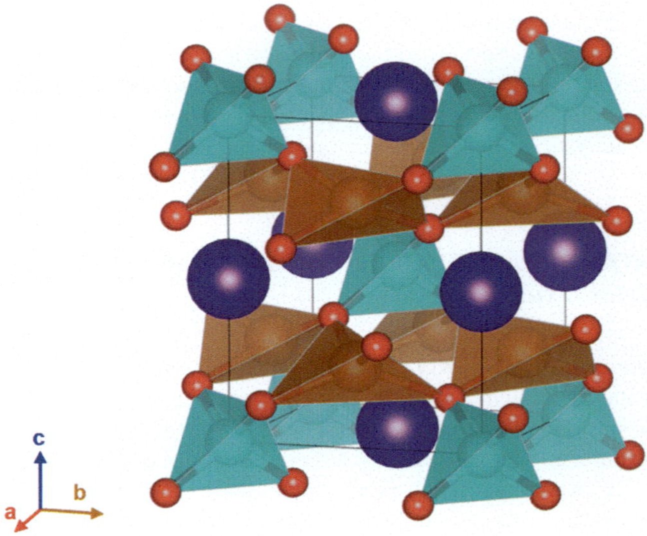

Fig. 10.11 Crystal structure of Li_2CaSiO_4. Crystal structure is shown as a clear view of the coordination environments of the Li and Si ions. Purple balls, Ca^{2+} ions; brown balls, Li^+ ions; blue balls, Si^{4+} ions; red balls, O^{2-} ions

References

1. Getsoian A, Theis JR, Paxton WA, Lance MJ, Lambert CK. Remarkable improvement in low temperature performance of model three-way catalysts through solution atomic layer deposition. Nat Catal. 2019;2(7):614.
2. Farrauto RJ, Deeba M, Alerasool S. Gasoline automobile catalysis and its historical journey to cleaner air. Nat Catal. 2019;2(7):603.
3. Misumi S, Yoshida H, Hinokuma S, Sato T, Machida M. A nanometric Rh overlayer on a metal foil surface as a highly efficient three-way catalyst. Sci Rep. 2016;6:29737.
4. Yoshida H, Kakei R, Kuzuhara Y, Misumi S, Machida M. A comparative study on TWC reactions over Rh thin films and supported Rh nanoparticles under lean conditions. Catal Today. 2019;332:245.
5. Ueda K, Tsuji M, Ohyama J, Satsuma A. Tandem base-metal oxide catalyst: superior NO reduction performance to the Rh catalyst in NO–C_3H_6–CO–O_2. ACS Catal. 2019;9(4):2866.
6. Royer S, Duprez D, Can F, Courtois X, Batiot-Dupeyrat C, Laassiri S, Alamdari H. Perovskites as substitutes of noble metals for heterogeneous catalysis: dream or reality. Chem Rev. 2014;114(20):10292.
7. Babier Jr. J, Duprez D. Reactivity of steam in exhaust gas catalysis I. Steam and Oxygen/steam conversions of carbon monoxide and of propane over PtRh catalysts. Appl Catal B: Environ. 1993; 3:61.
8. Royer S, Duprez D. Catalytic oxidation of carbon monoxide over transition metal oxides. ChemCatChem. 2011;3(1):24.
9. Hirvi JT, Kinnunen TJ, Suvanto M, Pakkanen TA, Norskov JK. CO oxidation on PdO surfaces. J Chem Phys. 2010;133(8): 084704.
10. Li Y, Yu Y, Wang J-G, Song J, Li Q, Dong M, Liu C-J. CO oxidation over graphene supported palladium catalyst. Appl Catal B: Environ. 2012;125:189.

11. Hosseini M, Barakat T, Cousin R, Aboukaïs A, Su BL, De Weireld G, Siffert S. Catalytic performance of core-shell and alloy Pd–Au nanoparticles for total oxidation of VOC: the effect of metal deposition. Appl Catal B: Environ. 2012;111–112:218.
12. Mars P, van Krevelen DW. Oxidations carried out by means of vanadium oxide catalysts. Chem Eng Sci. 1954;3:41.
13. Wang C, Gu X-K, Yan H, Lin Y, Li J, Liu D, Li W-X, Lu J. Water-mediated Mars-Van Krevelen Mechanism for CO oxidation on Ceria-supported single-atom Pt1 catalyst. ACS Catal. 2016;7(1):887.
14. Carlotto S, Natile MM, Glisenti A, Paul JF, Blanck D, Vittadini A. Energetics of CO oxidation on lanthanide-free perovskite systems: the case of Co-doped $SrTiO_3$. Phys Chem Chem Phys. 2016;18(48):33282.
15. Montini T, Melchionna M, Monai M, Fornasiero P. Fundamentals and catalytic applications of CeO_2-based materials. Chem Rev. 2016;116(10):5987.
16. Artini C. Rare-earth-doped ceria systems and their performance as solid electrolytes: a puzzling tangle of structural issues at the average and local scale. Inorg Chem. 2018;57(21):13047.
17. Yao HC, Yao YFY. Ceria in automotive exhaust catalysts I oxygen storage. J Catal. 1984;86:254.
18. Hosokawa S, Nogawa S, Taniguchi M, Utani K, Kanai H, Imamura S. Oxidation characteristics of Ru/CeO_2 catalyst. Appl Catal A: Gen. 2005;288(1–2):67.
19. Hosokawa S, Taniguchi M, Utani K, Kanai H, Imamura S. Affinity order among noble metals and CeO_2. Appl Catal A: Gen. 2005;289(2):115.
20. Lykaki M, Pachatouridou E, Carabineiro SAC, Iliopoulou E, Andriopoulou C, Kallithrakas-Kontos N, Boghosian S, Konsolakis M. Ceria nanoparticles shape effects on the structural defects and surface chemistry: implications in CO oxidation by Cu/CeO_2 catalysts. Appl Catal B: Environ. 2018;230:18.
21. Delimaris D, Ioannides T. VOC oxidation over $CuO–CeO_2$ catalysts prepared by a combustion method. Appl Catal B: Environ. 2009;89(1–2):295.
22. Delimaris D, Ioannides T. VOC oxidation over $MnO_x–CeO_2$ catalysts prepared by a combustion method. Appl Catal B: Environ. 2008;84(1–2):303.
23. Liotta LF, Ousmane M, Di Carlo G, Pantaleo G, Deganello G, Boreave A, Giroir-Fendler A. Catalytic removal of toluene over $Co_3O_4–CeO_2$ mixed oxide catalysts: comparison with Pt/Al_2O_3. Catal Lett. 2008;127(3–4):270.
24. Trovarelli A, Llorca J. Ceria catalysts at nanoscale: how do crystal shapes shape catalysis? ACS Catal. 2017;7(7):4716.
25. Wang W-W, Yu W-Z, Du P-P, Xu H, Jin Z, Si R, Ma C, Shi S, Jia C-J, Yan C-H. Crystal plane effect of ceria on supported copper oxide cluster catalyst for CO oxidation: importance of metal-support interaction. ACS Catal. 2017;7(2):1313.
26. Voorhoeve RJH, Johnson JDW, Remeika JP, Gallagher PK. Perovskite oxides: materials science in catalysis. Science. 1977;195:7.
27. Tanaka H, Misono M. Advances in designing perovskite catalysts. Curr Opin Solid State Mater Sci. 2001;5:381.
28. Peña MA, Fierro JLG. Chemical structures and performance of perovskite oxides. Chem Rev. 2001;101(7):1981.
29. Zhu J, Li H, Zhong L, Xiao P, Xu X, Yang X, Zhao Z, Li J. Perovskite oxides: preparation, characterizations, and applications in heterogeneous catalysis. ACS Catal. 2014;4(9):2917.
30. Xie X, Li Y, Liu ZQ, Haruta M, Shen W. Low-temperature oxidation of CO catalysed by Co_3O_4 nanorods. Nature. 2009;458:746.
31. Tian Z-Y, Tchoua Ngamou PH, Vannier V, Kohse-Höinghaus K, Bahlawane N. Catalytic oxidation of VOCs over mixed Co–Mn oxides. Appl Catal B: Environ. 2012;117–118:125.
32. Faure B, Alphonse P. Co–Mn-oxide spinel catalysts for CO and propane oxidation at mild temperature. Appl Catal B: Environ. 2016;180:715.
33. Dong C, Qu Z, Qin Y, Fu Q, Sun H, Duan X. Revealing the highly catalytic performance of spinel $CoMn_2O_4$ for toluene oxidation: involvement and replenishment of oxygen species using in situ designed-TP techniques. ACS Catal. 2019;9(8):6698.

34. Nishimura T, Hosokawa S, Masuda Y, Wada K, Inoue M. Synthesis of metastable rare-earth-iron mixed oxide with the hexagonal crystal structure. J Solid State Chem. 2013;197:402.
35. Hosokawa S, Jeon HJ, Iwamoto S, Inoue M. Synthesis of rare earth iron-mixed oxide nanoparticles by solvothermal methods. J Am Ceram Soc. 2009;92(12):2847.
36. Hosokawa S, Jeon HJ, Inoue M. Thermal stabilities of hexagonal and orthorhombic $YbFeO_3$ synthesized by solvothermal method and their catalytic activities for methane combustion. Res Chem Intermed. 2011;37(2–5):291.
37. Hosokawa S, Masuda Y, Nishimura T, Wada K, Abe R, Inoue M. Catalytic properties of Mn-modified hexagonal $YbFeO_3$: noble-metal-free combustion catalysts. Chem Lett. 2014;43(6):874.
38. Hosokawa S, Tada R, Shibano T, Matsumoto S, Teramura K, Tanaka T. Promoter effect of Pd species on Mn oxide catalysts supported on rare-earth-iron mixed oxide. Catal Sci Technol. 2016;6(21):7868.
39. Roy S, Baiker A. NO_x storage-reduction catalysis: from mechanism and materials properties to storage−reduction performance. Chem Rev. 2009;109(9):4054.
40. Granger P, Parvulescu VI. Catalytic NO_x abatement systems for mobile sources: from three-way to lean burn after-treatment technologies. Chem Rev. 2011;111(5):3155.
41. Liu G, Gao P-X. A review of NO_x storage/reduction catalysts: mechanism, materials and degradation studies. Catal Sci Technol. 2011;1(4):552.
42. Olsson L, Persson H, Fridell E, Skoglundh M, Andersson B. A kinetic study of NO oxidation and NO_x storage on Pt/Al_2O_3 and Pt/BaO/Al_2O_3. J Phys Chem B. 2001;105:6895.
43. Mulla S, Chen N, Cumaranatunge L, Blau G, Zemlyanov D, Delgass W, Epling W, Ribeiro F. Reaction of NO and O_2 to NO_2 on Pt: kinetics and catalyst deactivation. J Catal. 2006;241(2):389.
44. Bhatia D, McCabe RW, Harold MP, Balakotaiah V. Experimental and kinetic study of NO oxidation on model Pt catalysts. J Catal. 2009;266(1):106.
45. Tamai K, Hosokawa S, Okamoto H, Asakura H, Teramura K, Tanaka T. NO_x oxidation and storage properties of a Ruddlesden-Popper-type $Sr_3Fe_2O_{7-\delta}$-layered Perovskite catalyst. ACS Appl Mater Interf. 2019;11(30):26985.
46. Yoshiyama Y, Hosokawa S, Tamai K, Kajino T, Yoto H, Asakura H, Teramura K, Tanaka T. NO_x storage performance at low temperature over platinum group metal-free $SrTiO_3$-based material. ACS Appl Mater Interf. 2021;13:29482.
47. Hong Z, Wang Z, Li X. Catalytic Oxidation of Nitric Oxide (NO) over different catalysts: an overview. Catal Sci Technol. 2017;7(16):3440.
48. Peng Y, Si W, Luo J, Su W, Chang H, Li J, Hao J, Crittenden J. Surface tuning of $La_{0.5}Sr_{0.5}CoO_3$ Perovskite catalysts by acetic acid for NO_x storage and reduction. Environ Sci Technol. 2016; 50(12):6442.
49. Kim CH, Qi G, Dahlberg K, Li W. Strontium-doped Perovskites rival platinum catalysts for treating NO_x in simulated diesel exhaust. Science. 2010;327(5973):1624.
50. Tamai K, Hosokawa S, Onishi K, Watanabe C, Kato K, Asakura H, Teramura K, Tanaka T. Dynamics of the lattice oxygen in a Ruddlesden-Popper-type $Sr_3Fe_2O_{7-\delta}$ catalyst during NO oxidation. ACS Catal. 2020;10(4):2528.
51. Chen H-Y, Collier JE, Liu D, Mantarosie L, Durán-Martín D, Novák V, Rajaram RR, Thompsett D. Low temperature NO storage of Zeolite supported Pd for low temperature diesel engine emission control. Catal Lett. 2016;146(9):1706.
52. Ryou Y, Lee J, Cho SJ, Lee H, Kim CH, Kim DH. Activation of Pd/SSZ-13 catalyst by hydrothermal aging treatment in passive NO adsorption performance at low temperature for cold start application. Appl Catal B: Environ. 2017;212:140.
53. Zheng Y, Kovarik L, Engelhard MH, Wang Y, Wang Y, Gao F, Szanyi J. Low-temperature Pd/Zeolite passive NO_x adsorbers: structure, performance, and adsorption chemistry. J Phys Chem C. 2017;121(29):15793.
54. Beppu K, Hosokawa S, Asakura H, Teramura K, Tanaka T. Role of lattice oxygen and oxygen vacancy sites in platinum group metal catalysts supported on $Sr_3Fe_2O_{7-\delta}$ for NO-selective reduction. Catal Sci Technol. 2018;8(1):147.

55. Hosokawa S, Shibano T, Koga H, Matsui M, Asakura H, Teramura K, Okumura M, Tanaka T. Excellent catalytic activity of a Pd-promoted MnO_x catalyst for purifying automotive exhaust gases. ChemCatChem. 2020;12(17):4276.
56. Yoshida H, Yamashita N, Ijichi S, Okabe Y, Misumi S, Hinokuma S, Machida M. A thermally stable Cr–Cu nanostructure embedded in the CeO_2 surface as a substitute for platinum-group metal catalysts. ACS Catal. 2015;5(11):6738.
57. Ueda K, Ang CA, Ito Y, Ohyama J, Satsuma A. $NiFe_2O_4$ as an active component of a platinum group metal-free automotive three-way catalyst. Catal Sci Technol. 2016;6(15):5797.
58. Ueda K, Ohyama J, Satsuma A. Investigation of reaction mechanism of NO–C_3H_6–CO–O_2 reaction over $NiFe_2O_4$ catalyst. ACS Omega. 2017;2(7):3135.
59. Hirakawa T, Shimokawa Y, Tokuzumi W, Sato T, Tsushida M, Yoshida H, Hinokuma S, Ohyama J, Machida M. Multicomponent spinel oxide solid solutions: a possible alternative to platinum group metal three-way catalysts. ACS Catal. 2019;9(12):11763.
60. Beppu K, Demizu A, Hosokawa S, Asakura H, Teramura K, Tanaka T. Pd/$SrFe_{1-x}Ti_xO_{3-\delta}$ as environmental catalyst: purification of automotive exhaust gases. ACS Appl Mater Interf. 2018;10(26):22182.
61. Ozawa M, Kimura M, Isogai A. The application of Ce-Zr oxide solid solution to oxygen storage promoters in automotive catalysts. J Alloys Comp. 1993;193:73.
62. Krzystowczyk E, Haribal V, Dou J, Li F. Chemical looping air separation using a Perovskite-based oxygen sorbent: system design and process analysis. ACS Sustain Chem Eng. 2021;9(36):12185.
63. Bulfin B, Lapp J, Richter S, Gubàn D, Vieten J, Brendelberger S, Roeb M, Sattler C. Air separation and selective oxygen pumping via temperature and pressure swing oxygen adsorption using a redox cycle of $SrFeO_3$ perovskite. Chem Eng Sci. 2019;203:68.
64. Fujita K, Asakura H, Hosokawa S, Teramura K, Kobayashi M, Fujita K, Tanaka T. Oxygen release and storage property of Fe–Al spinel compounds: a three-way catalytic reaction over a supported Rh catalyst. ACS Appl Mater Interf. 2021;13(21):24615.
65. Johnson JW, Johnston DC, Jacobson AJ, Brody JF. Preparation and characterization of vanadyl hydrogen phosphate hemihydrate and its topotactic transformation to vanadyl pyrophosphate. J Am Chem Soc. 1984;106(26):8123.
66. Beppu K, Hosokawa S, Teramura K, Tanaka T. Oxygen storage capacity of $Sr_3Fe_2O_{7-\delta}$ having high structural stability. J Mater Chem A. 2015;3(25):13540.
67. Sugiura M, Ozawa M, Suda A, Suzuki T, Kanazawa T. Development of innovative three-way catalysts containing Ceria-Zirconia solid solutions with high oxygen storage/release capacity. Bull Chem Soc Jpn. 2005;78(5):752.
68. Sugiura M. Oxygen storage materials for automotive catalysts: Ceria-Zirconia solid solutions. Catal Surv Asia. 2003;7:77.
69. Urban S, Djerdj I, Dolcet P, Chen L, Möller M, Khalid O, Camuka H, Ellinghaus R, Li C, Gross S, Klar PJ, Bernd S, Over H. In situ study of the oxygen-induced transformation of Pyrochlore $Ce_2Zr_2O_{7+x}$ to the κ-$Ce_2Zr_2O_8$ phase. Chem Mater. 2017;29(21):9218.
70. Vieten J, Bulfin B, Senholdt M, Roeb M, Sattler C, Schmücker M. Redox thermodynamics and phase composition in the system $SrFeO_{3-\delta}$—$SrMnO_{3-\delta}$. Solid State Ion. 2017;308:149.
71. Vieten J, Bulfin B, Call F, Lange M, Schmücker M, Francke A, Roeb M, Sattler C. Perovskite oxides for application in thermochemical air separation and oxygen storage. J Mater Chem A. 2016;4(35):13652.
72. Demizu A, Beppu K, Hosokawa S, Kato K, Asakura H, Teramura K, Tanaka T. Oxygen storage property and chemical stability of $SrFe_{1-x}Ti_xO_{3-\delta}$ with robust Perovskite structure. J Phys Chem C. 2017;121(35):19358.
73. Beppu K, Hosokawa S, Demizu A, Oshino Y, Tamai K, Kato K, Wada K, Asakura H, Teramura K, Tanaka T. Striking oxygen-release/storage properties of Fe-site-substituted $Sr_3Fe_2O_{7-\delta}$. J Phys Chem C. 2018;122:11186.
74. Beppu K, Hosokawa S, Asakura H, Teramura K, Tanaka T. Efficient oxygen storage property of Sr–Fe mixed oxide as automotive catalyst support. J Mater Chem A. 2019;7(3):1013.

75. Antipov EV, Abakumov A, Istomin SY. Target-aimed synthesis of anion-deficient Perovskites. Inog Chem. 2008;47(19):8543.
76. Miura N, Ikeda H, Tsuchida A. $Sr_{1-x}Ca_xFeO_{3-\delta}$ as a new oxygen sorbent for the high-temperature pressure-swing adsorption process. Ind Eng Chem Res. 2016;55(11):3091.
77. Ikeda H, Tsuchida A, Morita J, Miura N. $SrCo_xFe_{1-x}O_{3-\delta}$ oxygen sorbent usable for high-temperature pressure-swing adsorption process operating at approximately 300 °C. Ind Eng Chem Res. 2016;55(22):6501.
78. Popczun EJ, Tafen DN, Natesakhawat S, Marin CM, Nguyen-Phan T-D, Zhou Y, Alfonso D, Lekse JW. Temperature tunability in $Sr_{1-x}Ca_xFeO_{3-\delta}$ for reversible oxygen storage: a computational and experimental study. J Mater Chem A. 2020;8(5):2602.
79. Yin Q, Kniep J, Lin YS. Oxygen sorption and desorption properties of Sr–Co–Fe oxide. Chem Eng Sci. 2008;63(8):2211.
80. Dou J, Krzystowczyk E, Wang X, Robbins T, Ma L, Liu X, Li F. A- and B-site codoped $SrFeO_3$ oxygen sorbents for enhanced chemical looping air separation. Chemsuschem. 2020;13(2):385.
81. Wright AJ, Palmer HM, Anderson PA, Greaves C. Structures and magnetic ordering in the brownmillerite phases, Sr_2MnGaO_5 and Ca_2MnAlO_5. J Mater Chem. 2002;12(4):978.
82. Motohashi T, Hirano Y, Masubuchi Y, Oshima K, Setoyama T, Kikkawa S. Oxygen storage capability of Brownmillerite-type $Ca_2AlMnO_{5+\delta}$ and its application to oxygen enrichment. Chem Mater. 2013;25(3):372.
83. Motohashi T, Kimura M, Inayoshi T, Ueda T, Masubuchi Y, Kikkawa S. Redox characteristics variations in the cation-ordered perovskite oxides $BaLnMn_2O_{5+\delta}$ (Ln = Y, Gd, Nd, and La) and $Ca_2Al_{1-x}Ga_xMnO_{5+\delta}$ ($0 \leq x \leq 1$). Dalton Trans. 2015;44(23):10746.
84. Tanahashi K, Omura Y, Naya H, Miyazaki K, Saito G, Kunisada Y, Sakaguchi N, Nomura T. Sr-doped $Ca_2AlMnO_{5+\delta}$ for energy-saving oxygen separation process. ACS Sustain Chem Eng. 2021;9:9317.
85. Ling C, Zhang R, Jia H. Quantum chemical design of doped $Ca_2MnAlO_{5+\delta}$ as oxygen storage media. ACS Appl Mater Interf. 2015;7(26):14518.
86. Iseki T, Tamura S, Saito M, Tanabe T, Motohashi T. Tunable oxygen intake/release characteristics of Brownmillerite-type $Ca_2AlMnO_{5+\delta}$ involving atomic defect formations. ACS Appl Mater Interf. 2021;13(45):53717.
87. Valldor M, Andersson M. The structure of the new compound $YBaCo_4O_7$ with a magnetic feature. Solid State Sci. 2002;4:923.
88. Chmaissem O, Zheng H, Huq A, Stephens PW, Mitchell JF. Formation of Co^{3+} octahedra and tetrahedra in $YBaCo_4O_{8.1}$. J Solid State Chem 2008;181(3):664.
89. Karppinen M, Otani HY, Fujita T, Motohashi T, Huang YH, Valkeapää M, Fjellvåg H. Oxygen nonstoichiometry in $YBaCo_4O_{7+\delta}$: large low-temperature oxygen absorption/desorption capability. Chem Mater. 2006;18:490.
90. Räsänen S, Yamauchi H, Karppinen M. Oxygen absorption capability of $YBaCo_4O_{7+\delta}$. Chem Lett. 2008;37(6):638.
91. Motohashi T, Kadota S, Fjellvåg H, Karppinen M, Yamauchi H. Uncommon oxygen intake/release capability of layered cobalt oxides, $REBaCo_4O_{7+\delta}$: Novel oxygen-storage materials. Mater Sci Eng: B 2008; 148 (1–3):196.
92. Parkkima O, Karppinen M. The $YBaCo_4O_{7+\delta}$-based functional oxide material family: a review. Euro J Inorg Chem. 2014;2014(25):4056.
93. Remsen S, Dabrowski B. Synthesis and oxygen storage capacities of hexagonal $Dy_{1-x}Y_xMnO_{3+\delta}$. Chem Mater. 2011;23(17):3818.
94. Klimkowicz A, Świerczek K, Kobayashi S, Takasaki A, Allahyani W, Dabrowski B. Improvement of oxygen storage properties of hexagonal $YMnO_{3+\delta}$ by microstructural modifications. J Solid State Chem. 2018;258:471.
95. Klimkowicz A, Cichy K, Chmaissem O, Dabrowski B, Poudel B, Świerczek K, Taddei KM, Takasaki A. Reversible oxygen intercalation in hexagonal $Y_{0.7}Tb_{0.3}MnO_{3+\delta}$: toward oxygen production by temperature-swing absorption in air. J Mater Chem A 2019; 7(6):2608.

96. Otomo M, Hasegawa T, Asakura Y, Yin S. Remarkable effects of lanthanide substitution for the Y-site on the oxygen storage/release performance of $YMnO_{3+\delta}$. ACS Appl Mater Interf. 2021;13(27):31691.
97. Klimkowicz A, Hashizume T, Cichy K, Tamura S, Świerczek K, Takasaki A, Motohashi T, Dabrowski B. Oxygen separation from air by the combined temperature swing and pressure swing processes using oxygen storage materials $Y_{1-x}(Tb/Ce)_xMnO_{3+\delta}$. J Mat Sci. 2020;55(33):15653.
98. Kondratenko EV, Peppel T, Seeburg D, Kondratenko VA, Kalevaru N, Martin A, Wohlrab S. Methane conversion into different hydrocarbons or oxygenates: current status and future perspectives in catalyst development and reactor operation. Catal Sci Technol. 2017;7(2):366.
99. Liu J, Yue J, Lv M, Wang F, Cui Y, Zhang Z, Xu G. From fundamentals to chemical engineering on oxidative coupling of methane for ethylene production: a review. Carbon Resourc Convers. 2022;5(1):1.
100. Keller GE, Bhasin M. Synthesis of ethylene via oxidative coupling of Methane I. Determination of active catalysts. J Catal. 1982;73:9.
101. Zavyalova U, Holena M, Schlögl R, Baerns M. Statistical analysis of past catalytic data on oxidative methane coupling for new insights into the composition of high-performance catalysts. ChemCatChem. 2011;3(12):1935.
102. Ito T, Lunsford JH. Synthesis of ethylene and ethane by partial oxidation of methane over lithium-doped magnesium oxide. Nature. 1985;25:721.
103. Arndt S, Otremba T, Simon U, Yildiz M, Schubert H, Schomäcker R. Mn–Na_2WO_4/SiO_2 as catalyst for the oxidative coupling of methane. What is really known? Appl Catal A: Gen. 2012; 425–426:53.
104. Kiani D, Sourav S, Baltrusaitis J, Wachs IE. Oxidative coupling of methane (OCM) by SiO_2-supported tungsten oxide catalysts promoted with Mn and Na. ACS Catal. 2019;9(7):5912.
105. Iwamatsu E, Moriyama T, Takasaki N, Aika K. oxidative coupling of methane over Na^+- and Rb^+-doped MgO. J Catal. 1988;113:25.
106. Aika K, Fujimoto N, Kobayashi M, Iwamatsu E. Oxidative coupling of methane over various metal oxides supported on strontium carbonate catalysts. J Catal. 1991;127:1.
107. Xu J, Zhang Y, Xu X, Fang X, Xi R, Liu Y, Zheng R, Wang X. Constructing $La_2B_2O_7$ (B = Ti, Zr, Ce) compounds with three typical crystalline phases for the oxidative coupling of methane: the effect of phase structures, superoxide anions, and alkalinity on the reactivity. ACS Catal. 2019;9(5):4030.
108. Matsumoto T, Saito M, Ishikawa S, Fujii K, Yashima M, Ueda W, Motohashi T. High catalytic activity of crystalline lithium calcium silicate for oxidative coupling of methane originated from crystallographic joint effects of multiple cations. ChemCatChem. 2020;12(7):1968.
109. Matsumoto T, Ishikawa S, Saito M, Ueda W, Motohashi T. Studies on activation factors for oxidative coupling of methane over lithium-based silicate/germanate catalysts. Catal Sci Technol. 2022;12(1):75.

Chapter 11
Metal Oxide Catalysts for the Valorization of Biomass-Derived Sugars

Daniele Padovan, Kiyotaka Nakajima, and Emiel J. M. Hensen

11.1 Introduction

11.1.1 Need of Renewable Resources

Renewable resources consist of solid matter or energy vectors whose lifecycle is comparable to the one of human life, implying that they can be virtually reutilized without being depleted. Unfortunately, our current energy consumption and materials production heavily rely on limited fossil resources. A great concern of utilizing fossil feedstock is the strongly negative impact on the climate, which is by far the most substantial challenge faced by humanity. Therefore, a transition from fossil resources to renewable and environmentally benign alternatives is imperative and urgent. Amongst all the kinds of renewable resources, organic biomass offers a wide scope of applications. Products derived from biomass can be utilized directly as energy vectors (*e.g.* by combustion) or they can be converted into other chemical compounds via chemical processing [1, 2]. The latter possibility is particularly attractive because it can in principle lead to similar building blocks as currently manufactured by the petrochemical industry, besides the possibility to produce completely new molecules with tailored properties for sustainable applications. Therefore, in the last decades, there has been a great drive both from the scientific community and the chemical industry to start shaping the concept of a biorefinery as a central

D. Padovan · K. Nakajima (✉)
Institute for Catalysis, Hokkaido University Kita, 21 Nishi 10, Kita-ku, Sapporo 001-0021, Hokkaido, Japan
e-mail: nakajima@cat.hokudai.ac.jp

E. J. M. Hensen
Department of Chemical Engineering and Chemistry, Laboratory of Inorganic Materials and Catalysis, Eindhoven University of Technology, 5600 MB, P.O. Box 513 Eindhoven, The Netherlands
e-mail: e.j.m.hensen@tue.nl

W. Ueda (ed.), *Crystalline Metal Oxide Catalysts*,
https://doi.org/10.1007/978-981-19-5013-1_11

processing facility of the chemical industry, in which biomasses are the main alternative to fossil feedstock for materials and energy production. This is challenging because it will require a profound change in the chemical industry. The main reason for this is the very different chemical constitution of fossil and biobased feedstock, which almost categorically excludes the retro-compatibility of biomass resources for feeding existing petroleum-based chemical manufacturing plants.

11.1.2 Biomass

Biomass comes in many forms. The choice of the type of biomass to be used as feedstock depends on many factors such as composition, availability, and ethical issues such as competition with food production. Therefore, exclusively non-edible and waste-derived types of biomass are preferred because of their lower environmental impact and to avoid ethical issues. Such type of biomass can be collected in large quantities from agricultural waste (mainly rice straw, sugarcane bagasse, and wheat crops) and the wood and paper industry [3]. The main organic resource derived from this waste is lignocellulose. There are mainly two ways in which biomass can be converted through thermo-chemical transformation. Biomass can be gasified to synthesize gas (syngas, CO and H_2), which constitutes a high-temperature treatment where the biomass is decomposed to its most basic gaseous components [4]. Syngas obtained in this way can be directly integrated with minimal effort into existing C1 chemical processing units, for instance through the Fisher-Tropsch reaction to produce liquid hydrocarbons [5]. While some challenges need to be solved to practically realize this approach, it has the disadvantage that the potentially beneficial useful molecular structures contain in biomass are lost. Therefore, the alternative approach to valorize biomass by chemical transformation into a variety of platform molecules is drawing widespread attention. Such an approach implies fractionation of biomass into smaller and simpler units that serve as feedstock of dedicated chemical processes to obtain useful chemical intermediates and products is potentially more attractive than the syngas route, as it gives access to brand new molecules which could not be obtained by conventional petrochemical routes. The atom economy of such processes is mostly also better because a significant part of the chemical functionalities in the feedstock can be retained, thus reducing waste. A challenge for this approach is that the fractionation and chemical conversion steps need to be carefully designed and developed based on different types of biomass and different products. Lignocellulose is a complex composite material comprising three main components: lignin, hemicellulose, and cellulose, whose ratio strongly depends on the origin of the lignocellulose. Lignin is generally abundant in woody biomass and can be characterized as a highly complex polymer build up from phenols with methoxy and propanol substituents [6, 7]. Due to the high complexity of lignin, recovery of its energy by combustion is preferred over chemical conversion, although many thermocatalytic routes are under development. On the other hand, cellulose and hemicellulose are very suitable resources for conversion into chemical platforms, with some processes already being employed in practice. Cellulose and hemicellulose are polymers made up of glucose and a mix of C_6 and C_5 sugars, respectively. Cellulose is the most abundant biopolymer on Earth, which is a linear chain of D-glucose units linked by $\beta(1 \rightarrow 4)$ glycosidic bonds. The strong hydrogen bonding between polymer chains confers a very robust structure to this material. Depolymerization of cellulose and hemicellulose can be typically achieved by hydrolysis of the glycosidic bonds, yielding free carbohydrates which can be converted into chemical platforms for further upgrading to liquid fuels, solvents, monomers, and other chemical intermediates [8–11]. Glucose is often considered the most representative lignocellulose-derivative compound and its conversion route into valuable platform molecules has been well established, comprising an array of different products with potential applications in diverse fields ranging from liquid fuel to polymers [12]. Scheme 11.1 summarizes some reaction pathways. The retro-aldol condensation reaction of glucose and fructose yields shorter C_2, C_3, and C_4 carbohydrates, which are precursors of lactic acid, the monomer of biodegradable polylactic acid (PLA) [13–15]. Hydrogenation of glucose leads to sorbitol, which can be dehydrated into isosorbide, an important additive for biopolymers [16–18]. Dehydration of glucose to 5-hydroxymethylfurfural (HMF) is probably the most studied process of glucose valorization. HMF is an important platform molecule with a wide scope of application. Hydrodeoxygenation of HMF produces 2,5-dimethylfuran (DMF), which can be used as a fuel additive [19, 20]. Rehydration of HMF leads to levulinic acid which can be converted to γ-valerolactone (GVL), a green solvent [21]. One of the more promising processes using HMF as a platform is its oxidation to 2,5-furandicarboxylic acid (FDCA), which is the monomer for polyethylene furanoate (PEF), an analog to the fossil fuel-derived polyethylene terephthalate (PET), the most common material of beverage bottles [22, 23]. Many strategies are employed for high-yield production of FDCA, in which one of the most attractive ones makes use of protected HMF to achieve a controlled and selective stepwise oxidation to FDCA, reducing by-products unavoidably formed from free HMF [24–27]. Therefore, the future of biorefinery relies on efficient glucose dehydration to HMF. Today, making profitable processes for industrial HMF production still represents a challenge, which is mainly due to the high reactivity of HMF, resulting in unsatisfactory yields. The main component of hemicellulose, C_5 sugars like xylose and arabinose, are also of great importance for the biorefining industry (Scheme 11.2). In a similar fashion to glucose, their dehydration can lead to another furanic compound, namely furfural, which is a key intermediate for a variety of useful products [28–30]. Furfural is a more stable molecule than HMF, giving it some advantages for practical implementation in the chemical industry. As for HMF, furfural is a pivotal platform molecule from which an array of different products can be obtained. For example, furan and tetrahydrofuran are common organic solvents [31–33]. An important reaction is an oxidation of furfural to furoic acid, a molecule that is used to manufacture pharmaceuticals, pesticides, and food [34, 35].

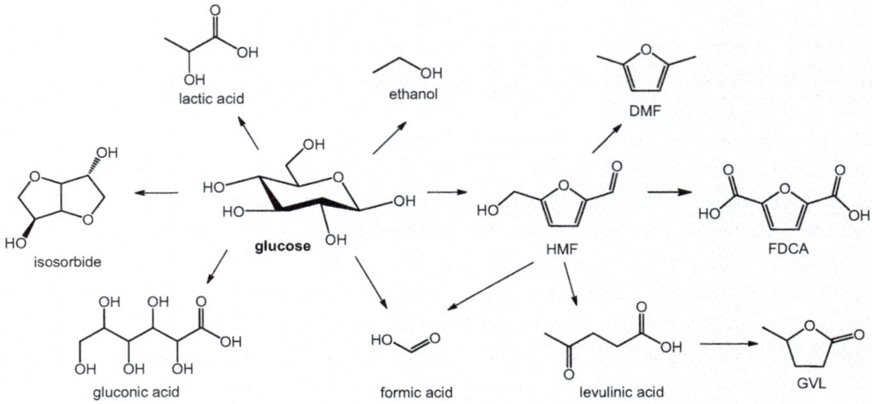

Scheme 11.1 Reaction network of important glucose-derived platform molecules

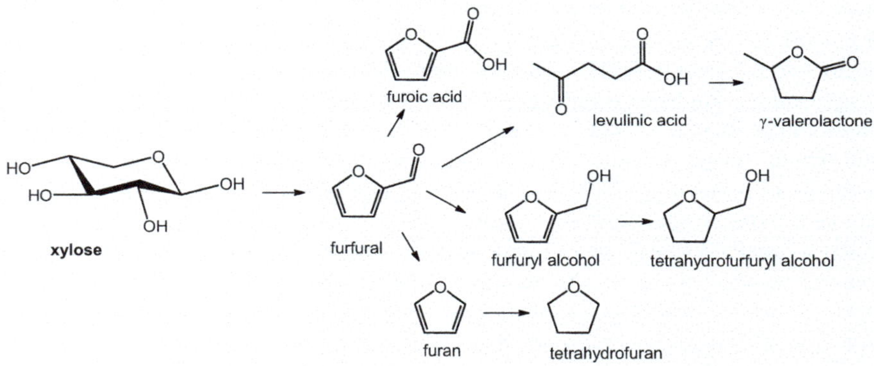

Scheme 11.2 Reaction network of xylose-derived production of important platform molecules

11.1.3 Metal Oxide Materials

Currently, most of the biomass-derived products are obtained through enzymatic industrial processes. Examples are the production of ethanol, succinic acid, lactic acid, and fructose. Despite being very selective catalysts, enzymes require strict and controlled conditions and typically mild reaction temperature, resulting in low productivity and large energy demand (e.g. due to the need to work in dilute conditions), affecting the cost of the final product. In order to make more cost-effective processes, biorefineries can adopt the use of thermocatalytic processes, which are the standard in petroleum refineries. In the petrochemical industry, heterogeneous catalysts are typically employed due to their robustness and stability besides facile separation of the products. In general, such benefits would also be important to large-scale processes for the conversion of biomass.

Lignocellulose as a feedstock can in principle be transformed into a large variety of products. Extensive reviews on biomass conversion and the role of different types of catalysts are available [36–41]. Here, we focus on the dehydration of monosaccharides into furanic compounds with glucose to HMF dehydration as the highlight to exemplify the role of alternative metal oxide catalysts to homogeneous catalysts. HMF itself is an important platform molecule with a wide array of downstream applications, while its efficient conversion from glucose is key to arriving at practical processes. Dehydration of glucose to HMF is normally conducted in water using acid catalysts. Brønsted acid catalysts are less effective and selective than Lewis acid catalysts. Nevertheless, the efficiency of Lewis acids in water is greatly hampered due to the formation of the stable Lewis acid—water adduct. One of the first Lewis acid catalysts developed for efficient operation in water was Sn-Beta catalyst [42]. The acid centers in Sn-Beta are tetrahedrally coordinated Sn atoms embedded in the zeolite framework. Owing to its high affinity for carbonyl groups, Sn-Beta has been employed successfully for glucose to fructose isomerization via intramolecular hydride shift [43] and for the formation of lactic acid via the retro-aldol reaction [44]. Despite its water tolerance, Sn species in Sn-Beta can leach upon prolonged reaction under hydrothermal conditions. This problem can be controlled to some extent by using a mixture of water and methanol as solvent at the cost of reducing the concentration of sugar in the stream, making this material less attractive for industrial applications [45, 46]. Furthermore, zeolites greatly suffer from deactivation by fouling, which is particularly challenging given the relatively large size of sugars and by-products formed during oligomerization such as humins.

In this regard, metal oxides including zeolites represent a broader class than only zeolites with potential applications in biomass conversion, mainly because of their tuneable acid and base properties [47–50]. Different from zeolites where the reactions take place inside narrow micropores, common metal oxides present an open surface in which the catalytic sites are usually easily accessible. The acid—base properties of their surfaces depend on many factors, such as the type of metal and its oxidation state, its coordination, crystalline structure, presence of defects and impurities, and so forth. Probe molecules such as CO, NH_3, pyridine, CO_2, etc. are often employed to determine the surface reactivity of such catalysts by analyzing the type and the strength of the interaction using spectroscopic techniques (IR, UV, Raman) [51–55]. A combination of such techniques combined with textural analysis and aspects such as crystal structure, and coordination environment of metals on the surface is essential to assess the relation between structure and performance. In this chapter, we will give an overview of biomass processes in which metal oxides are employed, giving a special focus on amorphous or low-crystalline metal oxides.

11.2 Lewis Acidic Amorphous Oxide and Determination of Structure–activity Relationship

11.2.1 Bulk and Surface Properties of Nb_2O_5

Dehydration of sugars is one of the most important reactions for obtaining chemical platforms. Niobium-containing materials have been known for many years as suitable acid catalysts and their acid character has been the subject of many studies [56]. Niobic acid is an amorphous solid material known for its unexpectedly high Brønsted acidity comparable to 70% H_2SO_4; it can be readily obtained by hydrolysis of $NbCl_5$ in water [57]. Acidic protons are stabilized as counterions of homopolyanions (Nb_xO_y) [58]. Niobic acid undergoes structural transitions during heat treatment. Calcination of niobic acid forms the TT-phase at around 400−550 °C, the T-phase at 550−800 °C, the M-phase at 800−1000 °C, and the H-phase above 1000 °C [59–61]. Along with changes in the crystalline phase, the surface area of Nb_2O_5 decreases from typical values of 150−200 $m^2\ g^{-1}$ after preparation and drying to less than 5 $m^2\ g^{-1}$ after calcination at 700 °C [62]. Such loss of surface area is also accompanied by a decrease in acid density, impacting the catalytic activity [63, 64].

Studies employing Raman spectroscopy have been carried out to elucidate the structure of Nb_2O_5 materials. Determination of active surface species on bulk Nb_2O_5 is proposed from supported Nb_2O_5 species on other oxides such as SiO_2, TiO_2, ZrO_2, and Al_2O_3 (Fig. 11.1) [65–68]. At high Nb-loading, polymeric Nb_2O_5 species consisting of octahedral NbO_6 units and tetrahedral NbO_4 units were detected with characteristic Nb–O−Nb vibration around between 550 and 750 cm^{-1}, and terminal Nb = O at around 980 cm^{-1}, respectively. The difference in bond length between lattice Nb−O and terminal Nb = O showed the presence of highly distorted NbO_6 units. However, at low Nb loading on SiO_2 only tetrahedral NbO_4 units were found with a characteristic vibration of terminal Nb = O species at 980−990 cm^{-1}. The lower coordination number of NbO_4 implies Lewis acidity which can be effective for reactions requiring an acid. NbO_4 tetrahedra are also present in the hydrated form on Nb_2O_5 and are responsible for Lewis acidity [69]. Identification of surface species is of critical importance as they are the main contributors to the catalysis, as opposed to bulk species.

11.2.2 Sugar Dehydration with Niobic Acid

Catalysts containing niobic acid have been largely employed for their Brønsted acidity, as Lewis acid sites are thought to be suppressed in the presence of water. The surface of amorphous niobic acid is composed of distorted NbO_6 octahedra and NbO_4 tetrahedra (Fig. 11.2). Surface octahedral units bear strongly acidic OH groups, while NbO_4 tetrahedra have a stronger Lewis acid character. We investigated niobic acid ($Nb_2O_5{\cdot}nH_2O$) for the conversion of glucose to HMF in water (Scheme 11.3) and

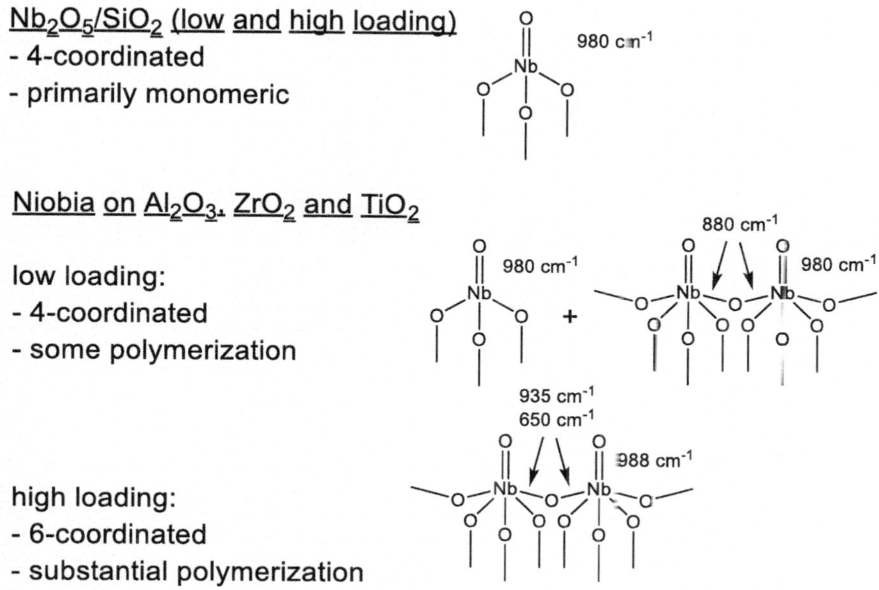

Fig. 11.1 Schematic drawing of the molecular structures present on supported niobia catalyst under dehydration conditions. Reproduced from reference [65]. Copyright 1999 American Chemical Society

compared its performance with reference catalysts (Table 11.1) [69]. Mineral acids (HCl and H_2SO_4) and Brønsted acidic resins such Amberlyst-15 and NafionNR50 lead mainly to the formation of formic acid and levulinic acid, which are formed upon rehydration of HMF. Indeed, it is well documented that strong Brønsted acids are effective for both dehydration and hydration reactions, making it very challenging to stop the reactions at the valuable HMF intermediate in the presence of water. Zeolites (H-Mordenite and H-ZSM-5) showed poor performance probably due to the slow diffusion of the sugar inside the zeolite micropores.

$Nb_2O_5 \cdot nH_2O$ also displayed a modest HMF selectivity at 12%, but it could be substantially increased to 52% when the catalyst underwent H_3PO_4 treatment (phosphate/$Nb_2O_5 \cdot nH_2O$). Despite the anticipated Brønsted acid character, the activity of $Nb_2O_5 \cdot nH_2O$ could be ascribed to the presence of tetrahedral NbO_4 units acting as Lewis acids. The role of Brønsted and Lewis acid sites could be discriminated by showing that the HMF yield did not change when all Brønsted protons of $Nb_2O_5 \cdot nH_2O$ were exchanged with Na^+ ions, supporting the view that the active species were the free Lewis Nb centers. These findings were exceptional as Lewis acid sites detected by IR investigations [70] had typically not been exploited for catalysis in water. Phosphate groups working as Brønsted acid were excluded for modified $Nb_2O_5 \cdot nH_2O$ as formic acid and levulinc acid formed by the decomposition of HMF were negligible. We also employed $Nb_2O_5 \cdot nH_2O$ for the dehydration of xylose to furfural in a biphasic systems [71]. Continuous extraction of furfural from

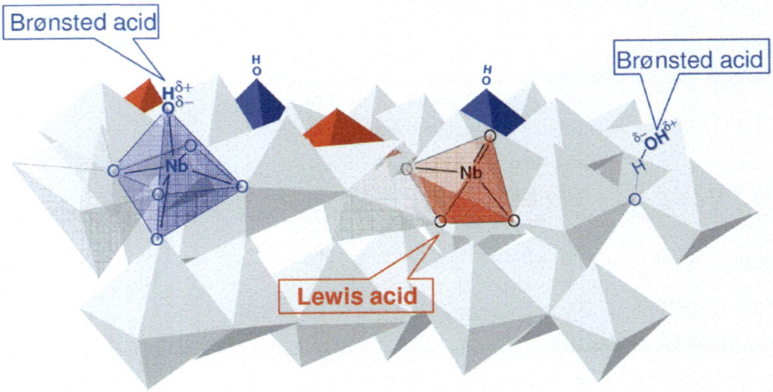

Fig. 11.2 Schematic structure of $Nb_2O_5.nH_2O$. Reproduced from reference [69]. Copyright 2011 American Chemical Society

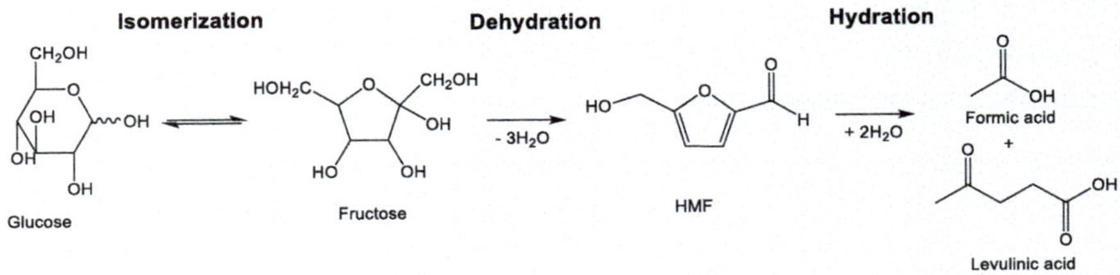

Scheme 11.3 Reaction pathways for dehydration of glucose to HMF via fructose isomerization. Reproduced from reference [69]. Copyright 2011 American Chemical Society

Table 11.1 Catalytic performance of several catalyst for the conversion of glucose to HMF in water. Reproduced from reference [69]. Copyright 2011 American Chemical Society

				selectivity %				
Catalyst	BA[a]	LA[b]	Conv.[c]	Fru[d]	HMF	FA[e]	LA[f]	unknown
HCl	9.9	–	100	0.0	0.0	5.7	27.1	65.5
H_2SO_4	22.4	–	100	0.0	0.0	8.4	56.4	35.2
Amberlyst-15	4.8	–	89	0.0	0.0	42.3	42.3	15.4
NafionNR50	0.9	–	65	0.0	0.0	9.8	35.4	54.8
H-mordenite	1.1	0.26	12	35.2	0.0	0.0	0.0	64.8
H-ZSM-5	0.15	0.05	34	0.0	0.0	3.8	0.0	96.2
$Nb_2O_5.nH_2O$	0.17	0.15	100	0.0	12.1	3.2	0.0	84.6
$Na^+/Nb_2O_5.nH_2O$	–	0.17	100	0.5	12.4	2.5	0.0	84.6
Phosphate/$Nb_2O_5.nH_2O$	0.04	0.11	92	0.8	52.1	2.6	1.2	43.3

Reaction conditions: 2 mL of water, 0.02 g of glucose, 120 °C, 0.2 g of catalyst. 3 h. [a] Brønsted acid amount (mmol g^{-1}), [b] Lewis acid amount (mmol g^{-1}), [c] Glucose conversion (%), [d] fructose, [e] formic acid, [f] levulinic acid.

the aqueous phase into the second organic solvent phase suppressed side-reactions [72, 73], reaching a final selectivity of 72% in a water/toluene mixture. $Nb_2O_5{\cdot}nH_2O$ displayed also a lower activation energy (83 kJ mol^{-1}) compared to homogeneous catalysts such as $Sc(OTf)_3$ (107 kJ mol^{-1}) and HCl (131 kJ mol^{-1}) for glucose dehydration.

Several spectroscopic techniques were employed to elucidate the origin of the acidity, with a particular focus on the strength and catalytic action of Lewis acid sites in the presence of water. Raman spectroscopy showed that the surface contained both NbO_6 octahedra (800–900 cm^{-1}) and NbO_4 tetrahedra (988 cm^{-1}) as shown in Fig. 11.3a. When water was adsorbed on $Nb_2O_5{\cdot}nH_2O$ the signal of NbO_4 tetrahedra disappeared and recovered after dehydration, showing the reversible formation of NbO_4-H_2O adduct due to weak interactions between the Lewis acid and water as a base. The presence of such a reversible NbO_4-H_2O adduct was also reported by Omata et al. during gas-phase cumene cracking in the presence of steam [74]. Lewis acidity was investigated using basic probe molecules such as CO and pyridine. Surprisingly, Lewis acid–base adducts between NbO_4 and the basic probe molecules were detected also for water-saturated NbO_4-H_2O, showing that part of NbO_4 tetrahedra could survive the presence of water. An estimated 20% of all Lewis acid sites on NbO_4-H_2O maintain their Lewis acidity after saturation with water (Fig. 11.3B).

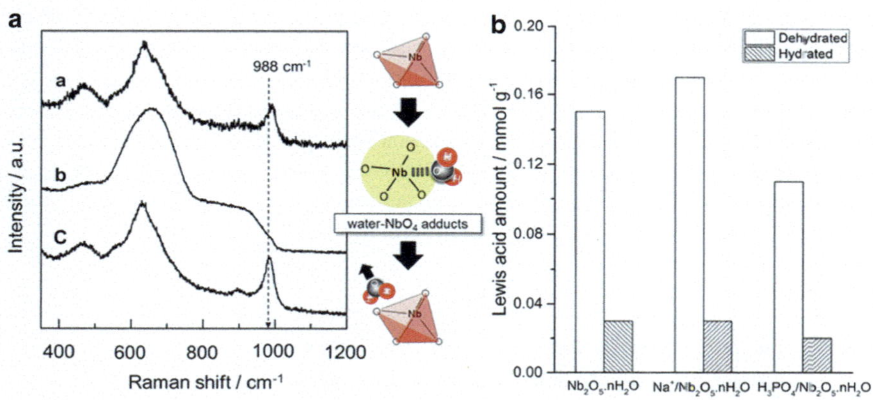

Fig. 11.3 **a** Raman spectra of $Nb_2O_5.nH_2O$ in a quartz cell. **a** $Nb_2O_5.nH_2O$ evacuated at 150 °C for 2 h to remove water. **b** Sample **a** exposed to saturated water vapor at 300 K for 3 h. **c** Sample **b** evacuated at 150 °C for 2 h. Reproduced from reference [69]. Copyright 2011 American Chemical Society. **b** Lewis acid amount calculated by Pyridine -IR for a series of dehydrated and hydrated $Nb_2O_5.nH_2O$ materials

11.2.3 Sugar Dehydration with Niobium Phosphate

Another extensively studied Nb-based material is niobium phosphate (here abbreviated NbP). Amorphous NbP can be obtained by precipitation of potassium niobate in a phosphoric acid solution. NbP remains amorphous even upon calcination at 800 °C and exhibits a high surface acidity up to 500 °C [75, 76]. In contrast, $Nb_2O_5{\cdot}nH_2O$ drastically loses acid properties upon heating at such high temperatures. NbP was applied as a catalyst for fructose dehydration, showing that HMF formation proceeded without the formation of rehydration products [77–79]. A reaction with a concentrated fructose solution (10 wt.%) exhibited 60% of fructose conversion and 89% of HMF selectivity [80]. Structural characterization of amorphous NbP by FTIR spectroscopy revealed that the surface is composed of NbO_6 distorted octahedral and orthophosphate tetrahedral units. Nevertheless, UV analysis indicated that low-coordinated Nb atoms present on the NbP surface work as catalytically active sites. XPS analysis confirmed that the electron density of the Nb surface atoms is lower than that of niobic acid, which suggests that the phosphate group might act as electro-withdrawing groups, thereby increasing the Lewis acidity of Nb.

Dehydration of glucose to HMF was found to proceed via isomerization and dehydration involving Lewis acid and Brønsted acid sites, respectively, in which isomerization was found to be the rate-determining step [81]. Consequently, several efforts were employed to tune the ratio of Lewis acid and Brønsted acid for NbP by modifying the synthesis conditions. Viera et al. synthesized several NbP catalysts and obtained a linear relationship between the rate of HMF formation and the Lewis acid/Brønsted acid ratio. The rate of HMF formation was proportional to the Lewis acid/Brønsted acid ratio, showing that Lewis acid sites played a central role in determining the overall reaction rate [82]. Similar results were obtained by Zhang et al. using mesoporous NbP, giving a glucose conversion of 68% with an HMF selectivity of 49% [83].

11.2.4 Sugar Dehydration with Amorphous and Low-Crystalline Ti-Based Oxides

Isolated TiO_4 tetrahedra can be obtained by grafting a Ti precursor on mesoporous silica SBA-15 (Ti/SBA-15) [84–86]. We assessed the Lewis acid character of Ti/SBA-15 with various Ti contents by performing the hydride transfer reaction of pyruvaldehyde to lactic acid and the Mukaiyama aldol reaction of benzaldehyde with 1-(trimethylsilyloxy)cyclohexene in water [87, 88]. These reactions successfully proceeded with the activation of carbonyl groups by the Ti-derived acid site of Ti/SAB-15 catalysts. CO adsorption FTIR experiments showed that Ti/SBA-15 forms a Lewis acid-CO adduct, even in the presence of adsorbed water, directly confirming the water-tolerant Lewis acidity of TiO_4 tetrahedra (Fig. 11.4). Notably, the same Ti-CO bond was observed in low-crystalline anatase TiO_2. The coordinated CO band on Ti/SBA-15 (2182 cm^{-1}) is higher than that on the anatase TiO_2 (2178 cm^{-1}), indicative of the differences in the Lewis acid sites between Ti/SBA-15 and anatase TiO_2. Nevertheless, the catalytic data show that anatase TiO_2 can also be used as a Lewis acid catalyst for biomass conversion.

TiO_2 is widely used in catalysis, both as a thermal catalyst as well as a photocatalyst. Besides, it has many applications in our daily life, such as white pigment in painting. Similar to $Nb_2O_5{\cdot}nH_2O$, low-crystalline anatase TiO_2 possesses TiO_6 octahedra and TiO_4 tetrahedra on its surface. However, distinct features of low-crystalline anatase TiO_2 compared to $Nb_2O_5{\cdot}nH_2O$ are the absence of Brønsted acidity and

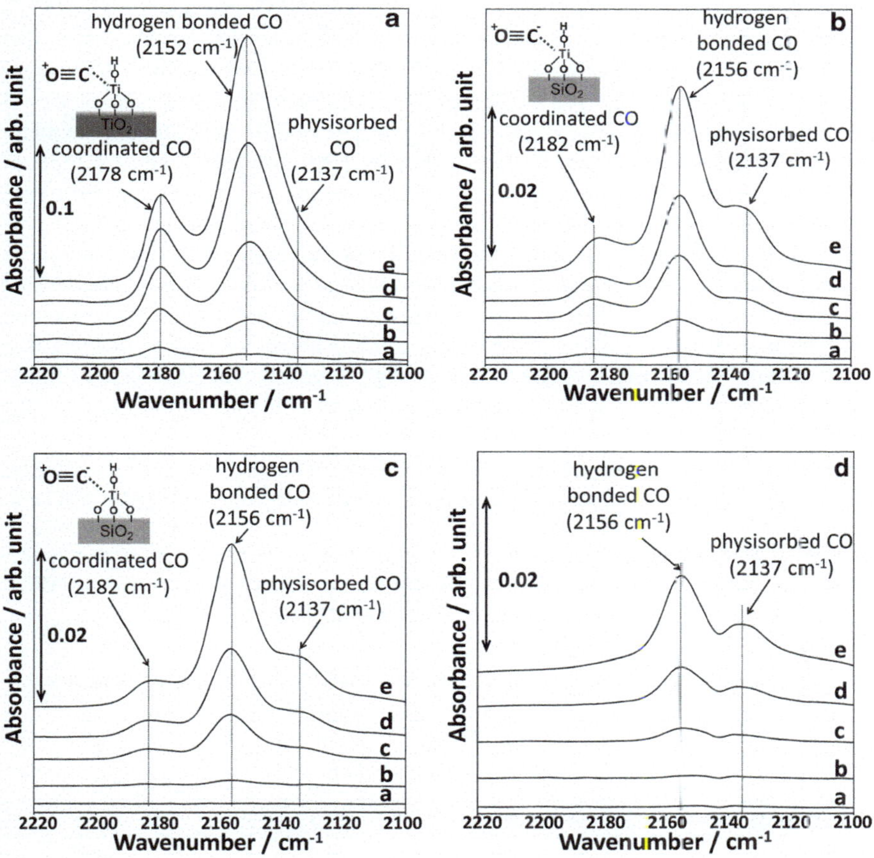

Fig. 11.4 FT-IR difference spectra of CO-adsorbed **a** hydrated TiO_2, **b** dehydrated Ti/SBA-15, **c** hydrated Ti/SBA-15, and **d** dehydrated SBA-15 at 100 K. CO pressures for spectra a, b, c, d, and e correspond to 0.1, 0.2, 0.5, 1.0, and 2.0 Torr, respectively. Reproduced from reference [87]. Copyright 2014 American Chemical Society

weaker Lewis acidity. The latter is potentially advantageous in aqueous-phase reactions because H_2O coordinated in TiO_4-H_2O adducts can be easily exchanged with other molecules including reactants [89].

Anatase TiO_2 was then employed for the conversion of glucose into HMF in a biphasic water/THF system (Scheme 11.3, Table 11.2) [90]. HCl converted glucose into HMF with 29% selectivity at 56% of HMF conversion. H_3PO_4 showed a very poor activity (glucose conversion 9%; HMF selectivity 1.2%). Two Brønsted acid resins (Amberlyst-15 and NafionNR50) also showed low HMF selectivity in water, suggesting that any Brønsted acid is ineffective for HMF formation under the given conditions. Homogeneous water-tolerant Lewis acids ($Sc(OTf)_3$ and $Yb(OTf)_3$) showed high glucose conversion, but the selectivity to HMF was also poor (<15%). As obtained for homogeneous Lewis acid catalysts, Lewis acidic anatase TiO_2 also gave high glucose conversion and poor HMF selectivity (9%), while HMF selectivity drastically improved to 81% after phosphorylation treatment, as also observed for niobic acid. Due to the negligible activity of phosphoric acid alone, phosphate groups on anatase TiO_2 are not the active sites for HMF formation. However, phosphate moieties immobilized on anatase TiO_2 surface play a vital role in improving the HMF selectivity. The difference between bare TiO_2 and phosphate TiO_2 was systematically examined. Pyridine adsorption FTIR experiments revealed that there is no difference in Lewis acid sites. Kinetic studies showed that the amount of HMF steadily increases during the reaction (2 h). However, glucose was quickly consumed at the initial stage of the reaction, with a small amount of fructose (<5%) as a possible intermediate (Fig. 11.5). This observation strongly suggests that HMF is not formed via the widely accepted fructose intermediate, but an unidentified intermediate for phosphate TiO_2.

Jadhav et al. proposed that HMF can be formed via 3-deoxy-glucosone (3DG) by step-size dehydration of glucose using homogeneous inorganic acids [91]. 3DG can

Table 11.2 Catalytic performance of several catalyst for the conversion of glucose to HMF in water. Reproduced from reference [90]. Copyright 2014 Elsevier

Catalyst	BAS[a]	LAS[b]	Conv.[c]	selectivity %				
				Fru[d]	HMF	FA[e]	LA[f]	unknown
HCl	9.9	–	56.2	0.0	29.4	0.0	0.0	26.9
H_3PO_4	10.2	–	9.2	0.0	1.2	0.0	0.0	8.0
Amberlyst-15	4.8	–	69.9	2.0	12.0	0.0	24.9	31.0
NafionNR50	0.9	–	28.7	0.0	3.1	0.0	0.0	25.6
$Sc(OTf)_3$	–	2.0	99.9	1.5	13.9	0.0	0.0	84.6
$Yb(OTf)_3$	–	1.9	89.4	0.0	10.6	0.0	0.0	78.8
TiO_2	–	0.24	99.9	0.0	8.5	0.0	0.0	91.5
phosphate/TiO_2	–	0.22	98	0.0	81.2	10.5	0.0	8.3

Reaction conditions: 0.2 mL of water, 1.8 mL of THF, 0.02 g of glucose, 120 °C, 0.05 g of catalyst, 2 h. [a] Brønsted acid amount (mmol g^{-1}), [b] Lewis acid amount (mmol g^{-1}), [c] Glucose conversion (%), [d] fructose, [e] formic acid, [f] levulinic acid.

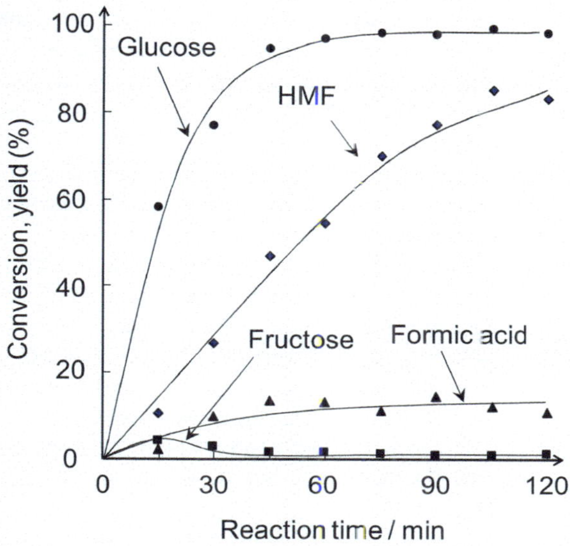

Fig. 11.5 Time courses for the conversion of glucose into HMF on H_3PO_4/TiO_2. Reagents and conditions: 0.05 g of catalyst, 0.2 mL of water, 1.8 mL of THF, 0.02 g of glucose, 120 °C. Reproduce from reference [90]. Copyright 2014 Elsevier

be obtained by dehydration of the C(3)OH moiety in glucose followed by tautomerization. However, the high reactivity makes 3DG very difficult to be identified during the course of the reaction. Tracer experiments using deuterium-labeled glucose were conducted with bare TiO_2 and phosphate/TiO_2 to clarify the reaction pathways for glucose dehydration to HMF [92]. Two different reaction pathways (isomerization/dehydration and stepwise dehydration) were proposed in HMF formation from glucose with an acid catalyst (Scheme 11.4).

For the isomerization/dehydration mechanism, the hydride step involved in the isomerization would shift the hydrogen from the C(2) to the C(1) position, resulting in the fructose intermediate fructose having the untouched deuterium atom in the C(1). The following dehydration step would have equal chances to retain or eliminate the

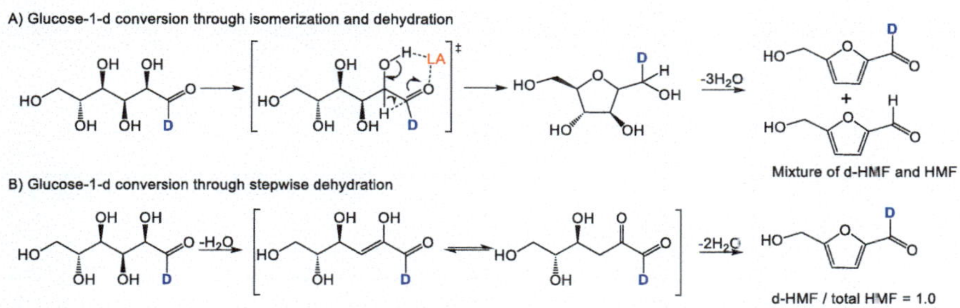

Scheme 11.4 Formation of HMF from deuterated glucose-1-d following (A) isomerization/dehydration mechanism or (B) stepwise dehydration. Reproduced from reference [92]. Copyright 2015 American Chemical Society

deuterium atom, resulting in a theoretical 50:50 mixture of H-HMF and D-HMF. In the case of the stepwise mechanism, the elimination of water from C(3) would leave the untouched deuterium atom in C(1) until its final conversion in HMF, leading to a theoretical 100% D-HMF yield. A reaction using glucose-1-d and phosphate/TiO_2 produced D-HMF with the D content of 98%. While direct detection of any intermediate could not be achieved under the reaction conditions, isotopic labeled experiments suggest that the stepwise dehydration mechanism via 3DG intermediate is the possible reaction pathway for phosphate/TiO_2. A control experiment using a homogeneous and water-tolerant Lewis acid ($Sc(OTf)_3$) gave a mixture of D-HMF and H-HMF with a D content of approximately 55%. This is in agreement with the isomerization/dehydration pathway expected for homogeneous Lewis acids. The active site on anatase TiO_2 surface, which is responsible for stepwise dehydration of glucose, was delivered successively using DFT calculations [93]. The dehydration of glucose into 3DG on abundant and unsaturated TiO_5 pentahedra on defect-free (101) anatase TiO_2 surfaces requires a very high energy barrier (207 kJ mol^{-1}), suggesting that the role of these sites is only marginal. The dehydration barrier was drastically reduced to 47 kJ mol^{-1} when an appropriate defective surface was constructed (Fig. 11.6). This surface contains Lewis acidic TiO_4 tetrahedra (Ti4c) with vicinal Ti–OH groups as conjugated Lewis base sites. Such conjugated Lewis acid–base pair sites were modeled and provided a good match with experimental observations made with FTIR spectroscopy of adsorbed $CHCl_3$. In contrast, the activation barrier for isomerization/dehydration of glucose with the same active site was calculated to be 115 kJ mol^{-1}, supporting that stepwise dehydration is the favored mechanism.

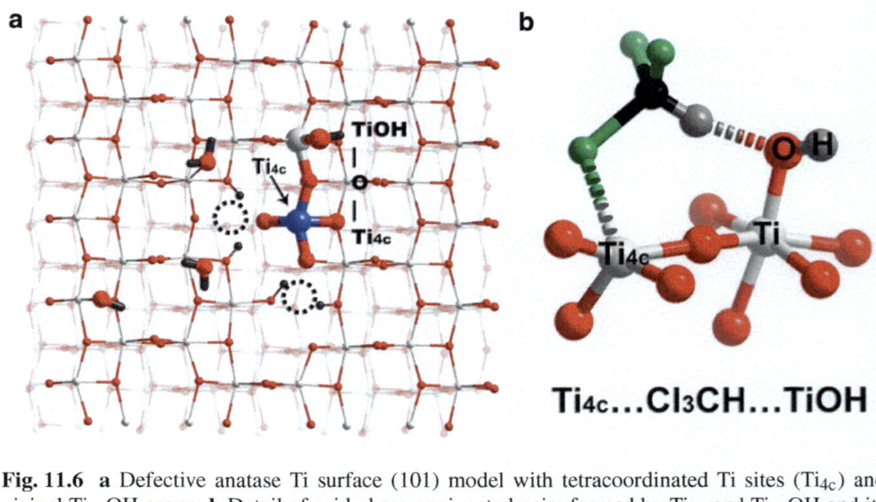

Fig. 11.6 **a** Defective anatase Ti surface (101) model with tetracoordinated Ti sites (Ti_{4c}) and vicinal Ti–OH group. **b** Detail of acid–base conjugated pairs formed by Ti_{4c} and Ti–OH and its interaction with a chloroform probe molecule. Reproduced from reference [93]. Copyright 2018 John Wiley and Sons

11.2.5 Limitations of Amorphous and Low-Crystalline Metal Oxide Catalysts

Although some amorphous or low crystalline metal oxides are active during sugar conversion, severe aspects limit their application in practice. Amorphous oxides are not thermally stable, because their textural and chemical properties drastically change by calcination at high temperatures. After extended catalytic cycles, accumulation of carbon deposition on the catalyst surface leading to deactivation will require such calcination. This is due to the deposition of insoluble matter, i.e., humins, on the catalyst [94, 95]. Removal of condensed organics is typically achieved by calcination in air (>400 °C) [96], which substantially changes the nature of metal oxides. Therefore, thermally stable and crystalline metal oxides are desirable as they can guarantee in principles higher durability simply by a conventional calcination treatment.

11.3 Crystalline Metal Oxides for the Conversion of Biomass Derived Compounds

11.3.1 Biomass Conversion with Crystalline Nb-Based Oxides

$Nb_2O_5{\cdot}nH_2O$ dramatically loses its textural and acid properties upon crystallization. As crystalline Nb_2O_5 is much less acidic than the amorphous precursor, it has not been explored for acid-catalysed reactions. Crystalline orthorhombic Nb_2O_5 synthesized by hydrothermal treatment of a water-soluble Nb precursor possesses a specific layered structure and heptagonal channels formed by NbO_6 and NbO_7 units (Fig. 11.7) [97, 98]. The surface of this material contains both Lewis acid sites and Brønsted acid sites. The Lewis acidity of crystalline orthorhombic Nb_2O_5 is different from that of $Nb_2O_5{\cdot}nH_2O$. The former retains its Lewis acidity even in the presence of water, while 80% of Lewis acid sites are deactivated with H_2O in $Nb_2O_5{\cdot}nH_2O$. This material was applied for obtaining lactic acid from dihydroxyacetone, in which consecutive dehydration and isomerization steps proceed. Crystalline orthorhombic Nb_2O_5 was highly active in such reactions compared to typical reference catalysts.

11.3.2 Biomass Conversion with Crystalline Ti-Based Oxides

Anatase TiO_2 is amphoteric with Lewis acidic TiO_4 tetrahedra and basic lattice oxygens, rendering this material suitable for a variety of acid–base catalyzed reactions including the dehydration of saccharides into HMF [99, 100]. Atanda et al. reported that phosphate anatase TiO_2 converts glucose into HMF with a yield of

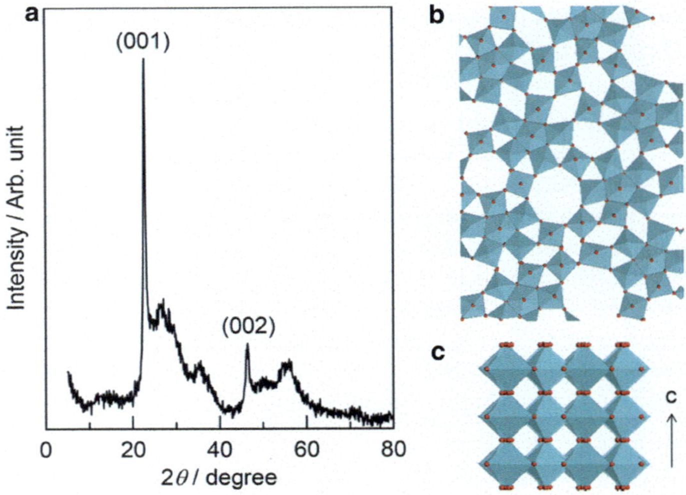

Fig. 11.7 **a** XRD pattern and proposed crystalline structure of orthorhombic Nb_2O_5 **b** in the a-b plane and **c** along the c-axis. Reproduced from reference [98]. Copyright 2018 American Chemical Society

81% at a temperature above 170 °C in a biphasic water/n-butanol system [101]. Bare TiO_2 was less efficient, providing an HMF yield of only 28%, highlighting the importance of surface modification with the phosphate group. Phosphated TiO_2 was applied to the direct conversion of cellulose to HMF [102]. However, the high crystallinity of cellulose greatly hampered its processability and only 6% of cellulose could be converted to HMF in a 4% yield. Conversion and yield could be gradually improved when the crystallinity and polymerization degree of cellulose were decreased by prior ball milling (29% conversion and 21% HMF yield) or acid-assisted ball milling (99% conversion and 75% HMF yield). This shows that the depolymerization of cellulose is the rate-determining step. Phosphated TiO_2 experienced a minor deactivation during HMF formation from glucose and cellulose. However, calcination could recover the initial activity. This regeneration by simple thermal treatment is an advantage of employing crystalline material compared to thermally unstable amorphous oxides. Despite these outstanding catalytic properties, anatase TiO_2 requires rather high reaction temperatures (approximately 170 °C) compared to low crystalline anatase TiO_2 (100–120 °C).

11.3.3 Biomass Conversion with W-based Oxides

The approach of using dopants to control the acid/base properties of an oxide host material was exploited by Hensen and co-workers for glucose dehydration to HMF. Tungstite (WO_3) is a layered oxide composed of an octahedral W coordinated to five

Table 11.3 Dehydration of glucose with Nb-doped WO_3 series. Reproduced from reference [103]

Catalyst	SA[a]	BAS[b]	LAS[c]	$Conv._{glu}$ [%]	Sel_{HMF} [%]
Nb_2O_5	184	143	88	100	16
Nb_1-WO_3	160	82	35	97	36
$Nb_{0.2}$-WO_3	63	8.7	5.9	86	40
$Nb_{0.1}$-WO_3	42	11.8	8.8	83	36
$Nb_{0.03}$-WO_3	26	6.8	8.7	79	32
WO_3	14	6.8	10.2	80	31

Reaction conditions:0.1 g of catalyst, 10 mg of glucose, 1 mL H_2O, 120 °C, 3 h. [a]Surface area (m^2 g^{-1}); [b]Brønsted acid sites ($\mu mol\ g^{-1}$); [c]Lewis acid sites ($\mu mol\ g^{-1}$).

lattice oxygen and a terminal water molecule. WO_3 is characterized by Lewis acid W centers and weak Brønsted acid sites. WO_3 doped with lower valence transition metals offers control of the amount of Lewis and Brønsted acid sites. By increasing the amount of Nb dopant Brønsted acid sites are formed at the expense of Lewis acid sites (Table 11.3) [103, 104]. The formation of Brønsted acid sites is ascribed to charge compensation of lower valence metals introduced in the tungsten oxide framework with acidic protons. Catalytic activity tests using these materials in the dehydration of glucose to HMF revealed that the optimum ratio of Nb to W was 1 to 5, namely $Nb_{0.2}$-WO_3. A lower amount of Nb dopant did not substantially improve the performance of the parent WO_3, while a higher amount led to a decrease in the HMF selectivity probably due to the higher content of the Brønsted acid site, leading to undesired side reactions. In addition, Nb doping has a positive effect on the surface area.

Kinetic studies with $Nb_{0.2}$-WO_3 showed that the rate of glucose conversion was higher compared to HMF formation. The fructose yield during the course of the reaction was always low, suggesting the rate-determining step is not the isomerization of glucose to fructose but the dehydration of a partially dehydrated intermediate formed from fructose. This is a peculiarity of this material, as other bifunctional Lewis-Brønsted acid system always proceeds with glucose isomerization as the rate-determining step. The slow dehydration steps are caused by the weak Brønsted acidity. DFT calculations confirmed that the introduction of pentavalent and tetravalent atoms in the WO_3 surface generates a high degree of surface deformation, which is beneficial for the isomerization step [105]. These calculations also highlighted that the introduction of Ti atoms increases surface deformation and lowers the isomerization barrier. FTIR measurements for pyridine-adsorbed Ti-WO_3 confirmed a higher density of acid sites compared to Nb-WO_3. However, the catalytic activities of these two materials were similar [106].

11.4 Conclusion and Outlook

This review shows that metal oxides are viable catalysts for biomass valorization. We introduced findings on the dehydration of sugars to furan compounds using metal oxides as catalysts. The controlling concepts in the production of furan compounds from hexose and pentose sugars are "water-compatible Lewis acid", "concerted catalysis of Lewis acid and Lewis base", and "the ratio of Brønsted acid and Lewis acid". Metal oxides composed of group IV, V, and VI elements such as Ti, Nb, and W are attractive for these reactions. Phosphoric acid treatment can modify the surface properties of these oxides, thereby enhancing their activity. This aspect needs to be understood better. Despite the importance of the resulting furan compounds, the overall reaction efficiencies are not yet satisfactory for industrial application. Further studies on the acid and base properties of crystalline metal oxides will be essential for the development of industrially relevant catalysts for biomass conversion.

References

1. McKendry P. Energy production from biomass (Part 2): conversion technologies. Bioresour Technol. 2002;83(1):47.
2. Clark J, Deswarte F. Introduction to chemicals from biomass. Wiley;2008.
3. McKendry P. Energy production from biomass (Part 1): overview of biomass. Bioresour Technol. 2002;83(1):37.
4. Rauch R, Hrbek J, Hofbauer H. Biomass gasification for synthesis gas production and applications of the syngas. WIREs Energy Environ. 2014;3(4):343.
5. Ahmad AA, Zawawi NA, Kasim FH, Inayat A, Khasri A. Assessing the gasification performance of biomass: A review on biomass gasification process conditions, optimization and economic evaluation. Renew Sust Energ Rev. 2016;53:1333.
6. Cao L, Yu IKM, Ruan X, Tsang DCW, Hunt AJ, Ok YS, Song H, Zhang S. Lignin valorization for the production of renewable chemicals: State-of-the-art review and future prospects. Bioresour Technol. 2018;269:465.
7. Galkin MV, Samec JSM. Lignin valorization through catalytic lignocellulose fractioning: a fundamental platform for the future biorefinery. Chemsuschem. 2016;9(13):1544.
8. Shrotri A, Kobayashi H, Fukuoka A. Cellulose depolymerization over heterogeneous catalysts. Acc Chem Res. 2018;51(3):761.
9. Chen P, Shrotri A, Fukuoka A. Synthesis of cello-oligosaccharides by depolymerization of cellulose: a review. Appl Catal A Gen. 2021;621: 118177.
10. Rinaldi R, Schuth F. Acid hydrolysis of cellulose as the entry point into biorefinery schemes. Chemsuschem. 2009;2(12):1096.
11. Van de Vyver S, Geboers J, Jacobs PA, Sels BF. Recent advances in the catalytic conversion of cellulose. ChemCatChem. 2011;3:82.
12. Chheda JN, Dumesic JA. An overview of dehydration, aldol-condensation and hydrogenation processes for production of liquid alkanes from biomass-derived carbohydrates. Catal Today. 2007;123(1–4):59.
13. Aida TM, Tajima K, Watanabe M, Saito Y, Kuroda K, Nonaka T, Hattori H, Smith RL, Arai K. Reactions of d-fructose in water at temperatures up to 400°C and pressures up to 100 MPa. J Supercrit Fluids. 2007;42(1):110.
14. Hu J, Yu F, Lu Y. Application of Fischer-Tropsch synthesis in biomass to liquid conversion. Catalysts. 2012;2(2):303.

15. Abdel-Rahman MA, Tashiro Y, Sonomoto K. Lactic acid production from lignocellulose-derived sugars using lactic acid bacteria: overview and limits. J Biotechnol. 2011;156(4):286.
16. Kobayashi H, Yokoyama H, Feng B, Fukuoka A. Dehydration of sorbitol to isosorbide over H-beta zeolites with high Si/Al ratios. Green Chem. 2015;17:2732.
17. Yabushita M, Kobayashi H, Shrotri A, Hara K, Ito S, Fukuoka A. Sulfuric acid-catalyzed dehydration of sorbitol: mechanistic study on preferential formation of 1,4-sorbitan. Bull Chem Soc Jpn. 2015;88(7):996.
18. Rose M, Palkovits R. Isosorbide as a renewable platform chemical for versatile applications—Quo vadis? Chemsuschem. 2012;5(1):167.
19. Roman-Leshkov Y, Barret CJ, Liu ZY, Dumesic JA. Production of dimethylfuran for liquid fuels from biomass-derived carbohydrates. Nature. 2007;447:982.
20. Hoang AT, Pandey A, Luque ZHR, Ng KH, Papadopoulos AM, Chen WH, Rajamohan S, Hadiyanto H, Nguyen XP, Pham VV. Catalyst-based synthesis of 2,5-dimethylfuran from carbohydrates as a sustainable biofuel production route. ACS Sustainable Chem Eng. 2022;10(10):3079.
21. Wright WRH, Palkovits R. Development of heterogeneous catalysts for the conversion of levulinic acid to γ-valerolactone. Chemsuschem. 2012;5(9):1657.
22. Casanova O, Iborra S, Corma A. Biomass into chemicals: aerobic oxidation of 5-hydroxymethyl-2-furfural into 2,5-furandicarboxylic acid with gold nanoparticle catalysts. Chemsuschem. 2009;2(12):1138.
23. De Jong E, Visser HA, Dias AS, Harvey C, Gruter GJM. The road to bring FDCA and PEF to the market. Polymers. 2022;14(5):943.
24. Kim M, Su Y, Fukuoka A, Hensen EJM, Nakajima K. Aerobic oxidation of 5-(hydroxymethyl)furfural cyclic acetal enables selective furan-2,5-dicarboxylic acid formation with CeO_2-supported gold catalyst. Angew Chem Int Ed. 2018;57(27):8235.
25. Kim M, Su Y, Aoshima T, Fukuoka A, Hensen EJM, Nakajima K. Effective strategy for high-yield furan dicarboxylate production for biobased polyester applications. ACS Catal. 2019;9(5):4277.
26. Coumans FJAG, Overchenko Z, Wiesfeld JJ, Kosinov N, Nakajima K, Hensen EJM. Protection strategies for the conversion of biobased furanics to chemical building blocks. ACS Sustainable Chem Eng. 2022;10(10):3116.
27. Wiesfeld JJ, Kim M, Nakajima K, Hensen EJM. Selective hydrogenation of 5-hydroxymethylfurfural and its acetal with 1,3-propanediol to 2,5-bis(hydroxymethyl)furan using supported rhenium-promoted nickel catalysts in water. Green Chem. 2020;22(4):1229.
28. Brownlee HJ, Miner CS. Industrial development of furfural. Ind Eng Chem. 1948;40(2):201.
29. Ricciardi L, Verboom W, Lange JP, Huskens J. Production of furans from C5 and C6 sugars in the presence of polar organic solvents. Sustain Energy Fuels. 2022;6(1):11.
30. Mariscal R, Maireles-Torres P, Ojeda M, Sadaba I, Lopez Granados M. Furfural: a renewable and versatile platform molecule for the synthesis of chemical and fuels. Energy Environ Sci. 2016;9(4):1144.
31. Zhang W, Zhu Y, Niu S, Li Y. A study of furfural decarbonylation on K-doped Pd/Al_2O_3 catalysts. J Mol Catal A Chem. 2011;335:71.
32. Lejemble P, Gaset A, Kalck P. From biomass to furan through decarbonylation of furfural under mild conditions. Biomass. 1984;4(4):263.
33. Kou H, Masatoshi S, Katsushiro F, Yosio O, Tominaga K. Ring transformation of oxygen containing heterocycles into nitrogen containing heterocycles over synthetic zeolites. Chem Lett. 1974;3(5):439.
34. Douthwaite M, Huang X, Iqbal S, Miedziak PJ, Brett GL, Kondrat SA, Edwards JK, Sankar M, Knight DW, Bethell D, Hutchings GJ. The controlled catalytic oxidation of furfural to furoic acid using $AuPd/Mg(OH)_2$. Catal Sci Technol. 2017;7(22):5284.
35. Pinna F, Olivo A, Trevisan V, Menegazzo F, Signoretto M, Manzoli M, Boccuzzi F. The effects of gold nanosize for the exploitation of furfural by selective oxidation. Catal Today. 2013;203:196.

36. Hara M, Nakajima K, Kamata K. Recent progress in the development of solid catalysts for biomass conversion into high value-added chemicals. Sci Technol Adv Mater. 2015;16(3):1.
37. Ong HC, Chen WH, Farooq A, Gan YY, Lee KT, Ashokkumar V. Catalytic thermochemical conversion of biomass for biofuel production: a comprehensive review. Renew Sust Energ Rev. 2019;113:109266.
38. Ennaert T, Van Aelst J, Dijkmans J, De Clercq R, Schutyser W, Dusselier M, Verboekend D, Sels BF. Potential and challenges of zeolite chemistry in the catalytic conversion of biomass. Chem Soc Rev. 2016;45:584.
39. Zhu S, Wang J, Fan W. Graphene-based catalysis for biomass conversion. Catal Sci Technol. 2015;5:3845.
40. Serrano DP, Melero JA, Morales G, Iglesias J, Pizarro P. Progress in the design of zeolite catalysts for biomass conversion into biofuels and bio-based chemicals. Catal Rev. 2018;60(1):1.
41. Fang R, Dhakshinamoorthy A, Li Y, Garcia H. Metal organic frameworks for biomass conversion. Chem Soc Rev. 2020;49:3638.
42. Corma A, Nemeth LT, Renz M, Valencia S. Sn-zeolite beta as a heterogeneous chemoselective catalyst for Baeyer-Villiger oxidations. Science. 2001;412:423.
43. Moliner M, Roman-Leshkov Y, Davis ME. Tin-containing zeolites are highly active catalysts for the isomerization of glucose in water. Proc Natl Acad Sci. 2010;107(14):6164.
44. Holm SM, Saravanamurugan S, Taarning E. Conversion of sugars to lactic acid derivatives using heterogeneous zeotype catalysts. Science. 2010;328:602.
45. Padovan D, Tolborg S, Botti L, Taarning E, Sabada I, Hammond C. Overcoming catalyst deactivation during the continuous conversion of sugars to chemicals: maximising the performance of Sn-Beta with a little drop of water. React Chem Eng. 2018;3(2):155.
46. Padovan D, Botti L, Hammond C. Active site hydration governs the stability of Sn-beta during continuous glucose conversion. ACS Catal. 2018;8(8):7131.
47. Ono Y, Hattori H. Solid base catalysis. Springer;2011. ISBN 978–3–642–18338–6.
48. Hattori H, Ono Y. Solid acid catalysis: from fundamentals to Applications, 2015, Pan Stanford Publishing Pte Ltd., ISBN 978–981–4463–28–7.
49. Vedrine JC. Heterogeneous catalysis on metal oxides. Catalysts. 2017;7(11):341
50. Takagaki A. Rational design of metal oxide solid acids for sugar conversion. Catalysts. 2019;9(11):907.
51. Tamura M, Shimizu K, Satsuma A. Comprehensive IR study on acid/base properties of metal oxides. Appl Catal A Gen. 2012;433–434:135.
52. Busca G. The surface acidity of solid oxides and its characterization by IR spectroscopic methods. An attempt at systematization. Phys Chem Chem Phys. 1999;1(5):723.
53. Huber S, Knozinger H. Adsorption of CH-acids on magnesia: an FTIR-spectroscopic study. J Mol Catal A Gen. 1999;141(1–3):117.
54. Lavalley JC. Infrared spectrometric studies of the surface basicity of metal oxides and zeolites using adsorbed probe molecules. Catal Today. 1996;27(3–4):377.
55. Yang C, Woll C. IR spectroscopy applied to metal oxide surfaces: adsorbate vibrations and beyond. Adv Phys-X. 2017;2(2):373.
56. Nowak I, Ziolek M. Niobium compounds: preparation, characterization, and application in heterogeneous catalysis. Chem Rev. 1999;99:3603.
57. Iizuka T, Ogasawara K, Tanabe K. Acidic and catalytic properties of niobium pentaoxide. Bull Chem Soc Jpn. 1983;56(10):2927.
58. Tanabe K. Niobic acid as an unusual acidic solid material. Mater Chem Phys. 1987;17(1–2):217.
59. Schafer H, Gruehn R, Schulte F. the Modifications of niobium pentoxide. Angew Chem Internat Edit. 1966;5(1):40.
60. Ko EI, Weissman JG. Structures of niobium pentoxide and their implications on chemical behavior. Catal Today. 1990;8(1):27.
61. Maurer SM, Ko EI. Structural and acidic characterization of niobia aerogels. J Catal. 1992;135(1):125.

62. Chan X, Pu T, Chen X, James A, Lee J, Parise JB, Kim DH, Kim T. Effect of niobium oxide phase on the furfuryl alcohol dehydration. Catal Commun. 2017 97:65–9.
63. Lebarbier V, Houalla M, Onfroy T. New insights into the development of Brønsted acidity of niobic acid. Catal Today. 2012;192(1):123.
64. Fernandez-Arroyo A, Delgado D, Domine ME, Lopez-Nieto JM. Upgrading of oxygenated compounds present in aqueous biomass-derived feedstocks over NbO_x-based catalysts. Catal Sci Technol. 2017;7(23):5495.
65. Burcham LJ, Datka J, Wachs IE. In situ vibrational spectroscopy studies of supported niobium oxide catalysts. J Phys Chem B. 1999;103(29):6015.
66. Jehng JM, Wachs IE. Molecular structure of supported niobium oxide catalysts under in situ conditions. J Phys Chem. 1991;95:7373.
67. Pittman RM, Bell AT. Raman studies of the structure of Nb_2O_5/TiO_2. J Phys Chem. 1993;97:12178.
68. Burke PA, Ko EI. Acidic properties of oxides containing niobia on silica and niobia in silica. J Catal. 1991;129(1):38–46.
69. Nakajima K, Baba Y, Noma R, Kitano M, Kondo JN, Hayashi S, Hara M. $Nb_2O_5{\cdot}nH_2O$ as a heterogeneous catalyst with water-tolerant lewis acid sites. J Am Chem Soc. 2011;133(12):4224.
70. Datka J, Turek AM, Jehng JM, Wachs IE. Acidic properties of supported niobium oxide catalysts: an infrared spectroscopy investigation. J Catal. 1992;135:186.
71. Gupta NK, Fukuoka A, Nakajima K. Amorphous Nb_2O_5 as a selective and reusable catalyst for furfural production from Xylose in biphasic water and toluene. ACS Catal. 2017;7(4):2430.
72. Roman-Leshkov Y, Chheda JN, Dumesic JA. Phase modifiers promote efficient production of hydroxymethylfurfural from fructose. Science. 1933;2006:312.
73. Romo JE, Bollar NV, Zimmermann CJ, Wettstein SG. Conversion of sugars and biomass to furans using heterogeneous catalysts in biphasic solvent systems. ChemCatChem. 2018;10(21):4805.
74. Omata K, Nambu T. Catalysis of water molecules acting as bronsted acids at lewis acid sites on niobium oxide. Appl Catal A-Gen. 2020;607: 117812.
75. Kurosaki A, Okuyama T, Okazaki S. Surface properties and catalytic activities of niobic acid treated with diluted phosphoric acid. Bull Chem Soc Jpn. 1987;60(10):3541.
76. Okazaki S, Wada N. Surface properties and catalytic activities of amorphous niobium phosphate and a comparison with those of H_3PO_4-treated niobium oxide. Catal Today. 1993;16(3–4):349.
77. Armaroli T, Busca G, Carlini C, Giuttari M, Raspolli Galletti AM, Sbrana G. Acid sites characterization of niobium phosphate catalysts and their activity in fructose dehydration to 5-hydroxymethyl-2-furaldehyde. J Mol Catal A Chem. 2000;151(1–2):233.
78. Carniti P, Gervasini A, Biella S, Aurox A. Intrinsic and effective acidity study of niobic acid and niobium phosphate by a multitechnique approach. Chem Mater. 2005;17(24):6128.
79. Carniti P, Gervasini A, Biella S, Aurox A. Niobic acid and niobium phosphate as highly acidic viable catalysts in aqueous medium: fructose dehydration reaction. Catal Today. 2006;118(3–4):373.
80. Carlini C, Giuttari M, Raspolli Galletti AM, Sbrana G, Armaroli T, Busca G. Selective saccharides dehydration to 5-hydroxymethyl-2-furaldehyde by heterogeneous niobium catalysts. App Catal A Gen. 1999;183(2):295.
81. Magno de Jesus Junior M, Fernandes SA, Borges E, Lobo Baeta BE, de Avila Rodrigues F. Kinetic study of the conversion of glucose to 5-hydroxymethylfurfural using niobium phosphate. Mol Catal. 2022;518:112079.
82. Vieira JL, Paul G, Iga GD, Cabral NM, Bueno JMC, Bisio C, Gallo JMR. Niobium phosphates as bifunctional catalysts for the conversion of biomass-derived monosaccharides. Appl Catal A Gen. 2021;617:118099.
83. Zhang Y, Wang J, Li X, Xia X, Hu B, Lu G, Wang Y. Direct conversion of biomass-derived carbohydrates to 5-hydroxymethylfurural over water-tolerant niobium-based catalysts. Fuel. 2015;139:301.

84. Berube F, Nohair B, Kleitz F, Kaliaguine S. Controlled postgrafting of titanium chelates for improved synthesis of Ti-SBA-15 epoxidation catalysts. Chem Mater. 2010;22(6):1988.
85. Kim TW, Kim MJ, Kleitz F, Nair MM, Guillet-Nicolas R, Jeong KE, Chae HJ, Kim CU, Jeong SY. Tailor-made mesoporous Ti-SBA-15 catalysts for oxidative desulfurization of refractory aromatic sulfur compounds in transport fuel. ChemCatChem. 2012;4(5):687.
86. Capel-Sanchez MC, Blanco-Brieva G, Campos-Martin JM, De Frutos MP, Wen W, Rodriguez JA, Fierro JLG. Grafting strategy to develop single site titanium on an amorphous silica surface. Langmuir. 2009;25(12):7148.
87. Shintaku H, Nakajima K, Kitano M, Ichikuni N, Hara M. Lewis acid catalysis of TiO_4 tetrahedra on mesoporous silica in water. ACS Catal. 2014;4(4):1198.
88. Shintaku H, Nakajima K, Kitano M, Hara M. Efficient mukaiyama aldol reaction in water with TiO4 tetrahedra on a hydrophobic mesoporous silica surface. Chem Commun. 2014;50(88):13473.
89. Nakajima K, Noma R, Kitano M, Hara M. Titania as an early transition metal oxide with a high density of lewis acid sites workable in water. J Phys Chem C. 2013;117(31):16028.
90. Nakajima K, Noma R, Kitano M, Hara M. Selective glucose transformation by titania as a heterogeneous lewis acid catalyst. J Mol Catal A Chem. 2014;388–389:100.
91. Jadhav H, Pedersen CM, Sølling T, Bols M. 3-Deoxy-glucosone is an intermediate in the formation of furfurals from D-glucose. Chemsuschem. 2011;4(8):1049.
92. Noma R, Nakajima K, Kamata K, Kitano M, Hayashi S, Hara M. Formation of 5-(hydroxymethyl)furfural by stepwise dehydration over TiO_2 with water-tolerant lewis acid sites. J Phys Chem C. 2015;119(30):17117.
93. Li G, Pidko EV, Hensen EJM, Nakajima K. A density functional theory study of the mechanism of direct glucose dehydration to 5-hydroxymethylfurfural on anatase titania. ChemCatChem. 2018;10(18):4084.
94. van Zandvoort I, Wang Y, Rasrenda CB, van Eck ERH, Bruijnincx PCA, Heeres HJ, Weckhuysen BM. Formation, molecular structure, and morphology of humins in biomass conversion: Influence of feedstock and processing conditions. Chemsuschem. 2013;6(9):1745.
95. Patil SKR, Heltzel J, Lund CRF. Comparison of structural features of humins formed catalytically from glucose, fructose, and 5-Hydroxymethylfurfuraldehyde. Energy Fuels. 2012;26(8):5281.
96. Wang S, Lin H, Zhao Y, Chen J, Zhou J. Structural characterization and pyrolysis behavior of humin by-products from the acid-catalyzed conversion of C6 and C5 carbohydrates. J Anal Appl Pyrolysis. 2016;118:259.
97. Murayama T, Chen J, Hirata J, Matsumoto K, Ueda W. Hydrothermal synthesis of octahedra-based layered niobium oxide and its catalytic activity as a solid acid. Catal Sci Technol. 2014;4(12):4250.
98. Nakajima K, Hirata J, Kim M, Gupta NK, Murayama T, Yoshida A, Hiyoshi N, Fukuoka A, Ueda W. Facile formation of lactic acid from a triose sugar in water over niobium oxide with a deformed orthorhombic phase. ACS Catal. 2018;8(1):283.
99. Martra G. Lewis acid and base sites at the surface of microcrystalline TiO_2 anatase: relationships between surface morphology and chemical behaviour. Appl Catal A Gen. 2000;200(1–2):275.
100. Chen HYT, Tosoni S, Pacchioni G. A DFT study of the acid–base properties of anatase TiO2 and tetragonal ZrO_2 by adsorption of CO and CO_2 probe molecules. Surf Sci. 2016;652:163.
101. Atanda L, Mukundan S, Shrotri A, Ma Q, Beltramini J. Catalytic conversion of glucose to 5-hydroxymethyl-furfural with a phosphated TiO_2 catalyst. ChemCatChem. 2015;7(5):781.
102. Atanda L, Konarova M, Ma Q, Mukundan S, Shrotri A, Beltramini J. High yield conversion of cellulosic biomass into 5-hydroxymethylfurfural and a study of the reaction kinetics of cellulose to HMF conversion in a biphasic system. Catal Sci Technol. 2016;6(16):6257.
103. Yue C, Li G, Pidko EA, Wiesfeld JJ, Rigutto M, Hensen EJM. Dehydration of glucose to 5-hydroxymethylfurfural using Nb-doped tungstite. Chemsuschem. 2016;9(17):2421.
104. Omata K, Matsumoto K, Murayama T, Ueda W. Direct oxidative transformation of glycerol to acrylic acid over Nb-based complex metal oxide catalysts. Catal Today. 2015;259(1):205.

105. Li G, Pidko EA, Hensen EJM. A periodic DFT study of glucose to fructose isomerization on tungstite ($WO_3 \cdot H_2O$): Influence of group IV–VI dopants and cooperativity with hydroxyl groups. ACS Catal. 2016;6(7):4162.
106. Wiesfeld JJ, Sommerdijk NAJM, Hensen EJM. Early transition metal doped tungstite as an effective catalyst for glucose upgrading to 5-hydroxymethylfurfural. Catal Lett. 2018;148:3093.

Chapter 12
The Rise of Catalysts Informatics

Keisuke Takahashi, Lauren Takahashi, Shun Nishimura, Jun Fujima, and Junya Ohyama

12.1 Introduction

The rise of catalysts informatics has changed how catalysts are designed and understood [1, 2]. Over the years, researchers in catalysis have conducted a tremendous degree of trial-and-error research through experiments. During such trial-and-error processes, researchers gain knowledge and experience that lead to the development of novel catalysts. Furthermore, the advancement of density functional theory calculations, commonly known as first principle calculations, has contributed towards unveiling the atomic scale mechanisms of catalyst activities as shown in Fig. 12.1. Thus, catalysts have typically been designed and understood through experimental and density functional theory perspectives.

In such situations, large amounts of catalyst data become available. There are several catalysts data sources available. One type of data is High-throughput experimental data. High-throughput experiments enable the ability to rapidly synthesize

K. Takahashi (✉) · L. Takahashi · J. Fujima
Department of Chemistry, Hokkaido University, North 10, West 8, Sapporo 060-8510, Japan
e-mail: keisuke.takahashi@sci.hokudai.ac.jp

L. Takahashi
e-mail: lauren.takahashi@sci.hokudai.ac.jp

J. Fujima
e-mail: jfujima@sci.hokudai.ac.jp

S. Nishimura
Graduate School of Advanced Science and Technology, Japan Advanced Institute of Science and Technology (JAIST), 1-1 Asahidai, Nomi 923-1292, Japan
e-mail: s_nishim@jaist.ac.jp

J. Ohyama
Faculty of Advanced Science and Technology, Kumamoto University, 2-39-1 Kurokami, Chuo-ku 860-8555, Kumamoto, Japan
e-mail: ohyama@kumamoto-u.ac.jp

W. Ueda (ed.), *Crystalline Metal Oxide Catalysts*,
https://doi.org/10.1007/978-981-19-5013-1_12

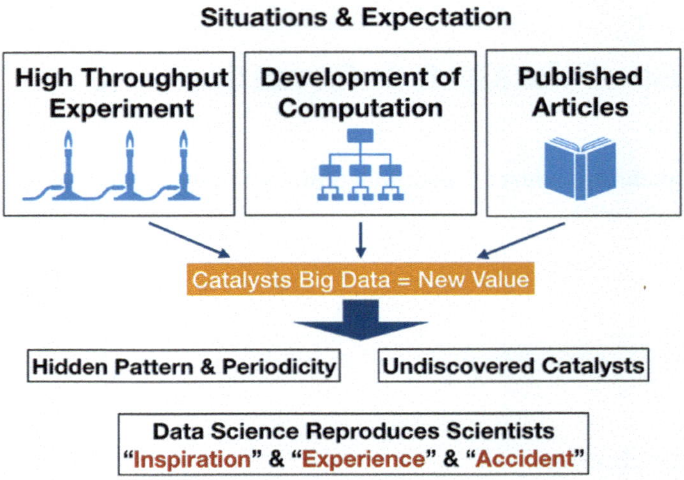

Fig. 12.1 Data in catalysts. Takahashi et al. The rise of catalyst informatics: towards catalyst genomics. ChemCatChem [1] Licence number5277780405983

catalysts and evaluate catalytic activities [3]. A second data source is the first principle calculation data. With the development of high-performance supercomputers, it has become possible to perform High-throughput calculations of metal surface reactions. A third data source is a literature and patent data. Large amounts of data generated by researchers are stored as data and often published in scientific journals, databases, and patents. Thus, large amounts of catalysts have become available for use. With this, it has been proposed that the design of catalysts can be achieved in principle if trends and patterns are identified in catalysts data [4]. In this chapter, an overview of catalysts informatics is introduced.

12.1.1 Concept of Catalysts Informatics

Catalysts informatics involves three key concepts: catalyst data, data to knowledge, and catalyst platform as shown in Fig. 12.2 [1]. It is important that those three components must be operated simultaneously in order to perform catalysts informatics. As mentioned previously, sources for catalysts data can range from experimental data, computational data, and literature data. Data to knowledge plays a major role in catalysts informatics. In particular, various data science techniques are implemented to extract knowledge from catalysts data. Lastly, a catalysts platform is the place where database and data science tools are provided for catalysts informatics. Here, further-detailed steps for catalysts informatics are introduced.

Catalysts informatics consists of several steps. The detailed steps are shown in Fig. 12.3. The first step is to set up an informatics environment. This is equivalent

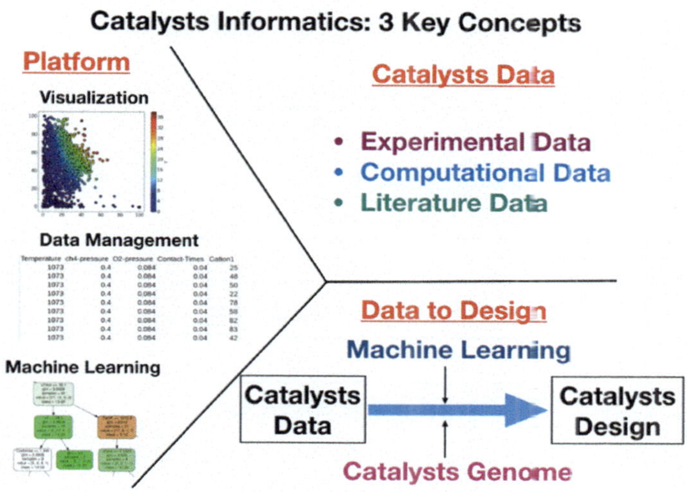

Fig. 12.2 3 key concepts in catalysts informatics. Takahashi et al. The rise of catalyst informatics: towards catalyst genomics. ChemCatChem [1] Licence number5277730405983

to setting up experimental devices. In general, catalyst informatics is recommended to be performed using the Python programming language under Linux OS. Python programming language offers various data science libraries including scikit-learn, python-pandas, and matplotlib. Those python modules contain basic data science environments whereas Linux provides such an environment. In particular, Ubuntu or Linux Mint within Linux comes with all necessary data science tools and python modules in their repository. Thus, Ubuntu and Linux mint play an important role in both server and desktop OS, respectively.

Once a data science environment is set, catalyst data must then be collected as shown in Fig. 12.1. Here it is important that depending on the target, appropriate datasets must be prepared or collected. When the catalyst data is ready, data preprocessing must be performed. Raw data generally contains outliers. Additionally, some data might use different units or may not be consistent. Text data must also be converted to numerical variables for machine learning visualization. Thus, data preprocessing, also known as data cleansing, must be performed. Once data is preprocessed, data visualization and machine learning are performed in order to design and understand catalysts.

Data visualization is the method used to unveil the patterns and trends present within catalyst data. In particular, data science provides techniques that can visualize multidimensional data. For instance, parallel coordinates and RadViz can plot multidimensional data. Such visualization gives a comprehensive understanding of data which would provide a hint towards what type of machine learning algorithm would be needed. Thus, data visualization plays an important role in catalyst informatics.

There are two types of machine learning, namely supervised machine learning and unsupervised machine learning. The fundamental idea of supervised machine

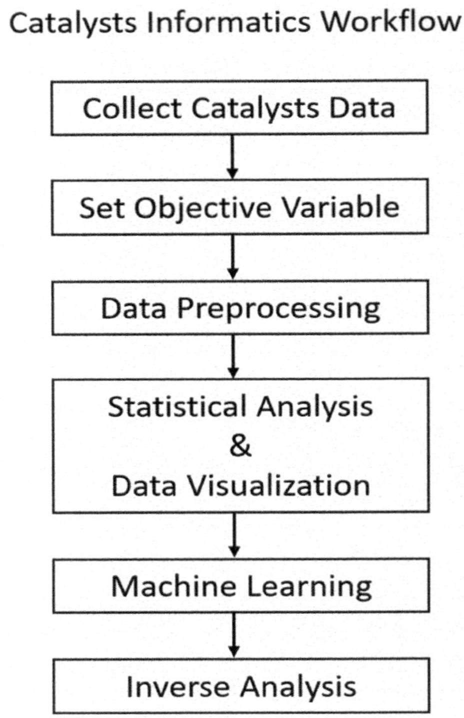

Fig. 12.3 Catalysts informatics workflow

learning is to solve the y = f(x) function where y and x are objective and descriptor variables, respectively. Here, objective variables can be yield, selectivity, conversion, or other factors that researchers choose to maximize the performance. X is the descriptor variable that represents the objective variable. In particular, there are various types of machine learning. It can be classified into linear and nonlinear supervised machine learning. One of the main tasks in catalysts informatics is to find appropriate descriptor variables with appropriate machine learning algorithms is the key. Once an appropriate train machine is ready, then inverse analysis can be performed by asking desirable descriptor variables to the trained machine where the machine returns the objective variables.

Unsupervised machine learning is used to find hidden groups in a data set. This is often called clustering. Principal component analysis, hierarchical clustering, and k-means clustering are commonly used unsupervised machine learning techniques. Unsupervised machine learning can find similarities in catalyst data, which often becomes an important factor in describing and understanding catalysts.

In addition, it must also be pointed out that Bayesian optimization plays an important role in catalyst data. In general, machine learning requires a large amount of data. However, it is not always possible to have the required amounts of catalyst data. In these situations, Bayesian optimization becomes an effective alternative. The underlying idea of Bayesian optimization is that the machine suggests experimental data points that could maximize or minimize the objective variable. In particular, Gaussian process regression in combination with acquisition function is used to predict the objective variable with expected improvement.

Gaussian process regression is trained using small data sets, in which the machine can predict objective variables similar to supervised machine learning. The difference is that Gaussian process regression provides the standard deviation which shows how far away the predictions are from the training data set. By using the acquisition function and Gaussian process regression, high expected improvement data points can be predicted. Thus, Bayesian optimization can contribute to minimizing the experimental procedures to maximize or minimize the objective variables.

12.1.2 The Role of Informatics in Catalysis

Catalysts informatics can be used to design and understand the catalysts [5]. Here, some case studies are explored.

Designing catalysts from High-throughput experiment data and literature data have been extensively investigated [3, 6, 7]. It has been demonstrated that High-throughput (HTP) experiment devices. HTP device enables the consistent data set which has major advantages in catalysts informatics. Such consistent data enable to design catalysts using machine learning.

Machine learning is also proposed to identify the active site of catalysts [8]. In particular, X-ray absorption fine structure spectroscopy data of Cu-zeolite is chosen for descriptor variables and methanol production is set as the objective variable. By using machine learning, it unveils that the Si/Al ratio, amount of Cu, and the local structure of Cu are found to be important factors for determining the methane to methanol reaction. Thus, combining spectroscopy and catalytic performance can reveal the underlying mechanism as well as identify key structural features. Thus, measurement data would also be a key component for catalysts informatics.

Those two case studies demonstrate that catalysts informatics can be not only used to design catalysts but also help to understand the mechanism. Thus, although catalyst informatics is a relatively new field, it demonstrates high effectiveness in catalysts design and understanding. In this chapter, further case studies and concepts are explored focusing on catalysts data, knowledge and design from catalysts data, and catalysts informatics platform.

12.2 Catalysts Data

12.2.1 The Growing Importance of Data

Data has grown to play an increasingly central role in modern material and chemical research [1, 9]. It is produced through studies, experiments, and calculations and consistently published daily across the globe. Data that is available for research purposes has increased exponentially due to both the increase in number of researchers that participate in material and chemical research as well as developments and advancements made regarding data creation, curation, and distribution [10]. The advent of computational science has also made a drastic impact on the research community where its introduction brought the possibility for researchers to approach material and chemical research in a manner that is not solely based on experiment or theory. Additionally, the continued evolution of modern science and the emergence of new perspectives and approaches to material and chemical research has led to the rapid development of data that is quite variable in nature. Data includes more than what one may typically associate with experiments or theoretical studies. While incorporating traditional types of data such as lab readings, imagery, observational notes, theory and equations, data now also includes computational models and calculation results—all of which are commonly stored in databases and published through outlets such as academic journals and patents. The increasing degree and accessibility of data, paired with concurrent advancements in computer science and artificial intelligence, has led to the emergence of a new perspective for chemical research: the perspective of data science.

Data science, increasingly referred to as the fourth paradigm of science, is an approach to research where data is the main focal point [11, 12]. Data science has quickly become popular for researchers to use because it allows researchers to utilize data science techniques such as machine learning to process data in levels far beyond what a researcher would be able to do on their own. By applying data science techniques to large data sets, it becomes possible to extract or otherwise uncover the knowledge that would otherwise be hidden within the data. With data science, or instance, it is possible to apply machine learning techniques in order to help understand any underlying trends that may be present within the data. Data science can also be used to aid in the search for descriptors, which are vital to understanding what a "material property" consists of and are of great importance for material design [13]. One particularly attractive aspect of data science is that it becomes possible to attempt to solve "inverse" problems such as predicting a series of conditions that may result in a desirable property. Applications of data science are quite diverse, and the possibilities continue to increase as research, computer science, and technology continue to develop.

12.2.2 Difficulties with Existing Data

Unfortunately, simply having access to large amounts of data does not mean that data science-centered research will immediately flourish. As seen in Fig. 12.4, materials data comes from many different sources, where each source has different types of data to report and interests in different types of information. While this provides a wide variety of data to work with, the variety in data itself becomes problematic when attempting to use the data for data science purposes. In fact, informatics-centered research is often presented with a myriad of issues when researchers attempt to use available data [14]. Such issues often can be reduced to issues with database quality, lack of industry standards, and the unconscious presence of biases.

Existing datasets vary in terms of quality. As there is no industry standard in place for researchers to follow, it is up to the researcher to decide what data is included and the manner in which it should be stored. As a result, databases are generally unique in their composition, naming formats, and overall design, and are also designed to fit the needs of the researchers that create them. In fact, researchers are responsible for constructing the databases for their data and its usefulness to other research groups will vary depending on factors such as comprehensive labeling, consistency, and degrees of thoroughness. For instance, one researcher may be extremely thorough and include many types of data while others may only include the bare minimum. Databases may also have a lot of missing data points, a factor that can, at times, render the database useless for reuse. It is also common to see such individualization resulting in databases that use terms and labels that are hard to decipher, are ambiguous, or else have multiple meanings/context that is dependent on the

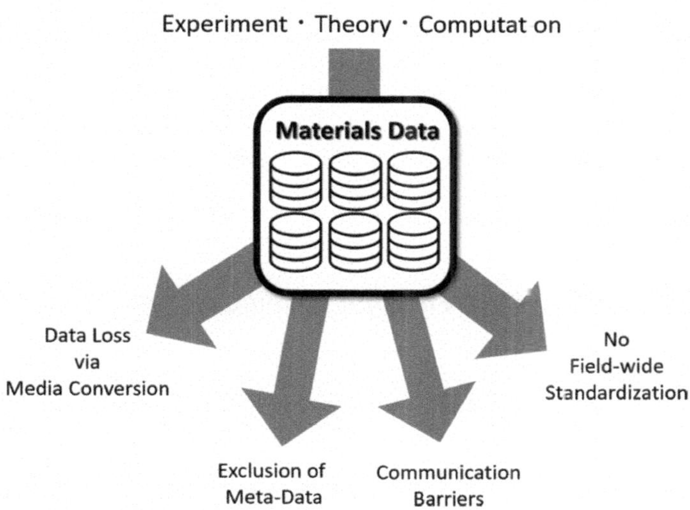

Fig. 12.4 Collecting materials data from multiple sources and the resulting difficulties [16]. Reprinted (adapted) with permission from Ref. [16]. Copyright 2022 American Chemical Society

researcher discipline that uses it. This can result in difficulties when translating the data. Given that it is not always possible to contact or otherwise communicate with the original creator of the database, researchers are often left to decipher the data on their own. The resulting difficulties in translating such data increase the probability of introducing human error into new databases and research. Finally, the lack of field-wide standards or guidelines leads to the presence of unconscious biases within databases. Researchers are often regarded as experts within their fields, and their research practices and interests are informed by their fields and years of experience. This can lead to unconscious biases when constructing databases where researchers will often make assumptions and decisions based on their experiences and understandings. This can result in data reporting where researchers are likely to only report data that is useful for their particular purpose. This becomes problematic when data is shared across research disciplines. Relying on the judgements of researchers can make databases unbalanced where certain types of data may be saturated or unbalanced and potentially useful data may not be included in the dataset as it does not directly address the needs of the original research project. Potentially useful solutions or insights may also be accidentally excluded because they are unconventional or assumed to be unlikely to occur. Thus, without some type of industry standard or guideline in place, these factors will continue to affect the quality and reusability of data and databases.

Imbalances such as these can negatively impact the success of data science techniques since data science techniques reflect the information found within the data. Preferences for certain types of data or assumptions held about particular topics may influence the structure of the database to favor certain outcomes. Additionally, researchers rely on their personal experiences and expertise when creating databases. Data that a researcher may judge to be "irrelevant" and is therefore excluded from the database may be vital information for researchers investigating topics that use the same data. Discarding negative data also hurts data science applications because many data science techniques learn trends and behaviors within data by analyzing the conditions of both the positive and negative outcomes. Without a means of addressing these issues, it will be difficult for informatics research to truly flourish and reach its potential.

12.2.3 Structuring Data Through Ontology

One possible method of addressing these issues is to implement ontology [15]. Traditionally, ontology is the study of existence. From a more technological aspect, ontology is a system that allows one to establish and define a domain, its objects, and the relationships between said objects. On the surface, ontology may not seem to be much different from existing database structures. However, unlike traditional databases, ontologies are constructed on the basis of description logic and formal semantics. By incorporating formal semantics via web ontology languages (OWL), it becomes possible to define vocabularies and data according to their relationship

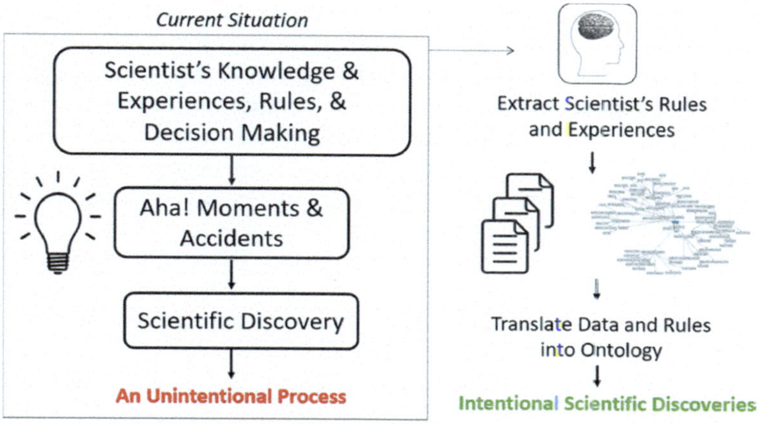

Fig. 12.5 Transforming the current path to scientific discovery into a process based on intent [16]. Reprinted (adapted) with permission from Ref. [16]. Copyright 2022 American Chemical Society

with other data and information and define them in a manner that allows machines to navigate data more intelligently. Ontology also provides the structure to define and incorporate metadata and other information such as annotations, alternative definitions, and additional observations made by the researcher. This makes it possible to expand data curation by providing the space to include additional layers of information beyond the initial raw data. Additionally, ontology can help add structure towards the scientific discovery process illustrated in Fig. 12.5 and can therefore help make the scientific discovery process more intentional.

The ability to include multiple layers of information with the original data, itself, is quite useful for the research community as it not only provides an expanded explanation of the data produced through research, but it also provides a clearer picture of the meaning behind the research being shared with the community. By defining the appropriate vocabularies and defining how they relate to each other, it becomes possible to define data in a way that reduces ambiguity. For researchers with different backgrounds or expertise, it becomes easier to translate what knowledge the database is attempting to convey. Ontology can not only help clarify what the original researchers wanted to share, but it also helps reduce time spent decoding and preprocessing data and, therefore, help reduce possible translation errors that may be introduced when preprocessing data.

Another benefit of incorporating ontology is its ability to combine multiple databases in a single space. In essence, multiple databases can be connected and accessed by machines via ontology if they share the same set of ontological rules. Provided that the ontologies are logically consistent, it becomes possible for machines to expand out and navigate multiple databases using the same set of ontological rules and to access large amounts of data at a fraction of the speed that a normal researcher would be able to if attempting to do it themselves. This, in itself, makes it attractive, especially when considering the costs in terms of time and resources when attempting to do the same task manually.

With descriptive logic as its foundation, ontology makes it possible to query large amounts of data using the definitions provided by the ontology. To query, one provides a series of definitions or restrictions and the ontology returns the data that matches. This system is particularly beneficial to research as it not only dramatically reduces the time it may take to find data that fulfill a specific set of conditions but also can be carried out over very large amounts of data.

In addition, ontologies are also able to consider inferences when navigating data. Inferences allow the machine to reorganize and reclassify data into different subclasses according to the provided definitions. Therefore, the machine is able to expand data classifications and definitions based without explicit instruction from a researcher. If terms are defined properly, it becomes possible for the machine to expand existing class membership and potentially uncover new or unique members of a class that may be otherwise left undiscovered if done manually by a researcher. It is reasonable, therefore, to believe that the same principle can be applied towards materials data and other materials or chemical databases, and potentially lead towards the discovery of new materials based on the rule defined by the ontology.

Ontology can become quite a powerful tool with its ability to combine multiple databases into a single space and navigate data based on the rules and definitions provided [16]. It not only provides the space to define datasets but also provides a space for researchers to also define their knowledge and understanding of data into a machine-accessible format. As seen in Fig. 12.6, researchers can query databases semantically, standardize material and chemical data, and restructure how materials can be classified. In doing so, not only is it possible to uncover new or unconventional materials, but it is also possible to better understand potential descriptors behind different material properties. Finally, with the ability to continuously add or refine definitions, it is possible to maintain and update existing databases with new scientific knowledge, allowing researchers to expand on existing knowledge with new findings.

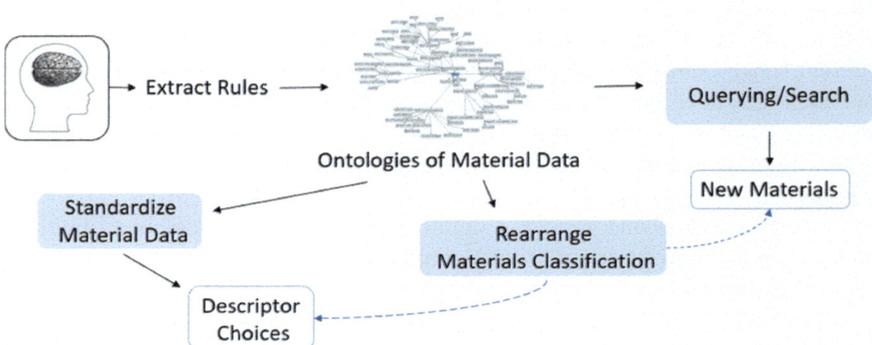

Fig. 12.6 Connecting to and updating existing knowledge through ontology [16]. Reprinted (adapted) with permission from Ref. [16]. Copyright 2022 American Chemical Society

In summary, it is essential for data to be organized, consistent, and easily accessible by both researchers and machines alike, especially as data continues to play an essential role in materials and chemical research. Issues such as lack of industry standards regarding data curation practices, terminology ambiguity, and underlying biases often make it very difficult to use existing databases without varying levels of data preprocessing, which also opens the possibilities for error when attempting to translate or restructure the data. By employing ontology, it becomes possible to address several of these issues within the same framework. Ontology restructures data and incorporates knowledge using description logic and web ontology languages. This makes it possible to define relations between different data and incorporate metadata in a manner that can be accessible by both researchers and machines, thereby helping to reduce ambiguity and translation errors. Additionally, with consistent definitions, it becomes possible to merge several databases into one location, making it possible for machines to access and query large amounts of data instantly. In doing so, it becomes possible to reveal new information and possibly unconventional solutions for queries thanks to the logic definitions the ontology provides, and thus can potentially revolutionize how material and chemical data is utilized.

12.3 Designing Heterogeneous Catalysts via Machine Learning

How the machine learning (ML) approach can aid scientists in developing high-performance catalyst? This query had been attracted many scientists for a long time. In particular, in homogeneous catalyst investigation, combined use of High-throughput (HT) screening experimentations and ML approach have successfully conducted as some answers. For examples, at the typical Pd-catalyzed reactions, the High-throughput experimentations with ML could be used for predicting the multidimensional potential space of the target reaction and predicting reaction performance with possible reagent and/or ligand combination [17–19]. This relies on the nature of homogeneous catalysis: *i.e.*, one is that design of active metal center and coordinating ligand is fundamentally crucial to its performance, and another is modern organic chemistry simply can be explained in accordance with the Hammet linear free-energy relationship [20, 21]. Accordingly, data-driven small molecular field regression in homogeneous catalysis in organic chemistry is widely demonstrated [22–24]. While, in the case of heterogeneous (solid) catalysis area, it's still a challenging subject to be popular because heterogeneous catalyst performance depends on many properties such as composition, size, morphology, crystallinity, support, additive and so on. Indeed, the identification of appropriate relationships between some physicochemical properties and their reactivity is a very complicated (multidimensional) subject. Clarifying the specific "active sits" on the surface needs long historical discussions with multiple characterizations and theoretical simulations.

For instance, since Prof. Haruta and Prof. Hatchings independently found the unique reactivity of dispersed gold on reducible metal oxides [25] and carbon-supported gold [26] respectively in 1980s, many researchers have attracted the gold-based catalyst. In particular, huge numbers of researchers have been studying gold-catalyzed CO oxidation at ambient temperature, and identified several hypotheses on the crucial factors of its catalysis such as cationic gold species, electronic structure in a cluster scale, bi-layer of gold in a critical thickness, a trace amount of co-water, effect of crystalline phases of oxide support and so on, for about 20 years. Then, Prof. Haruta insisted on one of the conclusions in 2011 [27] that the number of perimeter gold atoms directly related to the rate-determining step at the CO oxidation: *i.e.*, the adsorbed CO on the surface of gold reacts with the adsorbed oxygen at the periphery sites around gold nanoparticles via the Langmuir–Hinshelwood mechanism. Then, requirements to exhibit catalytic activity reported by various researchers are classified as shown in Fig. 12.7 [28]. However, there have been further discussions from several viewpoints on the importance of size and contact structure of Au nanoparticle, in which the O_2 activation step at the perimeter interface has still been debated. Moreover, in the development of further design of high-performance CO oxidation catalysts, different active centers such as Pd, Ag, Pt, Rh, Cu and/or these alloys have been further applicable. Their important parameters would be not exactly the same as that of monometallic Au cases. Accordingly, it is expected that determining the sufficient global descriptor and collecting the corresponding training data are really difficult even though at such a simple target reaction of CO oxidation.

Deep discussions on catalytic mechanism determining the descriptor at the target reaction need a lot of "costs" such as time and effort for clarification. Also, the present conclusion will be updated on the progress of analytical technology such as

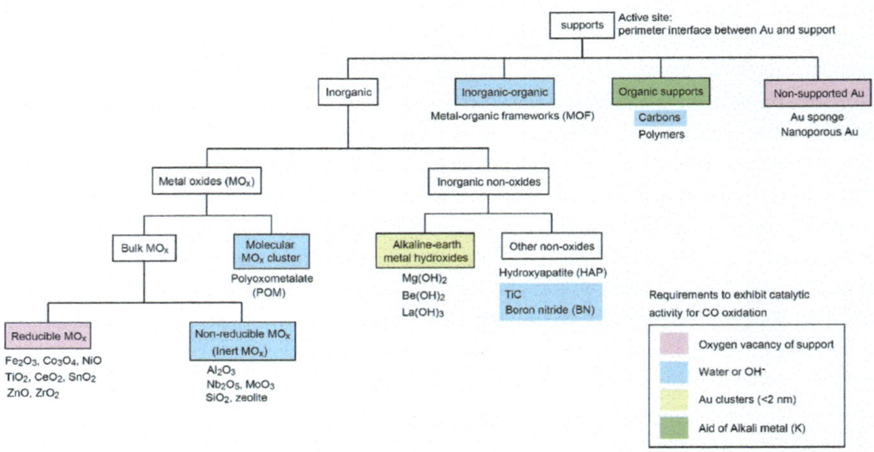

Fig. 12.7 Classification of supports for CO oxidation by their requirements for catalytic activity of gold [28]. Reprinted (adapted) with permission from Ref. [28]. Copyright 2022 American Chemical Society

operando spectroscopy. Therefore, fully understanding multiple networks between heterogeneous catalyst information (*ex.* components) and target performance (*ex.* yield of main product) implemented ML prediction has no realism and rational. In the simplified approach, a relationship between heterogeneous catalyst composite elements and its performance has been adapted to ML-derived catalyst investigation to predict targeting performance with complex none-liner models; such as artificial neural networks (ANN), random forest (RF), support vector (SV), Kernel ridge (KR), and Gaussian regression processes, and/or association rule mining. And then, the researchers attempt to detect meaningful patterns between input and target parameters. It has become the main discussion that how to design the appropriate regression model possessing a high score (>0.6) at the cross-validation of the constructed one, where the data were randomly split into 20 ~ 30% test and 70 ~ 80% trained data. In other word, constructing high-score models become one of goals even without experimental validation of predicted catalysts in somehow. However, there is an example made by the present authors that even if a well-defined RF model consisted with catalyst elements, reaction condition, and its reactivity based on 1868 catalysts data from literature, which possesses 73% score at the cross-validation check, it did not find any novel catalyst at its validations [29, 30]. It was expected that that is because the extraordinary points in learning data mis-leaded the process at ML prediction to predict outer area from the collected data. In other word, development of fine ML regression model and achievement of ML-driven new finding of undiscovered catalyst are not necessarily linear relationship; in particular, when it applied to predicting the outer area of experiment data field in heterogeneous catalyst at present situations.

Historically, one of the pioneer works in ML-derived investigation at heterogeneous catalysts area was examined by Prof. Hattori and co-workers with an ANN approach in 1990s [31–33]. They applied an ANN to elucidate product distribution (viz. selectivity) in oxidative dehydrogenation of ethylbenzene on a series of promoted and unpromoted SnO_2 catalysts [31]. Consistencies in which of an input layer including 9 parameters of physicochemical properties and reaction condition, a single hidden layer including 8 units, and an output layer including selectivities of 5 products made well matching between estimated selectivities and experimentally observe values. It was also reported that the acid strength of mixed oxides could be predicted by an ANN within appropriate experimental errors [32, 33]. One of important features on their examinations is that the ANN investigation can be used to imply the appropriate controlling factors on the catalytic activity by the trial-and-error cycles of the number of units in the input data [33]. At the same era, ANN applications on the analysis of NO decomposition [34] and propane ammoxidation to acrylonitrile [35] were explored by different groups. Subsequently, catalyst discovery by means of ML algorithms of not only ANN but also RF, SVR etc. have been applied in many applications in catalyst investigations and understanding of trends of data [1, 5, 36, 37]. However, the difficulty on prediction of outer area from the collected data points covered field is still a common issue at ML-based investigations. When the training machine tries to display an inexperienced (uncovered) area, the accuracy of prediction extraordinally decreases as it moves farther away from the covered data field as is the human. Moreover, the presence of multiple hidden patterns between input and target data also gave less adequacy.

The present authors have an idea that how to receive the meanings of the trends made by ML regression model is one of new ideas for ML-aided catalyst investigation towards different area from the collected one. For instance, oxidative coupling of methane (OCM) reaction had been found in 1980s [38, 39], however there has remained a catalyst performance issues for eco-friendly design of industrial plant for past 40 years. To overcome this issue, investigation of novel OCM catalyst has been demonstrated with ML assistance combined with experimental scientist's point of view [40]. One-hot encoding treatment was conducted on the open-source big data (40,330 data points with 350 types of catalysts) obtained at 700–1000 °C with a same High-throughput reactor, performance elucidation protocol, and same catalyst preparation methodology [3, 41]. SVR regression was implemented to increase the data impacts more elaborate. Then, the investigation of unique feature as lower temperature OCM catalyst derived from ternary-element-supported La_2O_3 catalyst was challenged. Importantly, by assuming that ML field may find the significant trend between input (viz. catalyst composite) and target (viz. C_2 yield), a different meaning of SVR regression field was tentatively defined for experiment validation: *i.e.*, predicted high C_2 yield values in SVR area would suggest the friendly element's combinations for OCM at the targeting support with a higher probability. When the high score elements are focused in the case of La_2O_3 support as SVR field, these may become a guide for superior element combinations as La_2O_3-based lower temperature OCM performance. Interestingly, among La_2O_3-based ternary element combinations predicted as TOP20 score at C_2 yield in the SVR field, 11 variations of ternary-element-supported La_2O_3 catalysts exhibited superior performances in comparison with La_2O_3 itself (none). The La_2O_3-derived unique lower onset temperature at OCM was enhanced as from 500 to 450 °C under $CH_4/O_2 = 2.0$ condition in the 11 catalysts in validation (Fig. 12.8). Though remained 9 validations were not matched with the assumption, it was more likely to effectively find undiscovered and novel catalysts with such *in-direct* ML toward different area of experimental data. It is expected that debating the common characters of these found catalysts would reveal the hidden features to improve the accuracy for direct ML prediction.

ML approach is good at finding important trends in data in big datasets. Instead of prediction, trend elucidation is another attractive property of ML. Understanding the fingerprint features in spectra are important matter in catalyst development, however it involves differences in interpretation among researchers. When the researcher tries to compare the big experiment data, a lot of efforts has been cared on manual comparison for discussions. Data science and ML can support such data analysis process as an identification toolkit. For instance, Timoshenko et al. developed ML-based determination of 3D structure of platinum nanoparticle stabilized on γ-Al_2O_3 with X-ray absorption near-edge structure (XANES) [42]. The training data was served with theoretical simulations using FEFF and FDMNES software, and then the X-ray absorption coefficient and the average coordination number of nanoparticles respectively set as input and output layer index at supervised ANN (Fig. 12.9). The research constructed a fine model describing complex nonlinear relationships

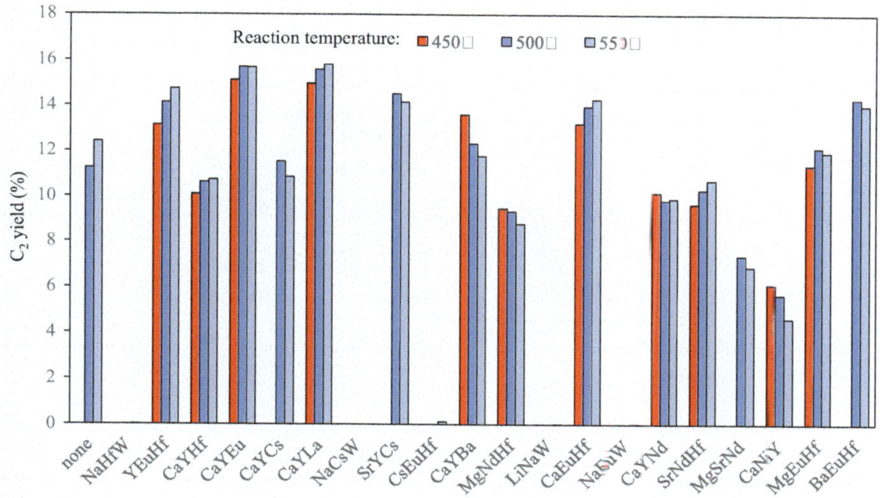

Fig. 12.8 Plots of C_2 yields over ternary-element-supported La_2O_3 selected from the viewpoints of high score of C_2 yield at SVR field derived from high-thoughput screening datasets, together with La_2O_3 support itself in $CH_4/O_2 = 2.0$. Ref. [40]. Copyright 2022 Royal Society of Chemistry

between XANES features data and structure of NP, and which predicted coordination numbers agree with the true values in tolerance error. Notably, this ANN/XANES can translate XANES spectra in nanostructures into real-space information of coordination environment such as size and shape of NP. It is expected that when it applied at *in-situ* analysis would be a powerful static analysis tool. It should also be mentioned that there are other reports demonstrated in experimental XANES spectra using ML assistances such as establishment of the reciprocal correspondence between XANES and geometry of Ni^{2+} site in the presence of gaseous adsorbates [43] and automatic oxidation threshold recognition of a given XANES feature without references [44]. Moreover, recent reach progresses at transmission electronic microscopy (TEM) technique using ML also become a hot subject: shape [45], a rich variety of defects [46], and the shape dependence of the strain distribution [47] are nicely identified by ML assistances. Those ML-aided analytical techniques of characterization data would be a beneficial milestone for catalyst development.

When there is a reasonable pattern between texture properties and catalytic performance, ML research is a powerful tool for researchers looking at unexpected areas. Zeolites are widely used heterogeneous catalysts in academic and industrial fields owing to its versatility of controllable topological and physicochemical properties. Li et al. developed gene-like stacking sequences of stacking layers for predicting, evaluating and identifying promising synthetic candidates with desired properties of ABC-6 zeolites, a family of industrially important catalysts build from the stacking of modular six-ring layers, among a large number of candidate structures (Fig. 12.10) [48]. To cover all possible staking sequences derived from every ABC-6 topology, a ternary numeral codes are implemented. For instance, the stacking sequences for

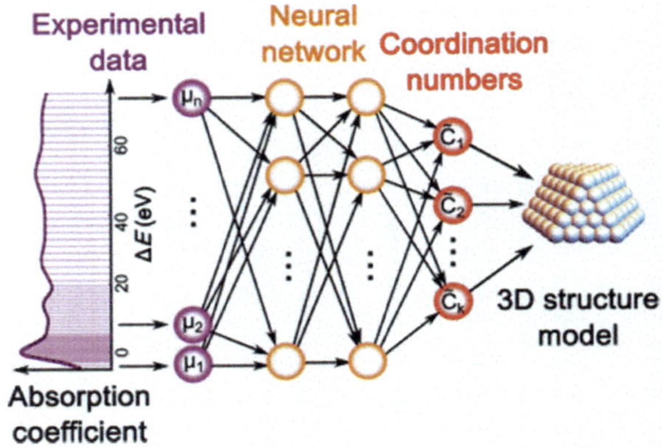

Fig. 12.9 Schematic view of artificial NN application for establishing the 3D structure of nanoparticle from experimental XANES [42]. Reprinted (adapted) with permission from Ref. [42]. Copyright 2022 American Chemical Society

cancrinte, sodalite, and chabazite are (AB), (ABC), and (AABBCC), respectively, when these three types of layers by layers are denoted with A, B, and C. It is noteworthy that 84,292 stacking sequences corresponding to all chemically feasible ABC-6 topologies as tectosilicates comprised of N stacking layers ($N \leq 16$) were enumerated on this method. Only 23 ABC-6 topologies have been found in the natural minerals or synthetic materials because this approach does not involve the strong host–guest interactions between ABC-6 cages and extra-framework species. However, it is expected that predicting and grouping on the enumeration of ABC-6 structures would be helpful to pre-screen: *ex.* for specific gas adsorption or separation applications. Genome-based identification and profiling have been adapted to investigation of materials including catalysts [36, 49, 50].

From the view of experimental researchers, it is still interesting demand that how to achieve the discovery of heterogeneous catalyst more effectively in comparison with the conventional manner with ML engineering; in particular, towards approaching "unexpected novel catalyst" which the researchers could not think of. One of the keys would be the identification of the appropriate numbers of adaptive descriptors to nicely explain the feature of the target catalysis. Fundamentally, (i) catalyst textural factors; composition, size, shape, crystallinity, strain, surface area, and dispersibility, (ii) reaction condition parameters; temperature, substrate concentration, stirring or feeding rate, amount of catalyst, and reactor design, and (iii) dynamic states under working condition; behaviour of oxidation state (redox), stability of particle (aggregation), and surface reconstruction, can be attributed to representation of catalyst reactivity. While, describing the element character has possible variety indexes of physicochemical nature such as atomic number, atomic weight, first ionization energy, electronegativity, covalent radius, ionic radius, atomic radius, valence electrons, redox potential, work function, melting point, conductivity, surface energy, and so on. In addition, it is notable that characterizations data such as XRD patterns [51]

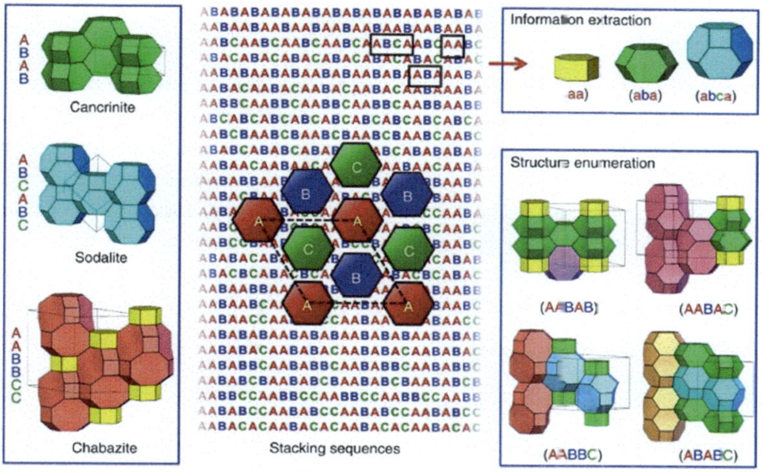

Fig. 12.10 Enumeration and interpretation of ABC-6 stacking sequences (Reproduced with permission from Springer Nature of [48])

and X-ray absorption fine structure (XAFS) features [8] effectively provided some trends of reactivity. If these multi-factors need to be revealed for whole understanding of the target reaction, that is no realism and excess trials on data collections would decrease advantage of more flexible and easier investigations based on ML. And also, it concerns misguiding the ML results owing overfitting. Accordingly, determining the meaningful descriptors to identify the target catalysts is really necessary skill for ML application in prediction of novel heterogeneous catalyst in high accuracy.

Fortunately, there are a lot of commonly available material databases and ML library are released one after another; named as NOMAD, Citrination, CatApp, Catalysis Hub, The Materials Project, TensonrFlow, scikit-learn, XenonPy, and/or CADS etc. Also, designs for High-throuhput screening experimentations for not only reactivity tests but also characterizations have gradually become common tools. These situations will decrease the cost on such featured data collection. Also, trans-disciplinary collaborations among not only data scientists and ML engineers but also additional theoretical scientists and experimental researchers gradually have been accelerated. Combining ML with DFT calculations allows prediction of *d*-band center and reactant gas adsorption energy [52], and graphic surface reaction networks on the basis of predicted kinetic parameters [53] are demonstrated. These trends are benefit for identification of appropriate descriptor for ML-based catalyst design [54]. It believes that new successful fusion at catalyst advancement for the desired catalytic reaction: ML-assisted searching the key factors from the physico-chemical nature and/or spectra from databases to identify the key components and structure, and the following experiments to validate exact reactive performance on the basis of ML-aids, are right there in the near future.

12.4 Catalysts Informatics Platform

In this section we introduce a Web-based integrated platform as an essential infrastructure for catalysts informatics. The platform provides three main functionalities: a repository for data sharing and publishing, an analytic workspace for exploratory visual analysis, and sharing pretrained machine learning models for catalyst property prediction. Such a platform helps decrease barriers to entry faced by researchers in catalytic chemistry when attempting to apply catalyst informatics towards data by providing analytical and visualization tools that can be simultaneously applied and easily accessed within a central space, thereby helping the advancement of catalyst informatics.

12.4.1 Concept of Platform

While Catalysts informatics shows much promise, it involves factors such as data construction, data management, visualization, machine learning, and a variety of other data analysis skills, thereby requiring a wide range of data science techniques and related knowledge. However, these tasks are not often collected or developed together, making it difficult to link data construction to catalyst design. Within the scope of materials informatics, various databases and platforms consisting of data generated from first principles calculations as well as data and codes related to computational materials science are available for use with examples ranging from the Open Quantum Materials Database (OQMD) [55], the Novel Materials Discovery repository (NoMad) [56], and Automatic Flow for Materials Discovery (AFLOW) [57].

Along these lines, web-based catalysis data platforms have also been developed with examples such as CatApp [58] and Catalyst Hub [59]. Unfortunately, these platforms lack data analysis functions and components, and more closely resemble data search engines with available visualization functions. Additionally, it is very difficult for experimental catalyst researchers that are unfamiliar with data management or software programming to use data provided by such data platforms. While some of the platforms provide data visualization functionalities, they are often very limited to a specific usage scenario. Hence, it is crucial to develop a platform that is accessible and is equipped with the tools necessary for catalyst informatics.

In order to help propel the development of catalyst informatics forward, it becomes necessary to provide an integrated platform where researchers can manage resources for catalyst informatics applications in a central location. In such a platform, researchers must be able to accumulate catalyst data and to successfully utilize the offered analysis and visualization techniques, including machine learning techniques, without requiring additional knowledge relating to computer programming or machine learning. All necessary resources including datasets, analysis methods, visualizations, and machine learning techniques should also be published and made accessible through the platform in a unified manner so that other researchers or other programs can reuse such resources for further analysis. Considering these factors, the developed platform aims to provide these functionalities in an integrated environment that is easily accessible to any catalyst researcher. Especially, the analysis application houses the core features of the developed platform. In order to provide users with a flexible and easy-to-use environment for analysis of heterogeneous data, coordinated multiple views (CMVs) [60] are employed.

12.4.2 Catalysts Acquisition by Data Science

Based on the discussion in the previous section, an innovative platform for a catalyst informatics platform, Catalyst Acquisition by Data Science (CADS) [61], is developed in efforts to address the needs of catalyst researchers that wish to apply informatics towards their research. The proposed platform provides three sub-applications: Data Management, Analysis, and Prediction. The platform allows researchers to upload local files containing catalyst data in the Data Management application. Once a file is uploaded to the platform, researchers can then analyze the underlying data in the Analysis application where various analysis methods are made available on the platform. These methods include not only simple visualization tools such as scatter plots but also complex analysis tools such as regression and clustering methods used in machine learning. Using trained models created in the Analysis application, researchers can then attempt to predict particular physical properties of catalysts. The platform is available on https://cads.eng.hokudai.ac.jp/.

When a user opens the Analysis application on CADS, a blank workspace is initially displayed. Here, the user selects data uploaded to the system in advance and adds components such as scatter plots and histograms to the workspace. With these components the user can conduct data analysis. Each component is equipped with a control panel for customizing the display content and analysis operation. For example, it is possible to set up which attribute values are assigned to the horizontal and vertical axes of the scatter plots, and which attributes are used as explanatory variables in regression analysis.

Figure 12.11 shows a snapshot of the actual analysis results. Here, methane oxidation coupling reaction data is used as the analysis target, and basic data such as C_2-yield and C_2-selectivity are visualized using "scatter plots" and "tables". Moreover, the regression result and the feature importance of explanatory variables used in the regression are calculated and displayed. In addition to this basic analysis, it is possible to select and highlight a part of the data by mouse operation on the part, and the selection operation is immediately propagated to other parts to highlight the corresponding data. For example, selecting data on the scatter plot will highlight data on the regression result view as well. Similarly, it is possible to automatically group data using the clustering component, color-coding the data points belonging to each group, and regression analysis using the regression component for only some of the data points segmented by clustering. In this way, exploratory visual analytics can be

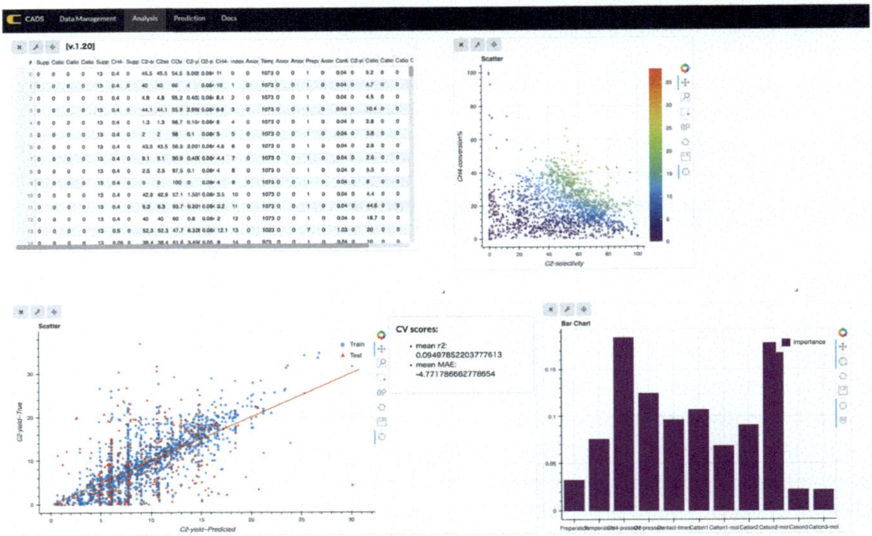

Fig. 12.11 The analysis tool on CADS. The workspace includes four components: Table, Scatter, Regression, and Feature Importance

performed by repeatedly adding and deleting components and changing the settings of the components, supporting users in understanding the characteristics of the data and approaching the various problems presented in the previous section.

References

1. Takahashi K, Takahashi L, Miyazato I, Fujima J, Tanaka Y, Uno T, Satoh H, Ohno K, Nishida M, Hirai K, Ohyama J. The rise of catalyst informatics: towards catalyst genomics. ChemCatChem. 2019;11(4):1146–52.
2. Medford AJ, Kunz MR, Ewing SM, Borders T, Fushimi R. Extracting knowledge from data through catalysis informatics. ACS Catal. 2018;8(8):7403–29.
3. Nguyen TN, Nhat TTP, Takimoto K, Thakur A, Nishimura S, Ohyama J, Miyazato I, Takahashi L, Fujima J, Takahashi K, Taniike T. High-throughput experimentation and catalyst informatics for oxidative coupling of methane. ACS Catal. 2019;10(2):921–32.
4. Nørskov JK, Bligaard T. The catalyst genome. Angew Chem. 2013;52(3):776–7.
5. Toyao T, Maeno Z, Takakusagi S, Kamachi T, Takigawa I, Shimizu KI. Machine learning for catalysis informatics: recent applications and prospects. ACS Catal. 2019;10(3):2260–97.
6. Kondratenko EV, Schlüter M, Baerns M, Linke D, Holena M. Developing catalytic materials for the oxidative coupling of methane through statistical analysis of literature data. Catal Sci Technol. 2015;5(3):1668–77.
7. Günay ME, Yildirim R. Knowledge extraction from catalysis of the past: a case of selective CO oxidation over noble metal catalysts between 2000 and 2012. ChemCatChem. 2013;5(6):1395–406.

8. Ohyama J, Hirayama A, Kondou N, Yoshida H, Machida M, Nishimura S, Hirai K, Miyazato I, Takahashi K. Data science assisted investigation of catalytically active copper hydrate in zeolites for direct oxidation of methane to methanol using H_2O_2. Sci Rep. 2021;11(1):1–10.
9. Takahashi K, Tanaka Y. Materials informatics: a journey towards material design and synthesis. Dalton Trans. 2016;45(26):10497–9.
10. Larsen P, Von Ins M. The rate of growth in scientific publication and the decline in coverage provided by Science Citation Index. Scientometrics. 2010;84(3):575–603.
11. Hey AJ, editors. The fourth paradigm: data-intensive scientific discovery, vol. 1. Microsoft research: Redmond, WA;2009.
12. Agrawal A, Choudhary A. Perspective: materials informatics and big data: realization of the "fourth paradigm" of science in materials science. APL Mater. 2016;4(5):053208
13. Ghiringhelli LM, Vybiral J, Levchenko SV, Draxl C, Scheffler M. Big data of materials science: critical role of the descriptor. Phys Rev Lett. 2015;114(10): 105503.
14. Edwards PN, Mayernik MS, Batcheller AL, Bowker GC, Borgman CL. Science friction: data, metadata, and collaboration. Soc Stud Sci. 2011;41(5):667–90.
15. Takahashi L, Miyazato I, Takahashi K. Redesigning the materials and catalysts database construction process using ontologies. J Chem Inf Model. 2018;58(9):1742–54.
16. Takahashi L, Takahashi K. Visualizing scientists' cognitive representation of materials data through the application of ontology. J Phys Chem Lett. 2019;10(23):7482–91.
17. Granda JM, Donina L, Dragone V, Long D, Cronin L. Controlling an organic synthesis robot with machine learning to search for new reactivity. Nature. 2018;559:377–81.
18. Ahneman DT, Estrada JG, Lin S, Dreher SD, Doyle AG. Predicting reaction performance in C-N cross-coupling using machine learning. Science. 2018;360(6385):186–90.
19. Allen C, Leitech DC, Anson MS, Zajac MA. The power and accessibility of high-throughput methods for catalysis research. Nat Catal. 2019;2:2–4.
20. Hammet LP. Some relations between reaction rates and equilibrium constants. Chem Rev. 1935;17(1):125–36.
21. Hammet LP. The effect of structure upon the reactions of organic compounds Benzene derivatives. J Am Chem Soc. 1937;59(1):96–103.
22. Fujita T, Winker DA. Understanding the roles of the "Two QSARs." J Chem Inf Model. 2016;56(2):269–74.
23. Sigman MS, Harper KC, Bess EN, Milo A. The development of multidimensional analysis tools for asymmetric catalysis and beyond. Acc Chem Res. 2016;49(6):1292–301.
24. Mukai M, Nagao K, Yamaguchi S, Ohmiya H. Molecular field analysis using computational-screening data in asymmetric *N*-heterocyclic carbene-copper catalysis toward data-driven in silico catalyst optimization. Bull Chem Soc Jpn. 2022;95(2):27 –7.
25. Haruta M, Kobayashi T, Sano H, Yamada N. Novel gold catalysts for the oxidation of carbon monoxide at a temperature far below 0 °C. Chem Lett. 1987;16(2):405–8.
26. Hatchings G. Vapor phase hydrochlorination of acetylene: Correlation of catalytic activity of supported metal chloride catalysts. J Catal. 1985;96(1):292–5.
27. Haruta M. Role of perimeter interfaces in catalysis by gold nanoparticles. Faraday Discuss. 2011;152:11–32.
28. Ishida T, Murayama T, Taketoshi A, Haruta M. Importance of size and contact structure of gold nanoparticles for the genesis of unique catalytic processes. Chem Rev. 2020;120(2):464–525.
29. Takahashi K, Miyazato I, Nishimura S, Ohyama J. Unveiling hidden catalysts for the oxidative coupling of methane based on combining machine learning with literature data. ChemCatChem. 2018;10(15):3223–8.
30. Nishimura S, Ohyama J, Kinoshita K, Le SD, Takahashi K. Revisiting machine learning predictions for oxidative coupling of methane (OCM) based on literature data. ChemCatChem. 2020;12(23):5888–92.
31. Kito S, Hattori T, Murakami Y. Estimation of catalytic performance by neural network—product distribution in oxidative dehydrogenation of ethylbenzene. Appl Catal A Gen. 1994;114(2):L173-8.

32. Kito S, Hattori T, Murakami Y. Determination of synergistically generated acid strength by neural network combined with experiment. Anal Sci. 1991;7:761–4.
33. Hattori T, Kito S. Neural network as a tool for catalyst development. Catal Today. 1995;23(4):347–55.
34. Sasaki M, Hamada H, Kintaichi Y, Ito T. Application of a neural network to the analysis of catalytic reactions analysis of no decomposition over Cu/ZSM-5 zeolite. Appl Catal A Gen. 1995;132(2):261–70.
35. Hou Z, Dai Q, Wu X, Chen G. Artificial neural network aided design of catalyst for propane ammoxidation. Appl Catal A Gen. 1997;161(1–2):183–90.
36. Le TC, Winkler DA. Discovery and optimization of materials using evolutionary approaches. Chem Rev. 2016;116(10):6107–32.
37. Goldsmith BR, Esterhuizen J, Liu J, Bartel CJ, Sutton C. Machine learning for heterogeneous catalyst design and discovery. AIChE. 2018;64(7):2311–23.
38. Keller GE, Bhasin MM. Synthesis of ethylene via oxidative coupling of methane: I determination of active catalysts. J Catal. 1982;73(1):9–19.
39. Hinsen W, Baerns M. Oxidative Kopplung von Methan zu C2-Kohlenwasserstoffen in Gegenwart unterschiedlicher Katalysatoren. Chem Ztg. 1983;107(7):223–6.
40. Nishimura S, Le SD, Miyazato I, Fujima J, Taniike T, Ohyama J, Takahashi K (2022) High-throughput screening and literature data-driven machine learning-assisted investigation of multi-component La2O3-based catalysts for the oxidative coupling of methane. Catal Sci Technol 12(9):2766–74. https://doi.org/10.1039/D1CY02206G
41. Nguyen TN, Nakanowatari S, Tran TPN, Thakur A, Takahashi L, Takahashi K, Taniike T. Learning catalyst design based on bias-free data set for oxidative coupling of methane. ACS Catal. 2021;11(3):1797–809.
42. Timoshenko J, Lu D, Lin Y, Frenkel AI. Supervised machine-learning-based determination of three-dimensional structure of metallic nanoparticles. J Phys Chem Lett. 2017;8(20):5091–8.
43. Guda AA, Guda SA, Lomachenko KA, Soldatov MA, Pankin IA, Soldatov AV, Braglia L, Bugaev AL, Martini A, Signorile M, Groppo E, Piovano A, Borfecchia E, Lamberti C. Quantitative structural determination of active sites from *in situ* and *operando* XANES spectra: from standard *ab initio* simulations to chemometric and machine learning approaches. Catal Today. 2019;336:3–21.
44. Miyazato I, Takahashi L, Takahashi K. Automatic oxidation threshold recognition of XAFS data using supervised machine learning. Mol Syst Des Eng. 2019;4(5):1014–8.
45. Yamamoto Y, Hattori M, Ohyama J, Satsuma A, Tanaka N, Muto S. Twinned/untwinned catalytic gold nanoparticles identified by applying a convolutional neural network to their Hough transformed Z-contrast images. Microscopy. 2018;67(6):321–30.
46. Ziatdinov M, Dyck O, Maksov A, Li M, Sang X, Xiao K, Unocic RR, Vasudevan R, Jesse S, Kalinin SV. Deep learning of atomically resolved scanning transmission electron microscopy images: chemical identification and tracking local transformations. ACS Nano. 2017;11(12):12742–52.
47. Aso K, Maebe J, Tran XQ, Yamamoto T, Oshima Y, Matsumura S. Subpercent local strains due to the shapes of gold nanorods revealed by data-driven analysis. ACS Nano. 2021;15(7):12077–85.
48. Li Y, Li X, Liu J, Duan F, Yu J. *In silico* prediction and screening of modular crystal structures via a high-throughput genomic approach. Nat Commun. 2015;6:8328.
49. Yuan W, He L, Tao G, Shreeve JM. Materials-genome approach to energetic materials. Acc Mater Res. 2021;2(9):692–6.
50. Takahashi K, Fujima J, Miyazato I, Nakanowatari S, Fujiwara A, Nguyen TN, Taniike T, Takahashi L. Catalysis gene expression profiling: sequencing and designing catalysts. J Phys Chem Lett. 2021;12(30):7335–41.
51. Corma A, Serra JM, Serna P, Moliner M. Integrating high-throughput characterization into combinatorial heterogeneous catalysis: unsupervised construction of quantitative structure/property relationship models. J Catal. 2005;232(2):335–1.
52. Takigawa I, Shimizu K, Tsuda K, Takakusagi, S. (2018) Chapter 3: Machine learning predictions of factors affecting the activity of heterogeneous metal catalysts. In: Nanoinfomatics. Singapore: Springer;2018. p 45–64.

53. Ulisse ZW, Medford AJ, Bligaard T, Nørskov JK. To address surface reaction network complexity using scaling relations machine learning and DFT calculations. Nat Commun. 2017;8:14621.
54. Nørskov JK, Bligaard T, Rossmeisl J, Christensen CH. Towards the computational design of solid catalysts. Nat Chem. 2009;1(1):37–46.
55. Kirklin S, Saal JE, Meredig B, Thompson A, Doak JW, Aykol M, Rühl S, Wolverton C. The Open Quantum Materials Database (OQMD): assessing the accuracy of DFT formation energies. npj Comput. Mater. 2015;1:15010.
56. Draxl C, Scheffler M. NOMAD: The FAIR concept for big data-driven materials science. MRS Bull. 2018;43:676.
57. Curtarolo S, Setyawan W, Hart GL, Jahnatek M, Chepulskii RV, Taylor RH, Wang S, Xue J, Yang K, Levy O, Mehl MJ. Aflow: an automatic framework for high-throughput materials discovery. Comput Mater Sci. 2012;58:218.
58. Hummelshøj JS, Abild-Pedersen F, Studt F, Bligaard T, Nørskov JK. CatApp: a web application for surface chemistry and heterogeneous catalysis. Angew Chem Int Ed. 2012;51:272.
59. Winther KT, Hoffmann MJ, Boes JR, Mamun O, Bajdich M, Bligaard T. Catalysis-Hub. org, an open electronic structure database for surface reactions. Sci. Data 2019;6:75.
60. Roberts JC. State of the art: coordinated & multiple views in exploratory visualization. In: Fifth International conference on coordinated and multiple views in exploratory visualization (CMV 2007). IEEE. P. 61–71.
61. Fujima J, Tanaka Y, Miyazato I, Takahashi L, Takahashi K. Catalyst acquisition by data science (CADS): a web-based catalyst informatics platform for discovering catalysts. React Chem Eng. 2020;5(5):903–11.

Chapter 13
Recent Advances in Density Functional Theory (DFT) and Informatics Studies on Metal Oxide Surfaces

Takashi Kamachi, Yoyo Hinuma, and Nobutsugu Hamamoto

13.1 Introduction

Computational chemistry is becoming an indispensable tool not only for experts but also for synthetic experimentalists, thanks to the development of quantum chemistry methods, the widespread availability of easy-to-use program packages, and dramatic advances in computing power. Especially for molecular systems of a hundred atoms, it has become relatively easy to analyze and corroborate experimental results such as reaction mechanisms and spectroscopic data. In solid surface chemistry, the concept of quantum chemistry was applied to surfaces in the 1980s, and important theories such as the d-band center emerged. However, there are still many limitations in analyzing important phenomena such as molecular adsorption and bond formation/cleavage on solid surfaces using realistic models, even with the current computer environment available. In this chapter, we review the advances of computational chemistry for surface science and outline its application on metal oxides surfaces and the future that we hope to achieve with comprehensive DFT calculations and machine-learning methods.

T. Kamachi (✉)
Department of Life, Environment and Applied Chemistry, Fukuoka Institute of Technology, Wajiro-higashi, Higashi-ku, FukuokaFukuoka 811-0295, Japan
e-mail: kamachi@fit.ac.jp

Y. Hinuma
Department of Energy and Environment, National Institute of Advanced Industrial Science and Technology (AIST), 1-8-31 Midorigaoka, Ikeda 563-8577, Osaka, Japan
e-mail: y.hinuma@aist.go.jp

N. Hamamoto
Faculty of Engineering, Department of Applied Chemistry, Sanyo-Onoda City University, 1-1-1, Daigakudori, Sanyo-Onoda, Yamaguchi 756-0884, Japan
e-mail: n-hamamoto@rs.socu.ac.jp

W. Ueda (ed.), *Crystalline Metal Oxide Catalysts*,
https://doi.org/10.1007/978-981-19-5013-1_13

13.2 DFT-Based Descriptors of Metal Oxides for Catalysis Informatics

The description of the catalytic activities as a function of a physical parameter has been a promising method of catalyst design, which can reduce extensive trial-and-error experimental testing. Important contributions have been made by employing surface properties related to catalysis such as d-band center [1], work function [2, 3], coordination numbers [4], strain [5], and surface energies [6] as descriptors of catalyst properties. In these efforts, the most widely explored examples involve establishing scaling relationships between the adsorption energies of reactants and intermediates using the d-band model [7], and between activation and reaction energies referred to as Brønsted-Evans-Polanyi relationships [8]. Based on the set of linear relationships, one can easily derive reactivity trends for a target catalytic reaction from the adsorption energy of reactants without time-consuming activation barrier calculations and the overall analysis of reaction steps involved in the course of the reaction. Nørskov and co-workers used this scaling relationships strategy and computational thermal and kinetic analysis to discover selective hydrogenation of acetylene with bimetal catalyst NiZn [9] and for CO_2 reduction with NiGa [10]. However, the computational techniques in determining electronic structures of complicated semiconducting and insulating surfaces and molecular adsorption on the surfaces are still under development, and hence in silico discovery of metal oxide catalyst is quite rare. This section shows recent examples of relationships between DFT-based descriptors and adsorption energy on metal oxide surfaces.

13.2.1 d-Band Center for Molecular Adsorption on Metal Surfaces

Since its development in the early 1950s, frontier molecular orbital (FMO) theory has provided a basis for explaining many aspects of chemical reactivity [11, 12]. According to Fukui's FMO theory [13] and the Woodward–Hoffmann rule [14], the selectivity of organic reactions is determined by both phase properties and the magnitudes of the coefficients of the reactant's highest occupied molecular orbital (HOMO) and lowest unoccupied molecular orbital (LUMO). FMO theory has successfully been applied to elucidate the selectivity of aromatic substitution and pericyclic reactions, among others. As an example, the observed preference for nitration at the 1-position on the naphthalene molecule can be easily understood by considering the HOMO coefficients. FMO theory has also been extended to solid and surface systems. Hoffmann and co-workers [15] used the concepts of FMO theory and of orbital interactions to intuitively explain the electronic properties of solid states and surface–adsorbate interactions. They have discussed the adsorption of small molecules and molecular fragments, such as CO [16, 17] and CH_3 [18, 19] on metal surfaces within a tight binding scheme based on the extended Hückel method.

The interactions between the frontier orbitals of these molecules and the d band of the metal surface make a major contribution to the adsorption process as well as to the energies of transition states on surfaces. In the Newns-Andersen model [20], couplings of an adsorbate state to sp and d bands of metal surfaces are considered to describe molecular adsorptions on the transition metal surfaces. Since the sp orbitals are spatially extended orbitals, they have large overlap and interaction with neighboring atoms, forming a band with a wide energy range, but the low density of states (DOS). On the other hand, d electrons, which are more strongly attracted to nuclei, have less overlap with the surrounding atoms, and form a band with a narrow energy width and a higher DOS. A molecular orbital level of the adsorbate decreases in energy and broadens upon interactions with the broad sp band (Fig. 13.1a), and the broadened adsorbate state splits into bonding and antibonding states upon further interaction with the narrow d band in a manner similar to that of orbital interactions in molecules (Fig. 13.1b). The adsorbate states broadened by the interactions with the sp band are energetically well below the Fermi level and are all filled with electrons, and therefore the contribution of the sp band to the molecular adsorption is comparable in energy across transition metal surfaces. In contrast, the adsorption energy of the molecule varies greatly from a transition metal to a transition metal due to the difference in the number of electrons occupying the antibonding state created by the interaction with the narrow d band. Since the antibonding state is always above the d-band, a higher d-band causes the antibonding state to rise as well, thus reducing the number of electrons occupying it and accordingly strengthening the bond between the metal surface and the molecule. Consequently, the center of the d band relative to the Fermi level is a good descriptor of molecular adsorption on the transition metals.

The d band center is linearly related to the adsorption energies of atoms and molecules and to the energy barriers for various reactions, and therefore the d band center has been used to identify and interpret trends in surface properties [7]. For example, Kibler et al. found a linear relationship between the d-band centers of

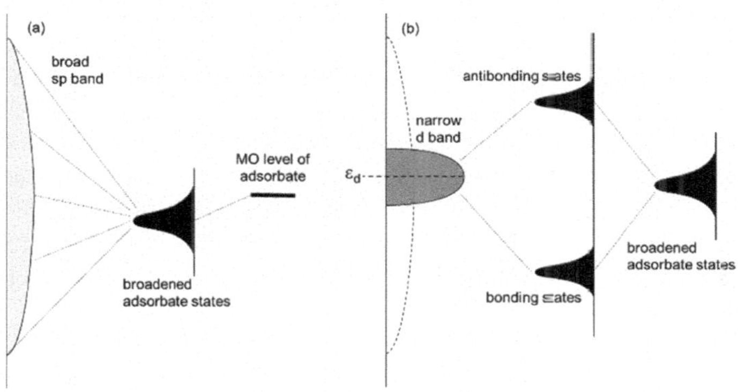

Fig. 13.1 **a** Coupling of an adsorbate state with the broad sp band of the metal surface, which gives rise to broadened adsorbate states **b** interactions between the adsorbate states and the narrow d band

various metal and alloy surfaces and experimentally measured hydrogen desorption potentials, which might help tune reaction rates by the modification of catalysts [21]. However, the d band center is not considered a useful descriptor for more complex metal oxides, and other properties of metal oxide surfaces have been featured to find correlations.

13.2.2 Scaling Relationships Between Adsorption Energies

Extensive studies of the adsorption of atoms and molecules on metal oxide surfaces have revealed that scaling relationships exist for the adsorption energies despite the structural and electronic complexities of metal oxide surfaces. Based on this relationship, the adsorption energies of various molecules on metal oxide surfaces can be predicted by using as descriptors the adsorption energies of atoms and simpler molecules such as H_2O, which can be easily obtained by DFT calculations. Thus, a database of adsorption energies on metal oxide surfaces is expected to serve as a basis for catalysis informatics, and many systematic studies on adsorptions of atoms and molecules have already been performed. Fernández et al. performed DFT calculations on the adsorption of AH_x intermediates and bare A atoms (A = O, S, N) on various transition metal oxide, nitride, and sulfide surfaces [22]. The same linear relationship was found to exist between adsorption energies of AH_x and A on surfaces of both metal oxides and pure metals. The trends in adsorption energies can be understood by using a modified version of the d-band model [23] and by "Fermi softness" [24], which is defined as the sum of density of states weighted by the derivative of the Fermi–Dirac distribution function at a certain non-zero temperature. Calle-Vallejo et al. described that the adsorption energies of O, OH, and OOH are linearly related to the number of outer electrons at the adsorption sites of metal, monoxide, and perovskite surfaces [25]. Although this effort found that the number of outer electrons can be a simple and efficient descriptor, van Santen et al. pointed out that experimental data on perovskite $LaXO_3$ compounds do not follow this trend but instead show double peaked volcano type behavior for the third-row reducible metal oxides, which is most likely to be due to high spin state stabilization of d^5 electron systems [26]. Recently, scaling relationships were reported to exist between the adsorption energies of various molecules bearing lone pair electron-containing atoms across a wide variety of materials [27]. It should be noted that adsorption energies on metal oxides surfaces are significantly affected by the presence of vacancies, dopants, co-adsorbed molecules. Therefore, further consideration is required in order to describe more precisely activites and selectivities of working metal oxides catalysts.

13.2.3 Oxygen Vacancy Formation Energy (E_{Ovac}) as a Descriptor for Catalytic Behavior of Metal Oxides

Defects play a key role in governing properties of catalysts. Various kinds of extended and point defects introduced into structures of catalysts largely govern their physical and chemical properties. Particularly, point defects in metal oxide surfaces such as oxygen vacancies have dominant effects on the properties of heterogeneous catalysts [28, 29]. Energies required to form the oxygen defects in metal oxide catalysts (E_{Ovac}) are typically used to rationalize and predict performances of these catalysts [30].

Oxygen vacancy formation in ABO_3 perovskites has been extensively investigated using DFT computations because it controls electrocatalytic activity [31–33]. By using DFT calculations, Chen and Ciucci systematically investigated the dependence of stability, and electronic and ionic conductivity on substitutions of A and B sites of $BaFeO_3$ [34]. They found that the E_{Ovac} is reduced by the presence of Na, K, Ca, Sr and Pb in the A site, and of Ni, Cu, Zn, and Ag in the B site. The E_{Ovac} were found to linearly correlate with the occupied O p band center, which is related to the basicity of the oxygen atoms. Emery et al. used high-throughput DFT to search for novel perovskites that promote thermochemical water splitting (TWS) [35]. They screened 5329 calculated perovskites on the basis of the E_{Ovac} and identified 139 potentially new candidates for TWS.

Kumar et al. studied oxidative coupling reactions of CH_4 on a range of metal oxide surfaces [30]. This reaction begins with the abstraction of a hydrogen atom from the strong C–H bond of CH_4 by lattice oxygen atoms, which leads to the formation of a CH_3 radical. They reported that this process takes place more readily for highly reducible oxides and that the activation energy for cleaving C–H bond is linearly correlated with the E_{Ovac}. The CH_3 radical intermediate is converted into desired C2 hydrocarbons by radical coupling or into undesired methoxy species by absorption to surface oxygen leading to the eventual formation of CO or CO_2. Thus, the adsorption of CH_3 radical regulates the selectivity for C2 hydrocarbon production. The adsorption energy of the methyl radical is also linearly correlated with the E_{Ovac}. Surfaces activating the C–H bond more easily bind the methyl radical more strongly, indicating that reactivity and C2 selectivity are essentially a trade-off. They proposed that metal oxides like TbO_x that can exist in multiple oxidation states could overcome the intrinsic competitive processes by changing the oxygen atmosphere.

Although theoretical studies on the formation of oxygen vacancies on metal oxides have been intensively conducted recently, research on surface defects is still rare and the numbers of investigated surfaces remain limited despite the obvious importance of oxygen vacancies on catalytic activities. Therefore, studies that provide E_{Ovac} values and reveal the physical factors determining E_{Ovac} are highly desirable. In this context, the author's group recently determined E_{Ovac} values of various semiconducting and insulating oxide surfaces using DFT calculations at comparable structure models and the same computational level [36]. A further investigation aimed at providing a deep understanding of the formation of oxygen vacancies found that the band gap, formation energy of bulk, and electron affinity of a metal oxides are important properties that govern E_{Ovac}. The reason why these electronic properties determine E_{Ovac} is that electrons enter the defect state(s) after the removal of oxygen, which could be in the conduction band for most cases. It is worth mentioning here that the use of ML for statistical analysis to assess the physical basis of this relationship would perhaps be more meritorious than predicting physical property values themselves [37]. Although limitations associated with a variety of metal oxide surfaces and computational accuracy exist, the information would be helpful for catalysis researchers.

13.2.4 Acid–Base Properties of Metal Oxides

Solid acids and bases have been employed to catalyze various reactions, and the role of the properties of these solids in governing catalytic performances has been investigated extensively [38]. An early computational study on the acid–base properties of the γ-Al_2O_3 surface found that the energies of occupied and unoccupied orbitals can be employed to rank surface Lewis acid and base sites, respectively [39]. Jenness et al. computed adsorption energies of alcohols, diethyl ether and water, and activation barriers for ethanol dehydration and etherification promoted by several Lewis acid sites on γ-Al_2O_3 surfaces [40]. The results showed that energies of unoccupied s band centers can be used as descriptors to evaluate the Lewis acidities of the Al^{3+} sites. This descriptor correlates the adsorption energies with barriers for ethanol dehydration in a quantitative manner, but it describes the barrier for ethanol etherification only qualitatively owing to the bimolecular nature of the process. Recently, the occupied p band center of O and unoccupied s band center of Al were used to evaluate the reactivity for C–H bond activation of CH_4 over γ-Al_2O_3 surfaces [41].

In a study of *n*-butane oxidation over vanadium phosphorus oxides, Wang et al. identified the active site for this reaction among 15 possible V = O and P = O sites in total [42]. Both V = O and P = O sites abstract an H atom from *n*-butane in what can be viewed as a proton-coupled electron transfer (PCET) process, in which V^{5+} and oxygens serve as electron and proton acceptors, respectively. Thus, the activity of the P = O sites is lowered by increasing the distance between V^{5+} and these sites. The results showed that a linear relationship exists between the centers of the energies of the P = O lone-pair bands and the energies for reaction at the P = O sites. Accordingly, the study demonstrated that energies of the s and p band centers are useful for rapid screening of potential active sites of metal oxide catalysts.

Brønsted acid sites (BAS) in zeolite micropores are formed by the protonation of oxygen atoms linking neighboring silicon and aluminum atoms for charge balancing. The presence of BAS in the confined environment of zeolite micropores is a crucial factor governing the catalytic properties of many zeolite catalysts [43]. Pidko and coworkers determined the acid strength of various zeolites by using adsorption energies of probe molecules such as NH_3, CO, pyridine, etc. [44]. The acid strength obtained by the adsorption energy calculations of NH_3 was found to correlate with the activation barrier of isobutene protonation in FAU zeolites. Matsuoka et al. screened

933,613 structures in a database of hypothetical zeolites to identify chemically feasible members that have strong Brønsted acidity [45]. Density, ring counting, and average ring sizes were used to eliminate unsuitable candidates. To assess Brønsted acid strength, a new structural descriptor called '*b*' was introduced, which is the distance between barycenters of the two triangles connecting O atoms linked to T atoms in the acid site. This parameter can be calculated for pure silica with low computational cost although the correlation between '*b*' and Brønsted acid strength is weak. They confirmed by using DFT calculations that 6 out of 12 final candidates contain a strongly acidic Brønsted site. This encouraging result suggests that even non-quantitatively accurate structural descriptors might be sufficient for screening purposes where speed is more important than accuracy.

13.3 Modeling of Metal Oxide Surfaces

We introduced in Sect. 13.2 that various properties of oxide surfaces obtained by DFT calculations can serve as descriptors for predicting molecular adsorption and catalytic activity. Once these descriptors are obtained with comprehensive DFT calculations for known materials reported so far or hypothetical materials expected to exist theoretically, a database can be constructed for machine learning to reasonably and effectively determine an optimal material for the desired functions such as reactivity, stability, selectivity, etc. To achieve this, in addition to a computing environment that can handle a large number of DFT calculations of high computational cost, it is essential to construct a high-throughput DFT calculation system, which can automatically solve various problems that may occur in actual DFT calculations (such as SCF convergence errors) and sequentially obtain and summarize the target descriptors extracting necessary data from output files. In addition, it is important to develop theories and algorithms that can automatically generate computational models that can reproduce the nature of the structure and electronic state of metal oxide surfaces and that can be practically handled using currently available computational resources. However, most DFT calculation program packages for materials are designed for three-dimensional periodic systems, and various innovations are required to calculate two-dimensional metal oxide surfaces with diverse structures and electronic states. Thus, the creation of a reasonable model of individual metal oxide surfaces remains a challenge and comprehensive analysis of metal oxide surfaces is a formidable task. In this section, we present an overview of concepts and algorithms for an automatic generation of metal oxide surfaces, which is vital in the building of DFT-based databases for catalysis informatics.

13.3.1 The Polarity of a Surface

What is polarity, and why is polarity important in the first place? All objects that we encounter have an outer boundary, which is an interface with something else than the "inside" of the object. Yet, we very frequently discuss the property of macroscopic objects using "bulk" that does not have an outer boundary and extends infinitely. Examples of such properties are density and heat of formation, where the surface, or more precisely, atoms near the surface have an extremely small contribution compared to those from atoms far from the surface. Consequently, the properties of the inside dominate the properties of the object. In contrast, some properties, such as the coefficient of friction, is governed by the surface, and details of the surface must be elucidated to discuss those properties in detail.

The shape of a crystal reflects microscopic surface characteristics. For instance, growing a table salt (NaCl) crystal is a staple experiment in science clubs, and interestingly, the grown salt crystals almost always form a cube. Natural quartz gems come in the form of a hexagonal rod that is pointed at a certain angle. Crystal growth is an atom-by-atom process and is a suitable topic for atomic-level computational modeling.

For simplicity, let us consider an infinitely spreading slab crystal with finite thickness. This crystal has only two surfaces. The surface energy of a crystal, in J, can be obtained by taking the heat of formation of this crystal and subtracting the heat of formation when this crystal is part of bulk, but this value diverges to infinity. The surface energy is better expressed after normalizing by surface area, and the unit is J/m^2. We note that the surface energy is for the two surfaces combined. The surface energy of one side can be obtained *only* when the two surfaces are exactly symmetrically equivalent. Another issue is charge distribution. If the charge is distributed such that there is an electric field penetrating the slab, the electrostatic potential at both sides of the slab is not the same. The electric field in a parallel capacitor is proportional to the net charge at the plates, and the potential difference is the electric field multiplied by the slab thickness. Uneven charge between the two surfaces of a slab can therefore lead to an extremely large potential difference, thus surface charge distribution is a very important topic.

Tasker classified the polarity of ionic compound surfaces into three distinct types [46]. Atom planes in a Tasker's Type 1 surface are neutral with both anions and cations, while those in a Tasker's Type 2 surface are charged and arranged symmetrically such that there is no dipole moment perpendicular to the unit cell. On the other hand, Tasker's Type 3 surface is charged with a perpendicular dipole moment. Gale and Rohl [47] separated Type 2 surfaces into Type 2a, while Goniakowski et al. [48] considered the concept of a dipole-free bulk unit cell, which is a bulk unit cell that may involve incomplete layers and does not have a dipole moment along a given direction. The frozen bulk termination is polar if the surface cannot be obtained by simply piling up dipole-free bulk unit cells and nonpolar if at least one dipole-free bulk unit cell exists that leaves the surface region empty. Stengel [49] used Wannier ion charges, which are derived from Wannier orbitals belonging to each ion and are typically formal charges, to determine the dipole-free bulk unit cell. Goniakowski and Noguera [50] defined the weakly polar surface as a surface such as the $SrTiO_3$ (100) surface.

Hinuma et al. [51] proposed a new categorization of surface (slab) polarity based on crystallography. One important feature is that information on the spatial charge is unnecessary, hence there is no need to analyze whether each layer of atoms is neutral, investigate what the nature of each bond is, or assign charge densities to each layer of atoms. All surfaces are either *polar* or *nonpolar*. A surface is polar when it is impossible to cut out a slab that looks identical when viewed from either direction normal to the surface (no identical termination). Surfaces that are considered nonpolar in Hinuma et al. [51] are exactly nonpolar by virtue of symmetry.

Nonpolar surfaces are further categorized into three types in Hinuma et al. [51]. A surface is "*nonpolar type A*" if the surface is not polar and each layer of atoms is stoichiometric. This is a stricter requirement than the Tasker type 1 surface that could have charge neutrality based on formal charge in every layer, for instance $SrTiO_3$ (001). The remaining surfaces can be categorized into "*nonpolar type B*" and "*nonpolar type C*" surfaces. The surface is the former if the boundaries of the dipole-free bulk unit cell, which is the repeat unit with no dipole moment perpendicular to the surface, lies between layers of atoms, while the surface is the latter if the boundaries must lie on layers of atoms. A nonpolar type C slab cannot, by definition, be simultaneously nonpolar and stoichiometric when simply cleaved from bulk. However, a nonpolar and stoichiometric slab can be obtained by reconstruction of the surface, for instance, by removing half of the atoms on the topmost layer on both sides.

Figure 13.2 shows prototypes of the four categories as well as issues arising in high-throughput calculations. All nonpolar type A surfaces, including the rocksalt (100) surface, are Tasker type 1 surfaces because stoichiometric layers are, by definition, charge neutral. On the other hand, nonpolar type B surfaces, such as the fluorite (111) surface, are generally Tasker type 2 surfaces, although some exceptions exist. Nonpolar type C surfaces are mostly categorized as Tasker type 3 surfaces, and a representative example is the rocksalt (111) surface. This surface of MgO and related compounds typically undergoes octopolar reconstruction [52] where 3/4 of atoms in the topmost layer and 1/4 of atoms in the next layer are removed. Some nonpolar type C surfaces such as the cubic perovskite $SrTiO_3$ (001) surface, which is terminated on each side by either a SrO or TiO_2 plane, may be considered nonpolar Tasker type 1 based on the Wannier charge approach [49]. Polar surfaces according to the crystallographic definition are generally Tasker type 3 surfaces, where a typical example is the wurtzite (0001) surface. However, a polar surface according to Hinuma et al.'s definition could be considered a Tasker type 1 surface in some cases. For example, if we take perovskite $SrTiO_3$ and move each TiO_2 layer by a fixed distance in the same direction along the *c*-axis without moving the SrO layers, this symmetry breaking changes the polarity of the (001) surface from nonpolar type C to polar. However, as all SrO and TiO_2 planes are neutral based on formal charge, the surface can still be regarded as a Tasker type 1 surface. It is possible, as Tasker pointed out using the fluorite (111) surface as an example, to choose a termination such that a surface is polar (type 3) even though it is also possible to leave a nonpolar (type 2) surface [46]. The surface-dependent reconstruction mechanism must be identified when investigating a nonpolar type C surface, so extra care is needed for high-throughput calculations of such surfaces.

In a nutshell, barring some exceptions, nonpolar type A surfaces are Tasker type 1 surfaces, nonpolar type B surfaces are Tasker type 2 surfaces, and nonpolar type C and polar surfaces are Tasker type 3 surfaces. An exactly nonpolar surface can be obtained for some Tasker type 3 surfaces but not in some, and the strength of Hinuma et al.'s categorization is the ability to describe the possibility of exact nonpolar surface formation.

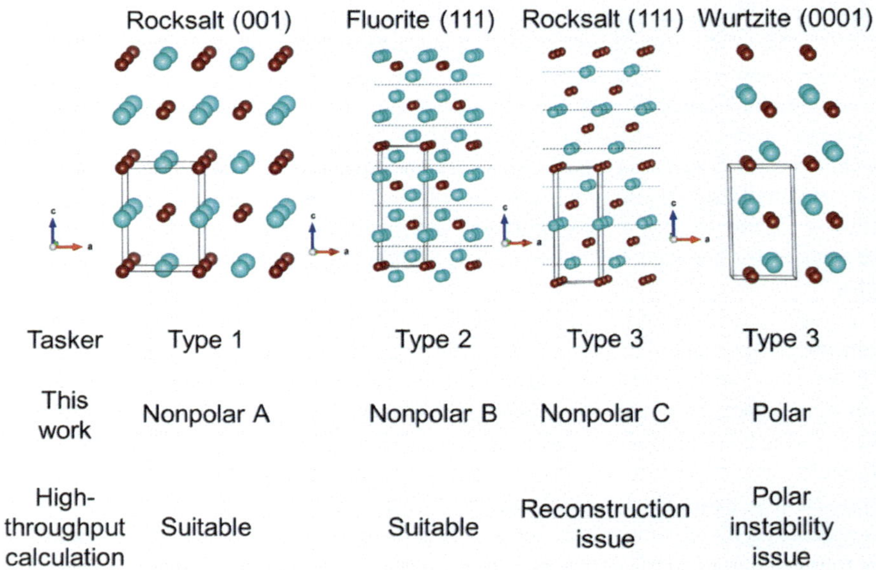

Fig. 13.2 Definition of slab polarity types based on Tasker [46] and this work as well as issues in high-throughput calculations. Small brown circles and large blue circles indicate cations and anions, respectively. Dotted lines in fluorite (111) and rocksalt (111) indicate the repeat unit where the macroscopic dipole moment perpendicular to the surface is absent in a slab of this repeat unit. One can alternatively choose to place the repeat unit boundaries on cation planes instead of anion planes in rocksalt (111). Adapted with permission from Ref. [51]. Copyright 2016 Elsevier

13.3.2 Automated Generation of Nonpolar Slabs

The procedure to generate nonpolar type A and B slabs with Miller index (*hkl*) is briefly outlined here with minimum use of equations. The reader is referred to Ref. [51] for details. A primitive cell where the **a** and **b** basis vector spans the (*hkl*) plane is generated first, which is the (*hkl*)-primitive cell. Isometries, which are symmetry operations that keep the crystal invariant, are searched using symmetry search software. Isometries of a certain form "flips" the crystal in the direction perpendicular to the (*hkl*) plane. Examples of such symmetry operations are inversion, mirror, and two-fold rotation. The forms of such isometries are used to identify potential slab centers, which are positions along the **c** basis vector (*z*-coordinate).

The (*hkl*) *n*-supercell is defined as the $1 \times 1 \times n$ supercell of the (*hkl*)-primitive cell. Transfer of *z*-coordinates between (*hkl*) *n*-supercells is very easy. The coordinate z_m in a *m*-supercell is transferred to z_n in a *n*-supercell as $z_m / m = z_n / n$.

Figure 13.3 shows (001) *n*-supercells of SnI_4 (space group $P\,\overline{4}\,3\,m$). There is a potential slab center (two-fold rotation along the b axis) at $z = 0.5$ in the leftmost (*hkl*)-primitive cell, or 1-supercell. The position of corresponding potential slab centers in is $(Z + m)/n$, where *Z* is the potential slab center in the primitive cell (0.5) and *m* is an arbitrary integer.

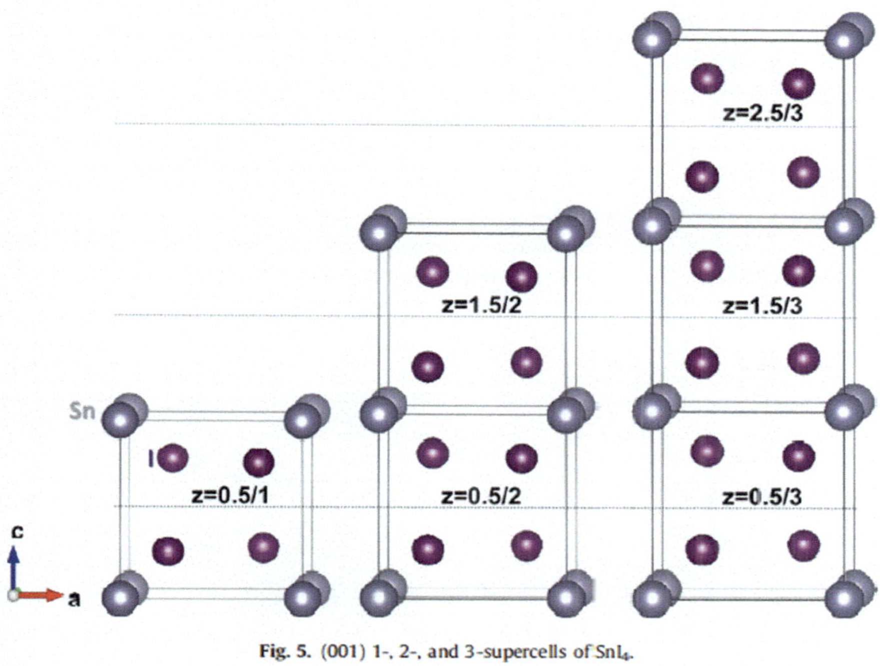

Fig. 13.3 (001) 1-, 2-, and 3-supercells of SnI_4. Adapted with permission from Ref. [51]. Copyright 2016 Elsevier

The "*unit layer thickness*" z_{nu} is defined as the minimum z_{nu} where a slab with boundaries $(z_{n-}, z_{n+}) = (z, z + z_{nu})$ is identical to one with $(z_{n-}, z_{n+}) = (z + z_{nu}, z + 2z_{nu})$ after any necessary in-plane rotation and/or translation. The polarity of slabs with various thicknesses in an (*hkl*) *n*-supercell can be derived by investigation of slabs in an (*hkl*) 3-supercell with slab thickness $z_{3t} = z_{3u}/2 = (z_{1u}/2)/3$ or $z_{3t} = z_{3u} = z_{1u}/3$ and slab center $z_{3c} = (z_{Pc} + 1)/3$ or $z_{3c} = (z_{Pc} + z_{1u}/2 + 1)/3$, where z_{Pc} is the position of the potential slab center in the (*hkl*)-primitive cell. Figure 13.4 shows the (100) 3-supercell of $BeSO_4$ (space group $I\,\bar{4}$). Here, $z_{1u} = 0.5$ and $z_{Pc} = 0$. Out of the four combinations of slab thickness and slab center, a cleaved slab with the combination $(3z_{3t}, 3z_{3c}) = (0.5,1)$ and (0.5, 1.5) leads to a non-stoichiometric slab, (1, 1) is a nonpolar type B slab, and (1, 1.5) is a nonpolar type C slab. The formation of a stoichiometric and nonpolar slab is therefore not guaranteed. Where to cleave surfaces to obtain a nonpolar and stoichiometric slab can be obtained by transferring slab boundary positions of an (*hkl*) 3-supercell, which results in a stoichiometric and nonpolar slab after cleaving, to a larger supercell while ensuring that the symmetry operation at the slab center remains the same.

Advanced topics on nonpolar slab generation are discussed below. Whether a certain orientation can form a strictly nonpolar slab is dependent on the point group and can be immediately identified by looking at the crystal form. A list of unique nonpolar orientations as well as a procedure to detect identical nonpolar terminations among the four combinations of $(3z_{3t}, 3z_{3c})$ is given in Refs. [54, 55]. Nonpolar type C slabs may be automatically generated by removing half of the atoms in the outermost surface of both sides using the procedure in Ref. [55]. This reconstruction is not necessarily guaranteed to be the most stable (cleaved surfaces without significant reconstruction is also not always the most stable surface termination, where one famous example is the Si (111) 7 × 7 reconstruction), however the existence of a procedure is better than nothing.

Models with step edges and terraces are attractive when trying to find surfaces that are more reactive. Strictly nonpolar models are necessary when the surface energy

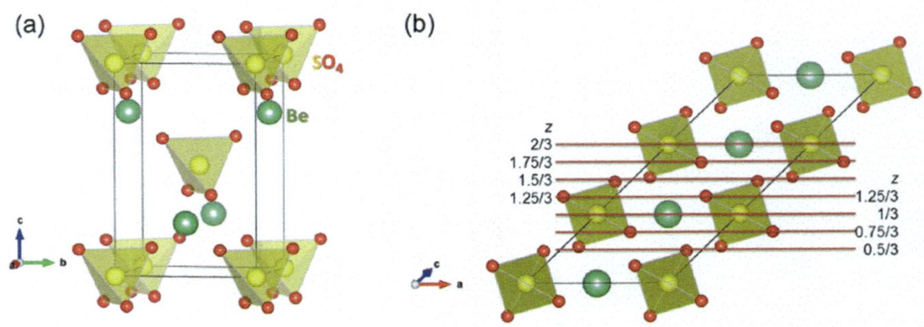

Fig. 13.4 Examples of nonpolar slab identification. **a** Unit cell and **b** (100) 3-supercell of $BeSO_4$. Green circles indicate Be and SO_4 are shown as tetrahedra. Lines in the 3-supercell are drawn at $z = 0.25/3$ intervals. Adapted with permission from Ref. [51]. Copyright 2016 Elsevier

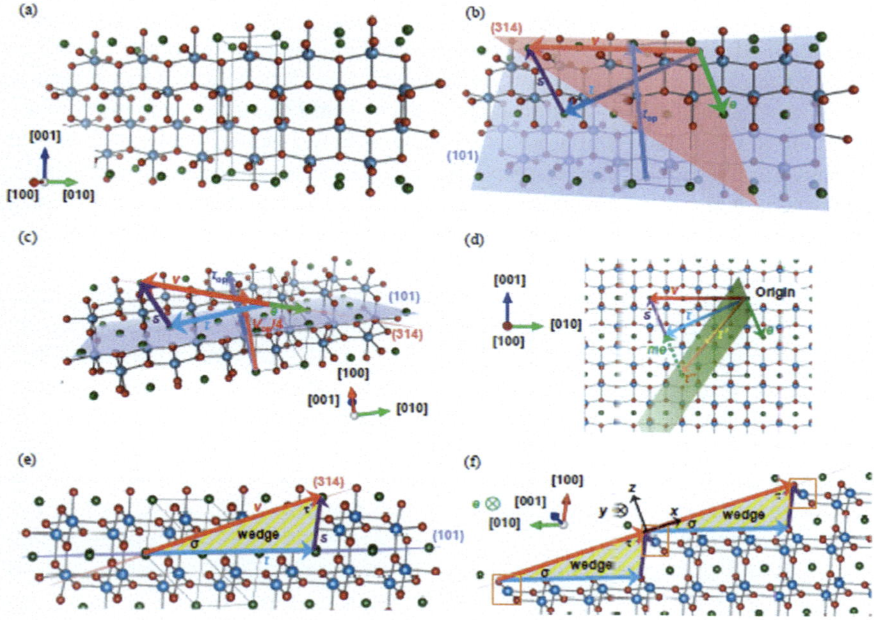

Fig. 13.5 Generation of an anatase TiO_2 step model. Adapted with permission from Ref. [55]

is required during analysis. A process to obtain nonpolar models with step edges, which is a painful process that requires the identification of many relevant high-index vectors and surfaces, is outlined in Ref. [55]. Figure 13.5 shows the process for obtaining a model of body-centered tetragonal anatase TiO_2 with step edges. Such a systematic procedure that can be applied to crystals with arbitrary symmetry is a necessity to investigate step-edge models beyond the simplest crystal structures.

As a side topic, quantitative detection of whether an atom is at the surface, or even better, distinguishment between a step edge, terrace, step bottom, or subsurface, is an important issue when considering adsorption and/or desorption sites on a surface. This can be accomplished by quantifying the solid angle of "open space" around an atom; details and applications are discussed in Ref. [54].

13.3.3 Applications of Surface Model Generation

These methods allowed us to perform calculations on surface properties over a wide variety of compounds. Although the IP of a material depends on the surface orientation and termination, the ionization potential of various surfaces of divalent [53] and trivalent [56] d^0 and d^{10} binary oxides showed a strong correlation with the atom size, namely $v^{-1/3}$ where v is the volume per atom [53]. The surface O vacancy formation energy (E_{Ovac}) of 32 semiconducting or insulating binary oxides covering 20 surfaces was calculated, and a good correlation between E_{Ovac} and the band gap as well as E_{Ovac} and the bulk formation energy was found [36]. Removal of O as a neutral species leaves two electrons behind at the surface. These electrons go to the conduction band but often end up in a defect state, and the size of the band gap is related to the position of the defect state above the valence band maximum. Consequently, a larger band gap typically results in a larger E_{Ovac}. The bulk formation energy is a measure of the bond strength, and a lower bulk formation energy (stronger bonds) tends to increase E_{Ovac}.

Within the same compound, a less unstable surface tends to have a lower E_{Ovac} because the surface is more susceptible to reconstruction. This was demonstrated in normal spinels containing zinc [57]. A lower E_{Ovac} suggests more reactivity, thus the formation of a very unstable surface could lead to a reactive catalyst. However, unstable surfaces are very difficult to attain experimentally. One possible reconstruction is the formation of a macroscopic facet formation. Let us consider the decomposition of surface A into surfaces B and C at an angle of θ_B and θ_C, respectively, with surface A. Denoting the surface energies of A, B, and C as E_A, E_B, and E_C, respectively, simple geometrical analysis shows that decomposition into surfaces B and C is more stable, although there is an increase in surface area, if $E_A \sin(\theta_B + \theta_C) > E_B \sin\theta_1 + E_C \sin\theta_C$. Identification of facet orientation pairs for a given orientation can be obtained using the algorithm in Ref. [58]. The stable and metastable surfaces of θ-Al_2O_3 and β-Ga_2O_3, which share the same monoclinic crystal structure, were studied in Ref. [59]. The identified surfaces are stable against macroscopic facet decomposition and reconstruction suggested by a genetic algorithm [60–63]. The E_{Ovac} can be lowered by slightly more than 1 eV if the corresponding metastable surface could be accessed [59]. Finding surface sites is a non-trivial task in complicated surfaces. The algorithm using the solid angle of open space around an atom allows quantitative identification of surface and subsurface sites [54].

When a metal nanoparticle is adsorbed on a support surface, the energy necessary to remove an anion from the surface is decreased if electrons can move from the support surface near the anion removal site to the metal nanoparticle, which is the electron scavenger effect [64]. The electron scavenger effect was found, when nanorods with large work functions were adsorbed on the In_2O_3 (111) surface when O was removed from sites near the nanorod. The (111) surface of the bixbyite structure is quite complicated. Initial guesses of metal adsorption sites were obtained by deriving the charge density isosurface above the surface. On-top, hollow, and bridge sites are points on the isosurface, namely the local maxima, minima, and saddle points, respectively, of the coordinate of the isosurface in the direction normal to the surface [64]. Manifestation of the electron scavenger effect was studied on experimentally known group 3, 4, 5 hydrides, nitrides, carbides, oxides, and sulfides [65]. The electron scavenger effect is more likely to happen in when the support has a smaller work function. Hydrides, nitrides, and carbides have a smaller work function than most elementary metals, strongly suggesting that the electron scavenger effect can be observed. Oxides and sulfides with lower cation valence tend to have a smaller work function. However, whether the resulting support will be a good catalyst depends on other factors. For example, a lower bulk formation energy will result in stronger bonds that will increase the anion vacancy formation energy. In contrast, a higher bulk formation energy means that the compound will be less stable. The geometry of the surface is also important because nanoparticles (and nanorods) are less likely to adsorb if the lattice parameters are incoherent because close-to-epitaxial adsorption cannot happen.

Surface models caused a revolution on grain boundary model generation. Conventional atomic-level grain boundary model generation relied on the existence of a 3D coincidence site lattice (CSL). Two grains are superimposed such that part of the lattice points share the same lattice, which is the CSL. Atoms are then removed as appropriate to form a model. Unfortunately, exact CSLs do not exist for cubic lattices, tetragonal lattices with rotation axis <001>, and hexagonal lattices with rotation axis <0001> [66], thus the conventional method cannot be used directly. However, atomic-level GB models useful for calculations intrinsically require 2D periodicity at the interface only and not a 3D CSL. A procedure to generate grain boundary models by stacking two slab-and-vacuum models is discussed in Hinuma et al. [67].

13.4 Molecular Adsorption on Metal Oxide Surface: Perspectives from Frontier Orbital Theory and Catalysis Informatics

As introduced in Sect. 13.2, DFT-based machine learning models for predicting molecular adsorption and chemical reactions on metal and metal oxide surfaces have been developed. In order to screen the most suitable materials from a large database to achieve desired functionalities by the machine learning methods, it is necessary to use as descriptors the properties of each material that are intrinsically related to the target variables such as yield and selectivity, and show resultant correlations. However, properties such as activation energy, which is difficult to compile into a database due to high computational costs, are usually not practical as descriptors, and more basic properties such as d band centers and defect formation energies are used. Although these are used to describe the metal and metal oxide surfaces, information on adsorbed molecules is also obviously important to achieve more accurate predictions. Molecular structures and electronic states are diverse so are their interactions with the surfaces and adsorption configurations of the molecules. In this section, we introduce our recent work on the frontier orbital theory of molecular adsorption on metal oxide surfaces and discuss the possibility of informatics using HOMO and LUMO levels and related properties of adsorbates as descriptors.

13.4.1 A Frontier Orbital Theory Study on Adsorption Behavior of Various Molecules on TiO_2 Surfaces

TiO_2 is utilized in numerous applications such as heterogeneous catalysis, photocatalysis, solar cells, batteries, dye-sensitized solar cell, lithium-ion batteries, pigments, and antireflection coatings. The broad interest in and applicability of TiO_2 are attributable to its nontoxicity, hydrophilicity, high natural abundance, and high stability under a wide range of conditions. Rutile ($P4_2/mmm$) and anatase ($I4_1/amd$) forms especially attract much attention for a variety of applications. Rutile is the most common naturally-occurring form and is widely used as a white pigment, while anatase is the most photocatalytically active polymorph [68, 69]. The structures and properties of these forms have been investigated in the field of surface science so as to improve the performance of TiO_2 as a catalyst. Surface energies, which can be easily obtained from DFT calculations, are one of the indicators to evaluate the phase stability and surface reactivity of catalytic systems. Judging from the computed surface energies of various surfaces, the (101) and (110) are the most stable surfaces in anatase and rutile, respectively, which is in good agreement with the observed predominance of these surfaces for each form [70, 71]. We chose TiO_2 as a prototype material for the comprehensive investigation of interactions between small molecules and metal oxide surfaces. For the anatase (101) and rutile (110) surfaces, we systematically investigated the adsorption of 41 small molecules (H_2, N_2, CO, CO_2, CH_4, NH_3, H_2O, H_2S, dimethyl sulfoxide, alkanes, alkenes, alkynes, aromatic compounds, alcohols, aldehydes, ketones, nitriles, carboxylic acids, amides, and amines) to identify important factors for molecular adsorption on oxide surfaces using machine learning methods [72].

13.4.2 Statistical Analysis of Molecular Adsorption on the TiO_2 Perfect Surfaces

The search for the most stable configuration with many initial structures revealed that the molecules prefer to adsorb on the top of the unsaturated fivefold coordinated Ti atoms on the anatase (101) surface. The adsorbates interact with the surface by various types of bonding, including chemical bonds between lone pair and Ti atoms, π-electron donations, and vdW interactions. The computed adsorption energies show that chemically hard molecules (N_2, CH_4, and CO) are weakly bound, while amines and pyridine (having lone pairs on their nitrogen atoms) are strongly bound. We searched for descriptors that are highly correlated with the adsorption energy of each molecule. As plotted in Fig. 13.6, the computed adsorption energies of the molecules under consideration are highly correlated with their HOMO levels but not at all with the LUMO ones with coefficients of determination (R^2) of 0.64 and 0.01 for the HOMO and LUMO, respectively. This result demonstrates that the HOMO of adsorbates is especially important for the molecular adsorption on the TiO_2 surface, which is also supported by the correlation map for descriptors and the adsorption energy (Fig. 13.6d). In addition, a higher R^2 value (0.87) between the adsorption energy and HOMO level can be obtained by dividing the molecules into two subgroups: hydrocarbons and heteroatom-containing compounds (Fig. 13.6b). This result might come from the fact that the main interaction with the TiO_2 surface for these two groups of molecules differs: dispersion forces for hydrocarbons and chemisorption on the Ti atoms for heteroatom-containing compounds. While the grouping is useful for obtaining the higher correlations, it should be noted that machine learning methods do not automatically discover the appropriate grouping in data and a deeper chemical understanding of the target is necessary.

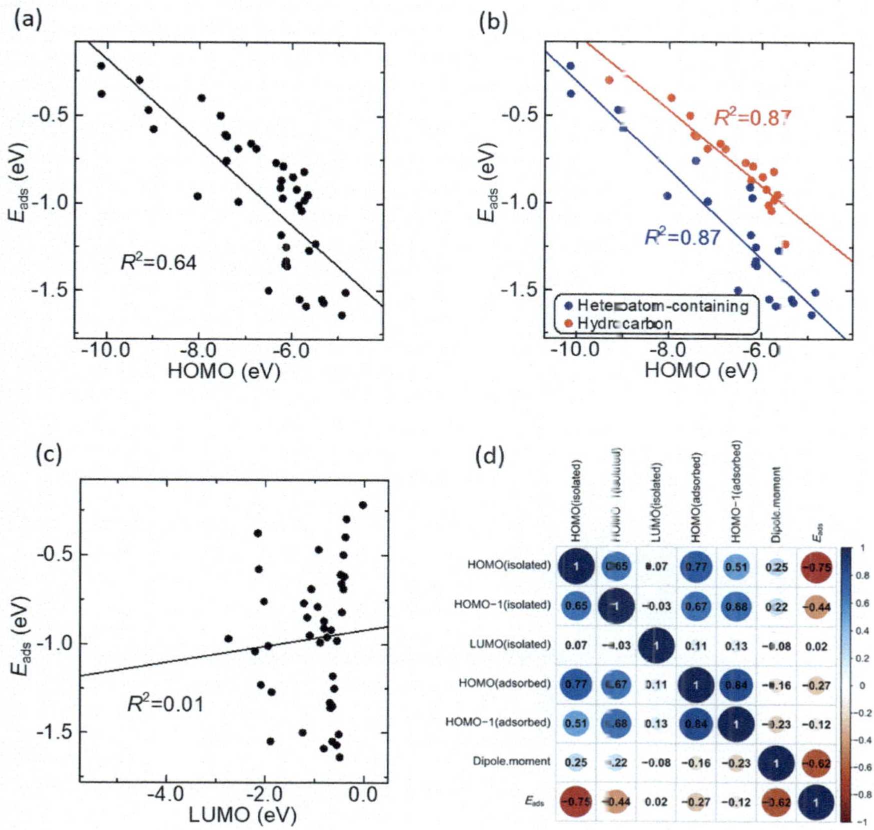

Fig. 13.6 Adsorption energies E_{ads} as functions of **a**, **b** HOMO and **c** LUMO levels for an anatase $TiO_2(101)$ surface. Adapted with permission from Ref. [72]. Copyright 2019 American Chemical Society

To obtain higher coefficients of determination, we generated a wide variety of descriptors to identify other important parameters associated with molecular adsorption using the DFT calculations and the Dragon program [73]. The ExtraTrees algorithm [74] was employed to rank the relative importance of the descriptors for the adsorption energy, and Fig. 13.7 shows the 10 most important descriptors for the prediction of adsorption energy based on nonlinear regression. As expected, the HOMO level of adsorbates is the most important descriptor. Interestingly, the dipole moment of the adsorbates plays the second most important role in predicting the adsorption energy. The dipole moment obtained from DFT computations was found to correlate with the adsorption energies but also showed a weak correlation with the HOMO levels (see Fig. 13.6d), which also supports the statistical significance of the dipole moment on the adsorption. These results can be understood by considering that electrostatic interactions are at work in molecular adsorptions in addition to orbital interactions. We expected HOMO-1 to be also important since there are reactions that are influenced by HOMO-1, but Fig. 13.7 shows that the HOMO-1 level is not an important descriptor in predicting adsorption energies. Figure 13.6d shows the HOMO-1 level is moderately correlated with the HOMO level, and hence the HOMO-1 data are statistically redundant in the prediction. In the FMO theory, molecular orbitals other than HOMO and LUMO are usually not paid much attention. One of the reasons why HOMO and LUMO are particularly important in FMO theory is probably that the information of HOMO-1 is already included to some extent in that of HOMO, so there is usually no need to consider the HOMO-1.

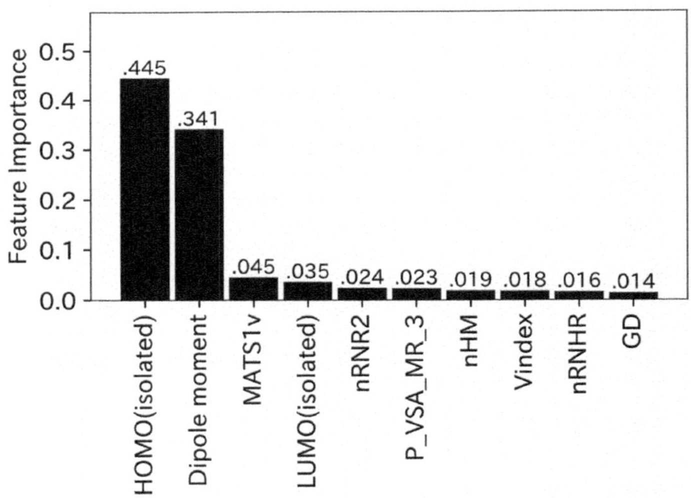

Fig. 13.7 Importance scores obtained using the ExtraTrees algorithm with 100 decision trees. Adapted with permission from Ref. [72]. Copyright 2019 American Chemical Society

13.4.3 Adsorption Behavior on the TiO_2 Surface with O Vacancy

The emergence of defects on a surface is well-known to affect various surface phenomena. In the case of metal oxides, the surface oxygen (O) vacancies lead to unique surface electronic properties and are expected to result in a higher catalytic activity not achievable on perfect surfaces. Such phenomena are reported sporadically but have not been comprehensively researched. In this section, we describe the investigation of the molecular adsorption on the surface with O vacancies and try to uncover the role of those defects in adsorption phenomena from the perspective of FMO theory [75].

The O defects enhance the reduction capacity of metal oxide surfaces, and therefore, all molecules studied bind more strongly to surfaces with defects than to clean surfaces. As shown in Fig. 13.8, both the anatase (101) and rutile (110) TiO_2 defect surfaces have localized defect states within the band gap at approximately −0.2 and −0.05 eV, respectively. Whereas the molecular orbital isosurfaces of the former show the localization of electrons at the defect site, those of the latter are localized at subsurface $\mathbf{Ti_{6c}}$ atoms. This difference in the electronic structures determines the adsorption configuration of the molecule on the defective TiO_2 surfaces: the molecules are adsorbed at the defect site on the anatase (101) surface, but on the rutile (110) surface they are adsorbed at $\mathbf{Ti_{5c}}$ sites rather than the defect site, as on the clean surface.

Figure 13.9 illustrates the correlation between adsorption energies and several descriptors on the basis of a simple linear regression analysis for the anatase TiO_2 (101) surface. The black line in this figure represents a regression line of the 41 nonradical molecules. The R^2 values for HOMO and LUMO are calculated to be 0.45 and 0.11, respectively, which shows that the electron-donating ability of adsorbates is more meaningful for adsorption phenomena than their electron-accepting ability in the adsorption of nonradical molecules, as discussed above. However, when a regression analysis (red lines) is performed including radical species (O_2, NO, and NO_2), the correlation between the adsorption energies and the LUMO levels becomes significantly higher as indicated by the R^2 value of 0.32 and 0.40 for the HOMO and LUMO, respectively. This might be because the LUMO also contributes to the adsorption of radical species on the anatase defective surface with the increased reduction ability. Actually, the molecular orbital isosurface for the defect state exhibits a bonding interaction between the 3d orbital of a Ti atom at the defect site on the surface and the π^* orbital of the $C=O$ bond of the adsorbed acetone molecule. These results can be understood intuitively by considering orbital interactions between the adsorbates and the clean or defective TiO_2 surface, as shown in Fig. 13.10. The valence and conduction states are derived primarily from O 2*p* and Ti 3*d* orbitals, respectively. In the course of the molecular adsorption process, the molecule approaches a vacant unsaturated coordinated Ti atom, but the extent of overlap between the FMO and the valence state remains close to zero due to the lack of d character in the valence state. Thus, the interactions between the LUMO

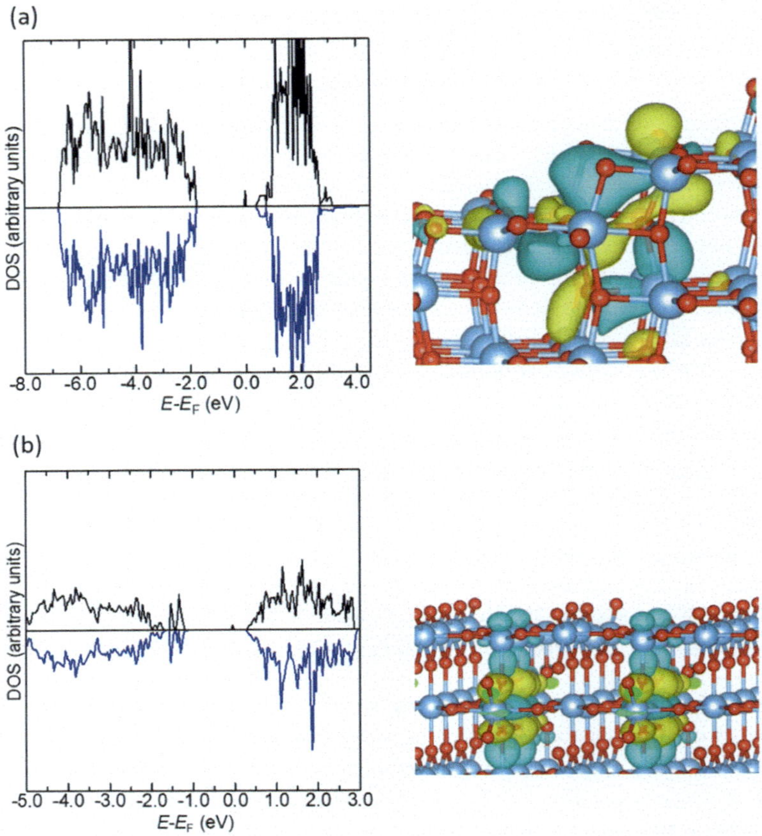

Fig. 13.8 Total DOS plots for **a** an anatase (101) and **b** a rutile (110) surfaces with an O vacancy. The up- and down-spins are colored black and blue, respectively. The energy values are measured from E_F. The molecular orbitals for the defect level are also shown. The red and blue balls represent O and Ti atoms, respectively. Adapted with permission from Ref. [75]. Copyright 2021 American Chemical Society

of the adsorbates and the valence states are negligible, and therefore, no correlation is observed between the adsorption energies and the LUMO in the case of the clean surface (Fig. 13.10a), and the molecular adsorption occurs through electron donations from the HOMO to the conduction states. On the other hand, the defect state lies within the band gap and is close to the LUMO in energy on the defective surface (Fig. 13.10b). Consequently, the radical species in particular form strong bonds with the defective surface.

Based on the result that not only the HOMO but also the LUMO of molecules play an important role in the molecular adsorption, we evaluated the effectiveness of the absolute hardness (η) and absolute electronegativity (χ) as descriptors in predicting adsorption energies because they include information on both HOMO and LUMO.

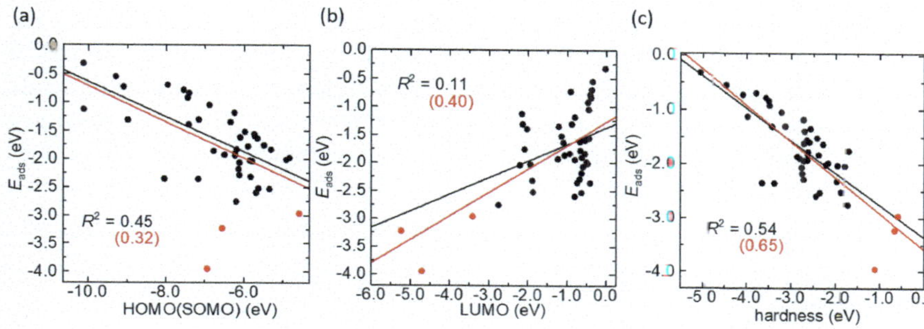

Fig. 13.9 E_{ads} as a function of **a** the HOMO level, **b** the LUMO level and **c** the absolute hardness for an anatase (101) surface with an O vacancy. While the black line represents a regression line for the 41 nonradical molecules adopted in Ref. [72], the red line includes O_2 and NO_x species. The black and red circles show the 41 nonradical molecules and three radical species (O_2 and NO_x), respectively. Adapted with permission from Ref. [75]. Copyright 2021 American Chemical Society

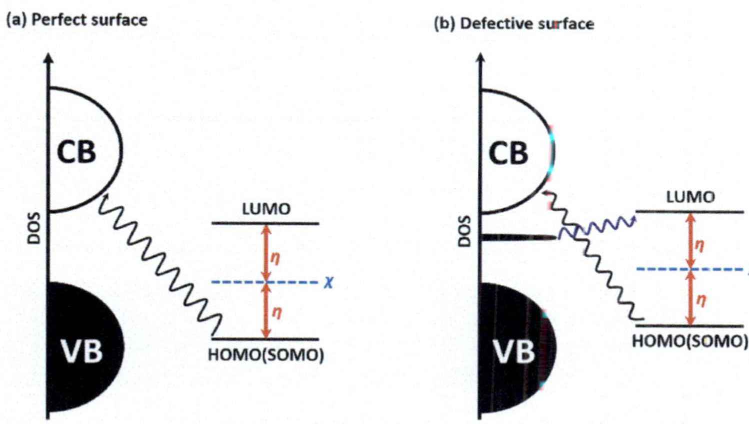

Fig. 13.10 Schematic representation of the orbital interactions between the HOMO (SOMO) and LUMO of an isolated molecule and a TiO_2 surface with or without an O vacancy. The major orbital interactions are illustrated by wavy arrows. Adapted with permission from Ref. [75]. Copyright 2021 American Chemical Society

Here, the η and χ values are defined on the basis of Koopmans' theorem

$$\eta \approx \frac{\varepsilon_{LUMO} - \varepsilon_{HOMO(SOMO)}}{2}; \quad \chi \approx \frac{\varepsilon_{LUMO} + \varepsilon_{HOMO(SOMO)}}{2}$$

where $\varepsilon_{\text{HOMO(SOMO)}}$ and $\varepsilon_{\text{LUMO}}$ are the energy levels for the frontier orbitals. As shown in Fig. 13.10, η represents one-half of the energy gap between the HOMO and LUMO levels, and χ corresponds to the midpoint of the HOMO and LUMO levels. In these two explanatory variables, the R^2 value for η is 0.54 (0.65) for the molecular group that excludes (includes) radical species (Fig. 13.9c), which is larger than that associated with the HOMO or LUMO. This suggests that η is a robust and generic descriptor for predicting adsorption energies on defective metal oxide surfaces. Finally, it is worthy of notice that the frontier orbital theory can provide a unified explanation of the adsorption of such a variety of molecules on solid surfaces with reasonable accuracy.

References

1. Ruban A, Hammer B, Stoltze P, Skriver HL, Nørskov JK. Surface electronic structure and reactivity of transition and noble metals. J Mol Catal A Chem. 1997;115:421–9.
2. Vayenas CG, Bebelis S, Ladas S. Dependence of catalytic rates on catalyst work function. Nature. 1990;343:625–7.
3. Shen X, Pan Y, Liu B, Yang J, Zeng J, Peng Z. More accurate depiction of adsorption energy on transition metals using work function as one additional descriptor. Phys Chem Chem Phys. 2017;19:12628–32.
4. Calle-Vallejo F, Tymoczko J, Colic V, Vu QH, Pohl MD, Morgenstern K, Loffreda D, Sautet P, Schuhmann W, Bandarenka AS. Finding optimal surface sites on heterogeneous catalysts by counting nearest neighbors. Science. 2015;350:185–9.
5. Kitchin JR, Nørskov JK, Barteau MA, Chen JG. Role of strain and ligand effects in the modification of the electronic and chemical properties of bimetallic surfaces. Phys Rev Lett. 2004;93:156801.
6. Zhuang H, Tkalych AJ, Carter EA. Surface energy as a descriptor of catalytic activity. J Phys Chem C. 2016;120:23698–706.
7. Nørskov JK, Abild-Pedersen F, Studt F, Bligaard T. Density functional theory in surface chemistry and catalysis. Proc Natl Acad Sci U S A. 2011;108:937–43.
8. Greeley J. Theoretical heterogeneous catalysis: scaling relationships and computational catalyst design. Ann Rev Chem Biomol Eng. 2016;7:605–35.
9. Studt F, Abild-Pedersen F, Bligaard T, Sørensen RZ, Christensen CH, Nørskov JK. Identification of non-precious metal alloy catalysts for selective hydrogenation of acetylene. Science. 2008;320:1320–2.
10. Studt F, Sharafutdinov I, Abild-Pedersen F, Elkjær CF, Hummelshøj JS, Dahl S, Chorkendorff I, Nørskov JK. Discovery of a Ni–Ga catalyst for carbon dioxide reduction to methanol. Nat Chem. 2014;6:320–4.
11. Fukui K, Yonezawa T, Shingu H. A molecular orbital theory of reactivity in aromatic hydrocarbons. J Chem Phys. 1952;20:722–5.
12. Fukui K, Yonezawa T, Nagata C, Shingu H. Molecular orbital theory of orientation in aromatic, heteroaromatic, and other conjugated molecules. J Chem Phys. 1954;22:1433–42.
13. Fukui K. Recognition of stereochemical paths by orbital interaction. Acc Chem Res. 1971;4:57–64.
14. Woodward RB, Hoffmann R. Stereochemistry of electrocyclic reactions. J Am Chem Soc. 1965;87:395–7.
15. Hoffmann R. How chemistry and physics meet in the solid state. Angew Chem Int Ed. 1987;26:846–78.
16. Glassey WV, Hoffmann R. A comparative study of the p(2 × 2)-CO/M(111), M=Pt, Cu, Al chemisorption systems. J Phys Chem B. 2001;105:3245–60.
17. Glassey WV, Hoffmann R. A molecular orbital study of surface-adsorbate interactions during the oxidation of CO on the Pt(111) surface. Surf Sci. 2001;475:47–60.
18. Zheng C, Apeloig Y, Hoffmann R. Bonding and coupling of C_1 fragments on metal surfaces. J Am Chem Soc. 1988;110:749–74.

19. Papoian G, Nørskov JK, Hoffmann R. A comparative theoretical study of the hydrogen, methyl, and ethyl chemisorption on the Pt(111) surface. J Am Chem Soc. 2000;122:4129–44.
20. Newns DM. Self-consistent model of hydrogen chemisorption. Phys Rev. 1969;178:1123–35.
21. Kibler LA, El-Aziz AM, Hoyer R, Kolb DM. Tuning reaction rates by lateral strain in a palladium monolayer. Angew Chemie Int Ed. 2005;44:2080–4.
22. Fernández EM, Moses PG, Toftelund A, Hansen HA, Martínez JI, Abild-Pedersen F, Kleis J, Hinnemann B, Rossmeisl J, Bligaard T, Nørskov JK. Scaling relationships for adsorption energies on transition metal oxide, sulfide, and nitride surfaces. Angew Chemie Int Ed. 2008;47:4683–6.
23. Vojvodic A, Hellman A, Ruberto C, Lundqvist BI. From electronic structure to catalytic activity: a single descriptor for adsorption and reactivity on transition-metal carbides. Phys Rev Lett. 2009;103: 146103.
24. Huang B, Xiao L, Lu J, Zhuang L. Spatially resolved quantification of the surface reactivity of solid catalysts. Angew Chemie—Int Ed. 2016;55:6239–43.
25. Calle-Vallejo F, Inoglu NG, Su HY, Martínez JI, Man IC, Koper MTM, Kitchin JR, Rossmeisl J. Number of outer electrons as descriptor for adsorption processes on transition metals and their oxides. Chem Sci. 2013;4:1245–9.
26. Van Santen RA, Tranca I, Hensen EJM. Theory of surface chemistry and reactivity of reducible oxides. Catal Today. 2015;244:63–84.
27. Kakekhani A, Roling LT, Kulkarni A, Latimer AA, Abroshan H, Schumann J, Aljama H, Siahrostami S, Ismail-Beigi S, Abild-Pedersen F, et al. Nature of lone-pair-surface bonds and their scaling relations. Inorg Chem. 2018;57:7222–38.
28. Puigdollers AR, Schlexer P, Tosoni S, Pacchioni G. Increasing oxide reducibility: the role of metal/oxide interfaces in the formation of oxygen vacancies. ACS Catal. 2017;7:6493–513.
29. Jia J, Qian C, Dong Y, Li YF, Wang H, Ghoussoub M, Butler KT, Walsh A, Ozin GA. Heterogeneous catalytic hydrogenation of CO_2 by metal oxides: defect engineering—perfecting imperfection. Chem Soc Rev. 2017;46:4631–44.
30. Kumar G, Lau SLJ, Krcha MD, Janik MJ. Correlation of methane activation and oxide catalyst reducibility and its implications for oxidative coupling. ACS Catal. 2016;6:1812–21.
31. Lee YL, Kleis J, Rossmeisl J, Yang SH, Morgan D. Prediction of solid oxide fuel cell cathode activity with first-principles descriptors. Energy Environ Sci. 2011;4:3966–70.
32. Curnan MT, Kitchin JR. Effects of concentration, crystal structure, magnetism, and electronic structure method on first-principles oxygen vacancy formation energy trends in perovskites. J Phys Chem C. 2014;118:28776–90.
33. Staykov A, Téllez H, Akbay T, Druce J, Ishihara T, Kilner J. Oxygen activation and dissociation on transition metal free perovskite surfaces. Chem Mater. 2015;27:8273–81.
34. Chen C, Ciucci F. Designing Fe-based oxygen catalysts by density functional theory calculations. Chem Mater. 2016;28:7058–65.
35. Emery AA, Saal JE, Kirklin S, Hegde VI, Wolverton C. High-throughput computational screening of perovskites for thermochemical water splitting applications. Chem Mater. 2016;28:5621–34.
36. Hinuma Y, Toyao T, Kamachi T, Maeno Z, Takakusagi S, Furukawa S, Takigawa I, Shimizu KI. Density functional theory calculations of oxygen vacancy formation and subsequent molecular adsorption on oxide surfaces. J Phys Chem C. 2018;122:29435–44.
37. Pankajakshan P, Sanyal S, De Noord OE, Bhattacharya I, Bhattacharyya A, Waghmare U. Machine learning and statistical analysis for materials science: stability and transferability of fingerprint descriptors and chemical insights. 2017.
38. Tanabe K, Hölderich WF. Industrial application of solid acid-base catalysts. Appl Catal A Gen. 1999;181:399–434.
39. Digne M, Sautet P, Raybaud P, Euzen P, Toulhoat H. Use of DFT to achieve a rational understanding of acid-basic properties of γ-alumina surfaces. J Catal. 2004;226:54–68.
40. Jenness GR, Christiansen MA, Caratzoulas S, Vlachos DG, Gorte RJ. Site-dependent Lewis acidity of γ-Al_2O_3 and its impact on ethanol dehydration and etherification. J Phys Chem C. 2014;118:12899–907.

41. Cholewinski MC, Dixit M, Mpourmpakis G. Computational study of methane activation on γ-Al_2O_3. ACS Omega. 2018;3:18242–50.
42. Wang P, Fu G, Wan H. How high valence transition metal spreads its activity over nonmetal oxoes: a proof-of-concept study. ACS Catal. 2017;7:5544–8.
43. Bhan A, Iglesia E. A link between reactivity and local structure in acid catalysis on zeolites. Acc Chem Res. 2008;41:559–67.
44. Liu C, Tranca I, Van Santen RA, Hensen EJM, Pidko EA. Scaling relations for acidity and reactivity of zeolites. J Phys Chem C. 2017;121:23520–30.
45. Matsuoka T, Baumes L, Katada N, Chatterjee A, Sastre G. Selecting strong Brønsted acid zeolites through screening from a database of hypothetical frameworks. Phys Chem Chem Phys. 2017;19:14702–7.
46. Tasker PW. The stability of ionic crystal surfaces. J Phys C: Solid State Phys. 1979;12:4977.
47. Gale JD, Rohl AL. The general utility lattice program (GULP). Mol Simul. 2003;29:291–341.
48. Goniakowski J, Finocchi F, Noguera C. Polarity of oxide surfaces and nanostructures. Rep Prog Phys. 2008;71: 016501.
49. Stengel M. Electrostatic stability of insulating surfaces: theory and applications. Phys Rev B. 2011;84: 205432.
50. Goniakowski J, Noguera C. The concept of weak polarity: an application to the $SrTiO_3(001)$ surface. Surf Sci. 1996;365:L657-62.
51. Hinuma Y, Kumagai Y, Oba F, Tanaka I. Categorization of surface polarity from a crystallographic approach. Comp Mater Sci. 2016;113:221–30.
52. Wolf D. Reconstruction of NaCl surfaces from a dipolar solution to the Madelung problem. Phys Rev Lett. 1992;68:3315–8.
53. Hinuma Y, Kumagai Y, Tanaka I, Oba F. Effects of composition, crystal structure, and surface orientation on band alignment of divalent metal oxides: a first-principles study. Phys Rev Mater. 2018;2: 124603.
54. Hinuma Y, Kamachi T, Hamamoto N. Algorithm for automatic detection of surface atoms. Trans Mater Res Soc Jpn. 2020;45:115–20.
55. Hinuma Y, Kamachi T, Hamamoto N. Auto-generation of corrugated nonpolar stoichiometric slab models. Mater Trans. 2020;61(1):78–87.
56. Hinuma Y, Gake T, Oba F. Band alignment at surfaces and heterointerfaces of Al_2O_3, Ga_2O_3, In_2O_3, and related group-III oxide polymorphs: a first-principles study. Phys Rev Mater. 2019;3: 084605.
57. Hinuma Y, Mine S, Toyao T, Kamachi T, Shimizu K. Factors determining surface oxygen vacancy formation energy in ternary spinel structure oxides with zinc. Phys Chem Chem Phys. 2021;23:23768–77.
58. Hinuma Y, Kamachi T, Hamamoto N. Automated identification of facet pair orientations. Mater Trans. 2020;6:1430–3.
59. Hinuma Y, Kamachi T, Hamamoto N, Takao M, Toyao T, Shimizu K. Surface oxygen vacancy formation energy calculations in 34 orientations of β-Ga_2O_3 and θ-Al_2O_3. J Phys Chem C. 2020;124:10509–22.
60. Oganov AR, Ono S. Theoretical and experimental evidence for a post-perovskite phase of $MgSiO_3$ in Earth's D" layer. Nature. 2004;430:445–8.
61. Oganov AR, Glass CW. Crystal structure prediction using ab initio evolutionary techniques: principles and applications. J Chem Phys. 2006;124: 244704.
62. Oganov AR, Lyakhov AO, Valle M. How evolutionary crystal structure prediction works-and why. Acc Chem Res. 2011;44:227–37.
63. Zhu Q, Li L, Oganov AR, Allen PB. Evolutionary method for predicting surface reconstructions with variable stoichiometry. Phys Rev B. 2013;87: 195317.
64. Puigdollers AR, Pacchioni G. CO oxidation on Au nanoparticles supported on ZrO_2: role of metal/oxide interface and oxide reducibility. ChemCatChem. 2017;9:1119–27.
65. Hinuma Y, Mine S, Toyao T, Maeno Z, Shimizu K. Surface activation by electron scavenger metal nanorod adsorption on TiH_2, TiC, TiN, and Ti_2O_3. Phys Chem Chem Phys. 2021;23:16577–93.

66. King AH, Shin K. Grain boundary structural transformations in hexagonal close packed metals. J de Physique Colloques. 1990;51:C1-203–8.
67. Hinuma Y, Kohyama M, Tanaka S. Boundary plane-oriented grain boundary model generation. Modell Simul Mater Sci Eng. 2022;30: 045005.
68. Diebold U. The surface science of titanium dioxide. Surf Sci Rep. 2003;48:53–229.
69. Hashimoto K, Irie H, Fujishima A. TiO_2 photocatalysis: a historical overview and future prospects. Jpn J Appl Phys. 2005;44:8269–85.
70. Vittadini A, Casarin M, Selloni A. Chemistry of and on TiO_2-anatase surfaces by DFT calculations: a partial review. Theor Chem Acc. 2007;117:663–71.
71. Perron H, Domain C, Roques J, Drot R, Simoni E, Catalette H. Optimisation of accurate rutile TiO_2 (110), (100), (101), and (001) surface models from periodic DFT calculations. Theor Chem Acc. 2007;117:565–74.
72. Kamachi T, Tatsumi T, Toyao T, Hinuma Y, Maeno Z, Takakusagi S, Furukaswa S, Takigawa I, Shimizu K. Linear correlations between adsorption energies and HOMO levels for the adsorption of small molecules on TiO_2 surfaces. J Phys Chem C. 2019;123:20988–97.
73. Kode srl. Dragon (software for molecular descriptor calculation), version 7.0.6;2016. https://chm.kode-solutions.net.
74. Geurts P, Ernst D, Wehenkel L. Extremely randomized trees. Mach Learn. 2006;63:3–42.
75. Hamamoto N, Tatsumi T, Takao M, Toyao T, Hinuma Y, Shimizu K, Kamachi T. Effect of oxygen vacancies on adsorption of small molecules on anatase and rutile TiO_2 surfaces: a frntier orbital approach. J Phys Chem C. 2021;125:3827–44.

The manufacturer's authorised representative in the EU is Springer Nature Customer Service Centre GmbH, Europaplatz 3, 69115 Heidelberg, Germany. If you have any concerns regarding our products, please contact ProductSafety@springernature.com

Printed and bound by CPI Group (UK) Ltd, Croydon, CR0 4YY
30/09/2026
02225315-0012